SCIENTIFIC INFORMATION SYSTEMS IN JAPAN

SCIENTIFIC INFORMATION SYSTEMS IN JAPAN

Edited by
Hiroshi INOSE
Computer Centre
University of Tokyo
Japan

NORTH-HOLLAND PUBLISHING COMPANY — AMSTERDAM · NEW YORK · OXFORD

ISBN: 0 444 86151 3

Published by:
NORTH-HOLLAND PUBLISHING COMPANY – AMSTERDAM · NEW YORK · OXFORD

Sole distributors for the U.S.A. and Canada:
ELSEVIER NORTH-HOLLAND, INC.
52 VANDERBILT AVENUE
NEW YORK, N.Y. 10017

Library of Congress Cataloging in Publication Data
Main entry under title:

Scientific information systems in Japan.

1. Information storage and retrieval systems--Science. 2. Science--Information services--Japan. I. Inose, Hiroshi, 1927-   .
Z699.5.S3S356   025'.065   81-1658
ISBN 0-444-86151-3   AACR2

PRINTED IN THE NETHERLANDS

PREFACE

The present volume is one of the fruits of the project study entitled 'Formation Process of Information Systems and Organization of Scientific Information' which was sponsored by the Ministry of Education, Science and Culture, Japan, through a Grant-in-Aid for Scientific Studies. The project study was carried out for three years, beginning in April 1976 and ending in March 1979, and involving more than 500 researchers from universities, colleges and scientific research institutions in Japan. The objective of the study was to carry out basic research for promoting the formation of information systems and the organization of information for a broad range of scientific disciplines. An interdisciplinary approach was therefore taken in which researchers from a large variety of disciplines were organized in 69 research units which were then categorized into five research groups, namely, input processing, structure recognition, storage and retrieval, systems approach and research trend analysis of scientific information. In addition, developmental activities were carried out on data base management systems, computer networks, input-output systems and system organization. Results of studies were published in Japanese in March 1977, 1978 and 1979. And in November 1979, a symposium, in which all the project members as well as nation-wide specialists participated, was held to summarize the project activities, and symposium proceedings were published in Japanese.

Although some of the results of the project study have been and will be published from time to time in foreign languages in various scientific journals, no publication covering the project study as a whole has ever been presented to the international scene. The intention of publishing this volume is therefore to provide international readers with a comprehensive view of the project study in the English language. In view of the available publication resources and of the necessity of describing respective topics in detail, however, the volume contains the results of only 29 study units selected by the researchers in each of the five categories, plus reports on developmental activities. A summary of the proposal for the organization of a nation-wide information system for scientific studies, which emerged in the course of the project study, is also included in this volume.

The project study as well as the publication of this volume would not have been possible without the enthusiastic support of many. With utmost gratitude, the editor wishes to acknowledge the oustanding leadership in project management provided by Professors Tosio Kitagawa, Takehiko Shimanouchi, Mamoru Hosaka, Toshiyuki Sakai, Yoshikazu Sawaragi, Hajime Tanaka and Atsunobu Ichikawa. Furthermore, the editor wishes to mention with gratitude that the project study could not have been accomplished without the enthusiastic efforts of all participating researchers and that the present volume could not have been completed without the dedicated efforts of the project secretaries, Professors Haruhisa Ishida, Takeo Yamamoto and Lecturer Hisashi Yasunaga. Last but not least, the editor wishes to express his gratitude to the Ministry of Education, Science and Culture, and in particular, to Messrs. Akira Tezuka and Takasi Okado, the former and present Director of the Research Aid Division, for their attentive support and encouragement.

Hiroshi Inose

Computer Centre
University of Tokyo

Final report of the 3-year (1976-1979) research project entitled —The Formation Process of Information Systems and the Organization of Scientific Information— and supported by the Ministry of Education, Science and Culture.

Organizing Committee

<Members>

Hiroshi Inose	University of Tokyo	Dept. of Electronic Engineering
Tosio Kitagawa	Kyushu University	Professor Emeritus
Takehiko Shimanouchi	Tsukuba University	Dept. of Information Science
Mamoru Hosaka	University of Tokyo	Institute of Space and Aeronautical Science
Toshiyuki Sakai	Kyoto University	Dept. of Information Science
Yoshikazu Sawaragi	Kyoto University	Dept. of Applied Mathematics and Physics
Hajime Tanaka	Hokkaido University	Dept. of Physics
Atsunobu Ichikawa	Tokyo Institute of Technology	Dept. of Systems Science

<Secretaries>

Haruhisa Ishida	University of Tokyo	Computer Centre
Takeo Yamamoto	University of Tokyo	Computer Centre
Hisashi Yasunaga	University of Tokyo	Earthquake Research Institute

Acknowledgement

A portion of this publication was produced by an on-line phototypesetter controlled by the HITAC-8800/8700 computer system at the Computer Centre of the University of Tokyo. The cooperation of the staff of the Computer center is gratefully acknowledged.

CONTENTS

Scientific Information Systems in Japan
H. Inose (ed.)

Introduction

Hiroshi Inose
Project Director

1. Outline of the Project Study

The project study entitled —The Formation Process of Information Systems and the Organization of Scientific Information— was established in April 1976 under the sponsorship of the Ministry of Education, Science and Culture through the Grant-in-Aid for Scientific Research.

As a result of the successful accomplishment of the preceding project study entitled —Advanced Information Processing of Large Scale Data over a Broad Area—, and in view of the urgent needs of organizing scientific information and establishing scientific information systems in this country, unanimous and enthusiastic request for organizing a new project study was raised from the researchers not only in information science but also in broad scientific disciplines and was endorsed by the Japan Science Council. In response to this request, the Ministry of Education, Science and Culture allocated in total 1,015,150,000 yen (approximately 4,400,000 U.S. Dollars) for the period beginning from April 1976 and ending in March 1979. The project study thus became one of the largest projects in basic scientific studies involving more than 500 researchers from universities, colleges and scientific research institutions in Japan.

The project was administered by the Computer Centre, University of Tokyo and directed by its director, Professor Takehiko Shimanouchi in its first year, and then by the author in its second and third years as a result of Professor Shimanouchi's retirement. To assist the project director in planning, directing and coordinating project activities, a Steering Committee was formed which consisted of eight members and three secretaries as listed in the front page. Reflecting the interdisciplinary feature of the project study, the Steering Committee members were a chemist, a physicist, three information scientists, three control engineers and an electronics engineer by profession.

The objective of the project study was to explore problems and solutions in organizing scientific information in broad range of disciplines, i.e., natural sciences, social sciences and humanities; and to promote research and development activities for the formation of information systems for scientific purposes. In view of very broad area to be covered and rather limited fund available, an interdisciplinary approach was taken in which research and development efforts were concentrated on selected topics while maintaining close coordination across disciplinary boundaries. For this purpose, more than 500 researchers invited from a large variety of disciplines were organized into 69 research units and 4 development committees. Research activities carried out by the research units as well as development activities by the development committees were coordinated by the Steering Committee.

2. Research Activities of the Project Study

The research units were categorized into five research groups, namely, input processing, structure recognition, storage and retrieval, systems approach, and research trend analysis. The titles, the leaders and their affiliation are listed in the Appendix A.

Research group A on input processing of information consisted of 17 research units. The activities of this group included the following :

(i) Automatization of information input and collection, evaluation and redundancy reduction of measured data.
(ii) Statistical handling and testing of original data.
(iii) Editing and input processing of bibliographical information.

Research group B on structure recognition of information consisted of 21 research units whose activities included the following :

(i) Pattern processing of visual and speech information.
(ii) Structural analysis in problem solving.
(iii) Structural recognition of languages.

Research group C on storage and retrieval of information consisted of 23 research units and included the following topics :

(i) Formation and operation of general-purpose and special-purpose information systems for scientific studies.
(ii) Development of data bases for visual, numerical and other forms of information.
(iii) Development of data bases for specific scientific disciplines, i.e., chemistry, physics, geology, astronomy, engineering, medicine, Japanese literature, ethnology, archeology, economics,etc.

Research group D on systems approach for information usage consisted of 4 research units and dealt with the following topics :

(i) Systems methodology for organizing large information systems.
(ii) Systems studies on international collaboration for scientific information.
(iii) Structural analysis on the utility of scientific information.

Research group E on research trend analysis consisted of 4 research units and included the following activities. :

(i) Methodology for research trend analysis and utilization of data bases for research trend analysis.
(ii) Research trend analysis by using relational graphs obtained from citations.
(iii) Utility analysis of scientific information.

3. Developmental Activities of the Project Study

Developmental activities were carried out by four development committees, i.e., the Data Base Management System Development Committee, the Network Development Committee, the Scientific Information System Development Committee and the Input-Output System Development Committee. For the purpose of coordinating the developmental activities and of reflecting the results to the realization of national scientific information system, the Development Planning and Promotion Committee was also set up. The members and their affiliation of these committees are listed in the Appendix B.

The Data Base Management System Development Committee analyzed the functions required for data base management systems for scientific information, explored the necessary approach for their development, implementation and operation, and planned the deployment of data bases with respect to scientific disciplines and geographical locations, by evaluating their availability and realizability.

The Network Development Committee laid out master plans for the realization of inter-university computer network in Japan, developed hardware and software for this purpose, performed feasibility demonstration and prepared manuals and other detailed documents. As the result, a network architecture tentatively named the N-1 Architecture was completed that provides remote job entry and time sharing services based upon such international standards as CCITT Recommendations X.25 and X.21, by way of circuit-switched and packet-switched digital data network DDX developed by the Nippon Telegraph and Telephone Public Corporation.

The Scientific Information System Development Committee explored the present state of information usage in various scientific disciplines and prepared the basic data necessary for system planning and organization of scientific information.

The Input-Output System Development Committee explored the basic specifications and systems planning specifically on input-output processing of Chinese characters, interactive graphic systems and input-output systems for visual information.

4. Steering Activities of the Project Study

The coordination of the entire research and development activities was the responsibility of the Steering Committee. The Steering Committee members therefore actively participated in all phases of the project activities, as the coordinator of research groups and as the chairmen or vice-chairmen of the development committees. The Steering Committee held 36 meetings during the project period for the purpose of planning, coordination and direction of the project. The committee organized annual and summer symposia in each year that involved all the research units and development committees to provide forums of discussions. The committee members also attended the meetings held by the research groups and subgroups from time to time to better understand the research progress and trend. In an effort to promote close and multi-lateral collaboration, the committee also organized, in the first year, the clusters that horizontally connect the research units sharing common interests but belonging in different research groups; and in the second year, the district liason activities that geographically connect the research units as well as the specialists in each district.

In view of the importance of reflecting the expertise and concept gained in the course of the project study to the national planning of scientific information systems in this country, the Development Planning and Promotion Committee which consisted of steering committee members plus data base and communications specialists, issued a draft proposal on the formation of national scientific information systems to be distributed for discussion, and taking into account of various comments, the report was completed and published in 1979.

5. Publication of the Project Study

Results of studies of the project were published in Japanese in March, 1977, 1978 and 1979 as the project's annual reports. In addition, development committees, research groups and research units published reports on their studies mostly in Japanese.

After the completion of the project study, the Ministry of Education, Science and Culture allotted a Grant amounting to 11 million yen (about 48,000 U.S. Dollars) for holding a symposium and publishing comprehensive reports covering the results of the project study. The symposium was held in November 1979 being participated by all the project members as well as nation-wide specialists, and the final report in Japanese was published and distributed as the proceedings of the symposium.

The publication of this volume was also made possible by the Grant. Although some of the results of the project study have been and will be published from time to time in foreign languages in various scientific journals, no publication covering the project study as a whole has ever been presented to the international scene. The intention of publishing this volume was therefore to provide international readers with a comprehensive perspective of the project study in English.

In view of the available publication resources and of the necessity of describing respective topics in detail, however, the volume contains the results of studies of 29 research units selected by the researchers in each of the five categories, plus the reports on developmental activities. A summary of the proposal made by the Development Planning and Promotion Committee for the organization of a nation-wide information system for scientific studies is also included. For information about the results of studies of the research units not included in this volume, the readers are asked to kindly contact with the leaders of respective research units whose affiliations are listed in the Appendix A.

Scientific Information Systems in Japan
H. Inose (ed.)

FORMATION OF SCIENTIFIC INFORMATION SYSTEM AND ITS IMPLICATION TO INFORMATICS

Tosio KITAGAWA
Emeritus Professor of Kyushu University
International Institute for Advanced Study of Social Information Science, Fujitsu Ltd, 17-25, Shinkamata 1-chome, Ohta-ku, Tokyo 144

The paper aims to give a methodological review of the Special Research Project (SRP). Sections 2 and 3 are devoted to an objective description of SRP, while Sections 4 and 5 reflect the subjective views of the present author regarding research directions and evaluation of research achievements. Section 6 gives an illustration of the implications of SRP research achievements to an development of informatics. Section 7 points out some remaining work to be performed for establishment of scientific information system in Japan. Appendixes give an brief explanation of keynotions introduced.

1. INTRODUCTION

The purpose of this paper is to give a methodological review to the research activities of the Special Research Project (SRP), "Formation Process of Information System and Organization of Scientific Information", being sponsored by the Ministry of Education, Science, and Culture of the Japanese Government, and to explain their implications to the development of informatics in the coming days. In Section 2, we shall start with a brief description of research organizations, and then we proceed in Section 3 to an introduction of research achievements which have been acquired up to the present time. These two Sections aim to give an objective description of the items concerned on the basis of several official reports given by the Coordinating Committee associated with SRP and official summary reports on research activities, each of which was written by the head of each Individual Research Unit (IRU). On the other hand, Sections 4 and 5 reflect the subjective views of the present author regarding research directions and evaluations of research achievements which have been established by SRP as a whole. Section 6 is devoted to an illustration of the implications of SRP to the development of informatics. The conclusion given in Section 7 points out the remaining work to be done by Japanese scientists in the coming days in order to realize a sound scientific information system in Japan in intimate cooperation with any worldwide scientific information system.

2. THE ORGANIZATION AND AIMS OF TWO SPECIAL RESEARCH PROJECTS

A brief description of two Special Research Projects is given in this section in order to provide general understanding about an outline of research organizations and research attitudes.

2.1 The Special Research Project "Advanced Information Processing of a Large Amount of Data over a Broad Area"

This Special Research Project started in April, 1973 and finished in March, 1976.
The full report [1] is given in Japanese amounting to more than 1,253 pages.

2.2 The Special Research Project "Formation Process of Information System and Organization of Scientific Information" (as in 1978, May)

"This Special Research Project (SRP) was started in April, 1976 and is expected to be finished in March, 1979. It has, at present, 69 Individual Research Units (IRU) and a Coordinating Committee Unit. The former IRU's are clssified into five Research Groups A, B, C, D and E; while the latter has five plan Promoting Committees, abbreviated as M, N, O, I and T committees, respectively. The total number of research workers engaged in our Special Research Project amounts to a little over 400, including scientists from various fields, far more than those engaged in the case of the previous Special Research Project explained in [1].
The intermediate reports in each year are given by two Japanese reports [2] and the report by the author [3]-[5], while the final report is completed after the termination of the SRP in 1979, November."

2.3 Methodology for analyzing scientific research activities from the standpoint of informatics

Since the beginning of the two Special Research Projects, the present author has been concerned with a fundamental consideration of the important features of research activities, with reference to the basic notions of the logic of informatics. In what follows, we shall give our summary views in terms of the five assertions, in order to make clear our philosophy and our strategy in promoting our present research. Specific references to the papers [3], [6], [7] are recommended for further interests.

Assertion 1. An organization scheme of scientific information should be adequately set up in order to secure an efficient scheme of their circulation and their sophisticated usage.

Assertion 2. An organization of scientific information should be performed through a deep analysis of real features of the scientific research process in terms of, and from the standpoint of, informatics; and hence, it should be recognized as the formation process of a scientific information system which contains, as its subsystems, various information processing processes, as shown in the Group Constructions in the second Special Research Project.

Assertion 3. A set of typical real examples of advanced information processing techniques carefully selected from various fields of research activities including natural sciences, social sciences, and humanities is essential to the promotion of our research. A deep analysis of these typical real examples can be, and should be, performed from the standpoint of informatics, including theories and techniques regarding pattern recognition, linguistic information processing, computer aided design, data base management design, learning process, deductive process and so on.

Assertion 4. It can be expected that an organization of scientific information and the formation of an information system will present to information scientists an assembly of well concentrated problems of advanced information processing, including knowledge engineering and software engineering.

Assertion 5. There are two broad classes of sciences, called discipline-oriented sciences and mission-oriented ones. Pure sciences belong to the first class, while sciences such as energy, matter resources, environment, medical care and education are mostly mission-oriented. A usage system of the scientific information system should be organized so as to be a cybernetical system. An intensive study should be done in respect to the formation process of specific types of information systems such as the observation-control supporting system and the statistics-management supporting system.

3. RESEARCH ACTIVITIES AND THEIR ACHIEVEMENTS IN SPR

3.1 Individual Research Units (IRU)

We can describe research activities and the achievements of 69 IRU's according to the convention using the following specific notations:

(1) Each IRU belongs to one, and only one, of the five Research Groups(A, B, C, D and E) and is numbered in a certain way.

(2) Science Area A set of the abbreviated notations, each of which denotes some science area, is used to show the science area(s) to which the research activities and/or research achievements of each IRU are intimately concerned.
The following abbreviation notations showing science areas are used in this subsection. (i) A=Astronomy; (ii) Arch.=Archaeology; (iii) B=Biology; (iv) C=Chemistry; (v) E=Engineering; (vi) Eco.=Economics; (vii) Env.=Environmentology; (viii) Ethn.=Ethnology; (ix) G=Geology; (x) Gen.=Genetics; (xi) H=Humanities; (xii) I=Informatics, Information Eng.; (xiii) L=Linguistics; (xiv) M=Mathematics; (xv) Med.=Medical Sciences; (xvi) Met.=Metallurgy; (xvii) Meter.=Meterology; (xviii) Phar.=Pharmaceutics; (xix) P=Physics; (xx) Pol.=Policy science; (xxi) S=Social sciences; (xxii) St.=Statistics; (xxiii) Sy.=Systems science; (xxiv) U=All science areas (universal applicability with no specifications)

(3) Information processing subprocesses

The whole process of information processing is understood to consist of the following seven subprocesses:
(P_1): input information processing
(P_2): preliminary information processing
(P_3): information structuring
(P_4): information storage
(P_5): information retrieval
(P_6): information circulation
(P_7): information usages.

An IRU whose research activities and/or research achievements are mostly concerned with (P_i) and (P_j), is denoted by (ij). Similarly for (i), (ijk) and so on.

(4) Information expression schemes The following notations are used to identify them.
(m): mathematical symbols and formulas
(n): numerical values due to observations and/or calculations
(s): statistics; (l): linguistic information;
(d): documental information; (f): figures;
(p): pictures; (v): voices and speeches

(5) Utility levels of data bases and/or fies

According to the proposal given by M-Committee, the Special Research Project has set up the following four utility levels: (see Fig 4.1 and Appendix I)
((1)): PRF type ((2)): ARF type
((3)): UDL type ((4)): CDL type,
where
PRF=Private Researcher File
ARF=Accumulative Researcher File
UDL=User Data Library
CDL=Center Data Library
whose explanation is given in the M-Committee Report.

(6) Ware types

The information resources which each IRU has already realized, and/or aimes to realize, can be classified into the following three wares:
[H]: Hardware [S]: Software [B]: Brainware
The terminology of brainware has been introduced by the present author in his paper [8], 1974, and has been extensively used by him in his papers [9], [10] and [7] since then.

(7) Keywords

To each IRU there is associated a set of keywords which specifies the characteristic features of its research activities and research achievements. It is noted that the set of keywords is not necessarily identical with the set of keywords chosen by the head of each IRU in his official report.
The research activities and the research achievements of each of all the 69 IRU's can be represented by the set of these notations, as we have given in our paper [22]. Here we shall give five examples for illustration.
Five representative examples are given in what follows.
A-1: Coordinate Research on Input Systems of Scientific Data and Information (Prof. T. Sakai, Kyoto Univ.) (B, Med., I)(12345)(p)((123))[HS]; Interactive image modelling system, Sequential similarity detection algorithm, Automatic monitoring, Frame picture.

B-1 Integrated Research of Formation Process of Scientific Information Systems (Em. Prof. T. Kitagawa, Kyushu Univ.) (I, M, L, St, Med., U)(1234567)(nsld)((1234))[HSB]; Brainware, Successive processes of statistical inferences, Generalized artificial grammar, MIR-RF, TEXT information Retrieval, New Interactive Statistical Analysis(NISAN).

C-1: Coordinate Research on the Organization of scientific Information (Prof. T. Shimanouchi, Univ. of Tsukuba) (C., I, U)(3456)((1234))[SBH]; Molecular vibration, Nuclear vibration, Magnetic resonance, Molecular Structure, Thermodynamic property, Central system, Unified treatment of independent databases.

D-1: Coordinate Research on Systems Science Approach to Utilization of Scientific Information (Prof. Y. Sawaragi, Kyoto Univ.) (Sys., E, U)(1234567)(nsd)((1)) [B]; Structural modelling and evaluation, Graph-theoretic approach, Educational Information System, DEMATEL, Multivariate analysis.

E-1: Comprehensive Study of Infological Research of Research Trend (Prof. H. Tanaka, Hokkaido Univ.) (I, St., E, P)(1234567)(nsd)((123))[BS]; Integrated usages of scientific documents, Structural analysis, Citation lineage, Successive processes of statistical inference, SCAT-IR system, Problem blocks, Representation of massive research trends, Clusterization.

3.2 Research Group Reports

In each Research Group, the IRU having the number 1, such as A-1, B-1, C-1, D-1 and E-1, has the responsibility of promoting smooth cooperation and of coordinating research activities in its respective Research Group.

3.3 Development Activities in Connection with Plan Promoting Committees

(1) M-Committee on data base management system (DBMS)

The objectives of M-Committee were assigned in May, 1977, to make clear the principal features of scientific data bases and to define the fundamental requirement specifications for DBMS which are both reasonable and realistic for an implementation starting from 1980 in University communities.

(2) O-Committee on the development of scientific information system

The objectives of the O-Committee were assigned in May, 1977, to specify the goals and the procedures for organizing scientific information and to establish a future plan for establishing scientific information systems in Japan.

(3) N-Committee on computer network system

The objectives of N-Committee were assigned in May, 1976, to conribute to the realization of an inter-university computer network system which was scheduled to start with general service from April, 1979.

(4) I-Committee on input/output system formations

The objectives of I-Committee were assigned in August, 1977, to solve information entry problems in general.

(5) T-Committee on general problems of coordination

The activities will be explained in the following Subsection 3.4.

3.4 Draft Plans for Establishment of Scientific Information Systems in Japan

In view of the basic objective of SRP, one of its most important jobs has been to prepare a draft plan for the establishment of scientific information systems in Japan on the basis of, and in view of, the research activities and the research achievements which have been established during the course of these five years since the first Special Research Project. In 1975, the Coordinating Committee published a draft plan for the establishment of a large scale scientific information system in its Japanese report, which was a continuation of the former Japanese report [11]. In view of our experiences accumulated through our research activities, and in accordnace with the recent technological innovation regarding computer network and the data base management system, we were in a position to revise our former tentative draft [11] so as to be in accordance with the recent requirements given in [12].
T-Committee is in charge of the task of organizing the preparation of draft plans in cooperation with the four Committees; that is, M, N, O and I Committees. The first tentative draft plan [13] was published in March, 1978. The final report for establishment of scientific information system was given in the November Meeting of the SRP in which the five definite proposals were presented.

4. STRATEGIC ANALYSIS OF RESEARCH ACHIEVEMENTS

The research achievements of our SRP can be observed from the two Annual Reports [2] of the Coordinating Committee and the final Report to be prepared in March, 1979. These Reports are to be based upon the official reports of sixty-nine IRU's, the five plan Promoting Committees, Research Group Coordinators and the Coordinating Committee. In this Section 4, we are not intending to present any summary of these objective description. Instead, we are going to present a strategic analysis of the research achievements by appealing to the virtual dynamical framework whose methodology is now under development by the present author [15], [16]. Three Subsections are set up in order to apply this methodology. These are namely, (i) starting research areas for SRP in Subsection 4.1, (ii) progress directions of the research activities of SRP in Subsection 4.2 and (iii) consolidation fo research activities in Subsection 4.3. The meaning of virtual dynamical considerations applied to an analysis of the research activity trends is explained in Section 5.

4.1 Starting Research Areas for SRP

Let us now explain the starting research areas for our SRP research activities at its beginning in March, 1976. For the sake of brevity, let us set up a set of seven items, each of which has two alternatives for a choice. Then the starting research area specification can be given by identifying a set of seven preference choices each of which is concerned with the two alternatives in the respective item. Our preference choice of A to B is briefly denoted by 'A to B' in what follows.
(1) scientific information to general social information
(2) discipline-oriented scientific information to mission-oriented scientific-technological information
(3) research information to education information
(4) reproduction of inforamtion to circulation of information
(5) generalization of information expressions to restriction of docummdntal information
(6) researcher files to data bases for common use

(7) knowledge as structured information to unstructured assembly of information

Here our preference of A to B implies (i) that more emphasis was placed on A than on B at the beginning of our research activities, (ii) that removal of our emphasis from A to B was not rejected if it was adequate and necessary in view of the progress of our research activities and (iii) that in all items, both A and B were coexistent in our research activities of SRP (as a whole) from the beginning and have been so during their progress.

4.2 Progress Directions of the Research Activities in SRP

The following six progress directions can now be pointed out in view of our experiences in research activities and through our discussions within SRP.

(1) Scientific information resources should have a development scheme as shown in Figure 4.1.

(2) Scientific information resources must be, in principle, distributed according to their origins and to the geographic considerations depending upon user distribution, except for the cases of highly universal scientific information resources which can, and should, be concentrated.

(3) Sophisticated structuring is required for data bases of scientific information in each specific science area in such a way that the knowledge system can be duly reflected in its construction.

(4) Scientific information systems should be designed so as to be evolving systems. In order to realize the specific requirements, this should be set up for the DBMS of scientific information storage.

(5) The construction of a computer network system is recognized as a crucial strategy in promoting the establishment of a scientific information system in Japan.

(6) A total system of the science and technology information system is required to establish intimate connection with the scientific information system which SRP is now proposing.

In what follows, we shall point out the three specific points which should be supplemented and promoted by the cooperative works within SRP. See (a), (b) and (c).

(a) Completion of two-dimensional connections regarding the research objects and the informatical methodologies

Each IRU has been developing some informatical methodologies mostly with reference to some specific science areas of its concerns. Such a specific reference to a science areas is indispensable in the establishment of each informational methodology in order to make its application feasible and, hence, to verify its fruitfulness. After that, however, further efforts should be performed, at least along the following two directions 1 and 2.

Direction 1. To endeavor to integrate various informatical methodologies, each of which has been applied to an assigned common science area with a certain high degree of success.

Direction 2. To gather all the informatical methodologies which have been developed by IRU's and which

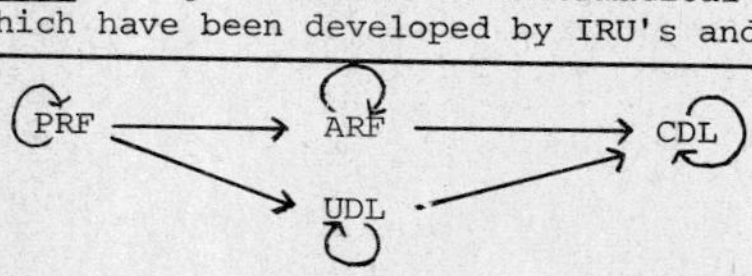

Fig. 4.1 Interrelation of evolutionarity among different utility levels of information storages.

may be recognized to contribute to an assigned common methodological aim, and then to arrange them in such a way that tells among others, which methodology should be in need according to a specification of the situation assigned.

Direction 1 may be coined as a system foramtion of informational methodologies applicable to the assigned science area, while Direction 2 may be coined as an integrated coordination of various methodologies, each of which has the same informational objectives. For instance, with regard to medical science areas, various informatical methodologies have been developed regarding picture image data processing, data base management systems, statistical software packages and so on. An investigation should be done as to whether (and how) all (or some) of them can be integrated for the establishment of a medical information system. In fact, such a development direction belongs to what we call Direction 1.

For instance, in Group E there have been several different methodologies; such as, (α) clustering procedures, (β) reference lineage representation, (γ) gravity center method for the common aim of finding our research trends in some research areas. Now we have to establish an integration of these research trends finding methodologies in such a way that a recommendation for combinations of these methodologies is given to the users of them. In fact, such an investigation belongs to what we call Direction 2.

(b) Informatical classification of scientific information systems Scientific information systems can be classified according to the condition, whether they are common to all science areas or they are specified to a certain set of specific science areas. In the latter case, scientific information systems can be nominated with respect to their specifications; such as medical information systems, chemical information systems and so on. Such a classification may be said to be very much pragmatic and useful because of the present state and the future plan of scientific information network systems. Nevertheless, we should proceed further to establish another classification of scientific information systems on the basis of the informatical standpoints, since such an informatical classificatory pattern will be more useful to further development of informatical approaches in building up an integrated scientific information system.

For instance, in the whole process of information processing, each subprocess ranging over (P_i) $(i=1\sim7)$ can, and should, be specified from the standpoint of informative logics in the sense of the present author [17]-[20]. Then we can introduce four-dimensional informational construction for the classification of a scientific information system to the following effect.

$$\begin{bmatrix} \text{1. field survey} \\ \text{2. laboratory experiment} \\ \text{3. study work} \end{bmatrix} \times \begin{bmatrix} \text{1. observation} \\ \text{2. measurement} \\ \text{3. statistics} \\ \text{4. calculation} \end{bmatrix} \times$$

$$\begin{bmatrix} \text{1. computation} \\ \text{2. control} \\ \text{3. communication} \end{bmatrix} \times \begin{bmatrix} \text{1. cognition} \\ \text{2. direction} \\ \text{3. evaluation} \end{bmatrix} .$$

For instance, meteorological information has the informatical characteristic features emphasizing their main functions;

$$\text{(field survey)} \cdot \begin{bmatrix} \text{observation} \\ \text{measurement} \end{bmatrix} \cdot \text{communication} \cdot \text{cognition}$$

which will be denoted as $1\cdot(12)\cdot3\cdot1$. Chemical loboratory information systems can be denoted by 2·(24)·1·1, while administrative information system 1·(13)·(13)·(123).

We can, and should, make clear the characteristic features of the existing information systems with respect to the four-dimensional construction just explained. However, it is noted that further development of some of these information systems will be expected to become more sophisticated, more knowledge-oriented, more powerful as decision-support systems and, hence, to cover almost all items given in the construction.

(c) Brainware formation as a consolidation of the whole research activities and research achievements of SRP

It is crucially important to our SRP to give a summary of research achievements by providing brainware formation, which is expected to be a consolidation of the whole research activities and research achievements of SRP consisting of sixty-nine IRU's, five Research Groups A, B, C, D and E, five Planning Committees T, O, M, I and N and one Coordinating Committee. The information resources consist of hardware, software and brainware. A consolidated brainware of our SRP aims to provide the following (α), (β) and (γ).

(α) A list of the hardware which SRP has provided to each IRU and to the Coordinating Committee Unit, respectively, and which should be available for common usage. This list will yield us directories for access to each individual hardware which is subject to the administration of the respective authorities under Ministry of Education, Science and Culture.

(β) A list of software resources which have been produced and/or maintained in the University or in the Institute to which each IRU belongs. These software resources are expected to be available for common usage of research workers under certain respective conditions.

(γ) A list of original papers, reports, memo notes and manuals on the use software resources which were published and/or delivered during these three years.

4.3 Consolidation of Research Activities

Detailed description of each IRU illustrated in Subsection 3.2 gives several observations which are important in yielding an objective basis when someone wants to give a check and review on all the research activities and research achievements of SRP. In what follows, two different kinds of Tables, 4.1-4.2, are given as summary observations.

(1) Table 4.1, (a), (b), (c), give us a set of connection matrices among the science areas and Research Group activities. Here, all of the science areas appealing in SRP are divided into five classes, (i) is defined under the specific consideration that the present SRP is basically informatics-oriented. The number in each cell of Table 4.1 gives how many times each symbol denoting its respective science area occurs in the explanations of all IRU's given in Subsection 3.1.

Table 4.1, (a), (b), (c), give us the following observations.

Observation 1. The applications of Group A research achievements are restricted to basic and applied natural sciences and are particularly remarkable in medicine and meteorology. The application fields for Group B activities and achievements are not specified clearly, and its intimate connections with E(Engineering) reflect merely the fact that most research workers in Group B come from the Faculties of Engineering. In Group C, chemistry and medicine are remarkable in comparison with other science areas.

Table 4.1 Research activities and achievements of SRP in view of the matrix Science Area × Research Group.

(a) Methodological area

Science Area	Research Group					
	A	B	C	D	E	Total
(M) Methodological A	13	29	11	6	6	65
(M-1) Informatics (I)	10	21	9	1	4	45
(M-2) Systems Sci.(SYS)	0	0	0	4	1	5
(M-3) Statistics (St)	2	1	1	1	1	6
(M-4) Mathematics (M)	1	4	1	0	0	6
(M-5) Linguistics (L)	0	3	0	0	0	3
Common Appl. Area (U)	3	10	3	2	2	20

Table 4.1 Research activities and achievements of SRP with reference to the matrix Science area × Research Group.

(b) Basic natural science and humanities area

Science Area	Research Group					
	A	B	C	D	E	Total
Basic Science	12	2	11	0	1	26
Astronomy (A)	0	0	1	0	0	1
Physics (P)	0	0	1	0	1	2
Meteorology (Meter.)	4	2	0	0	0	6
Chemistry (C)	2	0	6	0	0	8
Geology (G)	3	0	1	0	0	4
Biology (B)	3	0	1	0	0	4
Genetics (G)	0	0	1	0	0	1
Humanities in large	0	0	4	0	0	4
Humanities (H)	0	0	2	0	0	2
Archaeology (Arch.)	0	0	1	0	0	1
Ethnology (Ethn.)	0	0	1	0	0	1

Table 4.1 Research activities and achievements of SRP with reference to the matrix Science Area × Research Group.

(c) Applied Sciences and Social Science

Science Area	Research Group					
	A	B	C	D	E	Total
Applied Sciences	11	15	7	1	3	37
Engineering (E)	1	11	2	1	3	20
Metallurgy (Met.)	1	1	0	0	0	2
Medicine (Med.)	6	3	4	0	0	13
Pharmaceutics (Phar.)	1	0	0	0	0	1
Environment. (Env.)	0	0	1	0	0	1
Social Sciences in large	0	0	2	2	0	4
Social Soc. (S)	0	0	1	1	0	2
Economics (Eco.)	0	0	1	0	0	1
Policy Sci. (Pol.)	0	0	0	1	0	1

Table 4.2 Information storages provided by each Research Group in terms of utility levels.

UL Inform. Storage	Research Group					
	A	B	C	D	E	Total
((1)) PRF	11	3	4	4	1	23
((2)) ARF	0	0	0	0	0	0
((3)) UDL	2	1	1	0	0	4
((4)) CDL	3	1	0	0	0	4
((12)) PA	0	0	0	0	0	1
((123)) PAU	1	8	10	0	2	21
((1234)) PAUC	0	7	8	0	1	16
Total	17	21	23	4	4	69

Observation 2. The methodological bases for SRP are planned to be provided by informatics and systems science. The relative low percentage of informatics in Groups C and D and the localized existence of systems sciences within Group D show that methodological approaches in SRP have not been penetrating into all of the research activities. The relatively low percentages of the surrounding methodological approaches due to statistics, mathematics and linguistics show that few methodologicalprofound reformations and innovations were introduced in SRP research activities.
(2) Table 4.2 summarizes the utility levels of information storage provided and/or planned by each IRU. Table 4.2 shows that twenty IRU's, among the total sixty-nine IRU's, will provide scientific information resources to be stored and retrieved through some center; while the other twenty-three IRU's have been concerned purely with PRF levels. Most of the information storages with the level ((123)) in Table 4.2 are in the stage of planning, and their actual states may be said to belong to the level ((12)) at the present moment.

5. REVIEWS ON RESEARCH STRATEGIES AND CONSOLIDATION OF RESEARCH ACHIEVEMENTS

Our SRP has finished its activities in March, 1979, and now is going to complete a final report consolidating all research achievements within the period of one year; that is, until March, 1980. This Section is intended to contribute to an efficient organization in performing the jobs imposed upon our SRP by giving an orientation to fostering more intimate cooperation among the members of SRP by setting up a firm common goal.
Generally speaking, every SRP should have its general research objectives and research planning according to which the whole research activities should be coordinated. Our SRP has been sharply conscious of these requirements. With regard to the research strategies, each of which would give a decision on how to choose one of several alternative research activities in actual cases however, our SRP had not set down any principle to be applied to the decision-makings, except for the five Plan Promoting Committees. The research strategies of SRP (as they are called by the author) are virtual in the sense that they have been neither set up as a guide line before the research activities started, nor are the research workers even conscious of them during their reserach activities in each IRU. Nevertheless, the whole process of their research activities so far developed can be analyzed, understood and coordinated from a setup of the virtual framework. In this sense, we may call them research strategies which one may become conscious of in view of the actual experiences accumulated. In our present situation, we are in a position of being aware of the implications of the virtual research strategical considerations. It is a personal view of the present author that we can, and should, take advantage of this recognition in such a way that a set of research strategies now ought to be set up at the middle stage of the SRP activities. It is a characteristic feature of any scientific research activity, particularly within University communities, that not all of them can be strictly planned but that there is always something which should not, and can not, be predetermined and which should be left to learning by experiences accumulated through its research activities. The notion of virtual research strategies will be useful for the review and check procedure regarding the implication of the research activities and the research achievements. In fact, in this way we can find out the way to proceed and how to improve SRP activities in the coming days.

6. CONTRIBUTIONS TO DEVELOPMENT OF INFORMATICS AND SYSTEM SCIENCE

These can be summarized in the following six observations.
Observation 1. Generally speaking, the task of establishing a scientific information system and the organization of scientific information impose us with a lot of scientific and technical problems which can be and/or should be solved only through the development of information science and information processing technologies.
Observation 2. A development of knowledge engineering is urgently desired for the establishment of a scientific information system and the organization of scientific information by which to amalgamate various techniques having been developed in the area of artificial intelligence, pattern recognition and natural language analysis in such a way that their applications should now be coordinated with the common objectives.
Observation 3. A clear recognition of the urgent needs for development of software engineering methodologies in establishing a distributed system of various data bases and researcher files is being settled down among our research workers in view of their own experiences in belonging to our SRP.
Observation 4. The construction of researcher files and/or data bases requires the establishment and development of a new engineering area, titled brainware engineering, which seems to be crucially important in the construction of data bases in the areas of business and administration, as well as in scientific research areas.
Observation 5. One of the predominant tendencies in information processing technology in recent years is an introduction of the 4D system consisting of (α) Data Bases, (β) Data Communications, (γ) Data Entries and (δ) Data Utilities. In our SRP, each of these 4D technologies is intensively studied, as can be observed from the correspondence such as, (α) by the M-Committee and Group C, (β) by the N-Committee and some IRU's belonging to Group B, (γ) by the I-Committee, Group A and Group B and (γ) by the O-Committee, Group E and Group D. In this connection, we are sharply conscious of the urgent need for development of hardware technology in order to have an efficient information system available.
Observation 6. It is needless to say that the systems science approach is indispensable to our theoretical analysis of establishment, management and maintenance

of the scientific information system in connection with software engineering approaches. What we want to emphasize here, is that our SRP has made clear the importance of the roles and functions of researcher files and data bases in promoting the system science approach.

Apart from these six observations just given, the present author has been conscious of the need for an informatical analysis of scientific research acitivities which will be useful in evaluating all of the research activities of SRP. In his recent papers [6], [7], the present author gave this work on the basis of informative logics [17]-[19] and in view of experiences accumulated in the present SRP.

7. CONCLUSION

Our SRP should be subject to a checking and reviewing procedure by scientists, researcher workers and users in general who do not belong to SRP. Although the task of check and review does not authentically belong to the present author who is a member of SRP, he might be allowed to present his personal views on the check and review criteria. The scientific criteria may be said to be formulated in connection with the consolidation of research activities as illustrated in Subsection 4.3. From the pragmatic points of view, we have to point out several crucial problems which remain unsolved in Japanese societies. Although any solution to these problems is neither the sole responsibility nor the specific one to be imposed upon our SRU, we are still conscious of the needs of fostering our efforts within the responsibility of research workers in contributing to the following remaining works.

(1) An establishemnt of common understanding and substantial support of science communities, the Ministry of Education, Science and Culture and the Science and Technology Agency in order to realize our proposals of establishing a scientific information system in Japan.

(2) An establishment of an intimate connection of libraries in general, and university libraries in particular, with scientific information systems in our plan.

(3) An establishemnt of international communication and cooperation through scientific information systems.

(4) A clear specification of the roles and functions of the scientific information system within the general framework of the NIST plan in Japan.

(5) An establishment of informatical methodologies and technologies useful to various information systems in the areas of administration, education, medicine environment, material resources, traffic control and so on.

These problems are primarily concerned with national policies and are not directly oriented to scientific and technical research and development. Nevertheless, it is hoped that some of our research achievements may be useful when people have to encounter some of these problems and/or when research and/or development will be planned by any agency in order to solve some of these problems.

In conclusion, the present author would like to refer to his old paper [20] published in 1969, when his broad ideas of a scientific information system were presented. Although his characteristic emphasis may be said to be invariant for these ten years, the participation of the present author in two Special Research Projects leads him to a new prospect, as shown in his recent works [6], [7], [21].

Appendix I. The M-Committee Report on Data Base Management System in the present Proceedings specifies three levels of data storages PRF, UDL and CDL instead of four levels PRF, ARF, UDL and CDL. The M-Committee had started with an introduction of these four levels, but amalgamated to the three levels. According to the personal views of the present author it is worth while to emphasize the intermediate roles of ARF in the architecture of scientific information system in order to make our ideas more understandable.

Appendix II. There are a few concepts being introduced by the present author for informatical analysis of research activities and illustrated in the paper [6]. Since these concepts play a basic role in constructing the present paper, the author presents a list of Sections of the paper [6].

1. Informatical aspects of research activity components

1.1 General description of informatical aspects of research activities; 1.2 Research subjects; 1.3 Research methodology; 1.4 Research objects

2. Research methods

2.1 Information processing subprocesses in scientific research workes (1); 2.2 Fundamental features of data; 2.3 Simulator; 2.4 Generalized artificial grammar

3. Information usage subprocesses

3.1 Data and method dictionaries; 3.2 Role of computer in the usage of information Computer Aided Simulation Experiment (CASE)

4. Research resources

4.1 Research resources in connection with research activity components; 4.2 Scientific brainware

5. Planning of scientific information system in Japan

5.1 Plan methodology; 5.2 Basic process for planning work; 5.3 Responsibility of scientific research subjects; 5.4 Characteristic features of planning for scientific information system; 5.5 Preliminaries for draft plan of scientific information system in Japan.

Appendix III. The paper [23] discusses an informatical approach to models and architecture for distributed data sharing systems (DDBSS). In Section 6 knowledge engineering approach to DDBSS is given in the emphasis of (i) datalogical approach to generalization of data base management systems in connection with (a) data box (b) editing (c) structuring and (d) dbms and DBMS and (ii) generalized artificial grammar (GAG) approach to brainware engineering. General considerations on DDBSS are given in view of our experiences in SPR. In Section 7 relation calculus and content description of data in generalized artificial grammars are described. Furthermore mutual relations among three informative function cognition, direction and evaluation are discussed.

Appendix IV. The paper [24] explains the notion of generalized artificial grammars in the sense of the author. Applications of generalized artificial grammars to knowledge engineering approaches are explained with specific reference to (1°) abstract formation of scientific documents, (2°) informatical analysis of scientific research activities.

REFERENCE

[1] The Summary Report on the Special Research Project "Advanced Information Processing of Large Amount of Data over a Broad Area", sponsored by Ministry of Education, Science and Culture, during the period from April, to 1973, March, 1976 (in Japanese), Univ. of Tokyo Press, Tokyo, March, 1976, 1-1253.

[2] The Intermediate Reports on the Special Research Project "Formation Process of Scientific Information Systems and Organization of Scientific Information", (in Japanese) (1) The First Year Report, March, 1977; (2) The Second Year Report, March, 1978, The Coordinating Committee; (3) The Third Year Report, March, 1979, The Coordinating Committee.
[3] T. Kitagawa, "Formation process of information system in connection with organization of scientific information", Research Inst. Fund. Information Sci., Kyushu Univ., Research Report, 60 (1976), 1-21.
[4] T. Kitagawa, "The Special Research Project Formation Process of Information System and Organization of Scientific Information", (in Japanese) Joho Kanri, 20, 7(1977), 525-533.
[5] T. Kitagawa, The Special Research Project "Formation Process of Information System and Organization of Scientific Information" (in Japanese), Gakujitsu Geppo, 31, 5(1978), 33-46 (in press).
[6] T. Kitagawa, "Information Analysis of Scientific Research Activities", Research Inst. Fund. Information Sci., Kyushu Univ., Research Report, 81(1978), 1-51.
[7] T. Kitagawa, "A new trend in informatics with specific reference to scientific information systems", Presented to the FID Congress in Edinburgh, (Sept. 25, 1978).
[8] T. Kitagawa, "Brainware concept in an intelligent and integrated system of information", Research Inst. Fund. Information Sci., Kyushu Univ., Research Report, 39(1974), 1-20.
[9] T. Kitagawa, "Statistics and brainware in intelligent and integrated information system", Proc. 40th Session of International Statistical Institute, Warsaw, 1975.
[10] T. Kitagawa, "Structure formation of scientific brainware as a basis of data base management", Bull. Math. Statist., 17(1977), 55-64.
[11] Perspectives for establishment of large scale scientific informations system (Intermediate Report in Japanese), Part I, Introduction, The Special Research Project "Advanced Information Processing of Large Amount of Data over a Broad Area" sponsored by the Ministry of Education, Science and Culture, Report of the Coordinating Committee, 9(October, 1974), 1-127.
[12] T. Kitagawa, "Manual for preparing draft plan of scientific information system" (in Japanese), The Coordinating Committee, (May 4, 1977).
[13] Draft plan for scientific information system (in Japanese), The Special Research Project "Formation process of information system and organization of scientific information", sponsored by the Ministry of Education, Science and Culture, Report of the Coordinating Committee, 7(March, 1978), 1-60.
[14] T. Kitagawa, The Special Research Project "Formation Process of Information System and Organization of Scientific Information" (in Japanese), Joho Kanri, 20, 7(1977), 525-533.
[15] T. Kitagawa, "Neutral equation approach to recognition processes on the basis of reverberation and virtual dynamics", Presented to International Symposium on Mathematical Topics in Biology, (Sept., 10-12, 1978), Kyoto.
[16] T. Kitagawa, "An informatical formulation of generalized relational ecosphere on the basis of paired categories", Presented to International Conference on Cybernetics and System, (November, 1978), Tokyo and Kyoto.
[17] T. Kitagawa "The logic of information science" (in Japanese), Tokyo, Kodansha, Contemporary Ser., 200 (1969).
[18] T. Kitagawa "Three coordination systems for information science approach", Information Sci., 5 (1973), 157-169.
[19] T. Kitagawa, "Mathematical approaches in theoretical cybernetics and informatics", The 8th International Congress on Cybernetics, Namur, September 6-11, 1978, Assoc. Int. de Cybernetique, (1978), 267-284.
[20] T. Kitagawa, "Information science approaches to scientific information systems and their implications to scientific research", Research Inst. Fund. Information Sci., Kyushu Univ., Research Report, 3(1969), 1-41.
[21] T. Kitagawa, "The role of computerized information system in knowledge societies", Presented as keynote address in the 8th Australian Computer Conference, (September 29, 1978).
[22] T. Kitagawa, "Formation process of scientific information and its implication to informatics", Journal of Information Processing, 1(3), (1978), 111-124.
[23] T. Kitagawa, "Informatical approach to models and architectures for distributed data bases", Presented to Working Group C, IGDD, IRIA, Aix-en Provence 1979, May 16-18 (in press).
[24] T. Kitagawa, "Generalized artificial grammars and their implications to knowledge engineering approaches", International Institute for Advanced Study of Social Information Science, Research Report, No. 6, August, 1979, 1-29.

Scientific Information Systems in Japan
H. Inose (ed.)

GROUP STUDY ON THE PLANNING OF SCIENTIFIC INFORMATION SYSTEMS IN JAPAN

Takehiko SHIMANOUCHI

College of Computer and Information Science, University of Tsukuba, Sakura-mura, Ibaraki-ken 305

The Committee for Planning Scientific Information Systems in Japan (T-Committee) prepared two reports. The first report describes the details of the planning; the second summarizes the planning in the form of five proposals. The outline of the first and the chief details of the second are described. The five proposals deal with (1) technical developments in input-output facilities, database management systems and computer network systems, (2) developments in individual scientific fields, (3) the organization of information, (4) the promotion of national and international cooperation and (5) providing a foundation for promotion of system develpment.

1. INTRODUCTION

In Japan, and especially among the members of the present research group, the importance of planning scientific information systems in various fields of science is being emphasized. Demands for these systems from scientists in various fields are also increasing. This is the background for the formation and activity of the T-Committee. The first report of the Committee, "Planning of Scientific Information Systems in Japan", was published in March 1978 in preliminary form. The report was distributed to all the members of the present research group; they submitted many opinions, and a discussion meeting followed. A supplement to the report was then published. The final version of this report, 118 pages written in Japanese, was prepared and published in November 1979 and incorporated the earlier versions and the discussions at this meeting.

A second report, which summarized the T-Committee final report and other reports prepared in the research group, was prepared by Prof. T. Kitagawa and published in November 1979. The report, "Proposals for Planning Scientific Informaiton Systems" 23 pages in Japanese, deals with the problem, advances five proposals, and explains the details of each of them.

In the present report only the translation of the table of contents is given for the full T-Committee report; this appears here as Section 2. Essential points of the second report are given in Section 3.

2. CONTENTS OF THE FIRST REPORT, "PLANNING OF SCIENTIFIC INFORMATION SYSTEMS IN JAPAN"

3. PROPOSALS FOR PLANNING SCIENTIFIC INFORMATION SYSTEMS

3.1 Introduction

The report first defines scientific information as information of all kinds which is compiled, formed and utilized in the following steps of the development of scientific research:

(a) Accumulation of data and research materials obtained in the course of experiments, observations, field investigations and surveys
(b) Acceptance, evaluation and summarization of the above information
(c) Recognition of the information structure in relation to the research object
(d) Publication of the results in primary literature and in data depositories
(e) Secondary literature and review articles
(f) Inference of research trends.

One purpose of the present research is to analyse the process of scientific research, treating it as complex and higher-order information processing, and to study the methodology of forming information systems and also of organizing scientific information.

The present research will contribute to the progress of information science and system science. In addition, the results can be applied to the practical planning of scientific information systems in Japan.

Our report has been prepared from this standpoint, and we hope that it will be useful for policy-making by the government and by university people. The preparation of this report was useful also within our research group in the following ways:

(a) To utilize these results in the planning of scientific information systems
(b) To utilize this planning for guidance of research projects
(c) To provide a foundation for scientific research
(d) To emphasize the international significance of current research projects
(e) And to utilize this planning for future developments.

3.2 Basic understanding

The members of the present research group share common basic understanding with respect to the following points:

(a) The present situation in the organization of scientific information
(b) Characteristics of scientific information
(c) Demands for scientific information systems:
  (i) To transfer information obtained by individual scientists to the common resources
  (ii) To assign the obtained information to appropriate location within the whole system
  (iii) To disseminate the obtained information as an important resource in society
(d) Special situations
  (i) Members of non-profit organizations like universities and research institutes play principal roles.
  (ii) At present most scientific information is stored in libraries as books and journals.
  (iii) Scientific research is in principle international.
(e) Importance of an approach from a system science viewpoint.

3.3 Five proposals for the planning of scientific information systems

Proposal 1 (Planning for technical developments)

In the formation of a scientific information system, the techniques of information processing should be fully utilized, especially in the following respects:

(a) Input-output systems
Developments in automatic data input systems, man-machine interface systems, and data output systems adequate for various ways of dissemination, and the establishment of a flexible system of input-output standardization

(b) Database management systems
The processes of compiling, structurizing, storing and retrieving data should be automated by adopting the software and brainware*. DBMS's able to deal with all of the following files should be developed:
  Private researcher file (PRF)
  User data library (UDL)
  Center data library (CDL)
(The three files are explained in Section 3.4.1.2)

*This term is used for the knowledge stored by the information system.

(c) Computer network systems
Materials for scientific information are in most cases dispersed throughout universities and research institutes. In order to utilize these materials, network information systems should be formed in Japan and in the world. The backbone structure of the system should be supported by governments. The free-market system should be adopted for individual utilization.

Proposal 2 (Planning for development of systems in various fields of science)

Formation of the information network system in each field should consist of the following steps:

(a) The information committee is formed. It establishes the plan for organizing information, arranges the method of administrating the system, and cooperates with learned societies in the field and in other related fields. It should be supported by the main society in the field.
(b) The formation, dissemination and utilization of private researcher files in this field is promoted.
(c) The facilities for establishing user data libraries in this field are set up.
(d) The center data library in this field is planned.

Proposal 3 (Planning for organizing the overall information system)

(a) Establishment of Committees
A Committee dealing with the whole system is set up in the scientific information promotion center. Committees are also set up to deal with the systems for individual fields.
(b) Establishment of central organizations
These organizations are set up in regional computer centers, interuniversity research institutes, information centers in various fields and information centers in universities.
(c) Promotion of database projects
Projects are promoted by the Ministry of Education, Science and Culture and by other ministries and agencies. The obtained databases are stored in the computer files of these central organizations.

Proposal 4 (Promotion of national and international cooperation)

Mutual understanding and close cooperation among the Ministry of Education, Science and Culture, the Agency of Science and Technology, the Science Council of Japan, learned societies in many fields, and university and other libraries should be promoted. International cooperation with ICSU-AB, ICSU-CODATA and the information systems in various countries should also be promoted.

Proposal 5 (Providing a foundation for promotion of system development)

(a) Education and training of information specialists should be planned.
(b) Research projects in computer and information science should be promoted, and technical capabilities for the formation of scientific information systems should be maintained at the highest level.
(c) Arrangements for governmental support for database formation should be established.

3.4 Structure of individual factors of planning

(a) Structure with respect to the steps in information system planning
- (i) Development of scientific information
- (ii) Organization of scientific information
- (iii) Management and administration
- (iv) Dissemination of scientific information
- (v) Utilization of scientific information

(b) Structure with respect to the constituent parts of the information system
- (i) Training and assignment of specialists
- (ii) Information centers
- (iii) Hardware, software and brainware facilities
- (iv) Financial aspects

(c) Structure with respect to progress over time
- (i) Long-term and basic plans
- (ii) Short-term survey
- (iii) Annual plans

In the following the details of the proposals are given mainly from the viewpoint (a). In the practical procedure the other two viewpoints are also important.

3.4.1 Problems in technical developments (Proposal 1)

3.4.1.1 Input-output systems

(a) Basic factors
- (i) The information should be scientifically well-defined and needed by scientists. The planning for developing the information should be carefully prepared.
- (ii) The actual procedure of making the information machine-readable should be precisely described.
- (iii) Machine-readable information is not always machine-compatible. Procedures for producing machine-compatible information should be clarified. The machine-aided method should also be developed.
- (iv) The information users are in most cases the information producers. The method of including produced information in the system should be developed.
- (v) The role of the system administrator should be well-defined.
- (vi) The information system should reflect the status developments in performance, reliability, security and cost of the computer hardware and software.

(b) Factors raised by users
- (i) Number, cost-performance, and special abilities of the terminals
- (ii) Interactive method of training inexperienced users
- (iii) Effective use of the system corresponding to the experience level of the user
- (iv) User's manual
- (v) Training programs for users
- (vi) Acceptance of technical developments made by users
- (vii) Documentation of the system, information about system updating, etc., should be included in the system and be accessed rapidly and effectively by the users.

3.4.1.2 Scientific information management systems

(a) Factors needed for management systems

Updating system, retrieval system, output system meeting user demands, and processing and transfer of patterns and images

(b) Constitution of data-files in the system

An important characteristic of the formation of a scientific information system is that the procedure consists of the accumulation of the individual efforts of many scientists. Accordingly, the constitution of the data-file should be adapted to this characteristic. In the proposed system the following three kinds of computer files are developed:

(i) PRF (Private researcher file)
Scientists or groups of scientists compile and accumulate information in their individual fields of science. This kind of information is stored in the form of PRF. In this stage the procedure of organizing information is not always fixed. Accordingly, the structure and the utilization method should be flexible.

(ii) UDL (User data library)
The above PRF becomes stable after the information organization stage and grows into UDL. The UDL is offered to a wider range of users. In this stage, however, the system is still flexible. The details of the procedure of utilizing the system will often be improved. Accordingly, only a limited number of users are admitted; the purpose of opening the file to these few users is to improve the system.

(iii) CDL (Center data library)
The UDL grows into CDL after a certain period of time. The system is now fixed. The manual is carefully prepared. The use of the system is open to public.

3.4.1.3 Dissemination system

For the establishment of a successful scientific information system, the following three factors are important:

(a) Governmental support for the formation of basic databases
(b) Establishment of an easily accessible on-line data network
(c) A free-market structure promoting the formation of useful dissemination systems.

3.4.2 Problem in the development of systems in various fields of science (Proposal 2)

3.4.2.1 Basic problems

The following problems should be discussed and a consensus throughout the scientific community should be formed.

(a) The production and the processing of raw data
(b) The production of computer readable data
(c) The process of evaluating data or data systems
(d) The steps for opening data to the public and problems related to copyright

3.4.2.2 Policy for selecting fields to be promoted

Planning of scientific information systems in Japan should cover, in principle, all the fields of science. However, it is not practical for the scientists of all these fields to start simultaneously. In order to select the fields to be started, the following five criteria should be taken into account.

Criterion 1. Usefulness as scientic information
(a) Reliability of the contents
(b) Usefulness from the viewpoint of the speed of the transfer and the breadth of the dissemination
(c) Usefulness from the viewpoint of combination with other information
(d) Breadth of the field in which the information is used
(e) Usefulness of the information for solving problems

Criterion 2. Cost of preparing the system as PRF, UDL or CDL.

Criterion 3. General usefulness and the degree of opening to the public

Criterion 4. The degree of scientific and social expectation and the support of scientists in the field

Criterion 5. The existence of a group of scientists promoting the formation of the system and having capability and strong interest in its formation.

3.4.2.3 Cost of developing these systems

The cost of forming computer readable databases is estimated. It is not practical for present users of the database to pay the cost; governmental support is needed. In order to decrease the cost to be paid by users, the close cooperation among the scientists, information committees, learned societies and information centers is also needed.

3.4.3 Problem in planning for organizing scientific information (Proposal 3)

The organization of scientific information can only be performed by the cooperation of regional computer centers, interuniversity research institutes and university libraries. Regional computer centers should be equipped with large storage systems, operating systems and database menagement systems adequate for the acceptance of PRF, UDL and CDL. Interuniversity research institutes have special program packages and databases. They should be accessed by the remote terminals. University libraries should have computer-readable information systems for retrieval of primary and secondary literature. All of these three organizations should be connected by the information network and all the database should be used by the scientists all over the country.

3.4.4 Problems in the promotion of national and international cooperation (Proposal 4)

3.4.4.1 Concept of the scientific information promotion center

The committee advances a proposal for establishing a Scientific Information Promotion Center. The Center would plan, promote, and sometimes regulate scientific information activities and promote the use of the results.

The production and the preservation of PRF, the development and the introduction of data banks and the establishment of information centers in basic sciences would be planned in the Center.

The training of information specialists could also be performed in the Center.

3.4.4.2 Process of promoting the plan

The Science Council of Japan, the Ministry of Education, Science and Culture, the Agency of Science and Technology, and the learned societies in various fields should promote the plans they are preparing, and also should coöperate with each other so that their individual plans are well coordinated.

3.4.4.3 Role of university libraries in the formation of scientific information systems

University libraries have traditionally preserved books and journals for the use of scientists and students. In the proposed system of scientific information the knowledge is stored in computer files. These two are technologically different, but share the same role of preserving scientific knowledge for the user. Accordingly, they should be coordinated with each other, and be consolidated in the future.

Cooperation will help with three short-term problems:
(a) A considerable part of CDL consists of information from secondary-source literature. The primary-source literature is stored in the university library. Close cooperation is needed.
(b) University libraries all over the country could supply the space needed for high-performance computer terminals. This kind of cooperation is very useful for the formation of the scientific information network system.
(c) The cooperation is useful for development of library business automation.

3.4.4.4 International cooperation

Scientific information is international in nature and the necessity of cooperation is ever increasing. The information system in Japan should be constructed as a part of a global system, so that Japanese scientists can be active both as producers and users of the system. Cooperation with ICSU-CODATA, ICSU-AB, the EURONET system, and other international and national systems of the world should be promoted. Standardization in information processing, copyright problems, and economic problems are also important.

3.4.5 Problems in providing a foundation for promotion of system development (Proposal 5)

Proposal 5 is related to long-term contributions to the systems. Within education, we must emphasize two points: education of specialists for the system corresponding to education of librarians for the library and training of scientists in the formation of the system.

As for research, the study of decentralized systems of information processing, knowledge engineering, software engineering, natural language processing, and systems of pattern recognition is basic to the organization of scientific information.

Scientific Information Systems in Japan
H. Inose (ed.)

DEVELOPMENT OF SCIENTIFIC INFORMATION SYSTEMS IN JAPAN – REPORT OF O-COMMITTEE

Hajime TANAKA

Faculty of Science, Hokkaido University, Nishi 8-Chome, Kita 10-Jo, Kita-Ku, Sapporo

The activities of O-committee are reported for three years. Four problems are developed. The first is to catch the situation of the activities for scientific information in typical fields. The next problem is the manner to calculate the fund for making the data base. Furthermore the foundation of the Japan center for Promoting Scientific Information is proposed. This is the third one. Lastly the several policies are presented in order to encourage the activities of scientific information.

1. THE PURPOSE OF THE COMMITTEE, A RECORD OF ITEMS ON THE AGENDA DISCUSSED AND THE MEMBERS MAKING UP THE COMMITTEE

1.1 Objective

The objective of our Special Research Project is to scrutinize the formation process of an information system and to make a comprehensive study on the course of developmental research which is a most complicated and highly advanced information processing system. Based on this the methodology of the organization of scientific information and the development thereof, must be set up and while an effective social system of scientific information must be established, at the same time it must contribute to system science as well as to information science.

For this particular work a dual system consisting of a research groups and committee, cluster, was organized for the purpose of dealing with a multiple system. Out of necessity the research group had to choose between objectives and the method, thus the research group strove to overcome such limitations. In order to find an answer to the inevitability, combined sub-research, group clusters were formed or various committee groups were adopted which had a common objective or a common method.

The committee for plans and promotion of scientific information organization which was formed in the First year, and the Committee for development of scientific information which was formed in the Second and Third year are both of the same type of committee and both are generally referred to as O committee.

The objective and duties of the committees changed slightly in the First, Second and Third year in that the points stressed were slightly different. In the First year the objective and duty of the committee was in pointing out problematic points based separate investigation in various fields. In the Second and Third year with the aim of organizing a method by which various fields could be covered simultaneously, stress was laid on policy and formulation of a means by which to achieve this end.

The objective of the First year was to formulate and promote plans for the organization of scientific information in each branch of science.

i) To formulate plans for the organization of scientific information in each branch and each field.
ii) Liaison and adjustment of multiple plans for the organization of scientific information in each field and dicipline.
iii) To set up a circulation system for the utilization of organized scientific information in such fields that the organization is well advanced.

Especially in the First year because a research group for the field of sociology was lacking, in the functions of the committee this was included. The O committee which was set up in the Second and Third year, had its objective in clarifying the measures for organization of scientific information and to prepare future plans for the establishment of scientific information systems in Japan. To achieve this purpose the following points were taken to designate the duties of the committee.

Second year
a) To clarify the border lines of the various scientific information systems and to find a means of communication with the outside.
b) A survey on the present state of affairs and analysis of the systems that are in active use at present, including those that are being planned.
c) A preliminary study of plans for organizing scientific information and system analysis.

The purpose a) is mainly to study the relationship of the National Diet Library and JICST.

Third year
d) A preliminary calculation for the expenses entailed for the preparation of data base.
e) A survey of scientific information activity of scientific societies and associations.

1.2 On the organization

The organizing of scientific information is a category set up for the express purpose of dealing with the circulation of scientific information established in this particular research program. When used in a narrow sense, the organizing of scientific information can be defined as the structurizing the circulation and retrieving of scientific information according to the method used. However, generally it is taken as follows; the range of organizing is not limited to the goods and information and in considered as an entity in society and it includes the relationship with society.

Kitagawa in his "Guide lines for preparing a draft plan of a scientific information system" suggested that for the direction of organizing scientific information, he gave his B plans, when taken in a broadsense he seems to have clearly defined the context of organization. He further pointed out that the range of organizing his plans were, held within scientific data, users and institutions; he further pointed out that it should be a consolidation of "Data Organization Plan in a Scientific Information System","User Organization Plan in a Scientific Information System", "Institution Organization Plan in a Scientific Information System".

1.3 Course of Research

In the First year, the present state of affairs of the organizing of scientific information were investigated and discussed in the various fields of sociology, meteorogic data, nuclear fusion, solid physics, environmental science, pathology and pharmacology. At the same time as a problem related to the sciences as a whole organization of men of talent, financial costs and needs, and cooperative structure were discussed.

In the Second year the field in social science that remained uninvestigated, namely jurisprudence, political sciences, sociology were discussed. Further, an outline was given on the problems of scientific information. In Japan, at the same time to set fourth plans for organizing scientific information, typical fields of chemistry, mathematics, and medicine were chosen and the basis for plans and the extend of needs were taken into consideration.

At the same time a questionnaire investigation was conducted in order to make a data file. It was discovered that in order to produce a satisfactory data file, it was necessary to include not only the information retrieval system, but also it was essential to include widely such items as, numerical data, picture data, natural languages and mathematical equations. Further, the problematic points of a data center were discussed from an angle of production of Environmental Data Base, and from a point of view of Research Information Center for Controlled Nuclear Fusion. Still further along this line we outlined the possible relations between scientific information activity and interuniversity large computer centers. With the above discussions used as the basis ground, from both individual scientific fields and centralized structure, plans for scientific information were discussed.

In the Third year, as an actual pressing problem which can not be overlooked, the cost of data base production was assessed. For this purpose an intimate questionnaire investigation was conducted. At the same time another questionnaire investigation was conducted on the scientific information activity of scientific societies and associations.

The results of the above were printed and published as a report in 4 volumes.

a) The present state of affairs and problematic points in the organizing of scientific information. (Intermediate report of the committee for plans and promotion of scientific information organization) March, 1977.
b) Development Committee for Scientific Information System (O committee) report March, 1978 Comprehensive report 8
c) Development Committee for Scientific Information System (O committee) report March, 1979 Comprehensive report 14
d) Development Committee for Scientific Information System (O committee) report March, 1979 - Trends and Tendencies of Scientific Information Activity of Scientific Societies and Associations - Comprehensive report 15

1.4 Composition of the committee

The composition of the committee in the below mentioned years. ◎, ○, △, *, and + marks indicate respectively ◎ committee chief, ○ vice committee chief, △ manager, * resigned committee member, + committee member in Third year alone.

First year
◎Hajime Tanaka, ○Atsunobu Ichikawa, Choichiro Asano, *Tadashi Okamoto, Shu Ono, Kanzaburo Ganbo, Tairoku Kose, Yoshinosuke Terashima, Yoshiro Matsuda, Masaaki Naito

Second and Third year
◎Hajime Tanaka, ○Atsunobu Ichikawa, ○Choichiro Asano, △Haruhisa Ishida, Shu Ono, Shuji Takagi, Tairoku Kose, Yoshiro Matsuda, Yoshinosuke Terashima, *Masahiro Fujiwara, +Koichi Yamanaka, +Yasuhide Yukawa

2. PROBLEMATIC POINTS CONCERNING THE ORGANIZATION OF SCIENTIFIC INFORMATION

2.1 The Diversity of Scientific Information Activity

Scientific Information Activity is characterized by its diversity. Next we shall examine the various types of information, mode of expression, the method of store and the characteristics of the particular field.

The type of scientific information necessary and the mode of expression differs with each field, and due to the degree of necessity, the difference is remarkable. Fig. 1 shows the state of development of data base revealed by questionnaire investigation, and it shows the diversity of the information gathered.

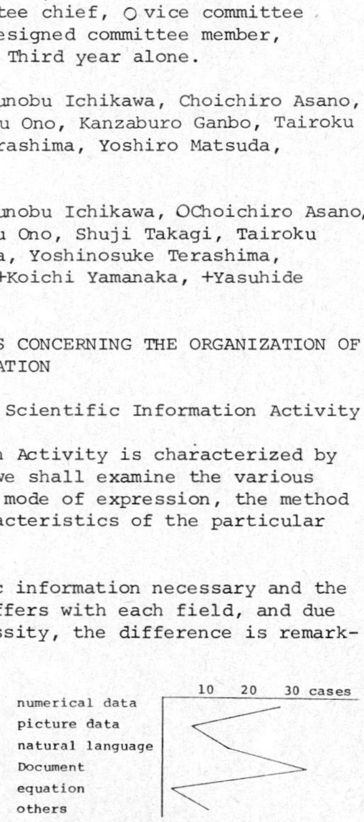

Fig. 1. Type of gathered data

Table 1 shows the results of the Third year. These factual data are gathered and stored with the same relative importance with respect to bibliographic data for document retrieval and fact data accompanied by numerical data and picture data. There are two types of data, namely, those which are out of necessity gathered and stored as time series data, and those which, are out of necessity, stored as up-to-data new data which are constantly replaced. Meteorology, environmental science, economy, political science, sociology, jurisprudence, human and cultural science largely come under the former while mathematics, physics, chemistry, pathology and pharmacology in most cases come under the latter.

			Produced	Organized	Usage	Total
Fact data	numerals		5	1		6
	English, Jap. Kana		1		1	2
	Chinese characters		1		1	2
	Figure and charts		2	1		3
	Total		9	2	2	13
Document Information	English, Jap. Kana	purchased	4			4
		produced	12			12
		combined usage	4	1		5
	Chinese characters					
	Figure and charts					
	Total		20	1	0	21
Sum Total			29	3	2	34

Table 1. Type of data

The present state of scientific information activity in each field, differs with its necessity and the degree of international demand. In the Second year, as typical fields sociology, mathematics, medical treatment were selected. With the state of affairs of each field as the sample material, plans were made for the promotion of scientific information activity and for the setting up of the entire policy. As may be seen in the points discussed hitherto and by grasping the present states in each field in the next paragraph, it may be seen that scientific information activity is diversified, so much so that it is impossible to select any particular one as a example. Regarding this, extreme care must be taken in setting up policies for promoting scientific information activity.

2.2 The state of affairs of each field which can be considered as typical

2.2.1 Mathematics

In the field of mathematics while the majority of research workers are not particulary concerned, there is a decided wish to avoid overlapping or redundancy in their work. In this respect the greatest service demand is in hard copy service; the majority of mathematicians seem to be satisfied with scientific information provided by referral service and hard copy service. As an international activity attempts to use data file has been temporarily given up.

As in the field of mathematics where scientific information activity is not necessarily high, it may be said that scientific workers of that field, rather than being interested in document retrieval are demanding scientific information service in various forms.

2.2.2 Social science

a) Data base of Document Information
In spite of the tremendous increase in the number of post war scientific papers, the data base necessary has not showed any advance both in Japan and in Europe and America. Technical difficulties seem to be rather scarce. With special regard to the yearly increase in the number of books while on one hand the MT method is being adopted in various countries with considerable progress, on the other hand on the vital old editions and classics hardly anything has been done.

b) Statistical Survey Data Base
It has been decided that regarding various investigations and statistical data in Japan, the Statistics Council based on the revision of by-laws and regulations of the Statistics Law on the basis of the recommendations of 1974 and 1977, has made progress in making available results recorded on magnetic tapes. As for production of data at the present day, we can see the activity of the Census of Establishment (every 3 years) and the National Population Census (every 5 years) which is a comprehensive list data of respondents and compiled by Bureau of Statistics. With regards to both, efforts are being made to solidify the center for supplying the frame of the survey. At the same time the production of data of time series data by the Economic Planning Agency are under production. As compared to this the data base of all Ministries and Bureau include secondary statistics gathered from the records of the affairs, and thus it cannot be expected that it will be made public. The data base of the time series data of the Japan Economic Research Center is becoming available as an on-line system and thus this is being utilized by some research groups of some of the universities. However, to be made available as a socio economic data system including factual data, it can not be expected in the near future.

c) Legal Information Data Base
In the field of social sciences, since legal information has its particular coding system and moreover since the store of information is large, there is no way to make it possible for computer processing, in Japan at least this has not been considered.

d) Socio-political Information Data Base
So-called field survey data by individual researchers, in most cases are done manually and not transformed to machine readable type, and system by which these can be systematically collected and stored has not been formed.

2.2.3 Chemistry

Chemistry is a field in which scientific information activity is high. In this field the development has advanced to where TOOL-IR can be utilized. It is noted that the utilization of TOOL-IR is the beginning of scientific information activity of large computer centers. Thus it can be said that large computer centers are to a certain extent suitable for scientific information service. In the field of chemistry the development of numerous data files are planned. And as in the case of crystal data base, there are some that can be put to actual use. Naturally there are some that are still in the planning stage and it can not be said that all data files can be put to actual use.

2.2.4 Medicine and Pharmaceutical Science

a) The organization of scientific data in pathology and pharmaceutical science, inasmuch as the demand is, out of necessity, extremely high is carried out as a natural course of events. Since both medical science and pharmaceutical science contribute to society directly, it may be pointed out that they have a strong organizational background. Especially, this feeling is strong in the field of medical science. In the case of pathology, based on specialists for the treatment of diseases with a firm basic medicine research organization in the background, conditions for making cooperation possible in the data processing of information are necessary. Storage with accessible input information, together with medical standardization as a common national resource, and with the development of methodology of data analyses is a problem of the future.

b) The organization of scientific information in this field is characterized by its long lasting basis. In pharmaceutical science the organization of scientific information is effectively carried out by combining the function of libraries and sub-library rooms. The main titles of research reports and medical journals and understanding of summaries or resumes are necessary, but the researchers are not directly concerned with the organizing of information and that they had a degree of success largely is because they took time to educate and train the librarians. In pathology likewise the organizing of scientific information was not the direct responsibility of the doctors who are the main research workers. Several trained medical examination technicians are hired and do supplementary research; thus by spending a little money for these people a strong long term organization is made possible.

c) In the field of pharmaceutical science, inasmuch as a high demand for scientific information exists, the recipients of the scientific information, out of necessity, must bear the expenses.

d) The store of scientific information and the increase in quality thereof, as seen in the case of pharmaceutical science is made possible by the frequency of use of scientific information by users. In the organizing of scientific information, whereas it is accompanied by discussions for the completeness of the contents and increase in quality thereof, as in the case of pharmaceutical science, as a general rule the quality and quantity is sparked by the utilizers and advances accordingly. It is rather that the preparation and establishment of the analytical methodology system of stored information groups is important.

e) In the field of pathology which is in close cooperation with the present committee, the first edition of the PADAMS Scientific Information was made available to the public and already at 3 or 4 large cumputer centers of universities, common use is being made. This is typical of the experimental work of the committee which they are exceedingly proud of.

2.2.5 Environmental Information at the Pollution Research Center

The problematic points in data file compilation of the production of Environmental Data Base are as follows.

a) The compilation of data production by various self-governing bodies are sporadic and are not standardized. (e.g. The method of measurement (method of analysis, method of sampling and the method of maintenance of the various Monitoring Station))

b) The method of data check adopted by the various self governing bodies are scattered and are by no means standardized.

c) The method of data format adopted by the various self governing bodies are scattered and are far from standardization

d) Attributed data (When, Where, Who, What, How) are insufficient.

The role played by the data center in the compilation of data files is to solve problems from a)∼d), and according to the need of users, to supply data in such a way that they can be smoothly and readily used by the users. When this work can be done in a fixed way and becomes routine, it becomes a fairly simple task. However, the first workers who decide what is required not only needs expert knowledge in the management of information, but also must have a specialist knowledge from the users standpoint. He must be able to use his creative knowledge to the full extent which is the same as a researchers knowledge and ability. The problem is to find a way to entice such knowledgeable and talented people and to get them really interested in the problems which requires enthusiasm and dedication to accomplish the work.

2.2.6 Setting up a Research Information Center for Controlled Nuclear Fusion

Recent nuclear fusion research is not limited to the field of plasma physics and engineering and is a comprehensive research which requires an extensive range of specialists knowledge and cooperation. In order to promote research, it is necessary to fully utilize a tremendous amount of research information covering a wide range of special fields.

With this in the background Fusion Research Association of Japan following a nationwide discussion of researchers, arrived at the decision to found an "Information Center for Research Planning of Nuclear Fusion" in the Plasma Research Institute of the Nagoya University. In the 1977 fiscal year the plan was officially started.

The objectives of the center are the collection, compilation, evaluation and dissemination of scientific information required for fusion research and the analysis and investigation of problems arising in the course of development of fusion research, thus making the results available to all the scientists concerned and contributing to the planning of fusion research programs. The plans as first conceived, including the 3 department conception will in the course of time be brought into being. At the time of this writing several educators of professor status have been assigned, and with the cooperation of all researchers in Japan task committees have been organized and their activity is beginning to show results.

The center runs an information dissemination service based on a copy service between National Universities and other methods by which information is collected on a bartering basis etc, at the same time with the cooperation of the Computer Center, Institute of Plasma Physics, the center will begin an information retrival service. The data base which will be used in the latter case, will include such information as INSPEC and will be gradually increased in scope and range to USGRA which is mainly made up of research reports of American National Governmental Institutes.

2.3 Necessary Condition for Organization of Scientic Information

The results of the investigation of the data base

Data Capacity	Developed (α)	to be added (β)	(α+β)	under development
under 1MB	5	6	2	8
over 1MB under 10MB	7	5	9	8
over 10MB under 50MB	11	1	8	1
over 50MB under 100MB	5	1	3	1
over 100MB under 800MB	4	3	5	1
Total	32	16	27	19

Table 2. The numbers of cases and data base capacity under development.

development carried out in the 3rd year are shown in Table 2. In Table 3 the data base production funds are shown by the number of cases.

			Gov. Research Funds	other sources	not recorded	Total
Fact Data	Numeric characters		4	3	1	8
	English, Jap. Kana		1	4		5
	Kanji characters		3			3
	Diagrams	purchased		1		1
		produced	2			2
		combined	1			1
	Total		11	8	1	20
Document Information	English, Jap. Kana	purchased	8	2	2	12
		produced	9	6	2	17
		combined	5			5
	Kanji characters			1		1
	Diagrams					
	Total		22	9	4	35
Grand Total			33	17	5	55

Table 3. Funds Source. Data Base Prod. Cases by Type of Information.

These Tables shown that the data base production are greater in number when funds are available. As may be seen that the latent possibility of data base production in the academic world is extermely high, and it shows that Government Research Funds is the largest source of funds available.

Researchers	4,5,12,15,28,30,23,177,21
Info. Staff	31,4,88,22,104,5,54
Assistants	70,18,23,18,78,31,41,62, 25,9,70,68,130,352

Table 4. Number of Required for Data Bace Production per 1 MB.

Many people are required for data base production. Table 4 shows the number of research members, information staff and assistants required for collecting and store of data for 1 MB. The figures are listed at random according to each data base. These information staff members and assistants are in most cases hired for the job and they are not steadily employed.

These facts indicate that in order to organize scientific information, the basic requirements for changing the latent possibility to actual possibility, are to raise funds and increase the member of workers.

2.4 Funds required for the production of data base

In order to calculate the funds required for production of data base, a flow chart which shows the work, necessary for the production of data base and a flow chart in which the necessary numbers of researchers, information staff and assistants at each step are given are required. The information was obtained by questionnaire investigation. There were 84 answers to the survey. The aggregate of the data base production work at each step came to 345 from the questionnaire answers. These are categorized into 4 categories, i.e., organizing, collection and analysis, mechanising of the data, and registering. These respective work numbers were considered and the work volume per 1 MB was calculated. Table 5 gives the results.

		Researchers	Information staffs	Assistants
collections and analysis	determine the location	x		α
	data selection and collection	5 ∿ 10	10 ∿ 20, x	1 ∿ 3
	editing, processing, analysis	5 ∿ 20		$0 \leq\!< 50$
mechanization of data	entering data			$100 r_c d_c + c$
	punch etc		α	$10 r_p d_p$
	correction, adding, verification	1 ∿ 10 or x'		1 ∿ 20
	conversion and input		1 ∿ 2	1 ∿ 2
registration	editing files			
	verify, proofreading, correction	1 ∿ 10	0 ∿ 20	5 ∿ 10
	adding records		α	
	test utilization	0 ∿ 10		
	secondary information storage		α	

Table 5. Per 1 MB work number, necessary for Data Base Production

Here x, x' and α represent uncertain work numbers, for example when the work number is from 1 to 20 at times it is 100. rp (r_c) the rate of information volume that requires punch work (transcription work), while dp (d_c) shows whether the punch work (transcription work) is complicated or not.

As shown in Table 5 to calculate the work volume of data base work, results may be obtained by the following rough guidelines.

1) First we must see whether the work number comes to 1000 or not in the following, namely, the researchers in investigating the location, researchers and information staff required for selection and compilation of data, assistants needed for data recoring, work number of researchers needed for correction, addition and verification. If the values are approximately like this, respective individual work numbers are approximately around these values.

2) If the work numbers are to be known in detail, the work numbers of the following stages must be estimated; determination of location; selection and collection of data; editing, processing and analysis; data recording; punch work etc; correction, adding, verification; conversion and input; editing files; verify, proof reading, correction; adding records;

test utilization secondary information store. These values are in all cases ≤ 20 or thereabouts.

2.5 International demand

In the field of physics, chemistry and nuclear power etc. the various work required in data base production must needs be divided internationally, and work allotment must be made. This trend is increasingly heightened and the expectation of scientific information activity in Japan and by Japan, seems to be extremely high. Recently the yearly publication of research papers in Japan in fields of mathematics, theoretical work in nuclear physics, material science, chemistry etc. have attained the 10% mark of the world. Judging from this, in order to respond to the international request in these fields at least, "we should organize 10 to 20% of the yearly volume of the published scientific information", at least this should be our objective. In order to set the volume of work that we are internationally responsible for, in organizing scientific information in other fields, we could use these figures as a reference.

3. PLANS FOR ORGANIZING SCIENTIFIC INFORMATION

As described in 2.1 in order to promote scientific information activity in various fields, it necessary to make plans to meet the actual conditions in each respective field. At the same time to equally promote scientific information activity in all fields, it would necessary to have some sort of central organization. The basic point of view for organizing plans, in order to suit the actual present state in each field, and to describe about "Japan Center for Promoting Scientific Information" (the Promotion Center), plans to fulfill the basic needs of these 2 points will be discussed.

3.1 Plans for organizing each various field

a) Field where scientific researchers have experienced data activity, and the activity attract attention of many researchers.

In such cases, plans for organizing the special field as a whole should be made, and efforts to promote comprehensively would seem most advisable.

b) Organizing plans which attract the heighest attention in a special area of a special field.

In such a case the actual organizing plans of a specialized field should be carried out by a research institute or by other research organizations and efforts should be made to establish a practical basis for scientific information of the special field.

c) A field where several dedicated researchers are striving for the development of scientific information.

Scientific information activity necessary for the researchers of this field should be sought out, and these should be promoted. It is necessary to see that the activity of such dedicated researchers is continued in the scientific information plans.

3.2 Japan Center for Promoting Scientific Information

a) Plans should be made to bring about the founding the Japan Center for Promoting Scientific Information as quickly as possible. The promotion cneter is a recognized organization specified in the attached table of the National University Establishment Law. Here in addition to the promotion of plans and drafting, Promotion, comprehensive coordination, promotion of utilization etc. we must make efforts to train talented people to organize scientific information and for work related to scientific information.

b) To describe this in more detail, the following should be promoted by the Promotion Center, that is, the attempts to develop, produce and plan "accumulative research files" by individual researchers in various fields and small groups, as well as the efforts to make a comprehensive adjustment of the data bank developing at present in Japan, and the one which is brought in from abroad. The Promotion Center also should promote the founding of Scientific Information Center covering all fields especially those of basic science and liaison work and coordination between each Scientific Information Center.

c) The Promotion Center is responsible for the training of specialists of scientific information and scientific information technical workers and at the same time they must set up a program for the training a pool of talented people to take part in scientific information activity.

d) The Japan Center for Promoting Scientific Information has the alone characteristics, its main function is in organizing and promoting scientific information including research and education. In this sense its activity is not in competition with the work and characteristics of Japan Science and Technology Information Center or the Diet Library. Scientific information activity hitherto has been carried out by various public bodies and special care must be taken to avoid competition with them. We must find the characteristics of the "Japan Center for promoting scientific information" and its starting point in the field of basic sciences

3.3 Basic Policies which will be the basic foundation

a) Funds for data base production will be supplemented by grant-in-aid for development scientific research.

As indicated, in 2.4 it is clear that the fact that scientific research funds are to be used for data base production is highly significant. When researchers take it upon themselves to produce data base, there is the case where production is done experimentally to find whether it is possible or to find problematic points in the production, and there is the case where it has not been decided whether the developed data base will be made public, at least whether the data will be made available to a limited number of researchers. The above are the 2 cases which can be considered. If we consider the character of the former development reseach and the latter in which we consider the character of whether to make public the results, we can consider for instance grant-in-aid for developmental scientific research funds for the former and we can consider for instance grant-in-aid for publishment of research report funds for the latter.

b) Maintenance of interuniversity large computer centers.

From a point of view of supporting the hardware of data base development, the role of computer systems of the interuniversity large computer centers and Interuniversity Research is exceedingly large.

The character of data base development for public use and the character of interuniversity large computer centers which are for common use, closely coincide with each other. The tendency must be actively supported in the future.

In this sense, as one means of supporting the development of scientific information data base, the maintenance of large computer centers for this purpose is indeed most appropriate and must be promoted strongly.

c) Maintenance of Scientific Information Data Base Management System (DBMS).

As a result of our present investigation it was clearly shown that the rate of using DBMS produced by oneself is large. This shows that the maintenance of DBMS which is easy to use as scientific information, is not developed and maintained very well.

The DBMS which is made easily by a limited amount of labor, generally is lacking in various points and can not be utilized widely and when we must consider that it is next to impossible to develop a well maintained DBMS for universal use by individual researchers and small group workers, we must remember that the early development and maintenance of DBMS suitable for universal use for scientific information is imperative.

Scientific Information Systems in Japan
H. Inose (ed.)

DEVELOPMENT OF INTERUNIVERSITY COMPUTER NETWORK IN JAPAN

Hiroshi INOSE

Chairman, Network Development Comittee

An experimental computer network for inter-university computer centers in Japan has been developed under the present project. The network features the use of the digital data network of Nippon Telegraph and Telephone Public Corporation which provides the circuit-switched service as well as packet-switched service. A protocol hierarchy, named the N-1 Protocol, has been developed to be compatible with both types of services. Such international standards as CCITT Recommendations X.25 and X.21 have been employed for physical, electrical, link-level and packet-level interfaces. The protocol hierarchy also includes, as its user level protocols, the remote job entry (RJE) protocol and the network virtual terminal (NVT) protocol. Field trials have been conducted between the University of Tokyo and the Kyoto University by way of the digital data network for the purpose of feasibility demonstration and system evaluation. The present paper outlines the system Configuration and experimental results with particular reference to user level protocols.

1. SYSTEM OUTLINE

The inter-university computer centers which exist in seven university locations (Sapporo, Sendai, Tokyo, Nagoya, Kyoto, Osaka and Fukuoka) have been providing services on a nation-wide basis to some 400 institutions of higher learning in Japan. The services provided are mainly local batch and remote job entry services, and to some extent, time sharing service. Although a number of universities and colleges have their own computational facilities, the service demand for the inter-university computer centers has been increasing not only quantitatively but also qualitatively year by year. Annual jobs which amounted to two million in the year 1977, increase at a rate of 20% per annum. Demands for such services as remote job entry, TSS and on-line file access are rapidly increasing.

For the purpose of realizing load-levelling and emergency back-up among computational facilities, and for meeting the demands for future services which generally fall in to the category of computer communications, the integration of inter-university computer centers by means of a communication network was planned. Prior to the implementation of an operational system covering seven inter-university computer centers, development of an experimental system for feasibility studies was started in 1974 through the joint efforts of the University of Tokyo, Kyoto University and the Nippon Telegraph and Telephone Public Corporation (NTT). The system makes use of NTT's Switched Digital Data Network DDX–2, which is scheduled to begin circuit-switched as well as packet-switched services in 1979.[1]

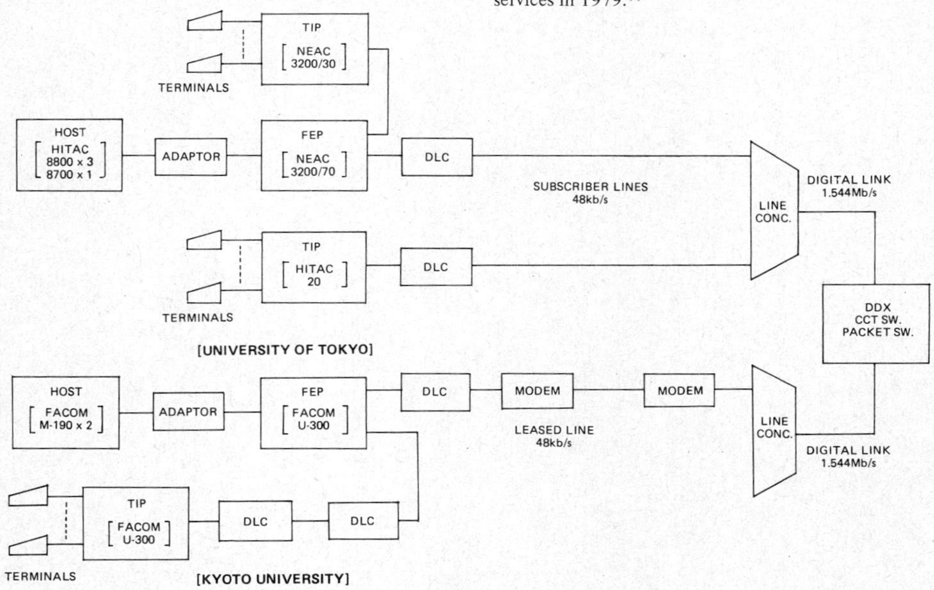

FIG. 1 Hardware Configuration

The decision to use switched data service was made on the basis of such considerations as that the network traffic being estimated to be rather low, especially in the initial service stage, in spite of the large variety of possibilities for using remotely located resources, and that the use of public data service permits smoother system expansion and relieves the computer centers to a large extent from installation, maintenance and operation of the network. It was also decided to employ circuit-switched as well as packet-switched services because the former is advantageous for load-levelling, remote job entry, file transfer and real time processing, and that the latter is preferrable for TSS, on-line file access and other interactive use of the resources. Hardware and software development, however, has been conducted to be compatible with both types of switching.

Figure 1 shows the hardware configuration of the experimental system. The HOST at the University of Tokyo Computer Center consists of three HITAC 8800 computers (each with an average instruction time of 0.18 μs) and one HITAC 8700 computer (with an average instruction time of 0.68 μs) with 8-Mbyte memory and 8400-Mbyte file. Whereas the HOST at the Kyoto University Computer Center consists of two FACOM M-190 computers (each with an average instruction time of 0.15μs) with 6-Mbyte memory and 5200-Mbyte file. To interface the HOST with the digital data network, a Front End Processor (FEP) is provided in each location.

interface and vice versa. The latter arrangement has been made for the purpose of feasibility demonstrations in the earlier stage of digital network service where not all of the seven inter-university computer centers are covered. For the purpose of feasibility studies on different modes of connecting TIPs, one of the TIPs in Tokyo has been connected directly to the digital network, whereas the TIP in Kyoto and the other TIP in Tokyo have been connected to the digital network by way of their FEPs.[2]

The software configuration of the system is shown in Fig. 2. As a part of the operating system of the HOST, the Network Control Program (NCP), Remote Job Entry (RJE) service processes and Network Virtual Terminal (NVT) service processes have been implemented. The NCP has such functions as establishing and terminating logical communication paths between local and remote processes, transferring data between processes and communicating with the NCP of the other HOST. The RJE service processes have the function of assisting the transmission of user jobs and output data. The NVT service processes have the function of assisting interactive conversation between the HOST and the remote terminals by way of a remote HOST or TIP. For the FEPs, the communication control monitors, HOST-FEP and FEP–NET communication control routines, event and data link control routines and packet handling and supervisory tasks have been developed. The FEP–NET communication control routines are capable of handling up to eight 48 kb/s lines. For the TIPs, the NCP, NVT service processes, TIP–NET communication control routines and terminal handling routines have been developed. The terminal handling routines are capable of handling up to 32 low speed terminals.[3]

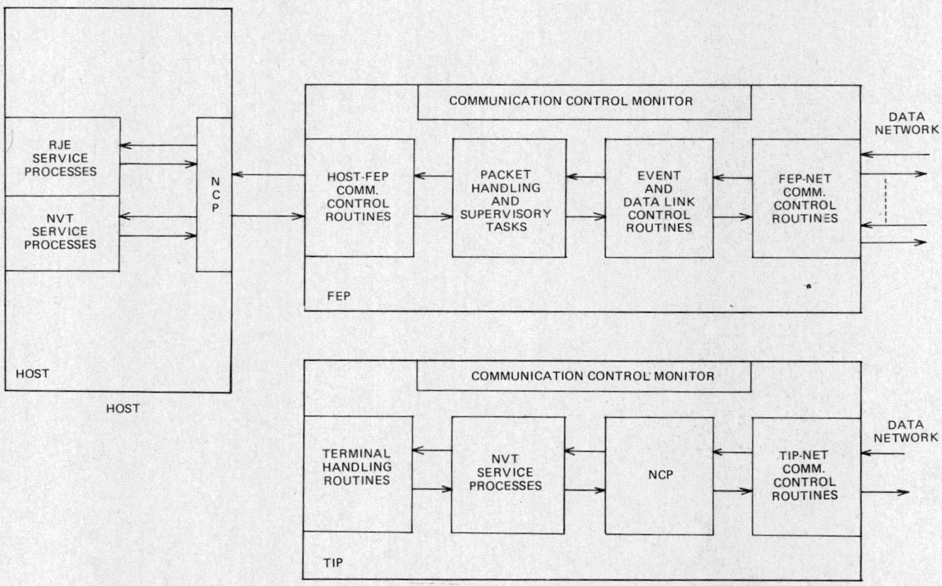

FIG. 2 Software Configuration

The FEPs are NEAC 3200/70 in Tokyo and FACOM U-300 in Kyoto, which are 16-bit-word minicomputers with 64-kbyte memory each. Adaptors and data link controllers (DLCs) are associated with the FEPs for interfacing with the HOSTs and the network, respectively. For the purpose of interfacing non-packet terminals with the network, the Terminal Interface Processors (TIPs) are provided which are minicomputers similar to FEPs.

Connection of the computer centers to the digital data network has been done in two different ways. The FEP and one of the TIPs at the University of Tokyo Computer Center have been connected by means of four-wire 48 kb/s base-band lines with the CCITT X.21 interface. The FEP at the Kyoto University Computer Center, on the other hand, has been connected by way of 48 kb/s carrier channels with the CCITT V.35 interface. A converter at the inlet to the ditigal data network converts the V.35 interface to the X.21

2. PROTOCOL STRUCTURE

The protocol hierarchy of the system is shown in Fig. 3. The user level protocols, namely, the Network Virtual Terminal (NVT) protocol and the Remote Job Entry (RJE) protocol, are at the top of the hierarchy. Then, the HOST–HOST protocol, the HOST–FEP protocol, the FEP–FEP protocol, the FEP–NETWORK protocol and the Data Link Control (DLC) exist in a descending hierarchical order.

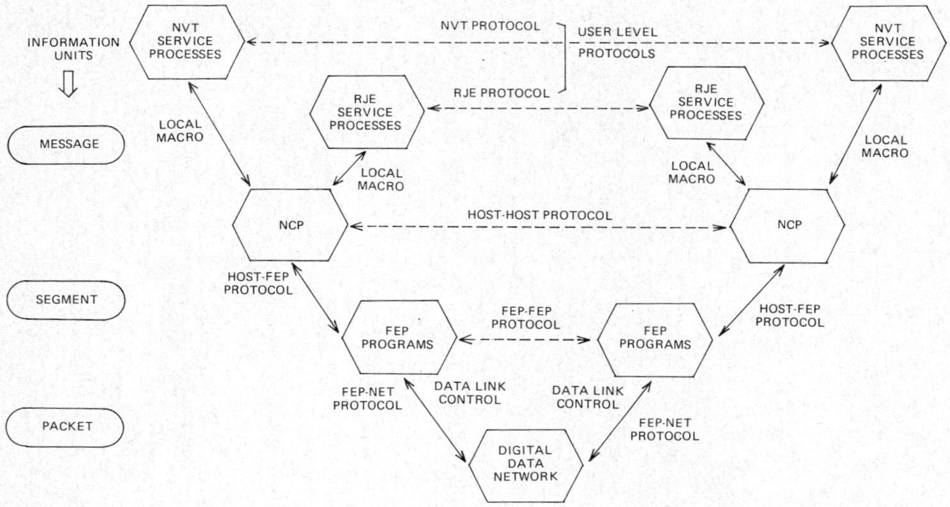

FIG. 3 Protocol Hierarchy and Information Units

The NVT protocol assists interactive conversation between a local HOST and a remote terminal by way of the remote TIP. The RJE protocol assists transmission between a local HOST and a remote job entry station by way of a remote HOST. The user level protocols are described in some detail in the next section.

The HOST–HOST protocol establishes and maintains communication between a sending and receiving process over a pair of logical links. A logical link is unidirectional and up to 256 logical links can be provided for each pair of HOSTs. In order to accomplish this, an NCP communicates with another NCP by means of control commands over a particular pair of links called control links.

The HOST–FEP protocol handles data transfer between a HOST and its FEP by means of a handshake procedure of transmitting commands. This avoids congestion in the FEP by controlling the transmission from its HOST. The FEP–FEP protocol is mainly for network supervision. If hardware trouble occurs in the data circuit, a FEP or a HOST, detection is made by either one or both of the FEPs and the result then is reported to a HOST or HOSTs for higher-level decision making.

The FEP–NET protocol differs depending on the types of service that are provided by the digital data network. In the case of circuit-switched service, the protocol establishes and releases a pair of physical links between a FEP and the network in accordance with the CCITT X.21 physical interface and signalling sequence. In the case of packet-switched service, on the other hand, the protocol is in accordance with the packet level interface as specified by CCITT Recommendation X.25. In this case, the protocol assigns a pair of logical links between a FEP and the network, structures the data and control information in the form of packets, and controls the transfer of packets over the assigned pair of logical links.

The data link control (DLC) is based on the High Level Data Link Control (HDLC) or the link access procedure as specified by CCITT Recommendation X.25. In the case of circuit-switched service, the DLC ensures the transfer of information over a pair of physical links between the FEPs. In the case of packet-switched service, the DLC takes care of data transfer over a pair of logical links between a FEP and the network.

For each level of protocols, information units are defined as shown in the figure. The information between the user process and the NCP is a message of indefinite length. The message is divided into segments, each of which has a maximum length of 2040 bits as is shown in Fig. 4. A segment consists of a segment header of 80 bits and data of up to 1960 bits. A segment header specifies the originating and terminating NCP addresses, the link number, the segment number, etc. The segment is an information unit between the HOSTs. In the case of circuit-switched service, the information unit between the FEPs in called a packet, which is identical to a segment. Whereas in the case of packet-switched service, the information between a FEP and the network is named an N-packet, consisting of a segment and a packet header as specified by CCITT Recommendation X.25.

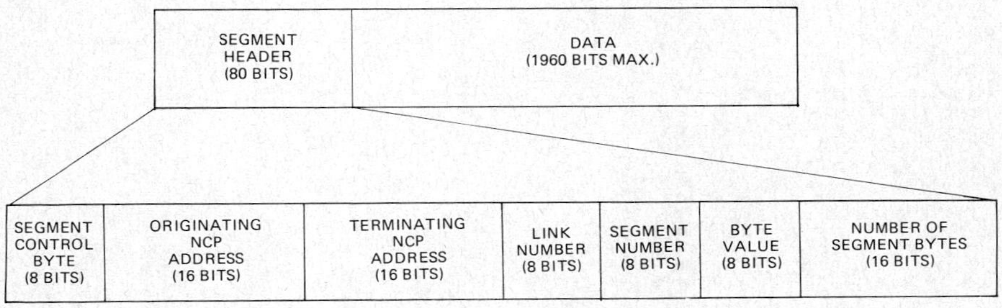

FIG. 4 Segment Format

In the case of circuit-switched service, a pair of physical links is established and released by the FEPs by initiating phases 1 and 5 of the data link control procedure. Once a pair of physical links is established, a plurality of logical links are assigned over it. This arrangement maintains compatibility with packet-switched virtual call service.

3. USER LEVEL PROTOCOLS

The details of a user level protocol depend largely upon the operating system of the HOST which provides the service. For example, such portions of a protocol as authorization of users, job creation, job inputting and job outputting are all different in each HOST. And it is unrealistic to modify and portion of an operating system which is operational. For this reason, user level protocols are organized in such a way as to include the minimum provisions which are needed to provide and receive service and yet to maintain sufficient freedom in choosing HOST-dependent factors. In this sense, the user protocols to be described are general enough, and may be considered as a framework for organizing detailed protocols which are to be defined when a specific HOST is given. Therefore, the users of the service have to take into account the HOST-dependent part of the protocol. In accordance with this framework, however, users can receive service from any one of the HOSTs by means of the same program.

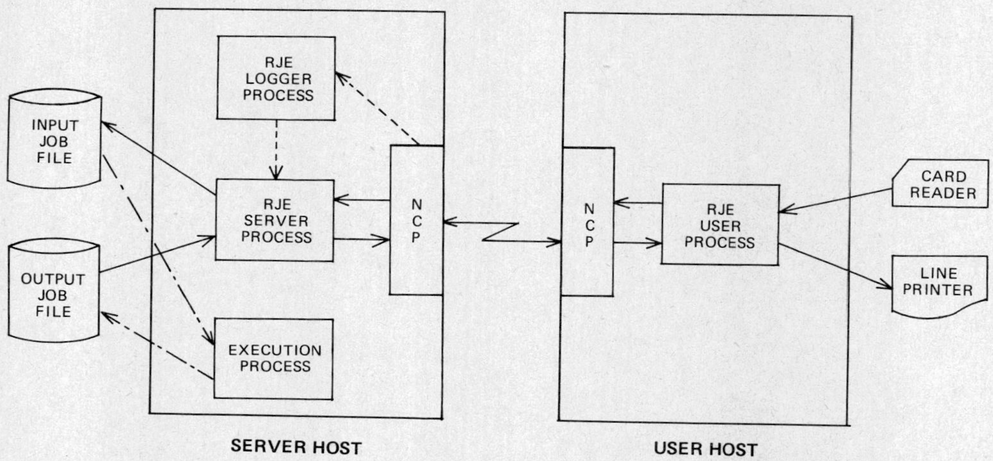

FIG. 5 Conceptual Structure of the RJE Protocol

(A) Remote Job Entry Protocol

The conceptual structure of the Remote Job Entry (RJE) protocol is shown in Fig. 5. Jobs from an RJE terminal are transferred, by way of the local HOST which is called here the user-HOST, to the remote host which is called here the server-HOST. The jobs are then processed in the server-HOST and sent back to the RJE terminal by way of the user-HOST. The sequence of RJE service is as follows.

1) The RJE user process in the user-HOST is activated in response to an RJE service request from a user.

2) The RJE user process establishes a pair of logical links to and from the RJE-server process in the server-HOST, in accordance with the HOST–HOST protocol of the NCPs of the HOSTs. The RJE logger process in the server-HOST assists in activating the RJE-server process and transfers information between the RJE-server and the RJE-user processes which is necessary to establish a pair of logical paths between them. This sequence is called the RJE initial connection protocol.

3) The RJE user and RJE server processes execute the jobs to be done and transfer the results of execution. The RJE command and RJE response to be described are used for the purpose of controlling the transfer and for performing various utility functions.

4) At the end of RJE processing, the RJE user process terminates the logical links to and from the RJE server process.

RJE commands and RJE responses are used to execute RJE processing. They are distinguished by the fact that a command from one party requests a response from the other party, whereas a response from one party does not request any response from the other.

A command may be sent from either the RJE user process or the RJE server process, and is transmitted over a pair of the logical links established between the RJE user and RJE server processes. The input and output data are also transmitted over the same links, so that, to distinguish the command and response, a message block with zero text length which is called the escape block is transmitted immediately following the transmission of a command or a response as is shown in Fig. 6.

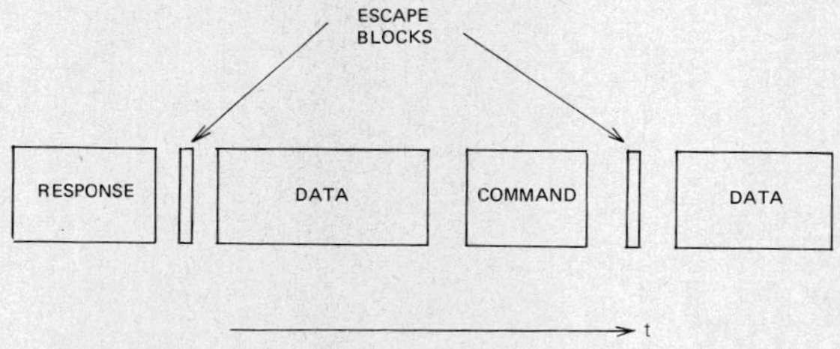

FIG. 6 Common Transmission of Data, Command and Response

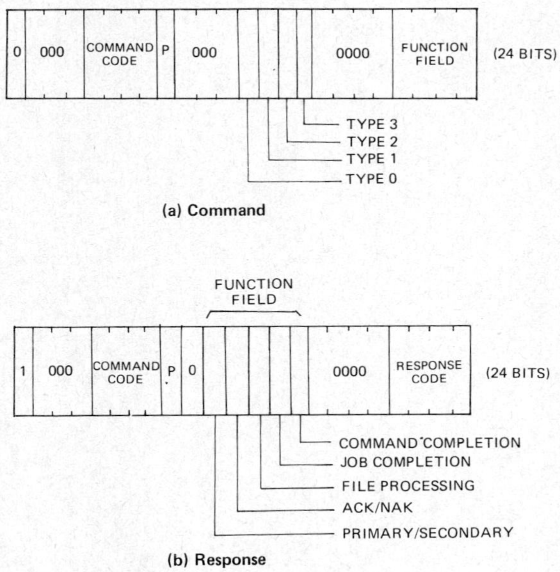

FIG. 7 The Formats of RJE Command and Response

Figure 7 shows the format of an RJE command and response. As is shown in (a), each command has a function field which defines additional functions. For example, the INPUT command has additional functions which specify the input device and the input file format. As is shown in (b), each response also has a function field which represents the result of processing associated with the command. For example, the response to the INPUT command has a function field that indicates the acceptance/rejection of the command.

Table 1. The RJE Command Set

Command Name	Command Code	Command Type	Direction	Meaning
USER	1	Type 1	User → Server	informs ID and password.
INPUT	2	Type 2	User → Server	requests data input.
OUTPUT	3	Type 2	User → Server	requests data output.
FILE	4	Type 1	Server → User	notifies characteristics of file.
CANCEL	5	Type 1	User → Server	requests to cancel processing.
BYE	6	Type 0	User → Server	requests to terminate processing.
ABEND	7	Type 0	Server → User	requests to terminate processing.
STATUS	8	Type 1	User → Server	requests to send job status.
EOF	9	Type 1	Both directions	notifies end of file.
MESSAGE	A	Type 3	Server → User	notifies to send message from server-HOST.

Table 1 shows the RJE commands set and their functions. The commands are categorized into the following four classes.

Type 0 command: A command which results in the end of RJE service if the response is received properly.

Type 1 command: A command which inhibits the receiver from further accepting command(s) or sending of command(s), before the receiver sends a response indicating the end of processing.

Type 2 command: A command which permits the receiver to accept further command(s) or to send command(s), after the receiver sends a response indicating the acceptance of the command.

Type 3 command: A command which permits the reciever to accept further command(s) or to send command(s) without sending any response to the command.

Among these commands, CANCEL, ABEND, STATUS and MESSAGE are optional.

The responses are categorized by the following factors.

Primary/Secondary: Before the reception of a primary response, the sender of the command is inhibited from transmitting the following command(s). Type 0 and 1 commands are responded to by primary responses. Type 2 commands are responded to by either primary or secondary responses. In this case, a primary response indicates the command execution and the result.

Acknowledge/Negative Acknowledge: Acknowledge is associated with responses indicating command acceptance or normal termination, and Negative Acknowledgement is associated with responses indicating command rejection or abnormal termination.

Processing Stages: In case the output job consists of a plurarity of files, the response indicates whether the job is in the command receipt level, in one of the file processing levels, in the job completion level or in the command completion level.

A command or a response may be followed by a parameter field that indicates such details as the output device format, the user's registration number, the result of a validity check, etc. When a parameter field is attached, bit P in the figure is set to one which is zero otherwise.

(B) Network Virtual Terminal Protocol

As is shown in Fig. 8, the Network Virtual Terminal (NVT) protocol consists of the server NVT program and the user NVT program that interface with the end process and the terminal, respectively. The server NVT program is located in the server HOST, whereas the user NVT program is located in the user Host or a Terminal Interface Processor (TIP) that accommodates the user's terminal. The NVT protocol is based upon the concept of the network virtual terminal (NVT) by which a standardized interface to terminal deivces and terminal-oriented processes is provided. This eliminates the need for the server HOST to keep information about the individual characteristics of physical terminals and terminal handling conventions. The NVT is assumed to be a bi-directional character device having a printer and a keyboard.

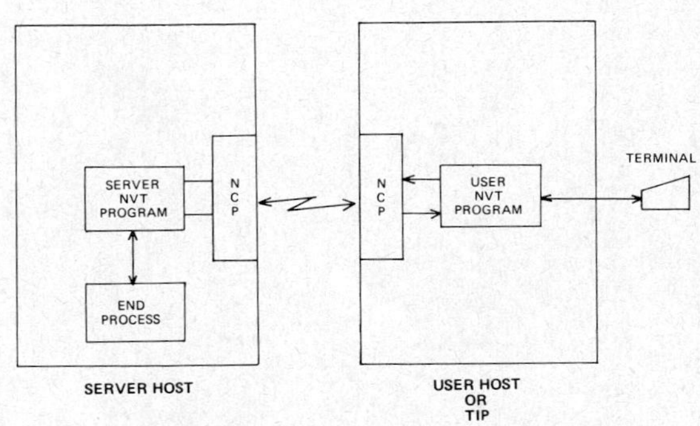

FIG. 8 Conceptional Structure of the NVT Protocol

The connection between a remote HOST and a local HOST or TIP is set up by the initial connection protocol. A pair of logical links thus established between the server NVT and the user NVT programs carry character blocks bi-directionally with interspersed control information. A character block consists of characters of less than 245 bytes. One control byte is included in a character block for the purpose of carrying commands on a piggy-back basis. This reduces the transmission of blocks that carry commands only. The control byte contains an identifier that indicates whether the block is with or without a command and whether it is the last block of the message. The commands are in hexadecimal code as shown in Table 2.

Table 2. Typical NVT Command Sets

Command Name	Command Code (hexadecimal)	Meaning
No Operation (NOP)	F 1	non.
Data Mark (DM)	F 2	notifies timing to restart subsequent action.
Break (BRK)	F 3	notifies actuation of break key.
Interrupt Process (IP)	F 4	requests to interrupt end process.
Abort Output (AO)	F 5	requests to abort output.
Are You There (AYT)	F 6	solicits response.

The HOST and the TIP have some support functions for assisting communication between a physical terminal and the user NVT program. The functions include those for format conversion, device control, etc., that are required to interface with the profiles of physical terminals.

4 EXPERIMENTAL RESULTS

Software development for this experimental system took three years. The amount of software for a HOST, namely, the NCP, RJE service processes and NVT service processes, is about 35 kilosteps. Whereas software for an FEP and a TIP amounts to about 12 and 28 kilo-steps, respectively. In addition to these, software for local test facilities has been developed. The effort required for the above software development was more than 100,000 man-hours at each of the locations.

An experiment on circuit-switched digital data network was conducted for RJE service. Remote jobs were entered from a terminal to its local HOST, transmitted over the circuit-switched network to the distant HOST, processed there, sent back over the network, and forwarded to the terminal by way of its local HOST. The overall element error rate for 48 kb/s data transmission through the network was in the order of 10^{-8}, and all transmission errors were detected by the cyclic redundancy check procedure. The maximum data transfer rate between the FEPs was 46 kb/s. The call setup time was in most cases less than 200 milliseconds, and the disconnect time was in the order of 10 milliseconds.

The system performance has been measured by directly connecting a test program (TP) to the NCP as shown in Fig. 9.

After transmitting a segment, the user-HOST has to suspend the transmission of the succeeding segment until after the reception of acknowledgement for the preceding segment from the server-HOST. As a measure of channel utilization, the segment transfer time, which is the time interval between the start of transmission of preceding segment and the start of transmission of succeeding segment, has been measured employing the circuit-switched network as well as the prototype packet-switched network (TL 1), and by changing the data length of segments and the number of active logical links. Some of the measured results are shown in Table 3 together with the utilization rate of physical line. It is observed from the table that the utilization rate of a 48 kb/s line can reach 98 %, the the segment transfer time is satisfactory even in such heavily loaded cases, and that a 48 kb/s circuit switched connection permits the establishment of as many as eight logical links each fully carrying segments with maximum length.

Table 3 Segment Transfer Time and Utilization Rate of Physical Line

		Circuit Switched Network		Prototype Packet Switched Network, TL1	
No. of Logical Links	Data Length (bytes)	Segment Transfer Time (ms)	Utilization Rate of Physical Line of 48 kbps (%)	Segment Transfer Time (ms)	Utilization Rate of Physical Line of 48 kbps (%)
1	10	69	9	101	6
	245*	115	39	186	24
8	10	280	16	273	18
	245*	354	98	367	97

*Maximum data length of a segment

By using the circuit-switched network and the prototype packet-switched network, the round trip transmission time of the ECHO command was also measured under the condition that no other information is transmitted between the HOSTs. The average round trip transmission time of the former was measured as 63 milli-seconds and that of the latter was 96 milli-seconds. Accordingly, it is estimated that the network delay of the latter is 17 milli-seconds on the average.

Performance of the packet-switched network TL 2, which fully adopts the CCITT Recommendation X.25, has been evaluated experimentally in terms of call set up and release times and packet delay under various packet throughput conditions. Some of the results are shown in Table 4. The table shows that, even in such heavily loaded cases as the packet throughput of 325 packets/sec., the packet delay of the network TL 2 is smaller than that of the network TL 1. It is also observed that the call set up and release times do not change significantly in accordance with the values of packet throughput.

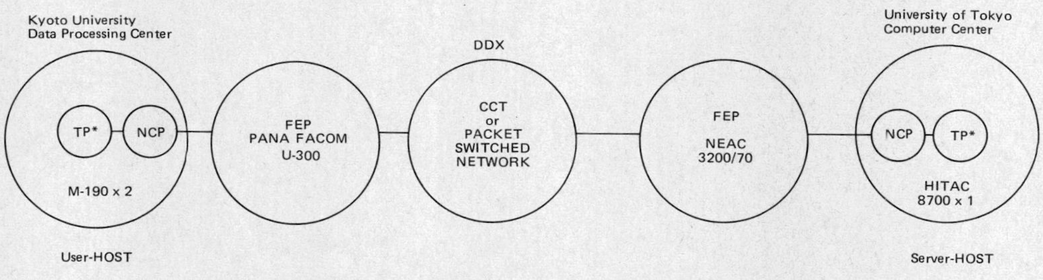

FIG. 9 Configuration for System Performance Measurement

*TP: Test Program

Table 4 Performance of Packet Switched Network TL 2

Packet Throughput (packets/second)	No-Load	170	325
Call Set Up Time (ms)	10	10	11
Call Release Time (ms)	8	9	11
Packet Delay (ms)*	4	5	6

* When the data length is 80 bytes.

Employing the packet-switched network, the performance of the network when used by TIPs has also been measured. Some of the results obtained when the network was used by the University of Tokyo TIP (HITAC-20) over a 48 kb/s line are shown in Table 5. The table shows, under various packet throughput conditions, the call set up and release times measured at the TIP as well as the packet delay for different data length.

Table 5 Performance of the Packet Switched Network When Used by a TIP (HITAC-20)

Packet Throughput (packets/second)		No-Load	250	400
Call Set Up Time (ms)		35	37	42
Call Release Time (ms)		28	31	37
Packet Delay (ms) for Data Packet Length (bytes)	5	18	19	18
	70	40	37	37
	250	100	100	98

The effect of background processing of a HOST to the network performance has been measured. Table 6 shows some of the results when the packet-switched network TL 2 was used by the University of Tokyo HOST with different loading. In the table, the normal-load condition is such that the HOST is loaded with the background processing of 13 local batch jobs, 50 local TSS jobs and 20 local RJE jobs. The over-load condition is such that all the computers of the HOST are fully loaded with 64 local batch jobs. It is known from the Table that, under normal-load condition, the utilization rate of physical line may reach close to 90 % while maintaining the segment transfer time below 0.2 sec.

Table 6 Network Performance vs. HOST Loading

Host Loading	No. of Logical Links	Data Length (bytes)	Segment Transfer Time (ms)	Utilization Rate of 48 kb/s Physical Line (%)
No-Load	1	10	91	5.5
	1	245	179	24.7
Over-Load	1	10	327	1.6
	1	245	755	5.7
Normal-Load	1	10	127	4.0
	1	245	236	18.6
	4	245	199	88.3

ACKNOWLEDGEMENT

The author wish to acknowledge with thanks the support of the Ministry of Education, Science and Culture through the Grant in Aid for Scientific Research entitled "Advanced Information Processing of Large Scale Data over a Broad Area" and "Formation Process of Information Systems and Organization of Scientific Information". The cooperation of a large number of people at the University of Tokyo, the Kyoto University, the Engineering Bureau and the Electrical Communication Laboratories of Nippon Telegraph and Telephone Public Corporation, Fujitsu Ltd., Hitachi Ltd. and Nippon Electric Co. are gratefully acknowledged. Particular thanks are due to the members of the Network Development Committee including Professor Toshiyuki Sakai, Vice Chairman and Dr. Masao Kato, Nippon Telegraph and Telephone Public Corporation.

REFERENCES

[1] H. Inose, et al.: Networking for Inter-University Computer Centers in Japan, Proc. 3rd Int. Conf. on Computer Comm., pp. 286–291 (August, 1976).

[2] H. Inose, et al.: Experimental Network for Inter-University Computer Centers in Japan, Proc. Int. Switching Symp. 1976, 441–4 (October, 1976).

[3] H. Inose, et al.: User Level Protocols for and field Trial on the Experimental Inter-University Computer Network in Japan, Proc. 4th Int. Conf. on Computer Comm., pp. 507–512 (September, 1978).

Scientific Information Systems in Japan
H. Inose (ed.)

STUDY ON METHODOLOGY OF INPUT AND OUTPUT OF INFORMATION SYSTEMS

Mamoru HOSAKA
Institute of Space and Aeronautical Science, University of Tokyo, Komaba, Meguro-ku, Tokyo

Activities of the I-Committee and summary of the symposia held by the committee are described. Necessity of establishing methodology and adopting interactive machine aided methods for production of organized data stored in information systems is emphasized. The symposia on machine aided document production and editing attracted many researchers. For English composing and editing, experiences of using available editing software packages, and design of a stand alone direct interactive system using a microcomputer were presented. For Japanese document processing, use of Kanji printers and various Kanji input methods were discussed. Research on Japanese language editor and study on statistics of Kanji characters used in scientific papers were reported. The other topics were: advanced input methods, standardization of Kanji codes, instrument calibration problems, collection of established software of image processing, and etc.

1. INTRODUCTION

This report describes philosophy and objectives of the I-Committee on development of input/output subsystems in scientific information systems, and summary of the symposia held by the committee during the period of the latter half of this research project. This committee was organized under the leadership of the supervisor group of the project in August 1977. Its objectives were (i) to exchange and disseminate knowledge about techniques and methods of input and output of data treated in scientfic information systems among researchers participating in the project, and (ii) to make general specification for the I/O subsystems, if possible.

The committee also took over the activities of the study groups on Kanji input and on the standard image, which started in the first year of the project. It expanded the scope of interest to include not only data conversion methods at interface of information systems, but also study on characteristics of data treated in the systems and on machine aided technology for organizing data. The committee thought that it was too early to develop general specification for I/O subsystem appropriate in scientific information systems, because sufficient knowledge and experiences had not yet accumulated for the specification to be defined. The committee tried to collect and disseminate information of project members' experiences, researches, requirements on various I/O processes, and also of new I/O technology which will be available in near future, by holding symposia on topics such as the following:

(i) Document processing with machine aid,
(ii) Kanji input and its processing systems,
(iii) Advanced I/O techniques and graphics,
(iv) Standardization in input and output fields,
(v) Related items to image input.

2. DATA IN SCIENTIFIC INFORMATION SYSTEMS

Scientific information systems have to treat various types of data, such as numerical data or text data in alphabetic or Japanese characters, figure and image data and etc. and to respond to non-simple queries about these data. The essential component of an information system is organized collection of data which are objects of retrieval responding to users' inquiries. These data do not exist naturally, but they are produced with intellectual labour of many people from great deal of raw information they collected. In order to produce organized information used in the systems, methodology of organization and frameworks of data description must have been established to meet the objective of a specific information system. They are usually important research items in informatics.

Then the following operations are to be applied to the collected data: verification and calibration of original data, detecting and correcting errors, extracting features and meanings, finding relations and structures in them, and making organized description of them according to the established methodology and frameworks. As quantity of data to be treated are great, without automatic means, cost and time for constructing a scientific information system become large. And as the activities need man's understanding of data with professional

* Chairman of the I-Committee.
Other members are: T. Sakai (Vice-chairman, Univ. of Kyoto), M. Onoe (Univ. of Tokyo), T. Motooka (Univ. of Tokyo), S. Fujiwara (Univ. of Tokyo), M. Takagi (Univ. of Tokyo), S. Tsuji (Osaka Univ.), M. Nagao (Univ. of Kyoto), C. Yoshimoto (Hokkaido Univ.) and K. Ikeda (Tsukuba Univ.).

knowledge, machine aided interactive methods should be applied for their organization with accompanying clerical works. A computer system used for this kind of informatic activities must have ability of good man-machine communication with high level intellectual interaction.

As the first step of data acquisition, raw data must be converted in machine readable forms on some media. If possible, it should automatically be made at the time when the data are acquired. Then these converted data are used as raw materials for the subsequent machine aided processing to produce organized information. Output functions from the computer is also important. Forms and types of print or display of texts and graphics must be acceptable for man's easy understanding. And when interactive functions are required, provision of methods of direct communication with output display or records is highly desirable.

In scientific activities of researchers, they are not only users of information systems, but also producers of raw information in their respective fields. They should make an effort to apply machine power to their research works, especially to their information processing tasks to reduce thier clerical labour and enhance their productivity as well as to produce machine readable data for the subsequent processing and organization. One of their time consuming tasks is that of document production such as writing reports, papers and letters. Most of these works require man's creative thinking and much of secretarial labour. As an interactive machine aided method is expected effectively applicable for these works, a machine for this purpose should readily be usable at one's disposal. This implies that the machine should be cheap, versatile and available any time at one's desk side. As hardware usable for this purpose has become cheap recently, researchers in informatics in the academic community should try to design and make such a system, taking into account their experiences in document making, which may be different from that in the business world.

The committee considered that, in this research project, problems of input and output were not only those of information form conversion at the interface between an information system and its outside world, but also those of methodology of constructing and maintaining organized information in the system. From this point of view, the activity of the committee was planned.

3. DOCUMENT PRODUCTION AND EDITING

Reports and scientific papers are the most important sources of scientific information. Making of these documents requires great amount of intellectual labour of researchers. Man's creative process to produce their contents cannot be replaced by the computer, but accompanying routine tasks such as gathering data, copying, arranging, clean writing and etc. can be aided by the machine, if a proper interactive computer system could be available for the author of documents. If these aiding process are to be used in document writing, the original text thus produced is also in machine readable form when writing is finished.

In our daily work in universities in Japan, there are so many miscellanious paper work duties to be performed other than writing scientific papers. If we could hire an able secretary or assistant, we could deal with most of the paper works efficiently and make time for more creative tasks and advanced study. But in reality in the university environments, it is almost impossible to hire an able assistant or to expect hasty rationalization of the management of labour. Accordingly, we should try for ourselves to reduce the pressure of miscellanious chores by making use of computer power just as we use the computer in research and laboratory works. There are obstacles to be removed to realize these kinds of secretary system. Some of them are problems of Kanji input and output, typing habits, probable high cost and inefficient available TSS terminals, especially, conservative way of thinking. Though problem of Kanji input and output, discussed in the next section, is important, English document production is also indispensable in our academic community. As the obstacles for making a secretary system for English typing seems not many, we should develop a system suitable for our use.

Japanese researchers in general are not good in English, and they consume very much time in writing their papers in English. Commercially available word processing machines are not always helpful for them. Because they are too expensive to be afforded by them and moreover they are useful only in final stages of document making. Situation is almost the same in TSS terminals if they want to use editing software packages produced in U.S.A. As there were many people in this project, who had strong interest in this English document processing and were eager to know the current state of the art and to use available systems, a symposium on this theme, held by the committee, attracted many researchers in universities.

A process of making original documents involves many cut and try procedures, such as plannig, seeking proper expressions, inserting and correcting words or sentences, reusing of deleted sentences or paragraphs, getting new ideas while examining, rewriting from the first etc. These procedures occurs usually even in our accustomed mother language. When in non accustomed foreign languages, we always experience difficulty in finding proper sentence and words for expressing our ideas, and in making many mistakes in grammer and spelling. If the machine could assist us in English composition to reduce the difficulties stated above, we hope it could have stored knowledge for English composition and could supply us examples of proper expressions or other useful remarks responding to our questions on translation or grammer. Of course, in the file there should be common knowledge useful for average researchers, but one could construct his own file in which his experienced knowledge could be added. Moreover the system must have the ability of easy handling of heavy correction and editing.

Display on CRT is not always adequate for interactive document processing, because correction of sentences on CRT replaces old ones, and comments or ideas brought forth during composition or correcting processes cannot be recorded near the proper location in desired style and format, and off-line examination of the written paper is sometimes more convenient.

In the symposium realization of the idea stated above was not seen, but one system pointing in this direction was reported. A business typewriter is connected to a microcomputer with a floppy disk device. A simple tablet is attached adjacent to the typewriter platten, so as to locate a position on the typed sheet on the tablet with a pen. The pen is used not only to point a location on the sheet, but is used for writing symbols indicating commands to be executed. The essential features of the system are that the inserted paper in the typewriter is the working area for composition and correction. This is the same process as in usual proof reading. Contents written on the sheet is always stored automatically in the file and formatted printing is also automatic, and its cost is low because of the laboratory assembled hardware and software. The transfer of the technology was highly desired.

Another items to be reported were as follows. Experiences on use of editing software package for program development in the work of document editing were reported by several users and also experience on output document format control with a program package, named ROFF, were reported. ROFF is written in high level language, it can be transfered easily, and though its commands for format control have to be inserted in appropriate position of the text to be printed, the results of printing are good. And also explanation of a phototypesetting machine introduced in a university computer center and its use experience was also reported. So long as editing and printing of texts are concerned, use of computer has just begun in the Japanese academic community.

As for an advanced system such as reported, we cannot expect computer makers or vendors outside of the academic community will develop and supply it to us at low price, because its main use is limited only in an academic circle. Development and technology transfer of such a system in universities should be encouraged.

4. PROCESSING OF DOCUMENTS IN JAPANESE

As a matter of course, in our country, necessity of document processing in Japanese is far greater than that in English. However, development in this area has been slow owing to difficulties in treating numerous Kanji characters by the machine. Recent progress in computer hardware has reduced the cost of storing Kanji patterns and Kanji texts, and made high speed Kanji printers practically feasible. With these situation, demand for use of Kanji in business systems has increased and various Kanji processing systems are being commercially available. But they are still expensive for use by academic researchers and not suitable for on-line text editing they need.

A comprehensive Japanese text editor is an indispensable tool for research on Japanese language processing. The current state of the arts in this area was reported in the symposium, and Japanese text synthesis in English-Japanese translation with this tool was explained. The editor has capability of processing two-byte data flexibly, retrieving the stored Japanese dictionary efficiently as well as using various techniques of language analysis properly. These research work is useful for syntastic and semantic analysis of Japanese documents in construction of data bases and also for planning a practical Japanese text editors. Experiences on use of computerized Japanese document printing system at Tsukuba Univ. was reported. So long as output printing of documents and figures is concerned, there are no problems in its speed and quality except for requirements of low cost and very high quality of printing.

As for input of Japanese text, attractive input devices and methods have not been realized yet. Currently available ones are full key-boards and pen touching menu tablets. Both of them require viewing on numerous keys or a fine-mesh menu, accordingly their input speed is slow and error-prone, and their operations are cumbersome. A study on typing of Japanese texts with touch typing was introduced by an ardent advocator of it. With intensive training, one's typing speed can be attained comparable to that of professional typist of English documents using ordinary typwriter. When secretaries with such typing skill will become easily available, efficiency of document making in business as well as in academic circles will be increased tremendously. We hope the time will come soon.

However, the touch typing method seems not suitable for layman such as researchers who want to use typing for themselves. Input of Japanese with Kana typing, though number of keys used is small, needs correct Kana-Kanji conversion which is not easy process due to many homonyms and requires confirmation of converted results. An advantage of this method is that touch typing can be applied without memorizing many special codes. Anyhow we still do not have a good input device of Japanese texts, which reseachers could readily use at their desk sides. We should make more efforts for inventing such devices.

Number of Kanji characters in common use is more than 2,000, but that used in a scientific paper in each specific field and of ordinary length is said about 500 to 600. This number is again comfirmed by a researcher who investigates several papers recently published in different categories. This property was utilized for making and displaying a Kanji input-menu on CRT view surface for individual use of a Kanji input system.

Standardization of Kanji code and their printing control code are important for transfer of Kanji processing programs and output printing. Current standardization activities of these items and the draft of control code standard were reported in the symposium.

5. ADVANCED INPUT TECHNIQUES

Human beings are accustomed to recognize intuitively two-dimensional representation of information. This may be attributed to the fact that man sees outside world through images projected on his retinas. The two-dimensional representation is natural way of expressing information for man, while the computer can only treat information which is expressed and stored one-dimensionally in it. Consequently there must be conversion between outside and inside expressions. When outside representation is semi-one-dimensioanl, such as usual texts, the conversion is not difficult and it can be made automatically, if proper techniques of character recognition could be applied.

As we use always two-dimensional expressions to describe relational, topological or geometric information, their automatic conversion for computer input and internal representation is very much desirable, but as their sizes and forms are variable and quantity of information attached to them are great, the automatic conversion is almost impossible. So man has to describe their forms amd meanings with a programming language to make the coresponding internal representation. Usually this process take time and efforts of man, and errors are apt to be introduced. As for output two-dimensional representation of information from the computer and format control of output printing, computer graphics methods have been established and there are few problems left for future research.

Research works on the input of two-dimensional expressions, in spite of their importance, have been few. People might think that use of two-dimensionality in an input method is ineffective, because pattern recognition by the machine have been applied only in very restricted objects arranged in a specified location and machine-understanding of expressions extended freely on a sheet of paper seems almost impossible.

However, if we investigate carefully structures and meanings of two-dimensional expressions, we can find that certain rules are applied in them, according to each specific field, and the expressions are, in a sense, symbols, which does not need precise geometric values. Then we have possibility to convey meaning of two-dimensional expressions to the computer, if we could indicate location of unit symbols, which can be automatically recognized, and hidden relations between adjacent or remote symbols, which can easily be understood by man. In the symposium, the current state of a research work based on the principle stated above was introduced, in which hand writing actions of making engineering drawings was taken into computer for recognition of its meanings. This was applied successfully to the input of the geometric model building and processing system. (Detailed description can be found in this proceedings.) As another special input method, an experiment on real time voice recognition was reported. A possibility of using voice with complement to the other input methods was suggested.

6. OTHERS

As research on image input was one of the main themes of the group A of this project, the committee did not directly treat it, but the following topics were presented in the symposia, which related to image input problems.

- (i) A future vision of iamge information systems, and its I/O and retrieval through communication lines.
- (ii) Definition of formats of standard images.
- (iii) Calibration methods of image input devices.
- (iv) Collection of portable image processing softwares with established techniques.
- (v) Necessity of standardized specification of performance for transducers which are produced and used in one main field, but may be used in other disciplines.

Importance of interface between man and machine and communication between them have been recognized by the committee members, but any unified guideline or specification for scientific information systems was not established. A report of the committee containing papers presented at the symposia was published and distributed to the project members.

Scientific Information Systems in Japan
H. Inose (ed.)

GROUP STUDY ON DATA BASE MANAGEMENT SYSTEM

Tosio KITAGAWA
Emeritus Professor of Kyushu University
International Institute for Advanced Study of Social Information Science, Fujitsu Ltd, 17-25, Shinkamata 1-chome, Ohta-ku, Tokyo 144

This Report is basically due to the final summary Report of the M-Committee written in Japanese and contained in the final Report of the Special Research Project. In the first four Sections the Committee activities during the whole three years are explained. Section 5 describes the principal guidelines for developing DBMS in scientific researches, where Section 5.1 gives our basic recognitions of characteristic features of informations in scientific researches and Subsection 5.2 states the five principles for developing DBMS in scientific researches under the present situation of DBMS development on the basis of recognitions given in Subsection 5.1. Section 6 gives an explanation regarding why and how we have summarized the surveyes and investigations of M-Committee during three years. There are several crucial conceptual frameworks, such as PRF, ARF, UDL, CDL, data box, knowledge base, restructuring, evolutionarity of data base, which should be introduced and investigated for an establishment of scientific system.

1. INTRODUCTORY

It was 1976 June 24th when the Coordinating Committee of the Special Research Project (SRP) initiated an establishment of the Plan Promoting Committee on Data Base Management System (DBMS), which has been abbreviated since then by the name of M-Committee. The M-Committee started with twelve members from our SRP including the Chairman Tosio Kitagawa, the Vice Chairman Mamoru Hosaka, and the General Secretary Seigo Kanō. The activity plans of the M-Committee were immediately set up which consisted of (1°) survey plan, (2°) study plan, (3°) problem solving plan, and (4°) cluster paln. According to these palns (1°)-(4°) several activities of the whole SRP regarding research activities on DBMS were planned, initiated, and organized. According to (1°) survey plan, several members of SPR were sent to Stanford University July, 1976 in order to learn about SPIRES developed in their campus, and others to United States and Europe in order to make an extensive survey regarding DB and DBMS being in use there. Due to our recognition that Japanese scientific communities had been behind the international progress in this area, our efforts had been so much devoted to (1°) and (2°) within the first year of SRP. In fact according to (2°) study plan, we had the Tutorial Seminare Course in which some of our Committee members as well as Associate Professors Setsuo Ohsuga and Takuya Katayama gave a series of lectures on the current textbook on DBMS, and to which research workers from every Individual Research Unit (IRU) attended for study. We organized the DBMS Cluster which consisted of research workers from more than twenty IRU's whose researches were particularly concerned with DBMS. The terminology 'Cluster' was used in comparison with the Research Group in the sense that the former should be organized not within an individual Research Group such as Group A, B, C, D and E but in combination of IRU's belonging to different Research Groups.
The substantial activities may be said to have started from the second year starting from 1977 April, but it is worth while to note that the preparations having been realized in the first year gave a solid foundation for the later activities of the M-Committee in the second and the third years. Two reports [1] and [2] were prepared regarding the first year activities of the M-Committee.

2. OBJECTIVES OF THE M-COMMITTEE

After the preparatory activities in the first year, a clear description of the objectives of the M-Committee, which was assigned by the Coordinating Committee, May, 1977, was (i) to make clear the principal features of scientific data bases and (ii) define the fundamental requirements specifications for DBMS which are both reasonable and realistic for an implementation starting from 1980 in University communities. It may be said that the problem to be solved by the M-Committee was set up in this way, and that the activities (3°) problem solving started from the second year.

3. MEMBERS OF THE M-COMMITTEE

Soon after the specification of the objectives of the M-Committee, the reorganization of the M-Committee was performed by which then seven members were newly appointed in June, 1977. Later in May, 1978 the M-Committee were argumented by an addition of other twelve research workers (denoted)* in the following list who had been active in theory and/or in practices of DBMS. The M-Committee had the following members in the last stage: (Chairman) Tosio Kitagawa; (Vice Chairman) Setsuo Ohsuga; (Secretary) Takeo Yamamota; (Members) Takuya Katayama, Kiyoshi Wadatsumi, Tosiyasu Kunii, Yuzuru Tanaka*, Masafumi Togashi*, Masamitsu Negishi*, Ryosuke Hotaka*, Tadao Saito*, Koichi Furukawa*, Kaname Kobayashi*, Bunji Kariya*, Yahiko Kambayashi*, Jun-ichi Tsuji*, Sei-ichi Uchinami*, Setsuo Arikawa*.

4. ACTIVITIES OF THE M-COMMITTEE

4.1 Activities of the M-Committee during the second year of the SRP (1977, April-1978, March)

It was on the occasion of the Summer Research Meeting of the SRP in July, 1977 when the M-Committee published a tentative view claiming the needs for introducing the four different utility levels of scientific data storages in terms of (1) PRF (Private Researcher

File), (2) ARF (Accumulative Researcher File), (3) UDL (User Data Library) and (4) CDL (Center Data Library) as an essential architecture of data storages for scientific researches and pointing out the possible development paths among these four data storages. Here we shall give our general ideas on concepts of PRF, ARF, UDL and CDL introduced in the second year.

(1) Private Researcher File (PRF) is a file of information resources which is (i) made for, (ii) managed by and (iii) belongs to the researcher himself. These information resources may cover various information expression types under various stages of information processing processes.

(2) Accumulative Researcher File (ARF) is constructed by the cooperation of a group of research workers as an accumulation of a set of PRF's, each of which belongs to the respective research worker belonging to the group. This can be realized only under the condition that every research worker in the group can cooperate to work under a common specification scheme by which input information should be subject to a prescribed sequence of an information processing process. It is noted that every member of the researcher group is both a producer and user of ARF. Both the ownership and the management rights belong to the researcher group. Some technical operations regarding DBMS may be handled by other persons.

(3) User Data Library (UDL) is a collection of information resources which is admitted to common use of the user group whose member is authorized by a certain set of conditions. This looks something like a club library in the sense that it can be used by a person if, and only if he(she) belongs to the club. Here, data is an abbreviation of any information resource. It is noted that a member of the user group is not obliged to be a producer of the data accumulated in UDL.

Because of the fact that the information producer group is not identical to the information user group, at least one of either the ownership or the management may belong to an authority which is not identical to either of these two groups. In the case where both the ownership and the management belong to an authority different from these tow groups, information circulation and usage will naturally tend to follow a current scheme provided by some DBMS. This is the reason why we call it UDL.

(4) Center Data Library (CDL) is simply a collection of data bases which are admitted to the common use of any applicant to the Center with equal opportunitiy. The Center has both the ownership and the management rights and establishes cetain universal authorization rules for applicants on the usage of data bases. Hereafter, we shall introduce the notion of utility levels to distinguish these four types of information resource storage.

(5) Various examples of these four utility levels can be observed from our description of IRU research activities in the form ((i))(i=1,2,3,4). Here, only one typical example of each utility level is given in the following Table 4.1. It is noted that there are several projects in our SRP which are now going to level up the utility.

Table 4.1 Examples of four utility levels.

Utility level	System name	IRU responsible
((1))PRF	MIR-RF	B-1
((2))ARF	GEODAS	C-10
((3))UDL	HUNDRED	B-18
((4))CDL	TOOL-IR	C-2

Four instance, during the three years several IRU's were now cooperating to establish ARF regarding informatics of which MIR-RF as a PRF would become its one constituent component.

During this second year period the Committee performed several surveys and investigations as well as initiations of meetings in the following matters:

(1°) Surveys on the existing DB within country and abroad

(2°) Surveys on the current DBMS being used within country and abroad

(3°) Research meetings on "Information Expression Types and Information Processings" in various fields of scientific researches, in cooperation with the O-Committee

(4°) Participations in each District Council Meeting being established in each of seven distincts (Sapporo, Sendai, Tokyo, Nagoya, Kyoto, Osaka, Fukuoka) ranging over the whole Japan and consisting of the researcher workers in their respective distinct.

Now let us give a brief explanation of each of these activities. Regarding (1°), the Committee sent to each Distinct Council a list of DB's being available in USA and Europe and requested members from each Distinct Council to give information about their demands and their views to these DB's. In the meetings mentioned in (4°), the Committee had the opportunities to hear from the participants regarding the present states of DB which were being either implemented or in preparation by themselves. Regarding (2°), the Committee made an extensive survey of the IR systems and DBMS now in current uses in the USA and Europe and prepared a list of check points for using some of these as components of our scientific information system. Our activities concerning (3°) were devoted to hear from each IRU member in our SRP who was preparing either researcher files or DB. The topics were naturally concerned with the following points based upon their information materials.

(i) The actual states and/or the requirements for information expression types including documents, numerical data, pictures, figures and so on.

(ii) Plans for DBMS in their own area and any requirement for DBMS.

On the basis of the reports regarding these points (i) and (ii) there were active panel discussions about the following topics:

(iii) Requirements for DBMS in scientific research — dialogues between users and makers

(iv) Characteristic features of DBMS in scientific research in comparison with those in business and administration views from the standpoints of DBMS designers.

In connection with (i) and (ii), there were reports on the following data bases:

(a) Protein structure data base (Osaka University) (Yasuoka and Isomoto)
(b) Molecular structure data bases (Tokyo University) (Tasumi)
(c) Nuclear reaction data base (Hokkaido University) (Togashi)
(d) Astronomical data base (Kanazawa Institute of Technology) (Terashita)
(e) HUNDRED system (Hiroshima University) (Ikeda)
(f) Present state of DB in social sciences (Hitotsubashi University) (Matsuda)
(g) Survey data (Osaka University) (Yasuda)
(h) Historical investigations on the prices of agricultural products in India (Tokyo University) (Matsui)

4.2 Informational functions of PRF, UDL, and CDL

The Coordinating Committee decided to publish a draft plan for establishment of scientific information system in Japan in the second year of SRP. The M-Committee cooperated with the T-Committee by presenting a draft plan for DBMS in scientific researches. Hereafter the M-Committee had adopted three types PRF, UDL and CDL by amalgamating ARL into either to PRF or to UDL. In what follows we shall give an English translation of the explanations in Japanese regarding the informational functions of PRF, UDL and CDL.
The essential objectives of scientific researches lie in building up a scientific knowledge-system in an unknown or uncultivated research area of sciences. There a common recognition shared by a majority of research workers in our SRP to the effect that the whole process of scientific researches can be analyzed in accordance to the information process consisting of seven subprocesses, namely, (P_1) input information processing, (P_2) preliminary information processing, (P_3) structuring of information, (P_4) storing of information, (P_5) information retrieval, (P_6) information circulation, and (P_7) information usages. Every research activity process must be originated from some creative ideas due to each individual research worker and/or to each group of research workers, and any research achievements being obtained from research activity processes is always subject to public evaluation and critique before they can be used as common information resources in scientific communities and/or in public societies in general. In view of these real pictures of research activities in scientific researches, it is quite understandable that there do exist various different types of data storages according to the different developmental stages of researches. In this connection we have set up three representative types of data storages (i), (ii) and (iii), whose informatical functions are explained in what follows.
(i) PRF (Private Researcher File) This is based upon storage of scientific informations which an individual research worker or a group of research workers has acquired, collected, and accumulated for his own or their own research purposes and which belong to the individual research worker or to the group respectively. Hence PRF is a file of information resources which is (1°) made for, (2°) managed by and (3°) belongs to the researcher himself or the group itself as we have mentioned before. This is the first characteristic feature of PRF. In view of the information obtained and the information processing processes leading to the storage of scientific information, an abduction ground for hypothesis formation can be formed and a new information will be amalgamated into the data storage, provided that the hypothesis is not rejected. Even in such a stage, however, research worker has not necessarily been in prossesion of (attained at) a well established knowledge system and therefore he is in the situation of being open to a reconfirmation of the hypothesis in view of any new information in the future, and hence it follows that there is always the possibility of reforming another hypothesis. The research worker, therefore, should be working in error and trial way because he has to deal with a scquence of hypothesis. In this sense PRF should be plastic in its structure and flexible in its usage. This is the second characteristic feature of PRF.
(ii) UDL (User Data Library) The development of PRF may have attained at the stage when its information expression types and its storage structure has been already fixed and stable as the consequence of research activities after the recursive and the evolutionary processes of error and trial. In such a stage we can expect that the objectivity of the information materials as common information resources among users has been secoured, the scope and the number of their users has been enlarged, and their usages as public data storages has been settled down. This is the stage when we can set up a UDL whose utility level is certainly higher than that of PRF.
In this way UDL will be realized through growth and/or improvement of PRF.
The characteristic features of data files can be summarized as follows. (a) In the first place an existence of user groups which are different from the data producers. (b) Secondly both data producers and user groups come from the same specific division of science. (c) Thirdly the structures of data files of UDL are required to be plastic enough to reflect research outcomes due to the research workers belonging to the group. The purpose of data sharings in the case of UDL is partly due to the sake of completion of data files, while the reason to limit to the group of specialists in the specific science area in the case of UDL comes from the needs for scrutinizing the whole process of information processing so as to secure right understanding and an efficient usage of data files which are still in the state of imcompleteness.
(iii) CDL (Center Data Library) By passing through the stages of (i) PRF and (ii) UDL, the scope of data storages becomes clear enough to be eligible to public usages by attaching to them a set of adequate notices for their usages. At this stage of data storages users are not requested to know the details of information processing processes by which the data storages have been established.
In the consequence data storages at this stage can be used by any applicant who obeys a certain set of notices being attached for their usages.

It is our recognition in SRP that the most essential characteristic feature of scientific information systems is to have to prepare not only (iii) but also (i) and (ii) by which to imply the process of organizing scientific information in themselves.

4.3 Activities of the M-Committee during the third year of the SRP (1978, April-1979, March)

(1°) Four Working Groups WG-1-4 were set up within the M-Committee in the following way:
WG-1: System analysis to DBMS (Saito*, Hotaka)
WG-2: Information structure in DBMS (Wadatsumi*, Kariya, Arikawa, Tanaka, Kobayashi, Tsujii, Togashi, Kambayashi, Ohsuga, Katayama)
WG-3: Cost evaluation in DBMS (Kunii, Uchinami*)
WG-4: Planning and acheduling of the M-Committee (Ohsuga*, Yamamoto),
where X* implies that X is the Chairman of the corresponding WG.
Each WG-i (i=1,2,3,4) presented to the Committee Meetings the reports regarding the subject assigned to the WG-i, which were then subject to free discussions in the Committee meetings.
(2°) Among several DB activities being performed by each IRU the following topics were subject to a series of deep discussions in the M-Committee with reference to the merits and the possible improvements of their respective DB's.
(a) GEODAS (Osaka University) (Wadatsumi)
(b) Representation, storage and use of knowledge in computer system (Tokyo University) (Ohsuga)
(c) Modeling of nuclear reaction data (Hokkaido

University) (Togashi)
(3°) Through the intensive discussions about these actual proposals regarding DBMS, the M-Committee entered into a deep consideration about the rigorous concept specifications of the three levels of utility of data storages called PRF, UDL, and CDL and about the mutual interconnections among these three levels. The consideration lead the Committee to the discussions how to build up a system of DBMS for dealing with these three levels of utility, PRF, UDL and CDL.
(4°) There were presented three reports each of which was devoted to the topics in serious concern of the M-Committee.
(a) Knowledge base vs data base (Tokyo University) (Ohsuga)
(b) Architecture of evolutionary data base (Tokyo University) (Kunii)
(c) Three levels in DB (Osaka University) (Uchinami)
(5°) After the intensive investigation activities explained in (1°)-(4°) the M-Committee established a set of agreements within its members so as to be able to prepare the final report to be presented to the third year annual meeting of the SPR scheduled March, 1979. The main results will be explained in the following Section 5.

5. PRINCIPAL GUIDELINES FOR DEVELOPING DBMS IN SCIENTIFIC RESEARCHES

The unified views which the M-Committee reached through the Committee activities explained in Section 4 regarding the topics of this Section consists of two components, as we shall explain in Subsections 5.1 and 5.2.

5.1 Basic recognitions of characteristic features of scientific informations

An establishment of DBMS in scientific researches requires a deep consideration on the following three features as basic recognitions of characteristic features of informations in scientific researches.
(1°) It is a characteristic feature of scientific informations that structure recognition in any scientific research is neither fixed nor stereotyped but flexible and evolutionary in general. In the consequence any DBMS in scientific researches is required to be flexible and evolutionary enough so as to admit dynamic restructuring of DB in an easy way.
(2°) Any DBMS in scientific researches should take into consideration the specific circumstances of scientific researches where advanced human intelligence plays its intensive roles and hence highly intelligent man-machine interface must be realized in the usages of information processing apparata and systems.
(3°) Due to their respective characteristic requirements to DB usage schemes being presented by each research area of sciences, there will be necessarily various different types of DB in scientific communities. These DB's will show a geographical distribution because a majority of DB's will be required to be existing at the locations where informations have either occured or been produced. In the consequence, the set of DB's in scientific researches is necessarily expected to be distributed in geography and in discipline.

5.2 Five principles of developing DBMS in scientific researches

Principle 1. A development of DBMS in scientific researches should be performed according to some multistage planning, because the very characteristic features of scientific researches would not admit any one-stage realization of the whole requirements of scientific researches to their DBMS.
Principle 2. It is realistic and adequate to start with the first stage of DBMS development in scientific researches by adopting some adequate existing DBMS in current use.
Principle 3. The second stage of DBMS development in scientific researches should aim to realize a relief of the following two specific constraints being posed by the DBMS in the first stage:
(1°) Structural constraints in dealing with input informations to DB
(2°) Structural constraints in dealing with output informations from DB.
In order to realize these objectives, we recommand to develop two independent subsystems, one for a formation of DB itself and the other for an access to DB, and then to establish an interface between these two subsystems.
Principle 4. In order to realize the requirements for restructuring of DB in scientific researches, different implementation procedures should be performed according to the different stages in the following way.
(1°) In the first stage when current existing DBMS will be adopted, an implementation of restructuring of DB is realized by means of specification language and generation.
(2°) In the second and the following stages, any PRF (Private Researcher File) and any UDL (User Data Library) contained in the DB system in scientific research area should have their own DBMS's, which we may and we shall call PRFMS (Private Researcher File Management System) and UDLMS (User Data Library Management System) respectively. In this context CDL (Center Data Library) has its own DBMS which may be specified by CDLMS (Center Data Library Management System). Each of these three DBMS must provide with the informational fnnctions called data box, editing, and structuring and restructuring.
Principle 5. The distinctions among PRF, UDL (User Data Library) and CDL (Center Data Library) should be described both from the standpoints of management and from those of restructuring feasibilities. The following three requirements are given in order to secure restructuring feasibilities:
(1°) Any structure of PRF is required to be palstic, because the real pictures in PRF are described as being in the stage of seeking for any structure, and hence naturally are subject to the possibilities of restructuring from an existing structure, which may be even drastic.
(2°) There are a comparatively few chances for restructuring of CDL, and, even in those cases, we can expect the amount of restructuring will be small in comparison with the usal cases of PRF.
(3°) An intermediate situation between (1°) PRF and (2°) CDL is expected to be valid for the cases of UDL.

6. EXPLANATIONS AND SUMMARY COMMENTS

Due to the M-Committee activities including enquete surveys to research workers in various science areas and intensive discussions with many DB user groups in scientific researches, which we explained in Section 4, the M-Committee had gradually reached to the recognition to the effect that any DB system aiming at a satisfication of the real requirements

from scientific research workers must have several essential features in its design philosophy which are not in a complete coincidence with those of the current existing DB systems.
(i) In the first place the expression types of data in scientific researches cover a very large variety ranging over documental, numerical, pictural, figurative factual and structural ones. (ii) Secondly the data in scientific researches require a high degree of plasticity in their respective structure. (iii) Thirdly they should be subject to a strict evaluation of data validies and efficacies including reliabilities. (iv) Fourthly we can not simply assume an existence of administrator organization who takes the responsibility of integrated management of the whole system. Furthermore (v) fifthly data in scientific researches as a whole are naturally distributed, and so on.
In view of these characteristic features of scientific information, there had been two opposite basic strategic view (a) and (b) within the M-Committee regarding how to proceed in realizing its objectives. The first one (a) was to emphasize an early realization of DBMS for science researches and hence to adopt some of current existing DBMS which had been originated from business application areas, even perhaps with a sacrifice of restricting the areas of applications to some extent. The second one (b) was to make clear a set of specifications which DBMS in science researches should satisfy. If the M-Committee would take the former attitude (a) and hence proceed to adopt a substitution of any DBMS in science researches by exclusive uses of current existing DBMS, the application fields would be severely limited both in their scope and in their informatical depth and the jobs would be reduced to a matter of techniques how to make efficient uses of current existing DBMS, which are certainly adequate to be handled by user groups, and are far from the authentic problem areas of system design of DBMS in any sense. There existed also another circumstance to be taken into consideration of the M-Committee. Before and during the three years of our Special Research Project, there had been no systematically organized research groups except the M-Committee for which we could expect an intensive investigations for establishment of a set of fundamental specifications for any DBMS in scientific researches. In view of these circumstances, the M-Committee had thought that it would be the responsibility of the M-Committee to prepare voluntarily in this early stage such a set of fundamental specifications for DBMS in scientific research activities, and in the consequence it became quite adequate and realistic to adopt the second basic strategic view (b), as we had decided.
The M-Committee was naturally conscious of the fact that a development of DBMS in scientific research, in which more difficult requirements have been posed than in those of current existing DBMS in business, would involve a large amount of efforts, time and costs far beyond the capacities of one special Committee associated with SRP. Therefore the M-Committee concentrated its efforts to a clearification of the principles for DBMS in scientific researches. Under this attitude of research the M-Committee investigated the informatical situations in scientific researches by taking into considerations both an analysis of user requirements to DB and DBMS and a recognition of the present states of art in software and hardware of computer technologies. The set of the five principles enunciated in Section 4 was presented with the definite object in mind aiming at an establishment of efficient scientific information system in Japan in the coming days as early as possible. Our sincere hope is that the set of these five principles can provide a framework for development of DBMS in scientific researches by which effective implementation will be performed in the coming days.

Let us give here further remarks as an explanation of our view points. We have realized, owing to the M-Committee activities, that there are two essentially different and somewhat contradictory requirements (α) and (β) for any DBMS in scientific researches. The one requirement (α) for a DBMS in scientific researches is that it should contribute to each individual research worker as a support of his research activities by providing tools in hardware, software, and brainware for information processing procedures ranging over its whole process ranging from an early stage of input information processing until a final stage of a publication of his research achievements. The other requirement (β) for a DBMS in scientific researches is that it should contribute to build up a public resource of information from the set of individual accumulations of information materials step by step in accordance with the progress of research achievements and finally should lead to an establishment of DB's which are capable to be open to the common usages of user groups and/or to the public usages in general. In realizing this requirement (β), it should be desirable that the loads to be imposed upon research workers who are producers of information are either none or admissible (negligibly small) to the majority of them in comparison with the scientific great benefits to be obtained by the usages of the DB system. These two requirements (α) and (β) are in reality not immediately compatible. In fact one of the main functional merits of our system formation consisting of three utility levels called PRF, UDL, and CDL is to bring these two requirements (α) and (β) to the situation where we can realize a satisfactory negotiation. The idea of the system formation consisting of the four levels were presented as early as in the second year, but an intensive effort by which to establish a theoretical foundation and a practical implication of our system formation was given by the members of the M-Committee in the last third year, as we have already mantioned in Section 4.

Let us now explain the principal direction of our investigation which were performed under the basic strategic view (b) of taking advantage of some current existing DBMS in order to realize an early implementation. It was rather natural in this view (b) to add a set of informatical functions which satisfy the first requirement (α) enunciated just now. These are (1°) restructuring functions of DBMS securing plasticity of data structures, (2°) editors admitting acquisition of input information in free form, and (3°) information acquisition and storage functions to be secured for raw and arranged informations in the name of data box. These three informatical processing functions are indispensable to the stage of PRF for the sake of individual research worker and/or small groups of research workers, and also useful to the stage of UDL for the sake of user group, because at these stages a certain degree of freedome from formal constraints imposed upon data structures are more or less an essential requirement for their usages. It is noted that the three file constructions in the name of PRF, UDL, and CDL had been introduced originally not with clear description of functional

distinction but rather with respect to the levels of utility, that is, the scope of publicity of these storages. In view of the arguments being performed within the M-Committee, we recognized that there were the needs for structural distinction of these three levels PRF, UDL, and CDL.
The problem happened to decide which of the following two assertions would expected to be valid.
Assertion 1. It is possible to have a common scheme of DBMS's which is valid to each of PRF, UDL, and CDL. Under a common scheme of DBMS's, the distinction among PRF, UDL, and CDL is essentially reducible to the simple distinction induced by management policy regarding the usages of them.
Assertion 2. There does not exist such a common scheme of DBMS's enunciated in Assertion 1, and we need each individual DBMS for each of PRF, UDL, and CDL respectively.
We had several meetings of serious discussions about the problem. Our M-Committee has reached three conclusions which summarized our discussions.
Conclusion 1. We have to recognize the actual possibility of having different structures between PRF and UDL.
Conclusion 2. In most cases we may assume the same structure of CDL and UDL. Nevertheless there will be more contraints of functions open to users in the case of CDL than in that of UDL, because in the case of CDL neither the contents of DB nor their structures are permited to the direct alternations due to user interfaces.
Conclusion 3. It is desirable to have a transformation scheme among a set of CDL's.
Let us explain first the reasons by which the Conclusion 1 was established. In the stage of PRF, the highest priority should be placed on the informatical function of the system by which to overcome any structural constraint of input information as far as possible. On the other hand in order to generalize the informational contents of PRF to those of UDL and CDL in a smooth way, it will be desirable to have DBMS which belongs to the same scheme of that of UDL and more hopefully to that of CDL. Nevertheless a management system of PRF, which we may call a PRFMS (Private Researcher File Management System), is in the direct interface with its user and has to make use of the CPU and the terminals belonging to him. Under such situations we have to expect that the scale and the capacity of these hardwares and softwares are too small to contain the whole set of the current existing DBMS. Because of these limitations of capacities of information processing facilities including memory capacities in the case of PRF, PRFMS has to be reduced rather to a minimal sufficient scale keeping the least minimal functions of DBMS.
Now let us give an explanation of Conclusion 2. In principle each of UDL and CDL has their respective level of publicity of data sharing and for any transition from UDL to CDL it is at least convenient, if not necessary, to have the same DBMS between these two systems. However there are still non-negligible differences between these two system. In most case UDL requires to have restructuring functions of data storages as well as partial structuring functions of input information. On the contrary in CDL such functions are not allowed to be open to users.
In Figure 5.1 an illustration is given to show the constructions of PRFMS, UDLMS (User Data Libarary Management System), and CDLMS (Center Data Library M. S.). It is noted that these terminologies were introduced in order to show the differences among management systems of these three data storages.

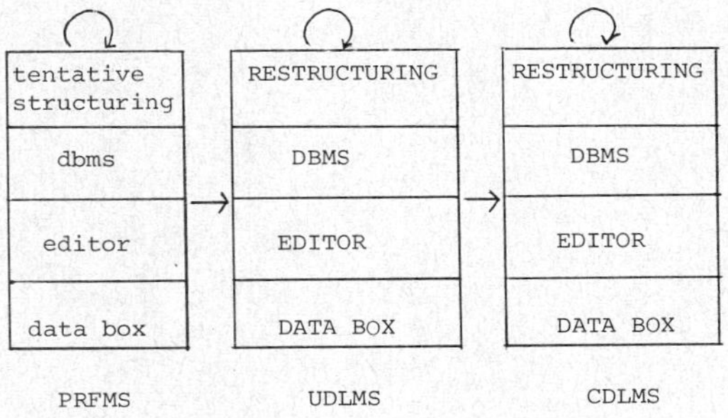

Fig. 5.1 Three levels of data base management systems

The shadows in CDLMS in Figure 5.1 indicate the fact that lockings are performed to the users.
Further technical discussions were performed in the M-Committee regarding the folowing items:
(a) Modularization of DBMS functions
(b) Data box function and editor by which input informations in any assigned form are acquired, processed, and transferred to the control of DBMS
(c) Adequacy of restructuring of files which are accumulations of experimental or survey data
(d) Procedures of avoiding restructuring
(e) Knowledge bases vs data bases
Although we have not succeeded in giving systems design of DBMS satisfying the requirements we have mentioned, we are self confident of having been able to give an important direction along which any scientific information system and its DBMS should be constructed. Further detailed discussions are given in the third year Report of the M-Committee [5].
Sections 3 and 4 of the Report [5] contain a description of the detailed discussions being performed in the M-Committee by the following contributions:
3.1. The three level construction of scientific information system (S. Arikawa and K. Wadatsumi)
3.2. PRF (M. Togashi and K. Kobayashi)
3.3. Design architecture for evolutionary data bases (T. Kunii)
3.4. Knowledge bases (J. Tsujii)
3.5. Standard translation form for data bases (R. Hotaka)
3.6. Construction of data base system in scientific researches (S. Ohsuga)
4. Operations and managements (T. Yamamoto)

7. REMARKS AND ACKNOWLEDGEMENT

This Report is based upon a series of the official M-Committee Reports mentioned in [1], [3]-[6]. As the Chairman of the M-Committee, I would like to express my hearty gratitude to our Committee members as well as many scientists within country and abroad who have been so cooperative to the M-Committee activities.
My sincere thanks are due to our Vice Chairman Prof. Setsuo Ohsuga who devoted himself to foster and to coordinate the M-Committee, and played the predominant roles in preparing our final Report. In connection with the contents of the present Proceedings, there are several papers on researcher file and data base management systems which will give the examples of PRF, ARF, UDL, and CDL. I might be permitted to remark that the paper of mine [7] in this present Proceedings will serve to give an infor-

matical implication of the M-Committee achievements in connection with the papers of mine quoted in the Reference of my paper [7].

REFERENCE

[1] A collection of survey reports in abroad in 1976, (in Japanese), Special Research Project 'Formation of Scientific Information Systems and Organization of Scientific Informations', Research Report, B-1, 1977, July, pp. 1-224.
[2] Reports of survey on SPIRES (in Japanese), Special Research Project 'Formation of Scientific Information Systems and Organization of Scientifc Informations', Coordinating Committee (unpublished).
[3] The Intermediate Report of the Special Research Project "Formation Process of Scientific Information Systems and Organization of Scientific Information" (in Japanese), (2) The Second Year Report, The Coordinating Committee, March, 1978, pp. 1-421.
[4] The Intermediate Report of the Special Research Project "Formation Process of Scientific Information Systems and Organization of Scientific Information" (in Japanese), (3) The Third Year Report, The Coordinating Committee, March, 1979, pp. 1-394.
[5] The report of the M-Committee on Data Base Management System (in Japanese), Report of Coordinating Committee, No.13, 1977, March , pp. 1-
[6] The Report of the M-Committee on DBMS (in Japanese), Report of SRP, 1979, November 5th-7th (in Press).
[7] T. Kitagawa, Formation of scientific information systems and its implication to informatics, (to appear in the present Proceedings)

Scientific Information Systems in Japan
H. Inose (ed.)

ORGANIZATION AND USE OF PRIVATE RESEARCHER FILES IN SCIENTIFIC RESEARCH WORKS

Setsuo ARIKAWA, Seigo KANŌ*, Tosio KITAGAWA** and Shun-ichi TAKEYA**

** Research Institute for Fundamental Information Science, Kyushu University 33, Fukuoka 812*
*** International Institute for Advanced Study of Social Information Science, Fujitsu Ltd, Tokyo 144*
Emeritus Professor of Kyushu University

This paper describes the systems we have studied and developed for these six years and discusses how they work for the organization and use of scientific information. The authors' fundamental philosophy of the scientific information systems is firstly given. The information storage and retrieval system MIR-RF based on the private researcher files and the automatic indexing system REKWEST are discussed. A one-way sequential search system using pattern matching machines and its three variants are described. Their powers are qualitatively evaluated. Finally a researcher file description language and its implication to such information systems are briefly discussed.

1. INTRODUCTION

The purpose of this paper is to describe one of the main subjects of the Research Unit "B-1: Consolidated Research on Formation Process of Scientific Information System" belonging to the Special Research Project (SRP). The central aim of our present study is to establish a cooperative relationship between individual researchers and the scientific information systems and, at the same time, to develop such systems for the scientific research workers.

The M-Committee of the SRP has proposed in 1977 the formation process which consists of the three levels; Private Researcher File (PRF), User Data Library (UDL) and Center Data Library (CDL). The proposal and opinion of the M-Committee are really consistent with our fundamental attitude and understanding which we had adopted in studying and developing our researcher file system MIR-RF (Multistage Information Retrieval system based upon Researcher Files) in the last Special Research Project during the period 1974-1976.

In fact we started with improving the MIR-RF system, which was at that time a PRF system for document retrievals, so as to work also for a UDL system and for an intellectual system as well. In the course of this study we have solved several fundamental problems and have newly developed TEXTIR (TEXT Information Retrieval) system and RFDL (Researcher File Description Language). We have also made several statistical observations on the word frequency, the sentence patterns and the types of errata in the references of scientific papers. Moreover, we have made and organized the common data for UDL such as the document files of papers from journals most frequently used in our study field, the books in our department library, and so on.

This paper concentrates on describing the developed systems and how they work for the organization and use of our private data. Section 2 discusses the philosophy of our information systems which will be realized in the later Sections. Sections 3 and 4 outline the first version of MIR-RF system and the REKWEST indexing system. Section 5 describes the newly developed one-way sequential search systems and gives a qualitative evaluation of the systems. In Section 6 we propose a notion of researcher file description language and its implications to our works.

2. RESEARCHERS AND SCIENTIFIC INFORMATION SYSTEMS

This Section explains our philosophy in designing scientific information systems to the effect that some cooperative relationship between the researchers and the scientific information systems should be established by reflecting the very fundamental fact that researchers are both producers and users of scientific information.

In most of the current scientific information systems, which are mainly for document retrievals in the level of CDL, the researchers are taken merely as users of the systems or receivers of information from them. The users can take parts in these systems only through their questions. This is due to the technical and economical performance consideration of the information systems for use by a great number of users. Hence we ought to be contented with this way of taking parts in the systems.

However we do need to find a way to make the researchers take more essential parts in the information systems. This is because the intrinsic mission of the scientific information systems is to assist the researchers in developing the next stages of their studies, and because the information searching activities of researchers are very subjective activities. As a matter of fact, a positive cooperation berween the researchers and the information systems is needed for clearing the way. So we shall mention some requirements from the researchers to the systems and vice versa, as some reciprocal relationship.

2.1 Requirements from Researchers to Systems

As just stated, the information searching activities of researchers are very subjective. Hence, first of all, the researchers should require:
1) That the researchers be able to reflect their individualities to the system.
If the researchers want this requirement to be satisfied, they are necessarily forced to cooperate with the system at least in preparing data. Their information searching activities are everyday jobs, but not, as a matter of fact, all of their research activities. Hence their second requirement should be

2) That the system be developed so that it does not become a heavy burden to the researchers, in preparing and making raw materials for data files. This requirement is also vital for continuation of the system. The following three additional requirements should be satisfied for the advanced uses of the information systems.
3) That the system be able to file and organize the researchers' private data which are lying scattered around them. The system should work as a workshop for the researchers' everyday activities. This function of the system contributes to organizing UDL and CDL.
4) That the system be able to use databases such as UDL and CDL which will cover the fields in which the researchers are interested.
5) That the system be able to store and to retrieve information not only on documents but also on numerical data and facts.

2.2 A Requirement from Systems to Researchers

It is well-known that the most essential and labouring work in managing information systems is to prepare, make and organize various kinds of data files. Hence the most essential requirement from the system to the researchers is
6) That they should cooperate with system when data for the system are prepared, made and organized.
Recently the number of journals in which papers have keywords and/or abstracts is gradually increasing. This is a stage in which editors of journals or authors of papers cooperate indirectly with the information systems. Also the researchers as readers of papers and books should further cooperate directly with the systems in extracting keywords and/or making abstracts from papers and books.

These jobs are usually entrusted to the specialists of documentation. However if the researchers want to reflect their own knowledge, view and insight in the information system, they should perform these jobs themselves according to their own thoughts.
The data presented by the researhcers will serve themselves. The accumulated data can lead to a UDL for a more general use, which will serve again for the researchers. In the new scientific information system, the researchers should keep in mind that they are not merely users of the system but the essential constituents of it.

3. MIR-RF SYSTEM

MIR-RF system is an information storage and retrieval system mainly for the use of researchers, which was developed by the authors [9],[10] at Kyushu Universisty. In this Section we give a short survey of the first version of the system which is a document retrieval system.

3.1 Researchers' Jobs

In order to maintain the document retrieval system for a long time by researchers, it is vital to minimize the researchers' efforts. There is not much to be done by the researchers for MIR-RF system. Their jobs are restricted only to the following: The first job is 1) to list the documents, in which the researchers feel interest, in such forms as the references and bibliographies in scientific papers and books. Every member of the lists should contain at least a) title, b) author(s), c) name of journal or publisher in case of book, d) year of publication, e) volume, number, page and language used. These lists of documents constitute the bases of data files of MIR-RF system. The second job of the researchers is
2) to extract keywords from and on books and papers, and organize them if necessary. Thirdly
3) the researchers can provide simple comments on books and papers.

The keywords can be organized into a tree structure, which represents the relations of 'equivalent to', 'lower than' and 'higher than' between the concepts, although concepts are usually organized into a kind of networks. The trees are represented as in Example 1, where the letters a, b,... mean the keywords and the strings 1, 11, 12,... indicate the tree representations of the keywords. By using the strings of letters we can also express our viewpoints, classifications and some attributes as well.

3.2 File Construction

From these raw materials three kinds of main files DF (document file), RF (researcher file) and KF (keyword file) are produced. DF is a set of data prepared in 1), 2) and 3), RF is a set of researchers' memos prepared in 2) and KF is a set of keywords extracted from titles of documents by the automatic indexing system REKWEST (REformed KWEST, see Section 4).

3.3 Multistage Retrievals

The researcher files are used for two purposes. One is to supplement keywords extracted from titles by REKWEST, and the other is to reflect researchers' individualities to the system. In principle it is possible to retrieve without using any researcher file. In such retrievals it has been observed in our experiment [9] that the precision rate becomes very high but the recall rate becomes extremely low. Hence it is really necessary to support REKWEST system. In general, such an indexing system is supported by a large scale thesaurus. The thesaurus is replaced by the researcher files in our MIR-RF system.

The basic retrievals by MIR-RF system are illustrated in Fig.1. The system is of multistage retrieval from two aspects. The first aspect is seen in tracing trees in a single RF. The other aspect is seen in Fig.1, that is, in use of two or more RF's.

MIR-RF system has two basic modes of retrievals. These modes, together with the other concepts of information systems, were formulated by using the mathematical concepts of relations and their compositions [9]. There we introduced two kinds of associations - the document associations and word associations. The system thus allows users to select some of these two associations.

Document associations are relations between pairs of documents based upon numerial nearness. The nearness of document pairs is defined by using three kinds of coefficients on the sets of keywords.

Word associations are, in turn, relations based upon several kinds of qualitative nearness of words. This retrieval mode is rather similar to the usual retrievals using a thesaurus. In our case the thesaurus is replaced by the multistage using of RF's. Three kinds of word associations E, H and L, namely 'equivalent to', 'higher than' and 'lower than', respectively, are used in the first version.

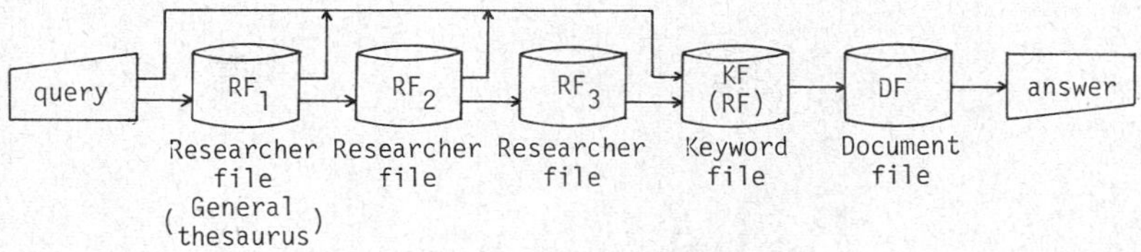

Fig. 1. Basic flow of MIR-RF retrievals

3.4 Researcher File Manipulations

MIR-RF system has a facility for manipulating and editing researcher files according to the requests by researchers [10]. This manipulation system consists mainly of two parts. One is to make unions of trees and the other is to make intersections of trees. Essentially, making the unions is joining fragmentary researcher memos (RM) or files together, while making the intersections is cutting off some extremely subjective words and branches from the trees.

Example 1. The followings are examples of a union and an intersection of two RM's.

$$\left\{\begin{array}{ll} 1 & a \\ 11 & b \\ 11 & c \\ 111 & d \end{array}\right\} \cup \left\{\begin{array}{llll} 1 & b & 2 & a \\ 11 & e & 21 & d \\ 111 & f & 21 & h \\ 112 & g, & 211 & i \end{array}\right\} = \left\{\begin{array}{llll} 1 & a & 1111 & i \\ 11 & b & 112 & e \\ 11 & c & 1121 & f \\ 111 & d & 1122 & g \\ 111 & h, & & \end{array}\right\},$$

$$\left\{\begin{array}{llll} 1 & a & 1111 & f \\ 11 & b & 1112 & g \\ 11 & c & 1112 & h \\ 111 & d, & 112 & e \end{array}\right\} \cap \left\{\begin{array}{llll} 1 & a & 112 & j \\ 11 & c & 1121 & e \\ 11 & i & 1122 & k \\ 111 & f, & & \end{array}\right\} = \left\{\begin{array}{ll} 1 & a \\ 11 & c \\ 111 & f \\ 112 & e \end{array}\right\}.$$

Now let R_{ijk} be a researcher file for researcher(s) k in a research group j of a field i. Then we can explain the process of editing our thesauri as follows:

$$R_{ij} = \bigcup_k R_{ijk}$$

is a thesaurus, i.e., an accumulation of researcher files of the group j of the field i. And

$$R_i = \bigcap_j R_{ij} = \bigcap_j \bigcup_k R_{ijk}$$

is a thesaurus of the field i. Lastly

$$R = \bigcup_i R_i = \bigcup_i \bigcap_j \bigcup_k R_{ijk}$$

is the general thesaurus of MIR-RF system at that time. Thus we can easily edit the general thesauri, which vary according to the lapse of time. The general thesauri are useful for visitors who have no RF in the MIR-RF system.

Schematically R_{ijk}, R_{ij}, R_i and R correspond to the notions of private researcher file (PRF), accumulated researcher file (ARF), user data library (UDL) and center data library (CDL), respectively.

4. REKWEST INDEXING SYSTEM

This section describes the automatic indexing system REKWEST. REKWEST is an abbreviation of REformed KWEST. And KWEST is an abbreviation of Key Word Extracted as a String of Terms. The underlying system KWEST of our REKWEST was introduced by T. Watanabe [1], in order to decrease the redundancy of the traditionally well-used indexing systems such as KWIC and KWOC. His idea of extraction of index terms as forms of word strings was accepted into our system not only because his proposition is agreeable but also because it can satisfy another point of view of ours. That is, we know and recognize by everyday experience that there is a trend that the more familiar the concepts become, the more frequently are they expressed in the forms of word strings without prepositions.

The REKWEST consists of three subsystems - the singularizer, the deleter of stop-words and the processor of quasi-stop-words.

4.1 Singularizer

We adopt a singularizer to our indexing system from two purposes: The first is to decrease the number of index terms and the second is to compress the size of the dictionary, i.e., the list of stop-words and quasi-stop-words.

English nouns are divided into those with regular plurals and those with irregular plurals. Hence a kind of a dictionary is necessary to process the latter. We have selected and summarized 30 suffixes of English words.

4.2 Stop-words

The stop-words in the usual automatic indexing systems such as KWIC are selected by taking into account of the frequency of occurrences of words. In our system, however, we did not adopt such a statistical criterion. The main reason is that we are afraid to lose meaningful words. We should, in our retrieval system, take care not to lose informations in order to answer varieties of researchers' requests.

In REKWEST system, from these viewpoints, we picked up about 80 words below: 1) prepositions except 'up', 'down', 'to', and, 'without', 2) articles, 3) co-ordinate conjunctions, 4) auxiliary verbs, 5) numerals from zero to ten, 6) personal pronouns, 7) interrogatives except 'how', 'what', 'which', and 'when', 8) pronominal adjectives, 9) some other adverbs and adjectives concerning degrees, and some ing-forms working participially and prepositionally.

4.3 Quasi-stop-words

A word is called a quasi-stop-word if it is deleted under a contextual condition. The essence and difficulty of indexing systems such as KWIC, KWOC and KWEST lie, in fact, in setting up a criterion for the stop-words. Under a loose criterion a trivial redundancy occurs, and under a rigid criterion a vital information loss occurs. In order to improve these systems without

appealing to statistical methods, it is necessary to develop at least syntactical considerations on titles. In general a more syntactical consideration is required, to construct some large and complicated dictionary. In our system, however, it was attempted to use the notion of quasi-stop-words and to make a step toward a syntactical consideration by keeping the size of a dictionary as small and simple as possible. The following three kinds of words were selected as quasi-stop-words:
1) 'to', 'when', 'how', 'what' and 'which'. These words are not deleted if they occur in forms 'when to', 'how to', etc., otherwise deleted.
2) words ending with 'ed' or 'ble'. These words are deleted if they occur in the right end of keywords, otherwise not deleted.
3) 'kind', 'note', 'study', 'remark', 'result', 'case', etc. These words are deleted if each of them occurs alone in the outputs of the deleter.

We give an example of the extracting processes by REKWEST. The title (A) is decomposed into the set (D) of keywords:

(A) ON THE LANGUAGES DEFINED BY SENTENTIAL FORMS OF CONTEXT-FREE GRAMMARS

⇓ (by the singularizer)

(B) ON THE LANGUAGE DEFINED BY SENTENTIAL FORM OF CONTEXT-FREE GRAMMAR

⇓ (by the deleter of stop-words)

(C) LANGUAGE DEFINED,
SENTENTIAL FORM,
CONTEXT-FREE GRAMMAR

⇓ (by the processor of quasi-stop-words)

(D) LANGUAGE,
SENTENTIAL FORM,
CONTEXT-FREE GRAMMAR.

4.4 Improvement of REKWEST

A main difference of titles from the usual sentences lies in the fact that in almost all titles verbs are transformed into nouns or neglected, and subjects are also neglected. In return, some prepositions are supplied. Consequently all the prepositions in titles play key roles to decide the syntax of titles. Hence we can say that some simple syntax of titles has been taken into consideration when deleting stop-words, since the stop-words mainly consist of prepositions.

In order to confirm this trend and to improve the system much more, we have made statistics on word frequency in scientific documents [20]. As a sample we have picked up three volumes of INSPEC tapes of Computer and Control Section, which covers a reasonable subfield of our study field. They include the titles and abstracts of about 8000 documents. Approximately the set of titles contains 67000 words in total and consists of 8200 different words, while the set of abstracts with titles contains 610000 words in total and consists of 20000 different words. Comparing the rates of the word frequencies in these two sets, we have statistically confirmed the trend.

5. TEXTIR SYSTEMS

Sequential searches based on pattern matching have mainly been studied for text-editing and symbol manipulations [2], [3], and have not been understood so practical for the information retrievals, because they, to say the least, require time proportional to the length of file to be scanned. However the newly invented techniques of pattern matching, together with the recent development and diffusion of hight-speed external memory units, are making it possible to apply the sequential searches to some practical information retrieval systems [4], [13].

We have developed a sequential search system TEXTIR (TEXT Information Retrieval system) at Kyushu University and at Edinburgh Western General Hospital in order to attain the following purposes:
1) Statistical study on natural languages, particularly on sentence patterns in abstracts of scientific documents.
2) Study on automatic indexing.
3) Development of information system for researchers' use.
4) Patient record system for doctors' use.
5) Support system for everyday business in laboratories.
We have constructed three variants of TEXTIR, which is based on Aho-Corasick's pattern matching machines [4]. They are different in retrieving powers but have two distinctive features in common:

a) Retrievals of records of the form $x_1 \ldots x_2 \ldots x_3 \ldots x_4$. Every variant of TEXTIR can retrieve records of that form, where the triple dots ··· mean "some string". The function is particularly useful for checking of some contextual conditions concerning the order of occurrences of keywords.

b) Independence of record formats. Our systems do not assume any definite formats on records or files to be scanned, but assume the whole file of records as one very long string of characters. We can freely set up the so called output delimiters like the usual keywords and can consider strings surrounded by the pairs of output delimiters as virtual records.

This Section briefly describes the three variants of TEXTIR and discusses their powers from viewpoint of formal language theory, and then briefly refers to the realization of the systems.

5.1 Preliminaries

We begin with a simple formalism of records, files and related concepts.

Let Σ be an alphabet, that is, a finite nonempty set of characters, Σ^* be the set of all words or strings over Σ including the empty word ε and Σ^+ be the set $\Sigma^* - \{\varepsilon\}$. Then a record is a string in Σ^* and a file is a subset of Σ^*.

When a file is organized or its elements are collected according to some definite rules, the set of rules can be taken as a grammar G and the file as a language L(G) generated by G. Usually the file is a set of collected samples from L(G) which possibly has infinitely many strings.

Similarly, retrieving some set of records which satisfy a query Q corresponds to making an intersection $L(G) \cap L(G')$, where G' is a grammar defined by the query Q. This understanding of the concepts will lead us to a new qualitative evaluation of the powers of one-way sequential search systems.

The class of grammars for queries is convenient if they have ability to define all reasonable subsets of L(G) or Σ^*. We call a one-way sequential search system R-universal if the class of all languages defined by the system and the queries to it contains the class of all regular languages, i.e., languages recognized by finite automata.

The Aho-Corasick's pattern matching machine is an efficient finite automaton which runs on text strings and detects all possible occurrences of keywords.

5.2 Queries

A query to TEXTIR consists of lists of input/output delimiters, keywords and logical formulae, and some other instructions such as selections of modes and object files.

Delimiters. TEXTIR makes use of input/output delimiters, which are strings in Σ^+ assigned to the variables A, B,..., Y. The output delimiters work for marking off some virtual records from a long text string, while the input delimiters work for telling the system when to evaluate the logical formulae. The system evaluates the formulae every time it detects input delimiters. According to a condition at the second output delimiter it displays the virtual record between the pair of output delimiters. The work and relation of the input/output delimiters are illustrated in Fig.2, where u_1, u_2, u_3 are output delimiters, n_1, n_2 are input delimiters and r_1, r_2

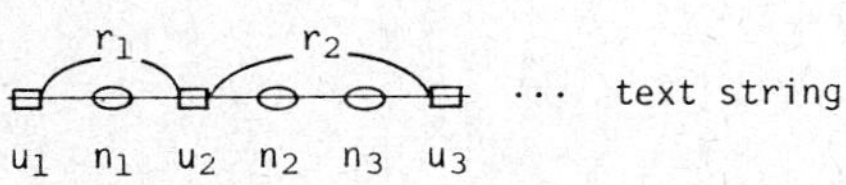

Fig.2 Input/output delimiters

are the virtual records. For example, in the usual retrieving, if the value of a logical formula is equal to 1 (true) at the output delimiter u_3, then the system will display the record r_2 as a reply to the query.

By virtue of these input/output delimiters we can use the system for very complicated retrievals if we are familiar with formats peculiar to the object files or text strings.

Keywords. We can use two kinds of keywords - the usual ones in Σ^+ and ones with some occurrences of triple dots. The keywords with triple dots such as

$$x_1\cdots x_2\cdots x_3\cdots x_4$$

work for defining or detecting virtual records of the form

$$\Sigma^* x_1 \Sigma^* x_2 \Sigma^* x_3 \Sigma^* x_4 \Sigma^*,$$

where x_1, x_2, x_3, x_4 are the usual keywords in Σ^+ without triple dots $\cdots$ as a substring. Thus the triple dots all together mean "some string" in Σ^*. Hereafter we use a symbol $\circ$ instead of the triple dots $\cdots$ in keywords to avoid a confusion.

The function of keywords of this type, which is not new in symbol manipulations [3], is particulary useful for checking some contextual conditions as to the order of occurrences of keywords in records.

The listing of keywords should be done by using variables A, B,..., Y issued by the system. We call the variables A, B,..., Y keyword variables.

Logical Formulae. A logical formula is a usual logical expression which is composed by using keyword variables A, B,..., Y, logical operators +(for 'or'), $\cdot$(for 'and') and $\sim$(for 'not'), and some pairs of brackets [and]. All keyword variables in the formula should be defined in the preceding keyword listing. The listing of logical formulae is done after the formula variables Z_1, Z_2,..., Z_{15} issued by the system.

5.3 Three Variants of TEXTIR

Three variants of TEXTIR - system **K**, system **W** and system **E** have been constructed. The basic queries and functions mentioned above are all common to these three systems. This Section describes other queries and functions proper to each variant.

System K. The system **K** is a standard version of TEXTIR, in which the output delimiters also work as input ones.

System W. As we shall see in Section 5.4, the function of the so called triple dots, which is one of the distinctive features of the system **K**, is more powerful than the sequential search system by Aho-Corasick. However **K** is not yet R-universal. In order to attain the universality and to enlarge the power much more, we have extended **K**.

The extended system **W** can restrict the universe Σ^* into some reasonably small subsets and make use of other variables than the keyword variables. The output delimiters can work properly.

1) Restrictions of universe. Queries in the system **K** define some subsets of Σ^*. The real Σ is the set of all characters used in computers. Hence it is not small. We sometimes want to restrict Σ within a small subset. We have realized this retriction in the system **W** by inhibiting the loop transitions in the initial state.

The system **W** has two retrieval modes {reset, noreset} related to the restriction. When the noreset mode is selected at the stage of querying and once the loop transition is taken place, the system ignores all keywords and input delimiters until it finds an output delimiter. By use of the noreset mode users can restrict Σ^* nearly into H^*, where H is the set of delimiters and keywords and is considerably small. The reset mode is the same as the implicit mode in the system **K**.

2) Use of formula variables in formulae. In the system **W** it is also possible to use the formula variables Z_1, Z_2,..., Z_{15} and a special variable @ called ε-variable in logical formulae and to indicate some formulae as temporary formula variables. The ε-variable @ plays a special role. Its value is true only when the system, leaving the last output delimiter, is about to start its job and becomes false after the system has started the searching for keywords.

The system evaluates the set of logical formulae in the order of Z_1, Z_2,..., Z_{15}, whenever it finds an input delimiter, and gives those virtual records between the last two delimiters as the outputs for the query when it finds an output delimiter and if the last value of the logical formula corresponding to the query is true.

The system clears all formula variables in use by Zi := ivz(i) whenever it finds an output delimiter. The value ivz(i) is determined at the stage of query editing as follows. Let A1 (= A), A2 (= B),..., An be keyword variables in use and Z1, Z2,..., Zm be formula variables in use as well, and let

$$Zi = f_i(@;\ A1,\dots,\ An;\ Z1,\dots,\ Zm\text{-}1,\ Zm)$$

for each i. Then the values of ivz are determined successively by

$$\begin{aligned} ivz(1) &= f_1(1;\ 0,\dots,0;\ 0,\dots,0,0) \\ ivz(2) &= f_2(1;\ 0,\dots,0;\ ivz(1),0,\dots,0,0) \\ &\vdots \\ ivz(m) &= f_m(1;\ 0,\dots,0;\ ivz(1),\dots,ivz(m-1),0). \end{aligned}$$

The extended mode of evaluating is automatically adopted if one of the following conditions is satisfied:

1) The ε-variable @ is used in some formulae.
2) There are some formulae $Zi = f_i$ in which a formula variable Zj with $j \geq i$ is used.

If none of them is satisfied, the system W works just like the system K. Hence a query such as

$$Zi = f_i(A1, A2, \dots, An)$$

in the system K is equivalent to a query

$$Zi = Zi + f_i(A1, A2, \dots, An)$$

in the system K. Hence the value of the right-hand Zi at time t is one at time t-1 and hence the variable works also as a memory. On the other hand, in a formula such as

$$Zj = Zi + f_j \qquad (i < j)$$

the variable Zi does not work as a memory.

System E. The system E, which is the final version of TEXTIR, is obtained from the system W by introducing several counters and by enlarging the operators in logical formulae. The counters are realized only by substituting A := A+1 and Z := Z+1 for A := 1 and Z := 1, respectively, and hence they count how often a keyword is detected in between a pair of input delimiters and how often a logical formula is satisfied in between a pair of output delimiters.

A logical formula in the system E is composed by using the keyword variables, formula variables and ε-variable @, integers, logical operators (|, ·, ∿), comparison operators (=, < =, > =, <, >, <>), arithmetic operators (+, *, -, ÷), and brackets [,].

In the system E a variable α is considered to be true if α > 0, and hence a logical formula Z is satisfied if Z > 0. The logical operators are interpreted as follows:

$$\begin{aligned} \alpha|\beta &= 1 && \text{if } \alpha > 0 \text{ or } \beta > 0, \\ \alpha\cdot\beta &= 1 && \text{if } \alpha > 0 \text{ and } \beta > 0, \\ \sim\alpha &= 1 && \text{if } \alpha \leq 0. \end{aligned}$$

5.4 Powers of the Systems

Now we consider the retrieving powers of the three variants of TEXTIR as well as Aho-Corasick system from a formal language theoretic viewpoint. The Aho-Corasick system A is also taken as a variant of our TEXTIR, although it has no function of the triple dots and the delimiters.

These systems can define many languages by various queries when we take files as Σ*. To make our discussion clearer we refer to an interpretation of these systems and queries.

A keyword assignment A = x in the system K as well as the system A can naturally be taken as a definition of a language

$$\|A\| = \|x\| \equiv \Sigma^* x \Sigma^*,$$

and $B = x_1 \circ \cdots \circ x_n$ in the system K as

$$\|B\| = \|x_1 \circ \cdots \circ x_n\| \equiv \|x_1\| \cdots \|x_n\|.$$

The interpretation can be extended to all formulae in the system K. A formula assignment $Z = f(\alpha_1, \alpha_2, \dots, \alpha_n)$, where α_i $(1 \leq i \leq n)$ are variables, defines a language

$$\|Z\| = \|f(\alpha_1, \alpha_2, \dots, \alpha_n)\|$$

by the following interpretation:

$$\begin{aligned} \|\alpha_i + \alpha_j\| &= \|\alpha_i\| \cup \|\alpha_j\|, \\ \|\alpha_i \cdot \alpha_j\| &= \|\alpha_i\| \cap \|\alpha_j\|, \\ \|\sim\alpha_i\| &= \Sigma^* - \|\alpha_i\|. \end{aligned}$$

A similar interpretation is also valid in the other systems W and E.

Let $C_A(\Sigma)$, $C_K(\Sigma)$, $C_W(\Sigma)$ and $C_E(\Sigma)$ be the classes of languages over Σ defined in the systems A, K, W and E, respectively, and by all possible queries, and let $C_R(\Sigma)$ be the class of regular languages over Σ. Then we have the following theorems on the powers of our systems. (See [16],[18] for proofs)

Theorem 1.

1) $C_A(\Sigma) = C_K(\Sigma) \subsetneq C_W(\Sigma) \subseteq C_E(\Sigma)$ if $\#(\Sigma) = 1$
2) $C_A(\Sigma) \subsetneq C_K(\Sigma) \subsetneq C_W(\Sigma) \subsetneq C_E(\Sigma)$ if $\#(\Sigma) > 1$.

From Theorem 1 we immediately have the following:

Corollary 2. The function of triple dots in the system K is independent of the logical operators.

Theorem 3. Let $C_K^{(n)}(\Sigma)$ be the class of all languages over Σ each of which is defined in the system K by a query with at most n uses of triple dots in a keyword. Then for Σ with $\#(\Sigma) > 1$,

$$C_A(\Sigma) = C_K^{(0)}(\Sigma) \subsetneq C_K^{(1)}(\Sigma) \subsetneq C_K^{(2)}(\Sigma) \subsetneq \cdots \subsetneq C_K(\Sigma).$$

Theorem 4. $C_W(\Sigma) = C_R(\Sigma)$ for any Σ.

Theorem 4 asserts that the system W and hence the system E are R-universal, and applicable to a syntax checker of finite state languages.

Theorem 5. The class $C_E(\Sigma)$ is properly contained in the class of recursive sets over Σ.

5.4 Implementations

The systems K and E have been realized in the small scale computer PANAFACOM U-400 in Fortran at Kyushu University, and the system W in GEC 2050 in Coral 66 at Western General Hospital. This Section briefly describes how the transition tables of pattern matching machines and the triple dots processor have been realized in the computers.

Transition tables. The transition diagram of the pattern matching machine is realized in an array of size $3 \times n$ if Aho-Corasick's algorithm is directly applied, where n is nearly equal to the number of states in the transition diagram. We have decreased the size to $2 \times n$ by deleting the row of goto function. The deleted goto function is easily recovered by a simple function goto(k)=k+1 for any state k. The original algorithm needs another big array as a working space at the stage of constructing failure function. We have deleted this working space while keeping the program in the same size and in nearly the same speed. And keeping the transition table in the same size, we have also deleted the redundant transitions as many as possible.

Triple dots processing. In principle it is possible to construct directly a finite automaton which accepts a language defined by $x_1 \circ \cdots \circ x_n$, but it forces us to abandon our present approach and to build a new theory. The directly constructed finite automaton will use many states. Hence we have adopted a method of marking the order of occurrences of sub-keywords x_1, x_2, $\cdots$, x_n at the stage of retrieving.

6. EXTENSION OF MIR-RF SYSTEM

In this Section we discuss an improvement and an extension of the first version of MIR-RF system.

6.1 Use of Citation Information

The first version of MIR-RF system has no efficient function to deal with the citation information, which is particularly useful for comprehending the research trends. As a matter of course, in the system it is possible to store the citation information in the form of researcher files with a tree structure and it is also possible to trace the citation relations by using the word-associations. Such files should include all references of the papers and books on which the researcher is making researcher memos. It will be tedious and troublesome for researchers.

Hence we have designed an automatic coder of papers and books. The code word consists of the following 12 characters:

AAAYYTTTJJVP,

where each sub-codes are determined as follows:
1) AAA is the initial three letters of the first author's family name. If the name is shorter than three, then add the initial letter of his first name.
2) YY is the last two digits of the year.
3) TTT is three initial letters of the first three words in the title, where the articles, prepositions and coordinate conjunctions are not counted in as the words. If this sub-code is empty, then recover the deleted words. If it is too short, then supply the lack from the last word. If it is still too short, then add the necessary blanks.
4) JJ is two initial letters of the first two words in the journal name or the publisher name in case of a book. This sub-code is determined just like in 3).
5) V is the last letter or digit of the volume number, and is a blank if missing.
6) P is the last digit of the start page, and is a blank if missing or in case of a book.

Example 2. The first three references of this paper are coded as follows:

WAT74KISDK65,
KNU74FPMTRO ,
FAR64SSMJA11.

The coder is much more easily implemented than the REKWEST system. It should be noted that the code words are independent of the code words already stored. Compared with the other code words of documents, ours are rather redundant. The redundancy keeps our coding nearly one-to-one. The coding has been designed according to the statistics on errata in the references of scientific papers [19].

Now the coded references can be associated with their entities in the DF and the RF by some word-association, if their entities are already in these files. This associating will work for making citation trees, and the trees will grow up each time the RF with these codes is added.

6.2 Researcher File Description Language

As we have seen, the MIR-RF and its improvement satisfy the requirements 1), 2), 3) and 6) of Section 2. The requirement 4) is gradually being satisfied in several university computer centers. Hence in so far as document retrievals are concerned, we have a prospect of achieving the scientific information systems proposed in Section 2.

The rest of the problems is to increase the precision rate much more and to satisfy the fifth requirement, which need to extend the MIR-RF system. For this purpose it needs to set up a more powerful description language than the trees and to enrich the contents of the RF's.

Requests for RFDL. We propose a researcher file description language (RFDL for short) to describe such researcher files. The RFDL should have the following properties:
1) Not so far from the natural language.
2) Powerful enough to describe the researcher files and comments.
3) Structually unambiguous.
4) Fit for highly intellectual processing.

Grammar of RFDL. The syntax of RFDL is described by a parenthesis grammar. The parenthesis grammar, introduced by R. McNaughton [5], is a well-conditioned context-free grammar and can represent the dependency relations between the words. Here we give two illustrative examples instead of the definition. The researcher memos a) and b) in the natural language are expressed as a') and b') in RFDL, respectively.
a) APPLICATIONS OF CONTROL THEORY TO GAMES.
b) IF X EQUALS Y AND Y EQUALS Z THEN X EQUALS Z.
a') APPLICATIONS OF (CONTROL (THEORY)) TO (GAMES).
b') IF ((X EQUALS Y) AND (Y EQUALS Z)) THEN (X EQUALS Z).

Researchers can easily learn the syntax of RFDL, since we do not set any essential limits on the natural language, except the parentheses which are used to show the dependency relations between the words. By means of these parentheses we can eliminate most structural ambiguities. The dictionary for RFDL is constructed from some collections of researcher files in a tree structure. In fact, it is a set of pairs (w,c), where w is a word with a tree code c, which can also represent some attributes of keywords as well as the usual conceptual relations. These tree codes for attributes will play an essential role in the interpretation of the sentences in RFDL.

The semantics of RFDL depends on how to use it in the information system to be constructed. We have not yet developed the semantics precisely, but we can assign meanings to the basic dependency structures [12].

6.3 Use of TEXTIR in Substitution for RFDL

We are now determining some standard and basic dependency structures by investigating the word frequencies and sentence patterns in the abstracts in the INSPEC tapes. The extended system we are designing will work as a highly intellectual knowledge system. However it will take much time to complete the system. Now one of the main purposes of developing the new system is to increase the precision rates when our system is linked to the large databases. Therefore we can use the TEXTIR in substitution for the system with RFDL.

As we have seen in Section 5, TEXTIR is quite suitable for retrievals from rather small databases. It has facilities for detailed retrievals. Thus TEXTIR is used for selecting really desirable data from private files which may be a selection from a large database through the MIR-RF system.

6.4 Basic File Structures

The extended system should also work as a kind of workshop for the researchers' everyday activities. Hence the system should be able to cope with variations of data. From this viewpoint we have designed the basic file structures and we are now implementing the extended system.

The files of our system are classified basically into two types, indexed files and sequential ones. Indexed files are realized by the multiway balanced trees, which allows varying-length keys.

Usual data are stored in the sequential files. A sequential file consists of some physical pages which are connected with bidireted pointers.

Thus we have designed the file structures paying our special attention to easier maintenance including file deletions and to more efficient use of memories.

7. CONCLUDING REMARK

We have outlined our basic principle of the scientific information systems, particularly of the PRF systems, and briefly discussed the MIR-RF system, REKWEST system, TEXTIR system and the extended MIR-RF system including the researcher file description language. For further details, the reader should be referred to the authors' publications [6] ~ [20] and also to their references.

The MIR-RF system was originally implemented in the mini-computer FACOM U-200 system and the extended version is now being implemented in PANAFACOM U-400. The extended MIR-RF system together with the TEXTIR system and the RFDL will work as a convenient workshop for the researchers' everyday research activities. By using the system the researchers will be possible to provide data for the UDL and receive data from the UDL.

REFERENCES

[1] T. Watanabe, "KWEST Index System - A New Attempt of Automatic Indexing.", Dokumen Kenkyu, 1974-06, 195-200, (1974), (Japanese).

[2] D. E. Knuth, J. H. Morris, Jr. and V. R. Pratt, "Fast Pattern Matching in Strings", TRCS-74-440, Stanford Univ. (1974).

[3] D. J. Farber, R. E. Griswold and I. P. Polonsky, "SNOBOL, A String Manipulation Language", J. ACM, 11, 21-30, (1964).

[4] A. V. Aho and M. J. Corasick, "Efficient String Matching: An Aid to Bibliographic Search", C. ACM, 18, 333-340, (1975).

[5] R. M. McNaughton, "Parenthesis Grammars", J. ACM, 14, 490-500, (1967).

[6] T. Kitagawa, "The Efficiency of Information Storage and Retrieval System for Scientific Research Activities", Res. Rept. Res. Inst. Fund. Inform. Sci., Kyushu Univ., 38, 1-21, (1974).

[7] T. Kitagawa, "Formation of Scientific Information System and Its Implication to Informatics", J. Inform. Processing, 1 (3), 111-124, (1978).

[8] S. Kanō and Y. Kai, "Optimal Reinforcement of Generalized Learning Processes by Means of Markov Decision Processes", Proc. HICSS-12, 2, 226-235, (1979).

[9] S. Arikawa and T. Kitagawa, "Multistage Information Retrieval System Based upon Researcher Files", Proc. 2nd USA-Japan Comp. Conf., 149-153, (1975), and also in Res. Rept. Res. Inst. Fund. Inform. Sci., Kyushu Univ. 51, 1-34, (1975).

[10] S. Takeya, "Researcher File Manipulation in MIR-RF System", Res. Rept. Res. Inst. Fund. Inform. Sci., Kyushu Univ., 55, (1975).

[11] S. Arikawa, S. Takeya and S. Kanō, "Users Guide, to MIR-RF System; (I) Presentation of Data, (II) Storage and Retrieval, (III) Manipulation of Researcher Files", Res. Inst. Fund. Inform. Sci., Kyushu Univ., (1976), (Japanese).

[12] S. Arikawa, "A Researcher File Description Language and Its Implication in Information Retrieval Systems", Preprint 20, COLING76, (1976).

[13] S. Arikawa, S. Takeya and M. Ishibashi, "On an Information Retrieval System Using Pattern Matching Machines", Proc. Inform. Processing Soc. Japan. 253-254, (1976), (Japanese).

[14] S. Arikawa and S. Takeya, "MIR-RF and a Generalized Pattern Matching Machine", Proc. INFORMATICA 77, 3 (126), 1-4, (1977).

[15] S. Takeya and S. Arikawa, "TEXTIR — A Text Information Retrieval System Using Pattern Matching Machines", Res. Rept. Res. Inst. Fund. Inform. Sci., Kyushu Univ., 83, 1-16, (1978).

[16] S. Arikawa, "A One-Way Sequential Search System and Its Applicability to Medical Information Processing", Proc. HICSS-13 (1980, to appear).

[17] S. Takeya, S. Arikawa and S. Kanō, "On Functions for Advanced Information Management in Laboratories", Proc. HICSS-13 (1980, to appear).

[18] S. Arikawa, "On Powers of Information Retrieval Systems Based on One-Way Sequential Searches", Res. Rept. Res. Inst. Fund. Inform. Sci., Kyushu Univ. (1979, to appear).

[19] S. Takeya, S. Arikawa and A. Okuma, "On an Investigation of References in Scientific Documents", Proc. Inform. Processing Soc. Japan, 699-700, (1977), (Japanese).

[20] S. Takeya and S. Arikawa, "On Words in English Abstracts", Proc. Inform. Processing Soc. Japan, 665-666, (1979), (Japanese).

Scientific Information Systems in Japan
H. Inose (ed.)

SCIENTIFIC INFORMATION SYSTEM FOR NUCLEAR PHYSICS RESEARCH

Masamoto TOGASI and Hajime TANAKA

Department of Physics, Faculty of Science, Hokkaido University, Nishi 8-Chome, Kita 10-Jo, Kita-Ku, Sapporo

A storage and retrieval system for charged particle nuclear reaction data (CPND) have been designed and inplemented. CPND is a typical example of raw data in scientific research. The system has an architecture of intermediate system between IR system and DBMS. Input data to the system are coded in a "data language" of free format. Doubly inverted file structure is incorporated into the system to represent set-element coupling and indexing. Implication of the system in context of scientific information system is discussed.

1. INTRODUCTION

In recent years, many journals which accept papers from nuclear physicists strictly limit the number of pages per paper to cope with the explosive increase in number of papers produced. As a result, only a few tables of experimental data, which are usually spacious, are picked up and the rest are omitted from accepted papers. These tables are reduced into some small graphs from which it is not easy to read out accurate values. Under these circumstances of difficulty against putting and catching detailed experimental data from journals, both experimental and theoretical nuclear physicists are eager to find new paths of access to numerical data to be established by utilizing computer files. On the other hand, organization of scientific information in general meaning is multi-staged processes. And, therefore, the whole system of scientific information consists of many subsystems each of which is designed to match with characteristics of information in each stage. One such subsystem would be the raw data management system which collects raw data in a field of science and roughly organizes them to prepare for further processing and analysis. The architecture of such a system and the nature of raw data from the informatical point of view have not yet been well studied.

Charged Particle Nuclear Data (CPND) in nuclear physics research would be a typical example of raw data in science. In this paper the newly designed information system for CPND, namely Nuclear Reaction Data File (NRDF), is proposed. How the architecture of the information system would be restricted by dealing with raw data has been investigated to give a general view. Matching the system with the CPND, an intermediate system between IR system and DBMS has been designed. A research project for CPND compilation in Japan was started in 1973. The first version of NRDF was designed and implemented in 1975 and was rather a prototype of NRDF. The system which is described in this paper is, thus, second version, namely NRDF2.

In this paper, architecture of the system and implication of it in context of scientific information system are shown. Section 2 gives a general view of CPND. Requirements to the system from points of view of scientific research and raw data management are pointed out in Section 3. In Section 4, architecture of the system is described from sides of information representation, information structure and information operation. In Section 5, how raw data has affected the design of the system is shown. Section 6 is the conclusion with future plans and some technical issues mentioned.

2. WHAT IS CPND?

Nuclear data is roughly divided into two main classes. One is nuclear structure data. This data represents attributes of nucleus, and types of the data are fixed and limited. The other is nuclear reaction data. Nuclear reaction is a kind of dynamic event in which possibly huge number of records are expected to be produced. Furthermore, number of reaction types increase according to increase in number of particles participating in the reactions. Nuclear reaction, in this context, is artificially caused by accelerated particle beam. And emitted particles which are detected by particle detection systems are observed. (Fig. 1)

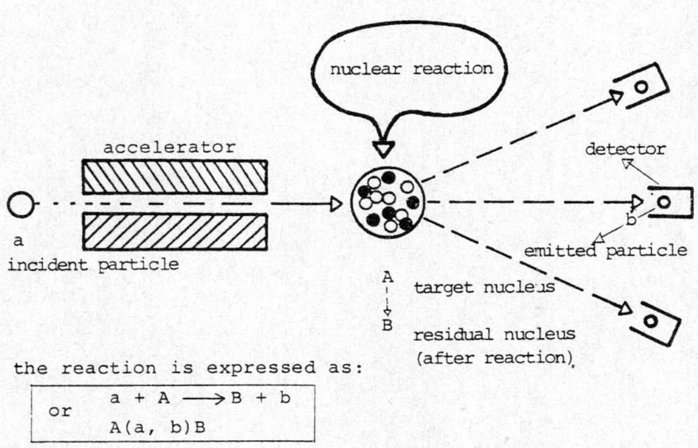

Fig. 1. Scheme of nuclear reaction experiments.

Various quantities about detected particles tells researchers what kind of reaction has actually happened. Probabilities of each type of reaction is usually represented as cross section, i.e. reaction cross section. This type of data is a typical one in nuclear reaction data. Nuclear reaction data produced in experiments using neutron as incident particle are called Neutron Nuclear Data (NND), which are usually used in application fields such as nuclear reactor design. Compilation of NND has been well organized so far. This is because of its application oriented character in which standardization of data can be done without spoiling value of data file. On the other hand, data from nuclear reaction caused by accelerated charged particle beam is called Charged Particle Nuclear Data (CPND). CPND is basic research oriented data. That is, CPND is produced and utilized in basic research environments of nuclear physics.

One characteristics of CPND is the variety of types of data. Number of charged particle which are available for incident beam of reaction are so many, including heavy ions in recent years, that it is possible to arrange various types of reactions, compared with NND, in experiments. Various aspects of nuclear reactions are observed and expressed as CPND according to the individual interest of the researchers. This variety refuses any rigid standardization. On the other hand, each unit of CPND must include a description of the individual condition of experiment. A set of observed values is not understandable without refering to this environmental information. That is, data and its environment are not separable in sense of informatics. This is another characteristics of CPND as raw data. It is worth to notice that the very locations of CPND production are fixed and limited in Japan. CPNDs are always produced at institutes where particle beam accelerators are installed. The number of such institutes in Japan is about 20. By monitoring the activities at these sites, all of the CPNDs in Japan would hopefully be covered.

3. REQUIREMENTS TO SYSTEM DESIGN

NRDF2 has two missions. One is to support basic research activities in nuclear physics from informatical side. And the other is to manage raw data produced in nuclear reaction experiments of various types. These two missions provide the outline of design of the system.

3.1 Research oriented system

Research activities in nuclear physics are not so well organized as activities of Data Base producers. Data to be input to the system, i.e. CPND, is produced in each specific context of work by individual researchers. And various kind of information is extracted from each specifically arranged experiments. These pieces of information are, usually, represented with arbitrary forms for ease of understanding by researchers. These data should be translated into machine readable data without cumbersome formatting or fixed field adjusting as so in case of specific and rigid input data formats. Free format data input is therefore desirable for the information system when many researchers deeply commit themselves. Input data should be easy to prepare for lowering psycological barrier for cooperation in the organization of CPND. Detailed information in CPND must not be cut off for the reason that some pieces of information could not be accepted in a specific format. Researchers expect to derive important information from files of the system by interpreting data and applying further processing on data. It is a question of individual researchers who use the file, which part of the data is important. Therefore, all parts of experimental data should be kept in the file from the point of view of system design. Description ability in the system is needed for incorporating various types of information including details of CPND. Flexibility is more important in the research oriented system rather than efficiency in processing and operation. On the other hand, simplicity of information structure helps researchers to understand what to do on retrievals of data. And the target of retrieval should be represented by simple form in the file. Basic mechanism of retrieval should be opened to researchers and should be simple enough for understanding what they are doing. It would be hopeless to expect researchers to be information specialists.

3.2 Raw data management system

CPND is a typical example of raw data in which many events are recorded as they are. By itself, raw data is not effective information for understanding object of observation. Further processing and reduction of information are expected to be applied on assembly of raw data. Management systems for raw data play the role of data intake and temporary storage in total architecture of a scientific information system. Thus, a file of raw data might resemble to files of materials in that individual case records are compiled and stored. Each material is interpreted by reading it when used. What is important in raw data management is that stored data is self descriptive. That is, sufficient information for understanding raw data should be associated with data itself. That is also needed for evaluation of raw data afterwords. In general, raw data is highly dependent to its production environment. In case of CPND, errors in data are deeply connected with performance of the observation system, such as particle detectors etc. Therefore description of this experimental condition should be associated with tables of observed values. Thus, raw data would be a complex of different types of information. The raw data management system should have appropriate framework in accepting complex information of raw data. On the other hand, many degrees of freedom to move point of view have been required for fine retrieval. That is, multiple indexing for units of raw data is needed to be incorporated into the system for CPND. Environmental information of CPND would be the main field of extraction of key information for indexing. These environments have two faces in the file: one is content and the other is label.

4. ARCHITECTURE OF THE SYSTEM

4.1 Information representation

4.1.1 Set representation of event records

A typical example of CPND is differential cross sections of specific reaction, associated with scattering angles. These data itself can be represented as a table consisting of 2 columns, i.e. scattering angles as independent parameter and differential cross section as dependent observed value. The table may well have data heading and data unit for each column, because there is no "type" of table to be catalogued in the system. This table of observed numerical data may be self-descriptive to some extent by adding non-numerical data on the location of the table. But the table does not hold sufficient enough information from the point of view of understanding and using the table for the purpose of research. Why? It is necessary to know what the type of nuclear reaction is, how the target nucleus was prepared, how the experimental conditions were set, and so on. This environmental information surrounding the table of observed values is needed to be associated with it. Usually, one run of the experiments gives a series of data table, in which environments to be recorded and added to the tables are slightly different from each other. For example, a series of reactions might be tested changing target nucleus, energy of incident particle and so on. There are, in general, overlaps between this environmental information. Overlapping parts might be common information about two or more data tables. (Fig. 2-A)

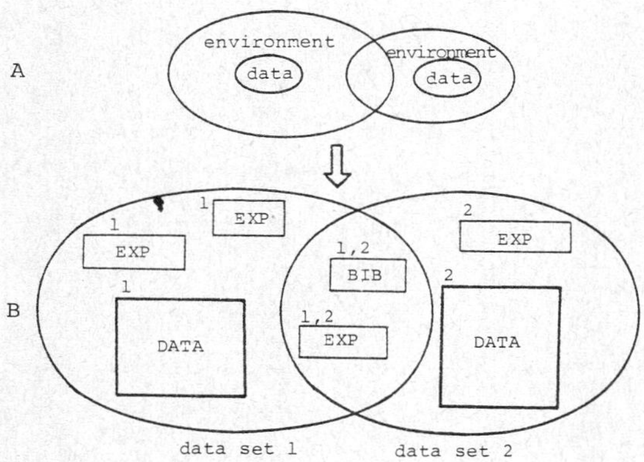

Fig. 2. Data set and section.

By dividing this information into some parts appropriately, one unit of CPND can be expressed as a set sharing some of its elements with other units. This set is called "data set". A data set holds a complete set of pieces of information needed for understanding the table. And what is essential to data set is that it is informatically independent from external data in elsewhere. Elements of a data set are called "section". Sections do not share any portions of themselves like data sets. By adding a control statement involving symbolic pointers to data sets, i.e. data set identifiers, each section points the data sets of which the section is to be an element. There are two kinds of contents of section: one is data table and the other is a set of data represented by a "data language". Sections which contain data table are called "DATA section" and they are cores of data sets. Data table itself is inserted in a DATA section as a sub-section for the reason that closely related information about the table should be located near to it being represented by the data language. A data table has data headings and data units in first and second line respectively. The number of columns of the table is implicitly represented by the number of headings. One characteristic feature of data tables in the system is that each data may be associated with, or itself is, symbolic pointers pointing other data and comments out of the table, as in the case of tables in published papers. Other types of sections are called "EXP section" and "BIB section", and generally they have no data tables. The EXP section holds information about the experimental condition, such as; reaction type, types of accelerator, and the description of the detection system for scattered particles. In the BIB section, bibliographic information such as data author and reference to a related paper which might have been published are recorded. Usually several data sets share a common BIB section, and the data sets might be produced in the same run of experiment. (Fig. 2-B)

4.1.2 Data language

In describing experimental conditions for each unit of CPND, data set, only a small subset of description terms are used. The subset varies from case to case. If all of the potential terms for description were introduced to each case, for the purpose of factoring out a common format, very "sparse tables" would be made up. To avoid this, compression of the sparse table has been done by picking up non-null elements of the table and redescribing them. Each non-null element is described by binding value and attribute name:

ATTR=VAL;

Attribute name is a kind of small environment for the value itself. Thus, direct binding of value and attribute name gives rise to informatical independence.

```
¥¥BIB,1,2,3;                                              <---------------- control statement
TITLE=/OPTICAL-MODEL ANALYSIS OF ELASTIC SCATTERING OF
3HE PARTICLES FROM 58NI AT 24.15, 27.64 AND 34.14MEV/;
PURPOSE=/TO DETERMINE THE OPTICAL MODEL PARAMETERS OF         BIB section
3HE SCATTERING/;ATH=(T.FUJISAWA'1',K.KAMITSUBO'1',T.
WADA'1',M.IGARASHI'2');INST-ATH=(2JAPIPC'1',2JAPTIT'2');
REF=JPJ;VLP=27(1969)278;RCTS=(58NI(3HE,3HE)58NI);
¥¥EXP,1,2,3;
RCT=58NI(3HE,3HE)58NI;ENR=X%;CHM=ELM;THK-TGT=0.65MG/
CM**2;BAC=SELF;POL-TGT=NO;ALGN-TGT=NO;ACC=(CYC);INST-
ACC=2JAPIPC;INC-ENGY-RANGE=(24.15MEV,27.64MEV,34.14MEV);
ERC-PRJ=0.2%;BEAM-INTNSTY=0.3UA;POL-PRJ=NO;DET-PARTICLE=      EXP section
(HE3);DET-SYS=(SI,CNTR-TLSCP);SOLID-ANGL=0.177MSR;ERS-
DET= 100KEV;ANL=(OPT-MODEL);PHQ=(XSECTN,DSIGMA/DOMEGA,
OPT-POTL-PARA);
¥¥DATA,1;
INC=ENGY=24.15MEV;EXC-ENGY=0MEV;DELTA-EXC-ENGY=0MEV;
J-PI=0+;
   ¥DATA;
     THTC     DSIGMA/DOMEGA    DELTA-DSIGMA/DOMEGA <--headings
     (DEG)       (MB/SR)            (MB/SR)        <-----units
     18.40       .1056E5            .3168E3
     21.02       .5212E4            .1564E3
     26.26       .1737E4            .5211E2
     31.49       .438E3             .1314E2                      DATA section
     23.64       .2882E4            .8646E2
                                                        data table

    127.44       .77E-1             .30E-2
    132.29       .55E-1             .42E-2          data points
    137.11       .41E-1             .20E-2
    146.71       .276E-1            .14E-2
    151.49       .310E-1            .16E-2
   ¥END;
¥¥DATA,2;
                 DATA section
¥¥EXP,3;
                 EXP section
¥¥DATA,3;
                 DATA section
¥¥END;
```

Fig. 3. An example of input data.

In other words, locality of information is gained. In this way, various aspects of experiments are described by a language. Other description functions of the data language in this system are: to get values together into one set and associate them with common attribute name;

ATTR=(VAL1,VAL2,VAL3, ····);

and to bind some attribute-value set, i.e. statement;

(ATTR1=VAL1,ATTR2=VAL2, ····);

Symbolic pointers can be attached to codes in statement:

ATTR1=VAL1'1'; ATTR2=VAL2'2';

Fig. 3 shows an example of input data.

4.2 Information structure

Here three types of files in the system namely, X1, X2 and X3, are introduced. X2 is the main file holding data sets. X1 is a collection of sections and X3 is inverted index. X1 and X3 play complementary roles to each other both interfacing X2. (Fig. 4)

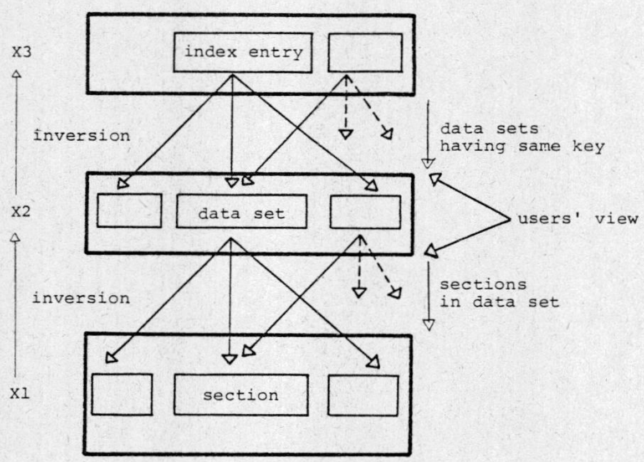

Fig. 4. Doubly inverted file.

4.2.1 X1

X1 is a collection of sections of input image. Input image means that sections are stored in X1 as character string including data tables which are, in that case, stretched into one dimension in row first order, and each data point delimited by blanks. An internal ID number is attached to each section. Symbolic pointers to belonging data sets of the section are also attached. Thus, one record of X1 has a form as follows:

<section ID><pointers to data sets><section body>

<section ID> is unique key to access the record.

There are no direct cross references among records in X1. Pointers to data sets are not in operational form for high-speed path tracing. This can also be said as to content of section body: statements coded in data language may have key information which is common in some sections, e.g. data author. But to realize this interaction between sections, it is necessary to "interpret" character strings, by scanning and syntax analysis on statements expressed in the data language.

4.2.2 X2

X2 holds data sets. Actually, each data set is represented and stored in a form of set of section IDs. Data sets, in general, have intersection with each other. It is not clear that there are data sets among collection of sections because realization of data sets is performed through the interpretation and processing of control statement located at the top of each section, i.e. set of data set IDs. In this connection, data sets, as a subset of collections of sections, are regarded as virtual ones. To realize units of information, i.e. data sets, and to make clear-cut separation between data sets, section IDs instead of section bodies are assembled. And this set of section IDs associated with data set ID makes up one record of X2. The general form of records in X2 is:

<data set ID><set of section IDs>

Both IDs are highly machine operational in that expansion from data set to section bodies are performed by locating sections in X1 through key-to address conversion mechanism. Data set ID is a unique key to locate the data set in X2. There might be various methods of representation for set of section IDs in X2 records. Some of set representation methods are proposed and compared with each other in context of index file design[1]. In this system both integer array and bit map are employed. These two methods are switched dynamically each other by criteria of which one saves more storage needed for representation of sets.

4.2.3 X3

X3 is inverted index to data sets, i.e. X2. Therefore one entry to the index points a set of data sets, in other words one subset of X2. Each subset pointed from index entries has the same key information abstracted in input process. A set of key attributes for which key extractions to be executed are defined at system generation time by original utility program. Form of records in X3 is:

<key information><set of data set IDs>

where key information varies according to key attributes, i.e. that may be name of data author, type of accelerator and so on. X3 is, exactly speaking, an aggregation of subfiles, each of which corresponds to a key attribute. Locating X2 records from an entry of X3 goes in the same way as in the case of expansion of data sets to sections. It should be noted that indexing is done for data sets not for sections. This means that key information actually extracted from a section may be registered two or more times when the section might be shared by some two or more data sets. Although resolving power of the index is weakened, i.e. sections are roughly sorted, quickness of searching data sets via X3 is expected at the sacrifice of non redundancy of X3.

4.2.4 Doubly inverted file

Interrelation of X1, X2 and X3 is shown in Fig. 4. Each record in inverted index X3 represents a subset of X2 regarding X2 as a set of records. Each element of the subset has somewhat the same key information in it. This is also the case for X2 in relation with X1. That is, every data set is a subset of X1. And each element of the subset has the same key information though it might be implicitly represented, saying it is related to an identical data table of observed values. This means that X2 is a kind of inverted file for X1. Thus, a collection of subsections on base set X1 makes X2, and a collection of subsets on base set X2 makes X3. This is read as "X1 is inverted into X2, and X2 is inverted into X3". For this reason, a file complex made up of X1, X2 and X3 is called a "doubly inverted file". Function distribution is well done in that X2 gives users a view, X3 assists retrieval process and X1 holds bodies of data.

4.3 Information operation

4.3.1 Input operation

Input processer of the system cuts input data into sections at first. At the same time, control statements which represent IDs of belonging data sets are extracted. These set of IDs together with section ID, which is issued by the system, are assembled through sections of input data. Next, each section is scanned and syntax of statements is checked by a data interpreter. The data interpreter also makes a temporary table of attribute-value sets. Ths table is looked up and elements related to index terms are extracted. By this stage, three kinds of information have been prepared: section bodies in form of character string, pairs of section ID and a set of data set IDs, and set of key information interpreted and extracted from sections. A pair of section ID and data set IDs represent "an element to sets" relation. And these relations are inverted into "a set of elements" relations. Using an element to sets relations of before inversion, key information extracted from each section are combined with a list of related data set IDs. Thus, three kinds of information prepared are:

<section body> ····· 1
<data set ID><set of section IDs> ····· 2
<key information><set of data set IDs> ····· 3

The formats of 1, 2 and 3 are the same as the formats of the records of X1, X2 and X3 respectively. This means that input data is equalized to the files. The last step of input operation is to merge these equalized input data into X1, X2 and X3 files. It would be worth to notice that the almost original image of input data are kept in X1 file. Therefore, reconstruction of X2 and X3 is possible enough even in case of destruction of these two files.

4.3.2 Output operation

The target of retrieval is data sets in X2. Using inverted index X3 for X2, extraction of data sets can be executed on X3 instead of searching in X2 directly. In X3, data sets have already been sorted and grouped by values of key attributes.

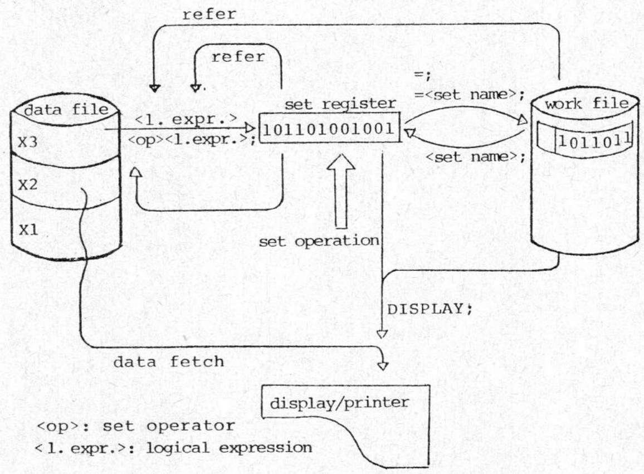

Fig. 5. Mechanism of the retrievals.

The system features two work areas for retrievals as shown in Fig. 5. They are set register and work file. A set of data sets extracted according to a query expression are moved into the set register as a set of IDs of data sets. These IDs are symbolic pointers to data sets in X2. Content of the set register may be moved into the work file with a name of set. This set in the work file can be pulled back into set register, later. A query expression addresses three kinds of sets. The first one is potential set implied by conditions for key attributes that the set should satisfy. This set will be realized after searching in X3. The second is content of set register. This set is represented in the expression by a specific symbol "#". The third one is the set kept in the work file. This set is called by its name. Syntax of the query expressions resembles to one for assign statements of Fortran except that the expression like A=B is read as that A is assigned to B. General form of query expressions is:

<logical expression>=<set name>;

terms in logical expression are of forms as follows.

(<key attribute><relational operator><value>)
<name of a set in work file>
#

where "#" addresses the set register. Syntax of logical expression is almost the same as the arithmetic expression in Fortran. A logical expression may have set operator at left most of it that means operation between the set register and evaluated expression except the operator. As shown in Fig. 6, each part of a query expression can be independently executed and the latest results are always in the set register.

A presented logical expression directs the system to evaluate it and load the set register the result. "=" sign says that store the set register to the work file. And if the sign is followed by a set name,

then the name will be attached to stored set in the work file. ";" symbol also has its own function more than being a delimiter, i.e. the sign direct the system to report hit count of each retrieval.
A query expression is translated into a sequence of lower level instructions, and then executed by an interpreter. There are two more commands for retrievals: DELETE command for deletion of elements in the work file, and DISPLAY command which displays a result set of retrieval. In execution of DISPLAY command, each data set in the result set is expanded into a set of sections through X2, and section bodies are fetched by reffering section IDs. Examples and timing data of retrievals are illustrated in Fig. 6.

```
                                              elapsed time
(PRJ=P)*(EMT=T)=PT;                           ----> 7.3 sec.
      10     ---------> hit count

(PRJ=P)*(EMT=D)=PD;                           ----> 4.6 sec.
      33

PT+PD=PTORPD;                                 ----> 2.8 sec.
      43

(INC-ENGY>=40)*(INC-ENGY<=70)=I40TO70;        ----> 4.3 sec.
      48

PTORPD*I40TO70*(EXC-ENGY<=4.0);               ----> 6.2 sec.
      18

=PTPD4070L4;                                  ----> 2.1 sec.
      18

*(EXC-ENGY<=2.0);                             ----> 5.3 sec.
      10

DELETE PTORPD;                                ----> 0.9 sec.

DISPLAY#;

 output of the latest result set
```

Fig. 6. Examples and timing data of the retrievals.

5. NRDF IN CONTEXT OF SCIENTIFIC INFORMATION SYSTEM

In the total system for scientific information, formation processes of integrated Data Base beginning with collections raw data through organization of information, should be supported. NRDF2 would partially share this load by managing raw data in nuclear physics research. And an experience with designing NRDF2 will give a general understanding of what raw data management is. What is important in raw data management for scientific research in context of integrated Data Base formation, is to accomodate sufficient information in machine readable form, and then prepare for further processing. Taking these interdependent roles of systems into account, it is illustrated in this section how the architecture of the system, i.e. NRDF2, has been affected by dealing with raw data for the purpose of assisting scientific research.

5.1 Informatical characteristics of raw data

Raw data which is produced under various experimental conditions is not informatically independent. Numerical data tables in CPNDs, though it is cores of CPNDs, do not explain what it is, by itself. The type of reaction and energy of incident particle, for example, should be associated with them to label the tables. Furthermore, detailes of experimental condition such as type and performance of accerelator or particle detection system is indispensable for evaluating original data to derive some other physical quantities. For example, the performance of the particle detector would affect the estimation of errors on observed values by the detector. Thus, raw data at stages of data production are highly dependent to environmental information so that a pair of observed values and environmental information should be associated somehow. On the other hand, a series of CPND is produced at a same time in a context of each specific work. These data have partially common environmental information, i.e. a part of experimental conditions. Sharing some parts of environmental information, say "valence information", each unit of raw data correlates with each other. Informatical characteristics of CPND are, thus, that incorporation of environmental information is indispensable in terms of informatical independence and that there are connections of independent units of information through valence information. And these would also be the case for other raw data.

5.2 Between IR system and DBMS

There are two major factors that characterize information systems: object information and use of data under that system. Input data of NRDF2 is given as a collection of sections. The sections are indirectly connected to each other through symbolic pointers. The object information has, therefore, inter-section structure in addition to intra-section structure. Although relationships between sections are not clear enough to factor out some types from that, useful information has to be extracted by structure traverse, i.e. navigation, based on the relationships. That is, a data set, for example, has to be found out by such a navigation in a file, if only a collection of sections are given to users. Navigation in cross refferenced files is usually supported by a Data Base Management System (DBMS). Thus, a kind of management system like DBMS is required in connection with representation and operation of interrelations among elements of data. On the other hand, from the point of view of utilizing files in research activity, only retrieval of event records is all that is necessary. Data analysis and finding out further information from raw data are of next stage of the total system. Event record in case of CPND is data table of observed values and its environment in one experiment. Retrieval of event records based on key information to be matched with one of data in the file must be supported by the usual Information Retrieval (IR) system.

From both requirements for storage and retrieval of raw data together, obvious conflict is emerged: Complexity of storage and simplicity of retrieval. What is resolution of these two? The question might be general one in raw data management system. Doubly inverted file is proposed to answer to that qestion. There are three files in NRDF2, i.e. X1, X2 and X3. X1 holds sections and is inverted into

X2 which holds data sets. X3 is inverted file to X2 and is used for data set retrieval. Data sets are found out from a collection of sections in input operation, and stored into X2. This can be interpreted as that structure traverse was executed in the stage of inversion of X1 into X2 to find out information units. Then records in X2 are retrieved in a manner of IR. Therefore, X2 plays the role of an interfacer between DBMS-like data representation and IR-like view of data. But neither DBMS nor IR system set up a file like X2 from the object information as the target of retrievals. And NRDF2 gives users a set of event records which are not visible at the stage of input data. Indexing and inversion of a file should be regarded as pre-search or say, pre-navigation, in the file giving views on data to users. And set representation of raw data is incorporated as a result of pre-navigation.

5.3 Data representation by language

Other characteristics of raw data correspond to environmental information about data table. In DBMS, common information structures are factored out as "schema", and relieve information storage from redundancy. This can be done because there is homogeniety to some extent in object information of DBMS. The individualism of information in raw data would not be accepted straightforwardly by DBMS. The concept of data language gives another way of data description. Data language utilizes syntax for data, and a variety of representation is available because of the synthetic nature of language. A data language describes states, entities and events while programming languages mainly describe actions to be taken by computer. A data language processer transforms data which is represented in a language into machine oriented form for internal processing, i.e. retrieval, navigation, and so on. This processer might be of compiler type or of interpreter type. Input data checking and indexing for updating X2 and X3 are executed by an interpreter in NRDF2.

6. DISCUSSION AND CONCLUSION

6.1 Implementation

System program of NRDF2 consists of 130 modules all of which are written in PL/I. The total number of steps of the programme is about 7000. The system has been implemented on a medium scale computer HITAC-8250 with 96KB main memory, two disks and two MT drivers. The total amount of work for development of the system is roughly estimated as 1 man*year. CPNDs have been compiled from 70 published and unpublished papers and the average number of data sets per paper was 7.2.

6.2 Technical issues

The input data processer has been originally developed, which includes syntax checking and translation into machine oriented object data. This is because of the introduction of data language in information representation. But techniques for these language processing have already been developed in programming language processer. Some of the input data processing modules in the system should be substituted by other ones, if general purpose language processing modules are to be available. Physical file management modules have also been originally developed in the system for logically direct access to variable length records. Individual record length varies from time to time by updating the file. Therefore, dynamic reorganization of the file is needed according to continuous growth of that file. From a logical point of view, file management modules in the system work like VSAM. These functions in file management would be essential in systems like NRDF2, and they should be substituted, in case of NRDF2, by VSAM or such other access method like VSAM. The information operation abilities of the system seems to be insufficient under expecting severe environments of practical use. On line data input with checking and editting, which is not yet available, would make the system more usable. Correction and deletion of stored data are batched for ease of implementation. These functions of information operation should be implemented and integrated into single "data editor" which works interactively with users.

6.3 Conclusion

Raw data seen as informatical object have characteristics of both separability and crossing. As event records, they are point-like and independent from macroscopic point of view. At the same time, they are structured bodies elements of which are dependent to each other from microscopic point of view. IR system supports the macroscopic view, and DBMS supports the microscopic view, respectively.
Raw data management system must support these conflicting two views together. Therefore, architecture of raw data management system should be of intermediate system between IR system and DBMS.

Doubly inverted file is the core of the system design, and supports zooming-in and zooming-out of a point of view towards stored CPNDs. The concept of data language for information representation has been useful in design of the system. Although the idea was not yet fully utilized, variety and individualism of raw data are well managed in the system by the synthetic nature of language. Introducing techniques of such as copiler writing system is expected for generalization of the idea. High descriptivity of data language also relieve raw data from leak and cutting-off important portions if information about environments. It should be noted that raw data is not data any longer without its environments.

Existing and well established techniques could take over many parts of the system. For example, they are language processer, VSAM like file management, inverter or index generator, set processer and so on. Raw data management system like NRDF2, therefore, can be constructed from common techniques facterd out from this experience in the system, with specific processing modules and parameters such as constants and code dictionary which are dependent to each specific field of science.

6.4 Future plans

Compilation of CPND will be continued keeping information representation unchanged. A little revision is expected according to growth of the file to speed up retrievals. Enhancement of information operation abilities are planned for the future to be capable of interactive operation. On the other hand, possibilities of on line data evaluation and feeding-back to the file have to be explored for further utilization of the system. This will make a turn-around system or, say, a self evolutional file.

ACKNOWLEDGMENT

The basic specification of the system which is developed in this work was suggested by the study group for CPND compilation. The members of the group are: Professors H. Ikegami (OSAKA Univ.), M. Kawai (KYUSHU Univ.), H. Ohnuma (Tech. Inst. of TOKYO), M. Yamada (WASEDA Univ.), M. Muraoka (OSAKA Univ.), T. Hasegawa (Univ. of TOKYO), Y. Abe (KYOTO Univ.) and the authors. The authors would like to thank Dr. T. Nojiri (OSAKA Univ.) for her cooperation in compiling the data in published papers and translating them into machine readable form.

REFERENCES

[1] J. Martin, "Computer data-base organization", Prentice-Hall, Inc., Englewood Cliffs, N.J., (1975).

Scientific Information Systems in Japan
H. Inose (ed.)

ACCUMULATIVE RESEARCHER FILE SYSTEMS AND MULTIDISCIPLINARY STUDIES IN GEOSCIENCES AND ENVIRONMENTAL SCIENCE

Kiyoshi WADATSUMI and Yasuo SHIMAZU***

**Faculty of Science, Osaka City University, Sugimoto-cho, Sumiyoshi-ku, Osaka*
*** Faculty of Science, Nagoya University, Furo-cho, Chikusa-ku, Nagoya*

This report describes results of a research project, sponsored by the Ministry of Education of Japan, to develop database systems for multidisciplinary studies and to apply them in both geosciences (subproject I)* and environmentology (subproject II)**. At the former, GEODAS (a geology oriented database system) was developed as a system based upon ARF (accumulative researcher files). ARF are sets of cooperative PRF (private researcher files) with distributed characteristics and locations in relation to that of researchers, who are both makers and users of PRF. Data loadings transfer ARF as sets of isolated file onto physical databases of GEODAS without any linkage. The relational join operation defines multipul logical databases, in virtual, based upon the fixed modules. A user can make query sets to the logical databases with the methods, which are nearly the same way that of the document retrieval system. At the latter, SMLESIS (a database system for seamles earth science) was developed also for multidisciplinary studies. Both descriptive and numerical data are accepted and managed. A significant part of numerical data is grid square data. Thesaurus oriented retrieval system is employed to guarantee a softness in which both tree and net logical thesauri are accepted. Query set is given as logical or arithmetic equations which are interpreted by a mini-compiler.

I. Database System for Multidisciplinary Studies in Geosciences - GEODAS

1. What is GEODAS ?

GEODAS is an abbreviation of "a geology oriented database system" and it was designed originally for the geosciences and recently have been extended its application to multidisciplinary studies of the space and time dependent occurrences on the earth. So that , it is also of "a general objective database system".

Throughout individual research activities, scientific data are being obtained and created in the form of PRF (private researcher files) in the fields not only of the geosciences, but also of the other natural and social sciences. In order to establish large scale scientific database systems, we must be able to make clear what are objects, roles and methods of organization and utilization of the systems.

At the present time, we are taking the method of bottom up organization based upon accumulative PRF in order to realize a system which includes private use of private database at the begining, common use of group database, general use of public database and finally multidisciplinary use of comprehensive databases.

In order to realize the concept, the prototype of GEODAS was developed and implemented on the small computer at Himeji Institute of Technology (Wadatsumi 1975 a,b, Wadatsumi and others 1976). The practical version of GEODAS has been implemented in the larger machine at the inter-university computer center of Osaka University (Wadatsumi 1977). The computer of the center is ACOS 77, Series NEAC 900 with OS (ACOS-6) installed by NEC (Nippon Electric Corporation), and which is comparable with IBM 370-168.

2. Design of GEODAS

GEODAS is composed of INQ (a information query system such as a generalized database management system of ADABAS type) implemented by NEC, accumulative researcher files, and generally accessible real-data dictionaries named MAP and CALENDAR (Wadatsumi 1979 a,b).

It has been required to establish a conceptional framework for the file organization of GEODAS by making clear through analysis the interrelationship among various kind of ARF. A working framework adapted in GEODAS is described as follows:

2.1 PRF and ARF

PRF (private researcher files) have been designed for each reseracher's applications and organized in a simple serial manner on computer readable files. Physical structure of PRF is essentially the same as that of logical files. Widely distributing characteristics and locations of PRF are closely related to that of each researcher. ARF (accumulative researcher files) are simple sets of spontaneously cooperative PRF, which include file structures of table, simple tree, and tree with repeating group.

ROWDAS (a row-wise case data system) is a system which consists of a logical set of records, files, and databases directly related to the paticular entites. The name of the system was derived from the view of a set of rows in data matrix (Wadatsumi 1978 a, b). Many files of local data has been accumulated as ROWDAS-ARF from observation, description and measurement of natural and social phenomena in the field.

On the other hand, COWDAS (a column-wise aspect data system) is a system which consists of a logical set of domains of aspect data, domain files, and domain databases, in which there are data related to the attributive properties. The data and their structures of COWDAS should be established after processing, evaluation and standerdization of ROWDAS. The real-data dictionaries, MAP and CALENDAR, are belong to COWDAS-ARF even in case they are in serial form. This means there are differences between the domain file and the inverted file.

Concerning of the relations between COWDAS and ROWDAS, mapping operation from COWDAS to ROWDAS means production of the case data in the real world. On the other hand, reverse mapping means reduction of the local data and/or production of knowledges in the logical world. According to this point of view, ARF are classified into two large groups, ROWDAS and COWDAS. Each of them is well associated with the other to dynamically assemble them into logical databases by means of INQ of GEODAS.

2.2 Data Loading (I,II) and Physical Databases

A general resercher, who is both a maker and a user of GEODAS, is expected to be able to take, in case by case, his positions from a file manager to a programmer or a enduser. He has mainly managed the some part of GEODAS in which there are his PRF and the other cooperative ARF.

Prior to data loading, schema information of input file, which is about data structure, type and attribute, should be described by researchers by means of DDL/FDL (a easy data and file description language) in nearly the same way of COBOL.

A meta-data dictionary keeps information of both the operation manuals and a set of the schemas of the previously created files. Every users can freely use the information of the dictionary at the time on operation through a user terminal.

Two steps of loading (I, II) make a pair of physical databases, which are both GMF (GEODAS master of files) and IRDB (a inverted relational database). These operations are able to be done automatically with the help of DML (a easy data manipulation language) or utility programs of INQ under the control of the information of schemas.

As the result of loading, ARF are transfered onto physical databases as a set of isolated file modules without any linkage such as the some of relational database. Addition and deletion of a file have no trouble to the other modules of the physical databases which have characteristics of open-ended modular system. So that, updating of GEODAS is able to do separately for each trouble file by a researcher who take the responsibility on it. Capacity of GEODAS is up to 300 databases including 63 files on each.

Each module of IRDB is of a inverted structure in order to permit easy operations for both association and rapid searching. On the other hand, each module of GMF has a linear standard structure to permit easy editing and output operations.

2.3 Definition of Logical Databases

Definition of logical databases (submodels of GEODAS) is carried out multifariously by the relational join operation between selected modules with a common variable. The variables related to the space and time are so common and fundamental through ARF that the modules are associated together dynamically around the real-data dictionaries, MAP and CALENDAR, which look just like cores of association.

The multiple logical databases, which are in virtual, can be consequently derived from the same fixed modules on the physical databases. So that, the physical storage organization of modules is independent of the logical views of users.

INQLIB is a meta-data dictionary in which subschemas of the logical databases are stored. A subschema of a logical database is accessable at the anytime. A user can alternatively change and use them from a enduser terminal in interactive mode.

2.4 Query Operation

Query operations of logical retrieval and editing output on GEODAS are able to perform really on a logical database with EQL (a easy enduser query language) in interactive mode, and within FORTRAN and COBOL applications in batch mode. Retrieval commands of EQL have predicates with comparison and Boolean operaters, which are very familiar at the information retrieval system as shown in the following example.

Query sets of the previous operations are also stored at GQLIB (a GEODAS query library) with catalog names. We can obtain a same result with only one selecting catalog name , so that a user has a great advantage to make a complex operation upon previous works.

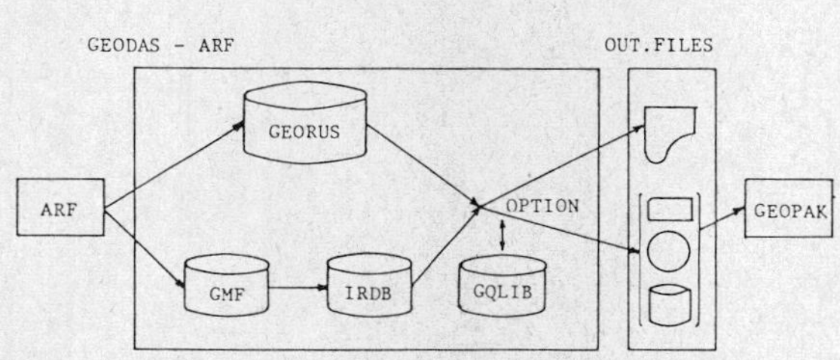

ARF : accumulative researcher files.
GEORUS : a GEO-thesaurus as a schemas dictionary.
GMF : a GEO-master of files.
IRDB : an all inverted relational database.
GQLIB : a GEO-query sets library.
GEOPAK : GEO-program packages.

Figure 1. Architecture of GEODAS-ARF.

2.5 Application Program Package

GEOPAK (a geology oriented application program package) is the system outside of GEODAS. Connection between GEODAS and GEOPAK is able to accomplish through the output files from retrieval operations of GEODAS. In this sense, GEODAS is a preprocessor to GEOPAK by which the selected data from GEODAS are analized.

GEOPAK consists of two subsystems, and one of them was implimented by Wadatsumi (1973) from the ROKDOK package developed by Loudon (1967-1974) together with some other programs. The other is from the domestic version of SPSS (statistical package for the social sciences: Nie et al. 1975), which has many programs useful in geology (Yamamoto and Nishiwaki 1977).

3. ROWDAS-ARF

ROWDAS-ARF on GEODAS are as follows:

(1) Earthquakes,JAPAN,1961-1972;Meteorological Agency.
(2) Surface geology,soil and geomorphology, Kinki, JAPAN; National Land Use Agency.
(3) Height on each 250m grids, Kinki; N. L. U. A.
(4) Chemical compositions of volcanic rocks, JAPAN; Geological Survey of Japan/Osaka City Univ.
(5) Active faults, Kinki,JAPAN; Osaka City Univ.
(6) Census data on each 1 km grid (mesh data), Osaka; Bureau of Statistics of Japan.

4. COWDAS-ARF

ROWDAS-ARF include generally the domains related to variables of both location and time. Some of the domains are widely sandardized, but the others are local and loose. So that, it is neccesary to make the dictionary in which the relation of the domains should be defined with the full set of real data. We should be able to concern many dictionaries in relation to the application of GEODAS. At present, the two dictionaries, MAP and CALENDAR, have been developing for the application as the space and time dependent database system (Wadatsumi 1979 a,b)

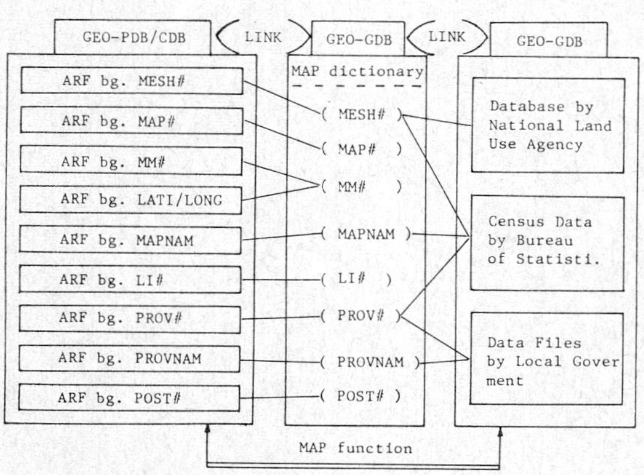

GEO-PCB : GEO private databases.
GEO-CDB : GEO common databases.
GEO-GDB : GEO general public databases.

Figure 2 A Schema of the Comprehensive GEODAS.

4.1 MAP

The Map dictionary consists of three files, the two of table relations and the other of a tree.

A. The Component Files as the Physical Modules.

(1) A normalized relation "D-ML".

D-ML (MESH#,MAP#,MAPNAM,MM#,LI#,LATI,LONG)
- MESH# (square codes downing to 1 km grids)
- MAP# (map codes downing to 1:25000)
- MAPNAM (map names)
- MM# (combination codes of MESH# and MAP#)
- LI# (locality index codes used by botanists)
- LATI; LT-NS, LT-DEG, LT-MIN (latitude of the lower left coners of grids)
- LONG; LN-EW, LN-DEG, LN-MIN (longitude of the same corners)

(2) A relation "D-PP".

D-PP (PROV#,PROVNAM,POST#)
- PROV# (province codes downing to virage and ward)
- PROVNAM (names of province)
- POST# (post codes)

(3) A tree "D-MP".

D-MP (MESH#, repeating PROV#)

B Map Models by the Logical Association.

1) Definition of MAP Models with the common domains.

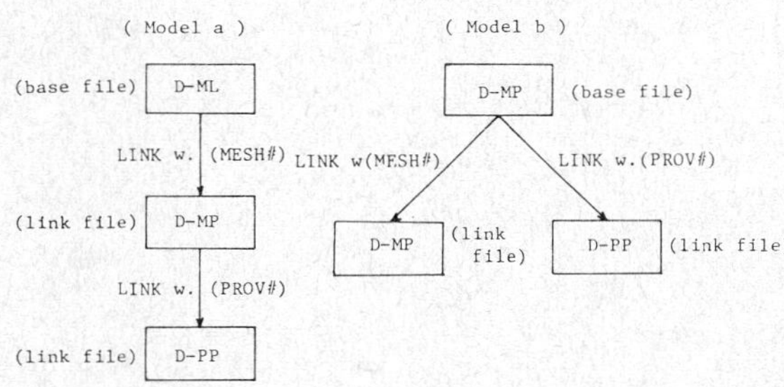

2) Functions of the dictionary.

The dictionary produces following functions;
(1) To refer the all domains of the dictionary one another,
(2) To add new domains to expand the dictionary,
(3) To join the domains of the dictionary together with that of PRF to expand them into wider PRF,
(4) To associate PRF around the dictionary to integrate them into a comprehensive database.

4.2 CALENDAR

It consists of three independent subsystems;
(1) D-YCAL is built upon a time axis of year unit.
(2) D-KCAL is of thousand (K) year unit.
(3) D-MCAL is of million (M) year unit.
Details are on the paper (Wadatsumi 1979 b)

5. Query Operation to the earthquake database (GEODAS-EQ)

```
TERMINAL READY

ACOS-6 TS2(R5.2) ON 10/29/79 AT 17.967
USER ID -6012950322
PASSWORD--
     ......................
     ......................
     ......................

     INQ DATA BASE RETRIEVE START

? FIELD GEODAS-EQ

  GEODAS-EQ     FILE  FIELD  DESCRIPTION

  LEVEL  DATA-NAME          ATTRIBUTE
  02       EQ-ID              PIC  X(8)  PKY.
  02       EQ-DATA.
   03      TIME.
     05    YEAR               PIC  9(4).
     05    MONTH              PIC  9(2).
     05    DAY                PIC  9(2).
     05    HOUR               PIC  9(2).
     05    MINUTE             PIC  9(2).
     05    SEC                PIC  9(4).
     05    ERROR-T            PIC  9(2).
   03      LOCATION.
    04     LONG.
     05    LONG-EW            PIC  X(1).
     05    LONG-DEG           PIC  9(3).
     05    LONG-MIN           PIC  9(2).
     05    ERROR-LONG         PIC  9(2).
    04     LATI.
     05    LATI-NS            PIC  X(1).
     05    LATI-DEG           PIC  9(2).
     05    LATI-MIN           PIC  9(2).
     05    ERROR-LATI         PIC  9(2).
    04     DEPTH              PIC  9(3).
    04     REGION             PIC  9(1).
   03      EQ-SIZE.
     05    MAGNITUDE          PIC  9(2).
     05    CLASS              PIC  X(1).
   03      MARK               PIC  X(1).
  FIEL  COMMAND  END
```

```
? RETR YEAR >= 1970
  1762        RECORDS FOUND

? SAVE 1
  SAVE  COMMAND  END

? RETR 7 <= MONTH <= 12 /1
  898         RECORDS FOUND

? SAVE 2
  SAVE  COMMAND  END

? RETR 130 <= LONG-DEG <= 140 /2
  464         RECORDS FOUND

? SAVE 3
  SAVE  COMMAND  END

? RETR 34 <= LATI-DEG <= 36 /3
  245         RECORDS FOUND

? SAVE 4
  SAVE  COMMAND  END

? RETR DEPTH GT 30 /4
  170         RECORDS FOUND

? SAVE 5
  SAVE  COMMAND  END

? RETR DEPTH GT 50 /4
  113         RECORDS FOUND

? SAVE 6
  SAVE  COMMAND  END

? RETR MAGNITUDE >= 4$ /6
  8         RECORDS FOUND

? SAVE 7
  SAVE  COMMAND  END
```

```
? DISP YEAR MONTH DAY LATI-DEG LONG-DEG DEPTH MAGNITUDE /7

 YEAR  MONTH  DAY  LATI-DEG  LONG-DEG  DEPTH  MAGNITUDE

 1970    07    14        36       139    060         40
 1970    10    07        35       140    060         41
 1970    10    30        36       139    060         49
 1971    08    12        35       140    060         48
 1971    08    31        34       140    060         41
 1972    12    04        36       140    060         40
 1972    08    11        35       140    060         43
 1972    09    23        34       140    060         41
  DISP  COMMAND  END

? TABLE
BB^7 J ?  YEAR BET 1960 1962 1964 1966 1968 1970 1972 /
BB^7 J ?  MAGNITUDE BET 0 10 20 30 40 50 60 70 80
BB^7 J ?

  YEAR
  1960 =< X1  < 1962
  1962 =< X2  < 1964
  1964 =< X3  < 1966
  1966 =< X4  < 1968
  1968 =< X5  < 1970
  1970 =< X6  < 1972

  MAGNITUDE
   0 =< Y1  < 10
  10 =< Y2  < 20
  20 =< Y3  < 30
  30 =< Y4  < 40
  40 =< Y5  < 50
  50 =< Y6  < 60
  60 =< Y7  < 70
  70 =< Y8  < 80

            X1      X2      X3      X4      X5      X6
     Y1    622    1065     696     237     347     315
     Y2      0       0       0       0       0       0
     Y3      0       0       2       1       3       0
     Y4      3       0      90     235     330     228
     Y5     48     295     376     527     821     432
     Y6     56     110     149     101     335     129
     Y7     13      21      26       6      41      54
     Y8      2       1       1       0      15      18

  TABL  COMMAND  END

? SORT YEAR ASC /7
  SORT  COMMAND  END

? SORT MAGNITUDE ASC /7
  SORT  COMMAND  END

? GRAP
BB^7 J ?  X(MAGNITUDE) Y(YEAR)
BB^7 J ?  /7
BB^7 J ?

             MAGNITUDE
          40.00     41.80     43.60     45.40     47.20     49.00
YEAR
            +.........+.........+.........+.........+.........+
   1970.00 I*                                                  I
   1970.00 I******                                             I
   1970.00 I***************************************************I
   1971.00 I*********************************************      I
   1971.00 I******                                             I
   1972.00 I*                                                  I
   1972.00 I*****************                                  I
   1972.00 I******                                             I
            +.........+.........+.........+.........+.........+

  ***   END OF LISTING   ***

  GRAP  COMMAND  END

? DONE

     INQ DATA BASE RETRIEVE END

SYSTEM ?BYE

**USED RESOURCE.....CPU=22SEC CON=30.1MIN LINE=NCU  T-ID=B9
**COST.  \97
$ N1041 17:23:57 DIS - CP
```

(___________; manual input)

6. Concluding Remarks

As a whole, GEODAS is considered to be; (1) a system of the distributed databases upon ARF, (2) an open-ended modular system at the physical storage level, (3) a relational database at the definition of the logical submodels, (4) an information retrieval system at the query operation, (5) a preprocessor system for GEOPAK. (6) a system with the logical capacity from one file database up to 300 databases including 63 files on each, (7) a progressive system covering from the small private use to the multidisciplinary use, (8) a information share system of not only the data, but also the user's views of both the logical databases and the query sets through the libraries, INQLIB and GQLIB.

Bibliography

Loudon,T.V.(1967) The Rokdoc Package. Reading Univ. geol. Rep. No.1

Loudon,T.V.(1974) Analysis of geological data using Rokdoc, a Fortran IV package for the IBM 360/65 computer. Rep. Inst. Geol. Sci., No. 74/1. 130 pp.

Wadatsumi,K.(1975 a) Basic design of the geology oriented data-base system (GEODAS) with scales of measurement in lattice theory as a schema attribute of data. Himeji Inst. Tech. Report, No.28A, 17-28 (in Japanese).

Wadatsumi,K.(1975b) Geology oriented data-base system (GEODAS) - background and system concept-. Geological Data Processing, No.1, 8-12 (in Japanese).

Wadatsumi,K.,Miyawaki,F.,Murayama,S. & Higashitani,M.(1976) GEODAS-DCRF: development of a relational data-base system and its application for storage and retrieval of complex data from researcher files. Computer & Geosciences, No.2, 357-364.

Wadatsumi,K.(1977) New specification of practical version of GEODAS as a DBMS to mix up well with each of the logical database models of a plate, relation, network, and hierarchy. Geological Data Processing, No.3, 45-62 (in Japanese).

Wadatsumi,K.(1978) Some problems in construction of geological databases and their application to geological correlation. Cenozoic Geology of Japan (Prof. Ikeb Memorial Volume), 217-229 (in Japa.)

Wadatsumi,K.(1979) Map function in the space and time dependent database system -an application of GEODAS (1)-. Geological Data Processing, No.4, 3-26 (in Japanese).

Wadatsumi,K.(1979 b) Calendar function in the space and time dependent database system -an application of GEODAS (2)-. Geological Data Processing, No.4,15-26 (in Japanese).

Yamamoto,K. & Nishiwaki,N.(1977) SPSS domestic version as GEOPAK, especially on the geologic programs added in it. Geological Data Processing, No.3, 1-7 (in Japanese).

II. Database system for multidisciplinary studies in environmental science - SMLESIS.

1. What is SMLESIS?

SMLES research group was organized in 1966 to enhance activities in earth science. SMLES is an abbreviation of "seamless earth science". Since the main method is system simulation, the name is also an abbreviation of "simulation earth science".

SMLES' basic stand point of view is that the earth is a seamless total system consisting of three major subsystems; non-biological subsystem (physico-chemical system), biological subsystem (ecological system) and human social subsystem (economy-culture system). Among the three, human social system has a special significance because of its impacts upon the environment of the terrestrial surface. Because of the importance of researches on environmental alterations, SMLES group expanded its research activities into environmental science which is called "environmentology".

Since various kinds and large amount of data should be dealt with in environmentology, a data management system by computer is essential. SMLESIS has been designed and implemented to satisfy the above need.

Recently, much about inter-disciplinary studies has been argued especially in environmental sciences. Those arguments emerges from the fact that environmental issues concern with various established disciplines. When an environmental issue emerges, usually a cooperative research group is composed by specialists from various disciplines. According to SMLES' view, however, an environmental system is a seamless total system which should be studied not partly by respective specialists but totally by every participants. Such type of research should be called "multi-disciplinary", instead of "inter-disciplinary".

Architecture and functions of database should depend largely upon for which type of research, inter-disciplinary of multi-disciplinary, the database is used. In multi-disciplinary studies, all members of the group should have an almost identical concept or philosophy about problems. Consequently, a database should reflect their overview, and it can be regarded as an expanded "researchers' file". It is also essential that such database is used by the member of a multi-disciplinary group to reach an identical concept. This role of a database is discussed in the following section.

The concept in environmentology should be validated through field data from actual environmental managements. Therefore a database for environmentology should be designed to be an effective tool for regional environmental management. For that purpose, interfaces of the database to administration and grass-root are carefully considered.

2. Design of SMLESIS

We already had various environmental database. For instance, EPA of USA has SAROAD (storage and retrieval of aerometric data), STORET (storage and retrieval system for water quality data), NEDS (national emissions data system), SEAS (strategic environment assessment system), *etc.* They are served for monitoring, search for technical source data and modelling. IRS (international referral system) of UNEP is to exchange environmental information. In Japan, almost all environmental information systems developed so far are for monitoring and controlling the air and water quality at local governmental level. National Land Use Agency, Japan, is developing a data bank which is used for planning at national level.

Under the above circumstances, a special database system is developed.

The present database system has both functions of model design and retrieval. A procedure is described as follows :

(A) design of model of logical thesaurus

Environmental researches and administration are exclusively interdisciplinary, and academic data and reports should often be referred from different view points of the original researcher. Consequently, a special thesaurus must be installed by which the user's philosophy or the environmental policy is expressed. In addition, the same term is often used for different definitions in different disciplines. Therefore, the thesaurus must have a function to disctinguish them. Because the specialists as well as data are scattered, it should have a function to trace the addresses of specialist systematically for each different project or environmental impact assessment.

Any project or academic research should have needs or image at the starting point. Here the image is used to represent an emotional feeling. Passing through procedures of brain storming and other methods, they become visualized in a form of logical thesaurus. This is nothing but a representation of system structure of the project, or a conceptional system model is developed.

(B) data file and information retrieval

The collected data are put into a file as a database. When a word or phrase (logical combination of words) is requested, the information is retrieved from the file referring to the thesaurus designed in (A). This is a process of discrepancy check between our philosophy (master plan) and the present pattern of activities (research activities or administrative activities). In a case of general information service, the thesaurus is developed by a process of statistical analysis of data to ensure the efficient retrieval. We have, instead, a loop of modifying the thesaurus by conversation with data.

(C) scenario and simulation

After passing through the fittness test, the obtained thesaurus will represent a master plan from a view point of our philosophy. The simulation can be carried out using filed data (both data and simulation program are included), and a scenario has a role of input conditions for simulation. It is remarked that not only a simulation procedure but also a modelling procedure are included.

The overall structure of the present system is shown in Fig.II-1, in which the procedures of planning and assessment are also shown.

In the present system, the major components are

(i) Descriptive data and their retrieval system

(ii) Grid data and their retrieval system

The "thesaurus" files are essential for these subsystems, because flexible data structure representing user's philosophy are installed in the database. They are described in the following sections.

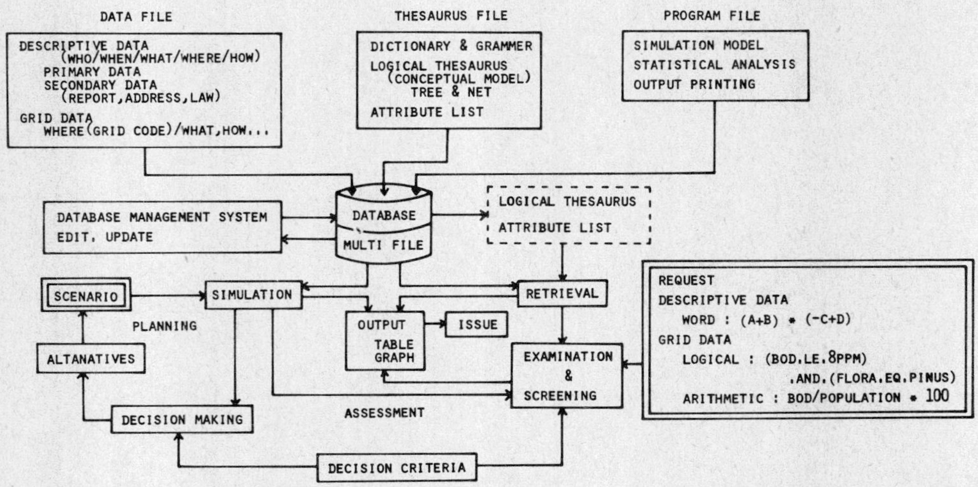

Fig.II-1 Database system to support regional environmental management

3. Descriptive Data

3.1 Types of Data

Descriptive data used for environmental database are as follows :

(A) Primary (academic) data
Typically they have the data structure as

WHO / HOW / WHEN / WHERE / WHAT
Dr.A - methodX - 1979.9.3 - riverα - BOD=10ppm

Field data by geological or ecological survey also have a similar structure defined by

WHERE / WHAT / WHEN / HOW / WHO

A significant point in filing these primary data into database is : "How it is evaluated?"

(B) Secondary data
They are classified into
i) report :

WHO	WHEN	WHAT	WHERE	HOW
author	date	title	journal	comments

ii) address of specialist :

WHO	WHEN	WHAT	WHERE	HOW
name	date of birth	research theme	address	background

iii) document (act and regulation code)
Sentenses of act and code have the structure as

WHAT	WHEN	WHERE	HOW
name of act	date of enactment	range of application	decision criteria

However, they are retrieved as

WHERE / HOW / WHAT / WHEN

for actual screening and approval of the individual projects. Consequently, the inverted file of the above form should be prepared.

It should be remarked that all the types of descriptive database have the same structure *i.e.* combination of WHAT, WHEN, WHERE, WHO and HOW. Therefore, all of them should be managed by a single program.

3.2 Descriptive Data Retrieval

Because of the rapid progress of environmental sciences and the diversity of the environmental issues, stress should be given on flexibility than speed of retrieval. To meet these, English (or Japanese) words, instead of codes or symbols to represent words or terms, are used for a query statement in the present system. For instance, the query

DIFFUSION * (SOx + NOx)

means that the data which contain one of SOx diffusion, NOx diffusion and their relating words should be retrieved.

4. Thesaurus File

Retrieval of data is performed by a help of thesaurus, which is composed by the following three files.

4.1 Dictionary and Grammer Book

Data are retrieved at a word level, and therefore, the dictionary to translate foreign words into Japanese for environmental technical terms should be installed. In addition, a declension table for nouns and adjectives is necessary.

4.2 Data Name List

The ordinary data name list and comments are used for identification and format generation.

4.3 Logical Thesaurus

The present system can adopt the thesaurus with tree and net structures.

(A) Tree structure
Indentation method is adopted to give a definition in logical thesaurus with tree structure.

(B) Net structure
To form the net thesaurus, a listing of simple binomial causal relations is the first step. The relations are classified into

A > B	A derives B
A < B	A is derived from B
A <> B	A and B are a two-way relation
A = B	A is synonimous with B
A / B	cut relation between A and B

Then the net structure is formed automatically, and the paths are searched by the following commands :

F(A >)	trace downward from A
F(A <)	trace upward from A (root search)
F(A > B)	find paths from A to B
F(A > A)	find loop containing A

This has also been extended into Net Structure Analyzing System of Causal Relation (NESSY) in which the delay time, transition probability and flux capacity to move from A to B are included.

Logical structure of thesaurus will be different from project to project. The present system can accept different thesauri which can be worked independently. It is remarked that a project team can be organized by retrieving the specialist address. In addition, programmed steps of training of project members with different backgrounds can effectively be achieved by using a flexible logical thesaurus.

5. Grid Square Data System

5.1 Types of Data in Master File

Major parts of the present database are the grid data system, where the following standard codes in Japan are employed.

(A) primary code (I2) 1 longitude x 40'latitude
(B) secondary code (I2) 7'30" long. x 5' lat.
(C) tertiary code (I2) 45" long. X 30' lat.
1km x 1km

Further,

(D) watershed code (I3)
(E) municipality code (I3) city, town, village
(F) boundary code (I2) prefectural boundary and coast to the open sea

are added for a convenience of retrieval.
Accepted data are : (1) topography, geology, land use pattern *etc.* (National information system by Geographical Survey and National Land Agency), (2) population *etc.* (National Census), (3) vegetation, national assets (Environmental Agency), (4) industry, traffic and other socio-economic activities.

5.2 Grid Data Retrieval

Various delicate evaluations are required in a decision making on environmental concerns. Therefore, query statements of "language" level which are expressed by arithmetic or logical equations, should be accepted to secure a soft and flexible retrieval. A "mini-compiler" is developed to interplete these query statements, of which specifications are as follows :

(A) Operators

priority		
1	**	arithmetic operator
2	*,/	arithmetic operator
3	+,-	arithmetic operator
4	.GE., .GT., .LE., .LT., .EQ., .NE.	logical operators
5	.AND., .OR.	logical operators

(B) Parameters

Xn: code of attribute name, which is given in a dataname list.
Yn: parameter for processing of data X , which is defined by *Yn*=f(*Xn*). (See (E) for details)
Zn: code of group name of attributes, which is defined by a thesaurus.

(C) Functions
The following functions are defined for group name (*Zn*),

SUM(*Zn*), MEAN(*Zn*), MAX(*Zn*), MIN(*Zn*)

by which a statistical analysis is possible. For instance, vegetation data with various species can be represented by one term as SUM(*Zn*).

(D) Printing instructions

MAP(*Yn*): map of the values of *Yn* is printed, where the maximum numbers of tones are ten.
CODE(*Yn*): code numbers of grids which "hit" the logical equation *Yn* are printed. Comments on the grid (point data), *e.g.* name of wild life to be protected, are also given.

(E) Query set and result
A query set has the following format :

```
Y1=SUM(Z1000)

Y2=1.0*(X114+Y1)+2.1*(X70+X78)
   ↓         |               |
 evaluation weight           |
             ↓               |
        quotation of previously defined
        parameter            |
                             ↓
                    data NO. in name list

Y3=(X15.GE.4).AND.(Y2.GT.5.5)
MAP(Y2)
CODE(Y3)
/END
```

Fig.II-2 shows a map of "low human activities and high degree of nature" in Aichi prefecture, where the query conditions are :

```
Y1=(X24.GE.5).AND.(X195.LE.50).AND.
   (X200.LE.500).AND.(X189.LE.1000)
Y2=Y1*(X24 + X195 + X2000 + X189)
MAP(Y2)
```

and

X24 : vegetation index defined by Environmental Protection Agency of Japan
X195: BOD load (kg/day)
X200: population
X189: energy consumption (kcal/day)

6. Concluding Remarks

As described above, the present SMLESIS is to be used by a small multi-disciplinary group whose size is up to ten. The size of data actually installed is about 10 MB. A large part of it is grid square data of environmental concerns in Tokyo and Aichi. About 4000 records of data of reports and documents are also contained and about 30 records are added monthly.

In some cases, large database is made for "inter-disciplinary studies" by collecting ready-made database from relating fields. Since such data have different types and forms which is inherent in the respective disciplines, the database made in such way is hardly to be effective. It should be re-emphasized that multi-disciplinary studies rather than inter-disciplinary studies should be encouraged for environmental sciences, of which former might be regarded as a new discipline. The database which is effective for such new discipline should reflect and represent the group's overview or philosophy even though its size is rather small.

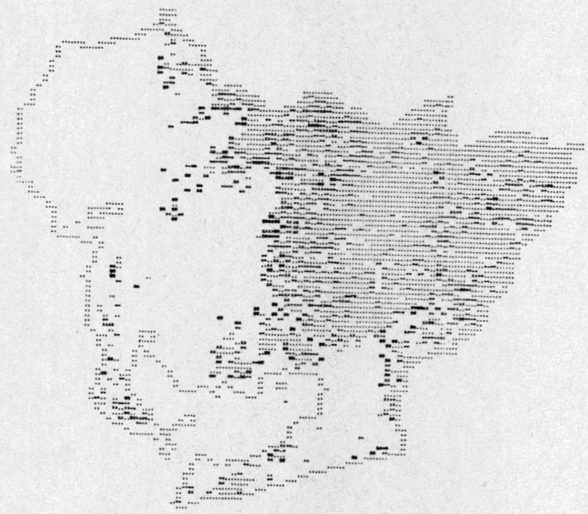

Fig.II-2 Low human activity and high degree of nature in Aichi prefecture

Bibliography

Hiramatsu,T. (1976) Thesaurus oriented documents retrieval system (THODORES) for laboratory database. Memoirs of Research and Development Division, Computation Center, Nagoya Univ., No2, 24-48.

Hiramatsu,T. (1978) Interactive net thesaurus processing system -- NESSY. Memoirs of Research and Development Division, Computation Center, Nagoya Univ., No4, 21-39.

Hiramatsu,T. & T.Fukumura (1979) Dynamic net structure analyzing system of causal relations (NESSY - III). Trans. Information Processing Soc. Japan (in press).

Shimazu,Y. (1977) Environmental management and information system. Kankyo Joho Kagaku (Environmental information science) 6 (1) 56-69 (in Japanese).

Shimazu,Y., T.Urabe and M.Uematsu (1978) Database for regional management. Proc. of International Conference on Cybernetics and Society in Tokyo, 704-708.

Manuals of SMLESIS *(in Japanese)*
(by Shimazu,Y., T.Urabe, T.Hiramatsu, T.Moizumi and M.Uematsu)

SMLESIS (I) (1976) Manual for information retrieval system in batch system, pp33.

SMLESIS (II) (1977) Introduction to grid data retrieval (GRIDREQ) and net thesaurus (NESSY) in Geological Data Processing 3, 9-16.

SMLESIS (III) (1978) Manual for grid data retrieval system in batch system and conversational system, pp87.

SMLESIS (IV) (1978) Manual for the system to process net thesaurus (NESSY-I) and the system to analyze net-structured causal relations (NESSY-II), pp40.

SMLESIS (V) (1978) Manual for the dynamic simulator of net-structured causal relations, pp37.

Scientific Information Systems in Japan
H. Inose (ed.)

INFORMATION RETRIEVAL SYSTEM AND SCIENTIFIC INFORMATION SERVICE IN HIROSHIMA UNIVERSITY

Hideto IKEDA and Sumiyasu YAMAMOTO
Computing Center, Hiroshima University, Higashi-Senda-machi, Hiroshima

A new bibliographic information retrieval system was developed in 1977 by the staff at the Computing Center, Hiroshima University. The new system is called HUNDRED (Hiroshima University New Document REtrieval and Dissemination) system. A major objective of the development of the HUNDRED system is to realize our theoretical work on the design of optimal balanced file organization scheme in an actual information retrieval system. On the other hand, utilization of the system, i.e., a literature search service to the University faculties and students, has been planned and carring out by the University Library members. In this paper, certain specific features of the system and the service, are described.

1. INTRODUCTION

A new bibliographic information retrieval system was implemented in 1977 by the staff at the Computing Center, Hiroshima University. This implementation was supported as a part of a research project on the "Organization of Scientific Information" by the Ministry of Education, Science and Culture of Japan. The new system is called HUNDRED (Hiroshima University New Document REtreival and Dissemination) system. It has been using for the dissemination of new bibliographic information about some selected fields to the University faculties and students. This service has been operating by the University Library under the technical cooperation of the Computing Center.

Our system HUNDRED has two specific features to be mentioned in comparison with other existing bibliographic information retrieval systems. One of the specific features is concerned with its storage structure. A new indexing method using multiple queries to an address transformation was designed and implemented in organizing indices in order to increase the performance of our bibliographic information retrieval system. The file organization scheme is called $HUBFS_1$ (Hiroshima University Balanced File-organization Scheme of order one). Another specific feature of the system is in its universal applicability to a number of bibliographic databases having various formats and contents. There are more than one hundred bibliographic databases prepared and distributed all over the world. Many bibliogrpahic information retrieval systems have been implemented to one of these particular databases. Some modifications about the main frame of the system are often necessary for the change of database. The main frame of our HUNDRED system, however, can remain invariant for such change of database except for setting of some parameters depending on the change of structures and usage.

During the course of the implementation of the HUNDRED, our University Library staff wanted to initiate a bibliographic information service for the University faculties and students by using currently available databases, namely, INSPEC by IEE and BIOSIS Previews by BioSciences Information Service. The system works on the computer (HITAC 8700) of our University Computing Center. Because of the limited hardware resources, especially on the capacity of disc packs, it is inevitable at this time to carry out SDI (Selective Dissemination of Information) service in batch mode. The system, however, is capable of performing many attractive functions in on-line mode. These functions will operate in full in the near future after the expected innovation of our computer system. Although, these on-line functions are now available for the users only in tuning up their personal profiles of sample bibliographic data provided for trial search is rather limited (ca 5×10^4), on-line trial-and-error of researchers and expert librarians are operating satisfactory. This strategy seems to be necessary and even inevitable for such bibliographic information service.

Serviceability of our HUNDRED system has been successfully improved by these service experiances and our efforts paid on the improvement of serviceability have been successfully reflecting on the wide-spread dissemination of the system among the University faculties and students.

2. BIBLIOGRAPHIC INFORMATION RETRIEVAL SYSTEM HUNDRED

One of the objectives of the implementation of the HUNDRED system is to realize our theoretical work on the design of optimal balanced filing schemes in an actual information retrieval system. The other is to meet with the desire of our Library staff to initiate a bibliographic information service.

Formally, the name HUNDRED is an acronym of the Hiroshima University New Document REtrieval and Dissemination system. It is, however, named in hopes to be able to manage more than one hundred bibliographic databases by a single system. The functions and their major characteristics of the HUNDRED system will be described in this section.

2.1 Functions of HUNDRED System

The system includes five major functions. The first is a system generation function. User can generate a retrieval system of a specified database by giving certain parameters. These parameters are

1) to describe the source record format,
2) to define methods for extracting access keys from records,
3) to define display formats of records,

4) to define conversion tables for code conversion if necessary,
5) to define stop words and
6) to define the database administraters and passwords for security control.

This function is performed by the system generator. It analize those parameters and creates a table to control a particular database system. This function realized by our technique is one of the most important functions of the system, since the main frame of the system can remain invariant with respect to the changes of databases by this technique. This can be regarded to be a realization of the principle of "data independency" in a database management system.

The second function is to store new records into the storage devices. In actual service, the new tapes arrive 1 to 10 time per month from the database supplier. Each time, the database administrater operates the system to store the new records in batch mode. The system generates three kinds of files, i.e., a text file, access key files and an address file. A text file is organized to store original records and access key files are organized to keep access keys. An address file is composed of every inverted list which consists of the address set of records pertinent to each access key. To add new records in the system is to update the above mentioned three kinds of files shown in Fig. 1. The system has high performance standard because of a new technique capable of organizing the address file. Table 1 shows the number of records, the number of keywords and the storing time which had been recorded with respect to BIOSIS previews.

The third is the on-line information retrieval function of a database. This function consists of various commands shown in Table 2. User can access every database by using these commands in on-line mode. Fig. 2 is an example of on-line search of BIOSIS previews.

The fourth is the preparation function of SDI lists with catalogue cards (option) in batch mode. This function operates each time at the arrival of new information tapes.

The last function is to maintain database systems. This function includes certain back-up commands of system files and listing commands of a database status.

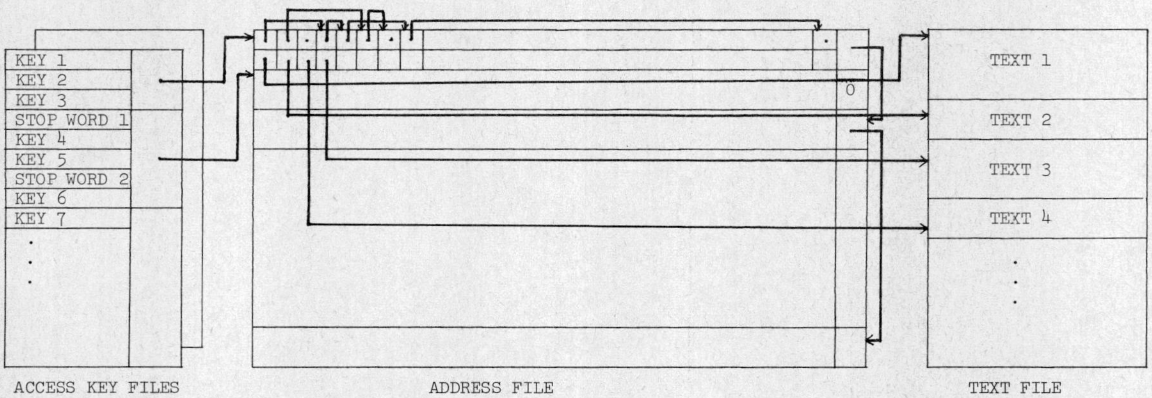

Fig. 1 Storage structure of the HUNDRED system

Table 1. Performance Statistics of the BIOSIS search system

	Tape No.	6305	6306	6307	6308	6309	6310	6311	6312
	Documents	6040	6040	6040	6040	6040	6040	6040	6048
	Documents (cumulated)	6040	12080	18120	24160	30200	36240	42280	48328
	Keys (cumulated)	34948	56838	75692	92266	108281	123621	137748	150953
CPU TIME	Parsing*	601	589	580	573	572	570	560	567
	Reducing**	878	956	1020	1053	1121	1158	1185	1244
	Storing***	292	299	309	298	300	304	298	307
	Total	1999	2105	2184	2204	2286	2333	2347	2434

* CPU time used for extraction of access keys from the records.
** CPU time used for construction of current access key files.
*** CPU time used for construction of current address file.

Table 2. HUNDRED on-line commands

No.	Command	Function
1	HUNDRED	to open a specified database
2	HSEARCH	to search records pertinent to keys specified and to create a document set
3	HBROWSE	to search records pertinent to keys specified and to display the number of pertinent records but not create any document sets
4	HAND	to create a document set obtained by the set-operation AND between two or more document sets
5	HOR	to create a document set obtained by the set-operation OR between two or more document sets
6	HDIF	to create a document set obtained by the set-operation DIFFERENCE between two or more document sets
7	HSAND	combined operation HSEARCH and HAND
8	HSOR	combined operation HSEARCH and HOR
9	HSDIF	combined operation HSEARCH and HDIF
10	HSTEXT	to create a document set extracted by text search of a specified document set
11	HSCOMB	combined operation of HSEARCH, HAND, HOR, HDIF and HSTEXT
12	HRESTRICT	to create a document set restricted with respect to a specified key such as publication year, tape no. and record type code
13	HDISPLAY	to display documents included in a specified set
14	HPRINT	to print documents included in a specified set to a lineprinter device
15	HQUERY	to display a query corresponding to a specified set as a formula
16	HCARD	to print documents included in a specified set on catalogue cards
17	HUSER	to register database users
18	HPROFILE	to register a user profile for the SDI service
19	HMENU	to display the list of all active database
20	HHELP	to display the command table
21	HEND	to exit from the HUNDRED system

```
WELCOME TO HITAC 8700/8800 TIME SHARING SYSTEM (HIRODAI C.CENTER) AT 16:10:23 ON
77-08-30
#EG304R4 TYPE IN 'JOB' COMMAND,PLEASE
//TOSHOKAN:JOB 1054821510,JC=T
#EG331R4 TYPE IN NUMERICAL PASSWORD FOR 84055451=8
#EG021I  PROFILE & COMLIB BEING USED
#EG330I  'JOB' COMMAND ACCEPTED ; READY AT 16:10:55 JSN=U00223 SYSTEM=OS7-05-01
//HUNDRED BIOSIS
WELCOME TO THE HIROSHIMA UNIVERSITY NEW DOCUMENT RETRIEVAL AND DISPLAY
(HUNDRED) SYSTEM VERSION-1
  DATABASE INFORMATIONS
    1. DATA-BASE     :BIOSIS  (7701-7709)
    2. TOTAL RECORDS:  54368 RECORDS
    3. UPDATE        :08/18
NAME?
@IKEDA H
USER ID ?
@0000000003
READY !
@/HS K.HEPATITIS,S.03200,S.86215,K.AUSTRALIA#,K.ANTIGEN#
  #  1(   97):KEYW=K.HEPATITI
  #  2(  834):KEYW=S.03200
  #  3( 7252):KEYW=S.86215
  #  4(  177):KEYW=K.AUSTRALI
  #  5(  311):KEYW=K.ANTIGEN
  #  6(    4):KEYW=K.ANTIGENE
  #  7(   91):KEYW=K.ANTIGENI
  #  8(  174):KEYW=K.ANTIGENS
@/HAND 1,2,3
  #  9(   69):OPER=  1*  2*  3
@/HOR 5,6,7,8
  # 10(  530):OPER=  5+  6+  7+  8
@/HAND 4,10
  # 11(    7):OPER=  4* 10
@/HAND 9,11
  # 12(    4):OPER=  9* 11
@/HD 12,1,2
-----------------------------------------------------------------------------------------

AUTHOR :NIK-AKHTAR B/KHAKPOUR M/PANAHI F/HESABI A M/
TITLE  :PREVENTION OF HEPATITIS TYPE B WITH SPECIFIC IMMUNE SERUM GLOBULIN.
        HUMAN AUSTRALIA ANTIGEN RENAL FAILUR HEMO DIALYSIS
JOURNAL:NPRNA/NEPHRON
BIBLIO :17 (5). 1976.402-406.
C-CODE :10064 10066 10068 10504 12100 12508 12512-14006-15001 15002 15506-22
        005 22018-33506 34504*36001 36006-
S-CODE :03200 86215
CIT-NO.:63015540
-----------------------------------------------------------------------------------------

AUTHOR :NOZDRIN-PLOTNICKI B/JANKOWSKI A/
TITLE  :ACITVITY OF COBALT ACTIVATED ACYLASE AND AUSTRALIA ANTIGEN IN CHILDR
        EN SUFFERING FROM VIRAL HEPATITIS.AGE/
JOURNAL:AITEA/ARCH IMMUNOL THER EXP
BIBLIO :24 (3). 1976.433-438.
C-CODE :04500 10010 10064 10069 10802 10808-12508 14006-25000-33506 36006*
CIT-NO.:63021577
-----------------------------------------------------------------------------------------
@/HEND
--LOGGING INFORMATION (77/08/30 : 16 37 - 16 40)-----------------
I DATABASE : BIOSIS      I  QUERY  :   6TIMES I FILES  :    12    I
I USER     : IKEDA H     I  KEYS   :   3      I CPU    :   20SEC  I
I UNO      : 0000000003  I  OUTPUTS:   2REC   I CHARGE :  160YEN  I
-----------------------------------------------------------------
END OF THE HUNDRED SYSTEM
//END
```

Fig. 2. Example of on-line search by HUNDRED system

2.2 Storage structure of HUNDRED system

Let Ω be a database or a set of all records. Every record ω in the stored database Ω has a unique accession number $a(\omega)$ together with its information contents. The accession number of a record is often realized by its address in a random access file. Let $\{A_1,A_2,\cdots,A_\ell\}$ be a set of ℓ attributes and consider the case where the information contents of every records can be characterized by a function

$$\delta : \Omega \longrightarrow 2^{\{A_1,A_2,\cdots,A_\ell\}}$$

where 2^S denotes the set of all subsets of a finite set S. If $\delta(\omega)$ includes the i-th attribute A_i, then the record ω is said to be pertinent to the attribute. In a bibliographic information retrieval system, those attributes which consist of keywords, author names and subject classification codes are used for access keys. In order to retrieve quickly those records pertinent to a query expressed by a predicate about access keys, it is familiar to organize in a bucket an index or an inverted list of accession numbers

$$I_i = \{a(\omega) \mid \delta(\omega) \ni A_i\}$$

is organized for each of the access keys. A key to address transformation technique can be used in this organization. The number of indices to be organized in this organization is, however, extremly large especially in the case of bibliographic databases.

A new organization scheme, called $HUBFS_1$ (Hiroshima University Balanced File-organization Scheme of order one) is used in our HUNDRED system. A c keys to one address transformation from $\{A_1,A_2,\cdots,A_\ell\}$ to the set of bucket addresses $\{1,2,\cdots,N\}$ defined by

$$\beta(A_i) = [(i + c - 1)/c]$$

is used and N buckets containing

$$B_j = \bigcup_{\beta(A_i)=j} \{(i,a(\omega)) \mid \delta(\omega) \ni A_i\} \quad (j = 1,2,\cdots,N)$$

are organized, where $[x]$ denotes the greatest integer not exceeding x. If $c = 1$, this scheme is identical with the inverted filing scheme of order one. In our HUNDRED system, $c = 51$ is selected. Our experimental experiance [4] shows that $c = 100$ is optimal in our actual circumstances.

When compared with the corresponding inverted filing scheme of order one, the scheme $HUBFS_1$ has several advantages in spite of some space overhead caused by storing i in addition to $a(\omega)$, i.e.,

(1) The number of buckets to be prepared by the system can be significantly reduced.
(2) Vacant memory space caused by the quite uneven distribution of keys in a database can be considerably reduced.
(3) Storing time can be reduced.

The scheme $HUBFS_1$ has originally been designed to reduce the redundant storage of accession numbers in a bucket in addition to the above-mentioned characteristic features [4]. Such a effort, however, is not adopted since duplicated storing of the same accession number is not so frequent in the case of bibliographic databases.

2.3 The present status and future of HUNDRED system

The present HUNDRED system controls several bibliographic databases as shown in Table 3 and is utilized to distribute scientific information to faculties and students at Hiroshima University.

On the other hand, an enhancement is in progress along with a replacement of the computer in April 1980. A new version is expected to have some additional functions. It will include capabilities for an on-line data inputing and updating function, a function of linking different databases and a function of handling KANJI-characters. We are planning to enhanced the system to includes library automation system, hospital automation system and university administration system.

3. LITERATURE SEARCH SERVICE AT THE LIBRARY

3.1 Brief history of the service

The literature search service using the HUNDRED system came into operation at the Library in October 1977. The service has been called HUNDRED service naming after the system. Prior to the start of the service, a survey was conducted using a questionaire to grasp the need of faculty members. Based on its findings, two kinds of databases, BIOSIS Previews and INSPEC, were selected. Currently the service is limited to SDI only in batch mode due to the limitation of the hardware, though the system has many attractive functions in on-line mode.

During the first two years, 1977 & 1978, the project has been supported by the Ministry of Education, Science and Culture of Japan and the operation was rather experimental. Since the third year, 1979, it has been fully operated by the University.

3.2 Characteristics of the service

The HUNDRED service has three distinguished characteristics in comparison with other services operating in our country.

The first is the fact that this is a cooperative venture by the University Computer Center and the University Library. Several functions needed to keep up with the service are clearly defined and assigned to the two separate departments as follows:

Computer Center.... Development, improvement and maintenance of the software, education of the library staff to operate the system.
Library........... On-line presearch for SDI service, output listings and their distributions, maintenance of stored data and publicity activities.

The second is the utilization of on-line capability of the system in presearch. In general, it would be important to refine expressed by profiles in the automated literature search, especially in the case of SDI service. Although our present service is limited in batch mode SDI only, on-line presearch function has been utilized to make profiles. Thus we can reduce time and cost for presearch. Whole operation processes for the service are shown in Fig. 3.

The third is concerned with the roll of the Library staff as intermidiary between users and the system. In this service, the Library staff members consult with researchers and then carry out presearch operations from a terminal to make queries. If the researchers give any negative response for retrieved information, the queries are adjusted or corrected and retrievals are retried. The Library staff members are likely better and experienced operators than the researchers. Searchers have up-to-date knowledges about.

(1) search system (mainly search commands),
(2) database characteristics (including how to use their search manuals),
(3) search strategy (how to make profile fit to the given requests),
(4) how to operate terminals.

The service has been successful with the current arrangement, in which the search is done by the experienced strategists. Steady growth in the number of users as shown in Table 4 demonstrates it.

3.3 Prospects of the service

Although the service has become fully operational from the beginning of this year, it has been obliged to be restricted only to the University staff members, because it involves the legal aspect of charging. It has been evident that many latent demands of researchers for the service also exist outside the campus.

To fulfil them and to make the most of our experiances, we are seriously looking into the expansion of the service area beyond the campus.

Table 3 Database controled by the present HUNDRED system

Database	Publication	Content	Note
BIOSIS Previews	BioSciences Information Service	life science	240,000 rec/year
INSPEC (A,B,C)	Institution of Electrical Engineers	physics, electronics & computers	150,000 rec/year
COMPENDEX	Engineering Index Inc.	technology	sample data 7,160 rec
SCI	Institute for Scientific Information	science	sample data 685 rec
MARC (books)	Library of Congress	book catalogue	sample data 2,966 rec
IMP	American Mathematical Soc.	mathematics	1973-1975 36,149 rec
PATELL	American Psychological Association	psychology	sample data 2,560 rec
HA	American Bibliographical Center-Clio Press	history	sample data 3,000 rec
GEOGRA	Hiroshima University	geography	sample data 416 rec
MATLIB	Hiroshima University	mathematics	1978- 700 rec
HUMARC	Hiroshima University	book catalogue	1978- 2,000 rec
STATAB	Kansas State University	statistics	1959-1975 25,000 rec

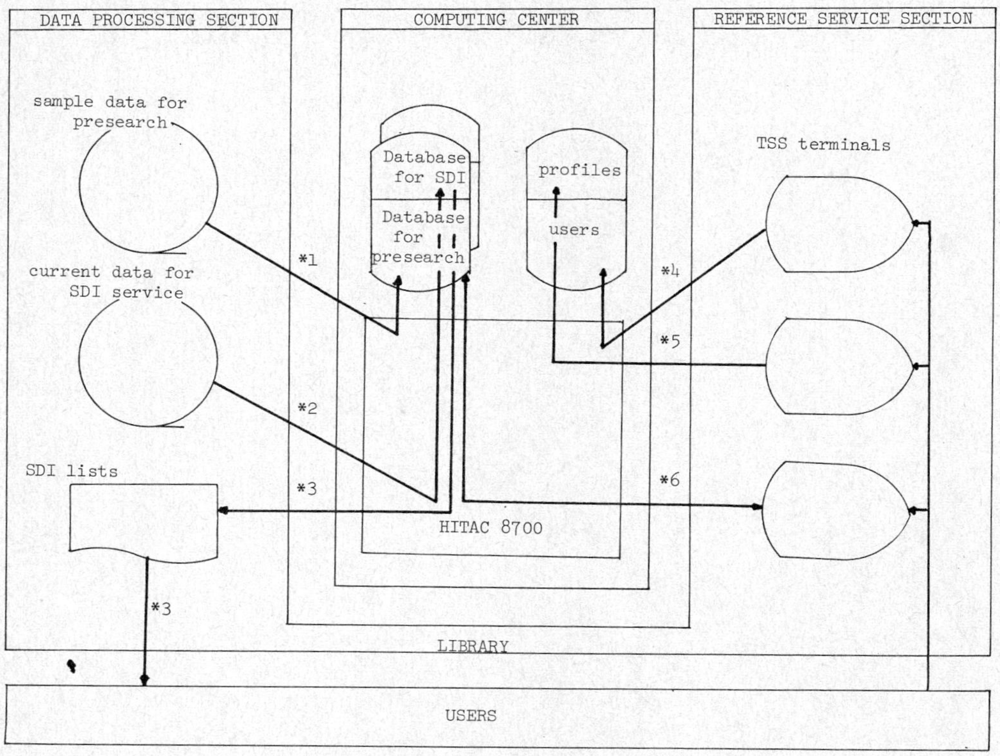

*1 storing of sample data for presearches
*2 storing of current data for SDI service
*3 retrievals for making of SDI lists
*4 registration of database users
*5 registration of profiles for SDI service
*6 presearches for making profiles

Fig. 3 Operation process of the HUNDRED service

Table 4. Growth of users in HUNDRED service

Database	year				1978												1979
	month	10	11	12	1	2	3	4	5	6	7	8	9	10	11	12	1
BIOSIS	No. of users	11	28	52	70	96	121	124	129	134	143	144	147	156	164	171	176
	No. of profiles	31	51	84	106	128	177	184	200	207	223	225	231	243	257	267	273
INSPEC	No. of users	7	12	13	24	25	25	26	28	28	30	32	32	35	38	39	39
	No. of profiles	9	14	15	28	30	30	32	36	37	41	46	50	56	61	63	63

ACKNOWLEDGMENTS

We would like to express our hearty thanks to Mr. Morioka and the Library staff of Hiroshima University for their contribution on the literature search service. We also wish to thank to Computing Center staff for the detailed programming of the HUNDRED system. We are grateful to Assistant Professor I. Morita of the Ohio State University for her valuable linguistic comments on this paper.

This study is supported in part by the Grant of Ministry of Education, Science and Culture of Japan.

REFERENCES

[1] S. P. Ghosh, "Data base organization for data management", Academic Press, New York (1977).
[2] H. Ikeda, "Evaluation of combinatorial file organization schemes", Proc. 3rd Int. Conf. on VLDB, 231-235 (1977).
[3] H. Ikeda, "Combinatorial file organization schemes and their experimental evaluation", Hiroshima Math. J., 8(3), 515-544 (1978).
[4] H. Ikeda and M. Naito, "Evaluation of a combinatorial file organization scheme of order one", Proc. Study on Scientific Database Management Systems, Edited by S. Ohsuga, 195-199 (1979).
[5] S. Tazawa, K. Ushio and S. Yamamoto, "Partite-claw-decomposition of a complete multi-partite graph", Hiroshima Math. J., 8(1), 195-206 (1978).
[6] K. Ushio, S. Tazawa and S. Yamamoto, "On claw-decomposition of a complete multi-partite graph", Hiroshima Math. J., 8(1), 207-210 (1978).
[7] S. Yamamoto, H. Ikeda, S. Shige-eda, K. Ushio and N. Hamada, "Design of a new balanced file organization scheme with the least redundancy", Information and Control, 28(2), 156-175 (1975).
[8] S. Yamamoto, H. Ikeda, S. Shige-eda, K. Ushio and N. Hamada, "On claw-decomposition of complete graphs and complete bigraphs", Hiroshima Math. J., 5(1), 33-42 (1975).
[9] S. Yamamoto and S. Tazawa, "Hyperclaw decomposition of complete hypergraphs", Proc. of the Colorado Conf. on "Combinatorial Mathematics and Optimal Design"", to appear (1978).
[10] S. Yamamoto and S. Tazawa, "Combinatorial aspects of balanced file organization schemes", to appear in Journal of Information Processing, (1978).
[11] S. Yamamoto, S. Tazawa, K. Ushio and H. Ikeda, "Design of a balanced multiple-valued file organization scheme with the least redundancy", to appear in ACM Transactions on Database Systems, (1977).
[12] S. Yamamoto, S. Tazawa, K. Ushio and H. Ikeda, "Design of a generalized balanced multiple-valued file organization scheme of order two", Proc. of 8-th ACM-SIGMOD Int. Conf. on Management of Data, (1978).
[13] S. Yamamoto, T. Teramoto and K. Futagami, "Design of a balanced multiple-valued filing scheme of order two based on cyclically generated spread in finite projective geometry", Information and Control, 21(1), 72-91 (1972).

Scientific Information Systems in Japan
H. Inose (ed.)

USES OF KNOWLEDGE BASES IN DATA BASE MANAGEMENT SYSTEMS

Setsuo OHSUGA

Institute of Space and Aeronautical Science, University of Tokyo, 4-6-1 Komaba Meguro-ku, Tokyo 153

This paper describes a knowledge based system named the KAUS(Knowledge Acquisition and Utilization System) designed to aid man in his intellectual works such as scientific research, decision making, engineering design and medical diagnosis. In particular, its subsystem named the SBDS(Structure Based Deduction System) is discussed in detail, which takes charge of handling both the knowledge base and the data bases. In this system, information is processed by being converted among four different forms, i.e., external representation such as natural language and graphic representation, program, data base and the representation of knowledge. Some important mappings between different forms and the deductive inference algorithm are defined.

1. INTRODUCTION

This paper describes a system named the SBDS(Structure Based Deduction System). This is a subsystem of a knowledge system named the KAUS(Knowledge Acquisition and Utilization System) that is being developed as an intelligent interactive system to aid man in his works such as industrial design, decision making, medical diagnosis and so on. The KAUS is expected to accept, understand and reply user's request presented in a natural-(like)-language and/or in a simple graphic form, or in other words, in man's everyday forms of representations. We call representations in these forms the external representations.

The SBDS takes charge of handling the knowledge bases and the data bases in the KAUS. It adopts, as its basic formalism for representation of knowledge, an extended form of the first order predicate logic because it is considered very suited for conversion to/from any of three major forms of information appearing in and arround the computer systems, i.e., (1) external representations, (2) data bases and (3) programs. Information is processed by being converted from one form to another among them. In the KAUS, all conversions are achieved via the representation of knowledge as is shown in Fig. 1.

Note here that there is no exact one-to-one correspondence among expressions in these different forms because of the difference in the expressive power of and redundancy contained in each representation. In general, the expressive power of the external representation is the largest among all. Therefore, it is often necessary to change a representation of information in the same formalism preserving its meaning until, at last, a representation is reached which has a corresponding representation in another form. This is the role of the deductive inference mechanism.

The advantage obtained by using the representation of knowledge is that it can be defined to have aspects very close to every other form so that the conversion to/from the other forms can be cleary defined.

Another advantage is in the fact that it can also be defined so that the inference algorithm can well be defined to it.

The mappings that the system should be provided with are those listed in Table 1.

In this paper, the outline of the SBDS is presented. The data bases viewed through representation of knowledge and the conversion algorithm from the latter to the former is discussed with a simple example.

2. REPRESENTATION OF KNOWLEDGE

From the discussion above and Fig.1, the requirements for the representation of knowledge are as follows:
(1) Conversion to/from external representation are possible and its algorithm can be obtained,
(2) Conversion to program is possible and its algorithm can be obtained,
(3) Conversion to/from data bases are possible and its algorithm can be obtained and
(4) Deductive inference algorithm can be well defined.

The basic configuration of the KAUS is as is shown in Fig.2 in which the first order predicate logic is adopted as the kernel concept of knowledge representation. We say it the kernel because, in the real system, it is extended towards higher order logic and also it is modified and combined with a concept of hierarchically structured universe to get high logical and physical performances. Since, however, our primary objective is to show how information is processed by being converted among different forms, we first discuss the simpler case of the first oder logic.

2.1 Representation of knowledge in the predicate form

We first show an example of representation of knowledge in the KAUS as follows:

$$(\forall x/PHYSOB)(\forall y/REAL)(\forall z/REAL)(\forall u/REAL)[(mass\ x\ y) \cap (vol\ x\ z) \cap (div\ u;\ y\ z) \Rightarrow (spc.grv\ x\ u)]$$

where each predicate has the following meaning ;
(mass x y): "mass of x is y",
(vol x y): "volume of x is z",
(div u; y z):"y/z=u" and

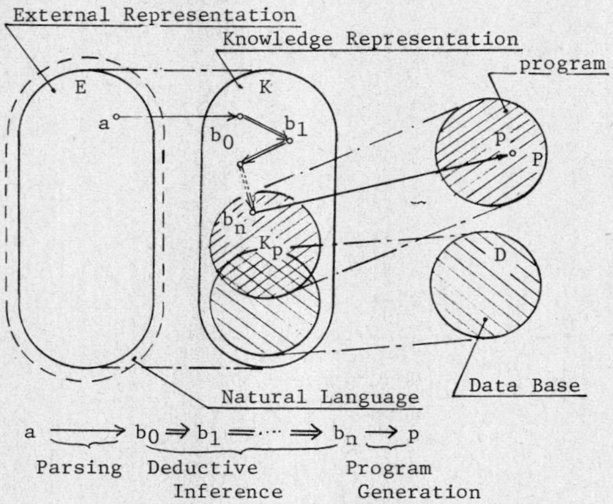

Fig.1 Relations among various representations

1	E→K : Parsing, Pattern Classification etc.
2	K→E : Output Generation
3	K→P : Program Generation
4	(K→D): Data Base Description
5	D→K : Generalization of Representation
6	K→K : Deductive Inference

Table 1 Necessary conversions

(spc.grv x u):"specific gravity of x is u". This formula is read as "for every physical object x and all real values y, z and u, if mass of x is y, volume of x is z and the quotient of y divided by z is u, then specific gravity of x is u." It gives the definition of "specific gravity". In general, definitions of concepts, mathematical theories, physical laws, facts and so on can be represented in the same style. The collection of such the formulas forms a knowledge base, and so, each formula as this example is called the knowledge element.

Note that, in this formula, a domain is explicitly specified to each variable and the formula is defined only over the specified domains of variables. It is very natural way of representation because these domains are brought in by the external words when a knowledge element is given by man in the form of external representation. In the KAUS, the logical structure of the matrix of a formula is representsed by an AND-OR tree as shown in Fib 3(b). Information on the variables are collected in a table called the variable table or V-table in short as is shown in the figure.

2.2 Knowledge structure

A universe is defined as the set of all objects referred in the system. Then a domain of variable in the predicate is one of its subsets. The universe itself forms a hierarchical structure based on the set-theoretical relations between sets in it.
In the KAUS this structure is represented by a graph

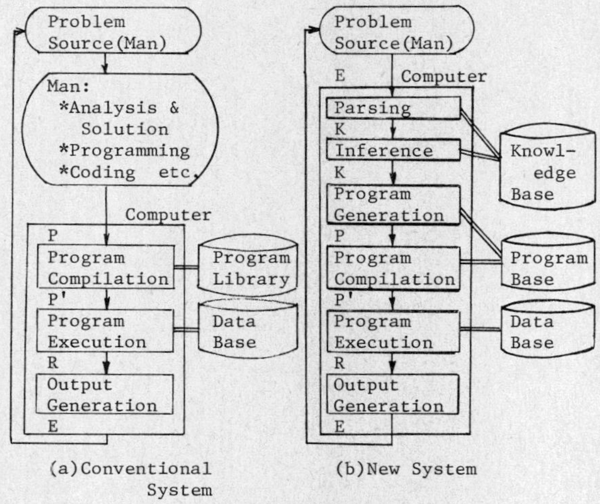

E; External Representation
K; Knowledge Representation
P; Program
R; Result

Fig.2 Conventional and new computer system

composed of nodes and two types of arcs. Each node represents a defined set and arcs represents two types of the set-theoretical relations between nodes; inclusion and disjointness. The structure thus formed is called the skeleton structure and is illustrated in Fig.3(a). To each node corresponds one or more external word(s). The association is made by a dictionary each entry of which contains a character string and a pointer to the corresponding node.

Every knowledge element is connected to one or more nodes in the structure in such a way that if the knowledge element has a domain X, then it is linked to the node X (ref. Fig.3 (b)).

2.3 Procedural Type Atom (PTA) and Non-procedural Type Atom

In a practical application, there may be a lot of knowledge elements in the system. Because the system must allow user to define and use his own predicate, the set of predicate symbols is kept open. However, some predicates of which predicate symbols are defined by some external words such as verbs, adjectives, prepositions and some nouns are defined in advance.

All atomic formulas or atoms in short are classified into two classes: procedural type atoms(PTAs) and non-procedural type atoms(NTAs). To each PTA a procedure is defined, which evaluates the logical value of the predicate as well as the value of some variable(s) in it. A few examples are as follows;

(add z; x y):"if x + y = z, then true",
(div z; x y):"if x/y = z, then true",
(g-than x y):"if x y, then true",
(sin y; x) :"if sin x = y, then true".

The set of these procedures forms the procedure base. Each procedure is evoked when a PTA in a formula becomes ready and is picked out for evaluation, that

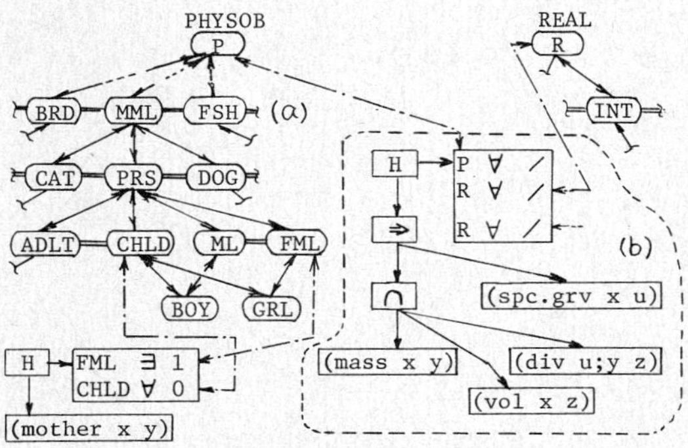

(∀y/CHLD)(∃x/FML)(mother x y)
(:"every child has mother")

I-Relation;
MML ⊃ PRS,CAT,DOG,---
PRS ⊃ ADLT,CHLD,ML,FML,--
CHLD ⊃ BOY, GRL,--
ML ⊃ BOY,--

D-Relation;
PRS ∧ CAT = ∅
PRS ∧ DOG = ∅
ML ∧ FML = ∅

BRD :BIRD
MML :MAMMAL
FSH :FISH
PRS :PERSON
ADLT:ADULT
CHLD:CHILD
ML :MALE PERSON
FML :FEMALE PERSON
GRL :GIRL
INT :INTEGER

Fig. 3 Structure of concepts and knowledge

is, a value has been given to every independent variable (underlined in the above examples) during the deductive operation as will be described in Sec.3. If a PTA includes one (or more) dependent variable(s), the result of the operation of the procedure defined to the PTA is substituted into it (them).

Besides these PTAs defined in the system, user can define his own PTA by writing and registering a procedure for it.

An NTA, on the other hand, is the one the logical value of which can not be obtained by itself. Its truthhood is proved if and only if it is implied by another formula that has already been asserted true.

3. DEDUCTIVE INFERENCE

3.1 Principle of deductive inference

Deductive inference algorithm is defined to the set of knowledge elements. There are two types of knowledge elements; simple assertion type written in the form of $A_1 \cap A_2 \cap --- \cap A_k$ and conditional type written as $B \Rightarrow C$ and read as "if B then C" or, in the logical term, "B implies C" where $B = B_1 \cap B_2 \cap --- \cap B_\ell$.

Suppose a query Q is presented. If it is proved that Q is implied by some A_i, $i \in \{1,2,\ldots,k\}$ of the assertion type knowledge element, i.e. $A_i \Rightarrow Q$, then Q is proved true. On the other hand, if Q is implied by the consequense of a conditional type formula, that is, $C \Rightarrow Q$, then the truthhood of Q depends on that of B. Hence Q can be replaced by B. We say that P satisfies the implicative condition to Q if either $P;A_1 --- \cap A_k$ and $A_i \Rightarrow Q$, $i \in \{1,2,--,k,\}$ or P; $B_1 \cap B_2 \cap -- \cap B_\ell \Rightarrow Q$ and $C \Rightarrow Q$.

In many cases, the query Q includes more than one atoms. Then, some atoms, say D_j, is selected out of them, and above test is applied to it. That is, if $A_i \Rightarrow D_j$ holds to the knowledge of the form $A_1 \cap --- \cap A_k$, then D_j is true. In this case, a label T is given to the node D_j in the AND-OR tree. On the other hand, if $C \Rightarrow D_j$ holds to the knowledge of the form $B_1 \cap --- \cap B_\ell \Rightarrow C$, then D_j is replaced by $B_1 \cap --- \cap B_\ell$ and the new AND-OR tree Q' is obtained to which the same procedure as above is repeated.

Thus the deductive procedure is formed of four major routines as shown in Fig.4 in which,

(1) SR(Selection Routine) selects some atom D_j out of atoms in the query,
(2) TIC(Test for Implicative Condition), given a query Q, searches and retrieves a knowledge element P that satisfies the implicative condition to the query,
(3) RR(Replacement Rule), given the query and the knowledge element P retrieved in the TIC above, replaces the old query by a new formula Q' which is obtained by modifying Q depending on P, and
(4) TT(Test for termination) evaluates the logical value of the query Q' above and determines if the process terminates.

3.2 SR

The SR selects an NTA in the query which includes the largest number of universally quantified variables or constants or both among all NTAs not evaluated yet. A variety of strategies can be applied at this stage to select one out of many (abbreviated here).

3.3 TIC

3.3.1 Test for the implicative relation

Suppose a couple of single atom formulas, say, P and Q, is presented. The TIC determines if the implicative relation, $P \Rightarrow Q$, holds between them by resolving the logical implicative relation into a set of set-theoretical relations between domains of each corresponding variables and as well into the condition on the orders of variabled in the prenex of each formula.

Let $P; (Q_{pi_1} x_{i_1}/X_{i_1}) \ldots (Q_{pi_n} x_{i_n}/X_{i_n})(F\ x_1 \ldots x_n)$ and
$Q; (Q_{qj_1}\ y_{j_1}/Y_{j_1}) \ldots (Q_{qj_n} y_{j_n}/Y_{j_n})(F\ y_1 \ldots y_n)$,

then $P \Rightarrow Q$ holds if

(1) for every pair of variables (x_i y_i), the condition shown in Table 2 holds, and
(2) there does'nt exist any such indices pair (i j) that (a) in P, the universally quantified x_i precedes the existentially quantified x_j, i.e. $P; \ldots .. (\forall x_i/X_i) \ldots (\exists x_j/X_j) \ldots (F\ x_1 \ldots x_n)$, while (b) in Q, the existentially quantified y_j precedes the universally quantified y_i, i.e., $Q; \ldots (\exists y_j/Y_j) .. .. (\forall y_i/Y_i) \ldots (F\ y_1 \ldots y_n)$.

This process is equivalent to the unification of the resolution principle.

Q_{pi}	$Q_{qi}(\overline{Q}_{qi})$	Conditions on Domains
∀	∀ (∃)	$X_i \supset Y_i$
∀	∃ (∀)	$X_i \wedge Y_i \neq \phi$
∃	∀ (∄)	Non Implicative
∃	∃ (∀)	$X_i \subset Y_i$

Table 2. Implicative rule

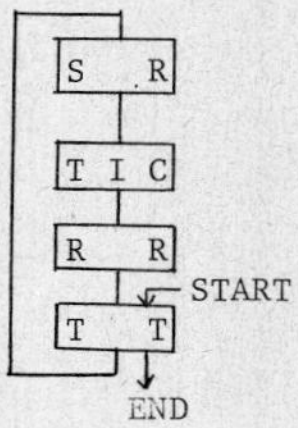

Fig.4 Deductive procedure

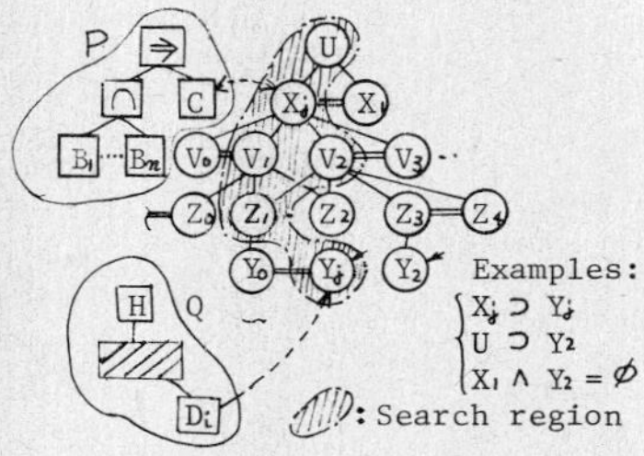

Fig.5 Searching a candidate and its test

	Q_{pi}	$Q_{qi}(\bar{Q}_{qi})$	$Q_{ri}(\bar{Q}_{ri})$	Domain of z_i
$z_i \in \mathbf{z}_0$	∀ ∀ ∃	∀ (∃) ∃ (∀) ∃ (∀)	∀ (∃) ∃ (∀) ∀ (∃)	Y_i $X_i \wedge Y_i$ X_i
$z_i \in \mathbf{z}_1$	- -	∀ (∃) ∃ (∀)	∀ (∃) ∃ (∀)	Y_i Y_i
	∃ ∀	- - - -	∀ (∃) ∃ (∀)	X_i X_i

Table 3. Rule for generating new formula

3.3.2 Searching the relevant knowledge element

The knowledge element that satisfies the implicative condition to the given query is effectively found in the knowledge base using the conditions in Table 2. Note that the set-theoretical relations in the table are the ones that are used to form the skeleton structure. Suppose some atom Di of the query is selected by SR and it contains a universally quantified variable y_j with the domain Y_j. If there is a knowledge element that satisfies the implicative condition to Di then its variable x_j at the corresponding position to y_j must also be universally quantified and its domain X_j must include Y_j (Table 2). All formulas satisfying this condition are connected to the node X_j in the skeleton structure, which can be reached by following the inclusion type arcs upwards starting from given Y_j (Fig. 5). Among these formulas, those of which the predicate symbols are the same with that of D_i have possibility to satisfy the implicative condition to D_i. We call such the knowledge elements the candidates to D_i. Since the skeleton structure expands rapidly down-word, the search region is considerably limited.

Next, suppose a candidate is found. Then all the remaining variables besides X_j in this formula must be tested for the implicative condition of Table 2. This time, both the query and the candidate are in hand and therefore, both domains of all corresponding pairs of variables are in hand. Then the set-theoretical relations between domains can be changed to the positional relations between corresponding nodes in the skeleton structure. Its algorithm is so simple that it can be realized in hard wired logic.
The TIC process is equivalent to the unification in the resolution principle.

3.4 RR

Suppose P of the form $B_1 \cap \cdots \cap B_\ell \Rightarrow C$ includes the set of variables, $\mathbf{x} = \{x_1 \cdots x_m\}$, of which the domain of each variable x_i is X_i, i=1, 2, ⋯, m. Let the single literal formula composed of the literal C and including only those variables contained in C be denoted $\hat{C}$. Suppose $\hat{C} \Rightarrow \hat{D}_j$ to D_j in Q of the form $D_1 \cap \cdots \cap D_\ell$ and Q includes the set of variables $\mathbf{y} = \{y_1 \cdots y_n\}$, $y_j \in Y_j$, j=1, 2, ⋯n. Then a new formula R is generated from P and Q according to the following rules of the RR. First, the matrix of R is obtained by replacing D_j in Q by $B_1 \cap \cdots \cap B_\ell$ of P. R inherits all variables included in P and Q except those which disappear by the replacement operation. Let those variables be divided into two classes: those included in C (and D_j) denoted $\mathbf{z}_0$ and the remainings denoted $\mathbf{z}_1$. Then the quantifier and the domain of each variable in R is obtained as shown in Table 3.

3.5 TT

The deductive procedure terminates either when the logical value of the query can be determined or when an AND-OR tree with no NTA is reached. An atom at the leaf of the AND-OR tree is labelled T meaning "TRUE" if it is evaluated as true while it is labelled F meaning "FALSE" if its negation is true. If no knowledge element could be found satisfying the implicative condition to the atom, then this is labelled U meaning "UNKNOWN". Then an intermediate node is evaluated using values of nodes below that by the rule shown in Fig.6. In this figure, the quantifiers at the left side and right side of slant symbol show the cases of the nodes being conjunctive node and disjunctive node respectively. Since the root node representes the query itself, if it is evaluated, then the deductive operation terminates. On the other hand, if there is no NTA remained in the query at all after the finite repetitions of deductive cycle, then the deductive process terminates too. Now, here is a question. Is every deductive process terminates in finite time?

Suppose a sequence of knowledge elements used for a deductive process for a given query is obtained. If, in this sequence, every predicate symbol once appeared in the consequence of a formula never appears in the premise of any of succeeding formulas, then the deductive process terminates in finite time because of the finiteness of the knowledge base. Resultant AND-OR tree with only PTA's corresponds to a program structure and it is not difficult to convert it to the program. To the contrary, if some predicate symbol ever appeared comes again in the succeeding or in the same formula, an endless repetition can happen to be. This is the case of recursive definition of knowledge element. Even in this case, however, if the recursive form is properly defined, the deductive procedure terminates in finite time. As an example, a recursive definition of the factorial function is given as;

(A) (fct 1 0)
(B) (∀x/I)(∀y/I)(∀m/I)[*(eql n 0) *(sub m:n 1) *(mult x:y n) (fct y m) = (fct x n)],

where (fct x n) means "n! = x" and * on the left shoulder of some atoms denotes that these atoms are PTA's. Its deductive process is shown in Fig.7 assuming that 2's factorial is asked by the query. The figure shows how the non-procedural form of the query is converted to the set of PTA's. Thus the deductive procedure acts also as the program generator.

n : node to be evaluated (∪ or ∩)
m_i: node below n (i = 1,2,....,k)

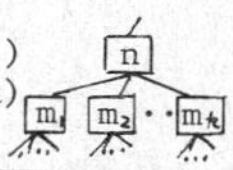

	(n:∩ / n:∪)		(n:∩ /n:∪)
a) if for ∀i	(m_i=T/m_i=F)	then	(n=T /n=F)
b) if for ∃i,s.t.	(m_i=F/m_i=T)	then	(n=F /n=T)
c) if for ∀i, for ∃j,s.t.	(m_i≠F/m_i≠T) (m_j=U/m_j=U)	and then	(n=U /n=U)

T: True, F: False, U: Unknown

Fig.6 Evaluation of nodes in AND-OR tree

4. ACCESS TO DATA BASES

4.1 Meaning and description of files

Thus we could realize the path from b_0 to p in Fig.1. Next we will discuss on the relationship between the representation of knowledge and databases. There are a number of different data models proposed within the database technology, each of which has own scheme for representing meaning. Here we will, among them, consider the relational model as an example.

In this model a relation is defined as a set(file) of n-tuple each of which is an instance of objects being in a relation f. Let i-th component in the n-tuple be confined to some set X_i. We say in this case, the i-th attribute of the relation f has the domain X_i. Then an n-tuple is considered as a point, in geometrical sense, in the Cartesian product space $X = X_1 x X_2 x$ ---- xX_n.

Suppose the set of all n-tuples in the relation f covers the whole space X. Then the relation f represents the same meaning as the single atom formula, $(\forall x_1/X_1)\ldots(\forall x_n/X_n)$(f $x_1 \ldots\ldots x_n$) .
In this case, a relation f in the relational database corresponds to the single atom formula in the logical form.
In general, however, the set is a proper subset of the product space. In this case a special procedural type atom denoted FGET is defined and used to describe the relation. It's format and meaning are,
(FGET R_i;x_1 x_2 ... x_n); "a tuple ($x_1 \ldots x_n$) is included in the file R_i (and retrieve it)". By using this,an arbitrary relation R can be defined and written as, $(\forall x_1/X_1)\ldots(\forall x_n/X_n)$[(FGET R ;$x_1 \ldots x_n$) ⇒ (f $x_1 \ldots x_n$)] that is read as "for every tuple ($x_1 \ldots x_n$)such that $x_1 \in X_1,\ldots,x_n \in X_n$, if it is included in a file R, then it satisfies the relation (f $x_1 \ldots x_n$)".

4.2 Basic database operators

A set of primitive operators is defined as is shown in Table 4 for the relational model. Note that the set of primitives are defined differently from that of the conventional $DBMS_s$ but the set has the same effect as the latter. For example, the conventional JOIN operation is formed, in this system, as a combination of an AND operation and two MULT operations.

In the ordinary database systems, a user manipulates the databases using the primitive operators. In oder for the knowledge system to be the intelligent interactive front end to the conventional database systems, it must be able to convert user's request represented in the knowledge form to a sequence of defined primitive operators.

```
Q:(∃x/I)(fct x 2)?:"What is 2's factorial?"

[0th] TT: Is there any atom that can be evaluated?
          --- No. Next
[1st] SR: D = Q (Select Q as D)
      TIC: (A) ⇏ D
           RIGHT HAND FORMULA OF (B)
                     (=(∀x/I)(∀n/I)(fct x n)) ⇛ D
           n ← z
      RR: ∼(eql 2 0)∩(sub m; 2 1)∩(mult x; y z)∩
           (fct y m)
       TT: ∼(eql 2 0)ⓐT, (sub m; 2 1)ⓐT, m ← 1
           then Q --> (mult x; y 2)∩(fct y 1)
          (eliminate atoms that have been evaluated)
[2nd] SR: D' = (∃y/I)(fct y 1)
      (Repeat similar procedure to [1st]'s and obtain)
      then Q --> (mult x; y 2)∩(mult y; z 1)∩(fct z 0)
[3rd] SR: D" = ( z/I)(fct z 0)
      TIC: (A) ⇛ D", z ⟵ 1, (fct 1 0)ⓐT
       TT:  Q --> (mult x; y 2)∩(mult y; 1 1)
            END
      where (---)ⓐT; (---) is labelled T,
            x ←α ; α is substituted to x,
            Q --> (Formula); Q is changed to (Formula).

(A) (fct 1 0)
(B) (∀x/I)(∀y/I)(∀m/I)(∀n/I)[∼(eql n 0)∩(sub m; n 1)
    ∩(mult x; y n)∩(fct y m) ⇛ (fct x n)
    where (fct x n); "n! = x"
```

Fig.7 Definition of factorial function and its evaluation through deductive operation

4.3 Generation of database access procedures

Suppose the deductive process terminates leaving a formula composed only of PTA's including FGET atom(s). Plural FGETs are then united to a single FGET to which the new file is defined through the set operations to the files in the original FGETs.

Suppose the formula Q is composed of N FGET atoms and represented as

$$Q\ ;\ (Q_1x_1/X_1)\ldots(Q_nx_n/X_n)[S\{(\text{FGET } R_j;\ x_{j_1}\ldots x_{jk_j})\}],$$
$$(j = 1,2,\ldots,N),$$

where Q_i denotes either ∀ or ∃ and $S\{Z_j\}$ represents a structure of AND-OR tree of this formula consisting of the literals Z_j, j = 1,...,N. Then it is transformed to the access procedure to the databases as follows:
(1) First of all, every FGET in the original formula is normalized. Let the set of all variables of this formula and the set of variables in the j-th FGET be $\mathbf{x} = \{x_1,\ldots,x_n\}$ and $\mathbf{x}_j = \{x_{j_1},\ldots,x_{jk_j}\}$ respectively. Then j-th file R_j in the j-th FGET is expanded by being multiplied by the domains of variables in $\mathbf{x} - \mathbf{x}_i$. It is denoted as $R_j^* = R_j \times (\mathbf{X} - \mathbf{X}_j)$ where $\mathbf{X}$ and $\mathbf{X}_j$ are the set of domains of variables in $\mathbf{x}$ and $\mathbf{x}_j$ respectively. Then the original (FGET R_j; $x_{j_1},\ldots,x_{jk_j}$) is equivalent to (FGET R_j^*; $x_1,\ldots,x_n$) in a logical sence provided the ordering of column in R_j is insignificant.
(2) After all FGETs were normalized, they are united to one FGET. A structure of set operations $\hat{S}\{\ \}$ is derived from the structure $S\{\ \}$ of the AND-OR tree by first replacing every conjunctive node by the set theoretical intersection and every disjunctive node by the set theoretical union and then replacing every FGET atom at the terminal node by the expanded file R_j^*, (j = 1,2,...,N). Then a sequence of set operations can be obtained from the structure thus obtained and it will produces a file R* defined in the space $X_1 \times \ldots \times X_n$. That is, $R^* = \hat{S}\{R_j^*\} = \hat{S}\{R_j \times (\mathbf{X} - \mathbf{X}_j)\}$.

Then,
$(Q_1x_1/X_1)\ldots(Q_nx_n/X_n)[S\{(\text{FGET } R_j^*;x_j,\ldots,x_{jk_j})\}]$ is logically equivalent to
$(Q_1x_1/X_1)\ldots(Q_nx_n/X_n)(\text{FGET } R^*;\ x_1,\ldots,x_n)$------(*A).

(3) Then this formula is transformed to a sequence of the basic operators listed in Table 4 using the following equations;

$$(Q_1x_1/X_1)\ldots(Q_{n-1}x_{n-1}/X_{n-1})(\forall x_n/X_n)(\text{FGET } R^*;x_1,\ldots,x_n)$$
$$=(Q_1x_1/X_1)\ldots(Q_{n-1}x_{n-1}/X_{n-1})(\text{FGET DIV}(R^*\ X_n);x_1,\ldots,x_{n-1})$$
$$(Q_1x_1/X_1)\ldots(Q_{n-1}x_{n-1}/X_{n-1})(\exists x_n/X_n)(\text{FGET } R^*;x_1,\ldots,x_n)$$
$$=(Q_1x_1/X_1)\ldots(Q_{n-1}x_{n-1}/X_{n-1})(\text{FGET PRJ}(R^*\ X_n);x_1,\ldots,x_{n-1})$$

These equations are applied to the formula until finaly the form of (FGET $\hat{F}$;ϕ) is reached where $\hat{F}$ is a sequence of basic operators applied to R*. $\hat{F}$ has value either 1 or 0 depending on whether the derived sequence of operations ends successfully, that is, ends without having resulted in the empty set (file) or ends unsuccessfully. Here, if $\hat{F}=1$, then the formula (*A) and, therefore, the original query is proved true, while, if $\hat{F}=0$, then it is untrue.

(4) Though $\hat{F}$ gives a mapping from the knowledge element to a database operation, this is not always an optimal one, but it can include redundant file operations and generate ineffective intermidiate files.

It is, therefore, optimized. The optimizing rules listed in Table 5 are used. These rules exchange the oder of operators in F to minimize the ineffective operations and are equivalent to that of [16].

<u>Example</u>

$$Q=(\forall x/X)(\exists y/Y)(\forall z/Z)[(\text{FGET } F_1;\ x\ y)\cap(\text{FGET } F_2;\ y\ z)]$$
$$=(\forall x/X)(\exists y/Y)(\forall z/Z)[(\text{FGET } F_1\times Z;\ x\ y\ z)\cap(\text{FGET } F_2\times X;\ x\ y\ z)]$$
$$=(\forall x/X)(\exists y/Y)(\forall z/Z)[(\text{FGET } F_1\times Z\wedge F_2\times X;\ x\ y\ z)]$$
$$=(\text{FGET DIV(PRJ(DIV(AND(MULT}(F_1\ Z)\text{MULT}(F_2\ X_1))Z)Y)X);\phi)$$
$$=(\text{FGET DIV(PRJ(AND}(F_1\ \text{MULT(X DIV}(F_2\ Z)))Y)X);\phi)$$

(5) In real problems, both FGET type atoms and ordinary PTAs may be included in an AND-OR tree. Then ordering of atoms is necessary because a PTA can be evaluated when a value is assigned to every independent variable and, if some inedpendent variable of a PTA is also the dependent variable of another PTA, then the latter PTA must precede to the former.

Note here that to execute a FGET type atom has the same effect as to assign values to variables x_i ,----,x_i included in the FGET. We call such the variables the evaluated variables. A PTA of which every independent variable is the evaluated variable can be evaluated. Inversely, it is possible o select minimum set of FGETs in the AND-OR tree such that the union of the set of all variables included in them and the set of already evaluated variables include the set of all independent variables of some PTA(s). To such the set of FGET atoms, the procedure (1) through (4) above are applied and to the end of the subsequence of file operations thus obtained the PTA(s) is appended. After then the set of the variables in those FGETs and the dependent variables of the PTA(s) are added to the set of evaluated variables. This process is repeated until all atoms are transformed to the sequence of operations.

(6) When the programming stage ends, the system enters the execution stage. Some variable(s) will be used to generate the answer and, therefore, must be saved to the end of the execution of the program. However, both DIV and PRJ erase, by definition, not only all the tuples that do not satisfy the conditions shown in Table 4 but also the key variables (and their values) irrespective of the variables being asked by the query or not, where by the key variable we mean the variables used as the divider in the DIV and the one used for projection in the PRJ. In order to save these variables, another set of oerators denoted as VPIV and VPRJ are provided. These are defined similarly to DIV and PRJ respectively except they do not erase the key variables but give them a special flag.

At the programing stage each variable is checked and determined if its values are to be saved or they can be erased at the execution stage. There are two cases of a variable to be saves. If it is included in the query then it should be saved until the answer is generated and (2) if a variable is included in some PTA as the independent variable but is not directly included in the query, then it is saved until the PTA is fabricated in the generated program.

5. SIMPLE EXAMPLE

The system thus obtained satisfies the basic requirements for it being an intelligent interactive system. To show how the system works, we will present a very simple example.

Suppose a set of knowledge elements(1) through (5) shown in Fig.8 exists in the knowledge base,where (1)and (2)represent facts such as "everything lighter than fluid float on it "and" the specific gravity of an object can be obtained by dividing the weight of the object by its volume" respectively.

The system is assumed to have three different files representing the relations holding between "sample

i) $\text{AND}(F_1,F_2) = \{(x_1^i\ldots x_n^i)/(x_1^i\ldots x_n^i)\in F_1\wedge F_2,\ i=1,2,\ldots\}$

ii) $\text{OR}(F_1,F_2) = \{(x_1^i\ldots x_n^i)/(x_1^i\ldots x_n^i)\in F_1\vee F_2,\ i=1,2,\ldots\}$

iii) $\text{MULT}(F_1,Y) = \{(x_1^i\ldots x_n^i,y^j)/(x_1^i\ldots x_n^i)\in F_1,y^j\in Y,\ i=1,2,\ldots,\ j=1,2,\ldots\}$

iv) $\text{DIV}(F_1,x_k) = \{(x_1^i\ldots x_{k-1}^i,x_{k+1}^i\ldots x_n^i)/(x_1^i\ldots x_n^i)\times X_k\in F_1,\ i=1,2,\ldots\}$

$\text{DIV}(\hat{X}_k,X_k) = 1$ (if $\hat{X}^k\supset X^k$) or 0 (if $\hat{X}_k\not\supset X_k$)

v) $\text{PRJ}(F_1,X_k) = \{(x_1^i\ldots x_{k-1}^i,x_{k+1}^i\ldots x_n^i)/(x_1^i\ldots x_n^i)\in F_1,\ i=1,2,\ldots\}$

$\text{PRJ}(\hat{X}_k,X_k) = 1$ (if $\hat{X}_k\wedge X_k\neq\phi$) or 0 (if $\hat{X}_k\wedge X_k=\phi$)

vi) $\text{CLP}(F_1,X_k) = \{(x_1^i\ldots x_k^i\ldots x_n^i)/(x_1^i\ldots x_k^i\ldots x_n^i)\in F_1,\ x_k^i\in X_k,\ i=1,2,\ldots\}$

vii) $\text{EXPD}(F_1,X) = \{(x_1^i\ldots x_n^i\ x^i)/(x_1^i\ldots x_n^i)\in F_1,x^i\in X,\ i=1,2,\ldots\}$

where
F_1 ; $\{(x_1^i\ldots x_n^i)/x_r^i\in X_r,\ 1\le r\le n,\ i=1,2,\ldots,N_1\}$
F_2 ; $\{(x_1^i\ldots x_n^i)/x_r^i\in X_r,\ 1\le r\le n,\ i=1,2,\ldots,N_2\}$

<u>Table 4 Primitive database operations</u>

(1) $\text{DIV(MULT(X,F),Y)} = \begin{cases}\text{MULT(X,DIV(F,Y))}, & \text{if } X\neq Y\\ \text{F}, & \text{if } X=Y\end{cases}$

(2) $\text{DIV(AND}(F_1,F_2)\text{,X)} = \text{AND(DIV}(F_1\text{,X),DIV}(F_2\text{,X))}$

(3) $\text{PRJ(MULT(X,F),Y)} = \begin{cases}\text{MULT(X,PRJ(F,Y))}, & \text{if } X\neq Y\\ \text{F}, & \text{if } X=Y\end{cases}$

(4) $\text{PRJ(OR}(F_1,F_2)\text{,X)} = \text{OR(PRJ}(F_1\text{,X),PRJ}(F_2\text{,X))}$

(5) $\text{DIV(OR}(F_1\text{,MULT}(X,F_2))\text{,X)} = \text{OR(DIV}(F_1\text{,X)},F_2)$

(6) $\text{PRJ(AND}(F_1\text{,MULT}(X,F_2))\text{,X)} = \text{AND(PRJ}(F_1\text{,X)},F_2)$

(7) $\text{AND(MULT}(X,F_1)\text{,MULT}(X,F_2)) = \text{MULT(X,AND}(F_1,F_2))$

(8) $\text{OR(MULT}(X,F_1)\text{,MULT}(X,F_2)) = \text{MULT(X,OR}(F_1,F_2))$

<u>Table 5 Optimizing rules</u>

(material) number and its (measured) weight","sample (material) number and its (measured) volume" and sample (liquid) number and its specific gravity" respectively. The formula (3) through (5) are descriptions of these files in the logical form. These are a portion of system's knowledge and user is not required to be aware of them. He can only ask the system like "Which sample material floats on all liquid sample?" (Q in Fig.8). Since there is no such data file (description) that contains data to reply this query directly, the system try to resolve the query into another formulas with equivalent meaning. "float on" is equivalent to a certain relation to hold between the specific gravities of two objects (Q1). The system tries to find answer to this new query but fails again because there is no data file on the specific gravity of the sample materials. Therefore, it resolves the query again and replaces the specific gravity of the sample material by its equivalent definition, i.e., the relation between the weight and the volume of the materials (Q2). This time, both weight and volume of materials are given in the files, and Q3 is obtained. Since Q3 contains no NTA at all, the deductive process terminates and a program including the file manipulations is obtained. This is converted to the database operations.

Fig.9 shows the results of this example obtained by our experimental system. This result is optimal in the sense that no ineffective file operation is achieved.

Knowledge:

$$(\forall x/P)(\forall y/F)(\forall z/R)(\forall u/R)[(\text{spc-grv } x\ z) \cap (\text{spc-grv } y\ u) \cap (\text{less-than } z\ u) \Rightarrow (\text{float } x\ y)] \quad (1)$$

$$(\forall x/P)(\forall y/R)(\forall z/R)(\forall u/R)[(\text{mass } x\ y) \cap (\text{volume } x\ z) \cap (\text{div } u;\ y\ z) \Rightarrow (\text{spc-grv } x\ u)] \quad (2)$$

$$(\forall x/P)(\forall y/R)[(\text{FGET } F_m;\ x\ y) \Rightarrow (\text{mass } x\ y)] \quad (3)$$

$$(\forall x/P)(\forall y/R)[(\text{FGET } F_v;\ x\ y) \Rightarrow (\text{volume } x\ y)] \quad (4)$$

$$(\forall x/L)(\forall y/R)[(\text{FGET } F_1;\ x\ y) \Rightarrow (\text{spc-grv } x\ y)] \quad (5)$$

Q: $(\exists x/P)(\forall y/L)(\text{float } x\ y)$; "which material floats on all liquid samples" (6)

Q_1: $(\exists x/P)(\forall y/L)(\exists z/R)(\exists u/R)[(\text{spc-grv } x\ z) \cap (\text{spc-grv } y\ u) \cap (\text{less-than } z\ u)]$
(from (1) and Q)

Q_2: $(\exists x/P)(\forall y/L)(\exists z/R)(\exists u/R)(\exists v/R)(\exists w/R)[(\text{mass } x\ z) \cap (\text{volume } x\ u) \cap (\text{div } v;\ z\ u) \cap (\text{spc-grv } y\ w) \cap (\text{less-than } v\ w)]$ (from (2) and Q_1)

Q_3: $(\exists x/P)(\forall y/L)(\exists z/R)(\exists u/R)(\exists v/R)(\exists w/R)[(\text{FGET } F_m;\ x\ z) \cap (\text{FGET } F_v;\ x\ u) \cap (\text{div } v;\ z\ u) \cap (\text{FGET } F_1;\ y\ w) \cap (\text{less-than } v\ w)]$
(from (3), (4), (5) and Q_2)

Fig.8 An Example of query processing

```
[EX/PU][AY/LIQ](FLT, $X, $Y)?
[1]   (#FGET   #WT    PU1    RNUM9    *WRK1)
[2]   (VPRJ    *WRK1  RNUM9  *WRK1)
[3]   (#FGET   #VOL   PU1    RNUM10   *WRK2)
[4]   (VPRJ    *WRK2  RNUM10 *WRK2)
[5]   (AND     *WRK1  *WRK2  *WRK3)
[6]   (EXPD    *WRK3  RNUM4  *WRK3)
[7]   (#DIV    RNUM4  RNUM9  RNUM10   *WRK3)
[8]   (PRJ     *WRK3  RNUM9  *WRK3)
[9]   (PRJ     *WRK3  RNUM10 *WRK3)
[10]  (#FGET   #SW    LIQ2   RNUM5    *WRK1)
[11]  (VPRJ    *WRK1  RNUM5  *WRK1)
[12]  (VPRJ    *WRK3  RNUM4  *WRK3)
[13]  (MULT    *WRK1  PU1    *WRK1)
[14]  (MULT    *WRK3  LIQ2   *WRK3)
[15]  (AND     *WRK1  *WRK3  *WRK2)
[16]  (#GT     RNUM5  RNUM4  *WRK2)
[17]  (PRJ     *WRK2  RNUM4  *WRK2)
[18]  (PRJ     *WRK2  RNUM5  *WRK2)
[19]  (DIV     *WRK2  LIQ2   *WRK2)
[20]  (VPRJ    *WRK2  PU1    *WRK2)
[21]  (OUT     *WRK2  *TTY)
```

Fig.9 Result of example problem

6 CREDIBILITY OF INFORMATION

In many cases, information in the real world contains ambiguity to some extent and also is not fully credible. In the KAU system, ambiguity is considered as a kind of incredibility. To utilize incredible information, its credibility must be evaluated.

There are two lines of approaches to evaluate the credibility. In the first approach the logical value of information and its credibility is inseparable and are evaluated at the same time. The multi-valued logic, the Fuzzy logic and so on belong to this class of approaches. The second method evaluates the credibility of information independent of logical value. The probability method belongs to this approach. According to the probability theory, a probability is defined to each (logical) event which is obtained as a consequence of logical operation. The KAU system adopts the latter approach because the former approach is anticipated to cause the system unnecessarily complicated. With the probability method, every logical formula is considered as an event and is assigned a real value in [0,1] as a credibility of the formula. The rule of evaluating the credibility decrement caused by the change of the formula at the deductive operation is defined. There are at least four causes of deteriorating the credibility for information. These are;

(1) incredibility of knowledge element itself because of ambiguity in the knowledge source,
(2) ambiguity implied in the conditional type knowledge element, that is, the uncertainty contained in B in $A \Rightarrow B$ even if A is certain,
(3) fuzziness existing in the set theoretical inclusion, $x \in X$, and
(4) error contained in each data in databases such as observation error.

The ambiguity implied in the conditional knowledge of (2) above is defined as the conditional probability and denoted P(B/A). Then, if the credibility of A alone is P(A), then that of B alone, P(B), is obtained by the rule of conditional probability as $P(A)\cdot P(B/A)$. The fuzziness of (3) is just the definition of the membership function in the fuzzy set theory. Though it is not the same as the probability in the strict sense, we regard them the same and apply the probability theory as an approximation.

Then the credibility of the given query is evaluated in parallel with the evaluation of its truth value in such a way that, each time information in the system is used in the deductive process, the credibility of the information is multiplied to update the intermediate value of the credibility of which the initial value was one. When the logical value of the query is determined, the current intermediate value gives its credibility.

7. EXTENSION OF FORMALISM

The first order logic is very suitable for representing knowledge, in partisular, for defining conversions, K-K, K-P and K-D in Table 1, as has been shown up to now. Its expressive power, however, is not good enough to allow rich class of the external representations. So we need to expand it. We first define the minimal set of allowable external expressions denoted here E* as follows;

(1) E* should be large enough so that every expression necessary for representing knowledge for intended application is to be included, and
(2) E* should be such that any user can easily understand without any special training how to express his knowledge or query by using it.

The fact that most knowledge element can be expressed in natural language tells us that system should be able to accept all basic syntactic patterns of language. Since every complex sentence can be analyzed to the logically connected set of basic sentences, the necessary condition of the system for representing knowledge is that the system can accept basic patterns of sentences and phrases.

In case of English, the simple sentences can be classified into about twenty-five classes of different patterns. Only one-third of those patterns can be represented in the framework of the first order predicate.

To expand the framework for representing the rich class of knowledge element, we extend the definition of atoms. So far, an atom is defined as an ordered set of simple variables preceded by the predicate symbol. The first extention is to allow the predicate symbol itself to be a variable. By this extention, a query of which the predicate symbol is the interrogated term can be represented. An example is " In what relation is A to B?" represented in a form ($\exists$x/RELATION)(x A B)?.

The second extension is to allow such an atom of which some positions are occupied not by simple variables but by some formulas. That is, the atom of the form (Variable$_0$, Variable$_1$/(Formula)$_1$,...,Variable$_n$/(Formula)$_n$) is allowed where Variable$_i$/(Formula)$_i$ means that either a simple variable or a formula can occupy i-th position. An example is ($\exists$x/PERSON)(know #Tom ((sick x)$\cap$(family x #Jim)));"Tom knows that someone of Jim's family is sick.".

Accompanying to these extensions in the formalism, the processing algorithm must also be modified. Especially to modify the deductive algorithm is inevitable. This modification in algorithm is not so difficult. But its foundation is on a deep theoretical consideration on the structure of the world treated in the system. Since the topics is too specific for the purpose of the paper, we only mention here that, according to the extention of the formalism, the TIC of the deductive algorithm is mainly affected. Let
$P;(Q_{pi_1} x_{pi_1}/X_{pi_1})\cdots(Q_{pi_n} x_{pi_n}/X_{pi_n})(x_1 x_2 \cdots x_m)$ and
$Q:(Q_{qj_1} y_{qj_1}/Y_{qj_1})\cdots(Q_{qj_n} y_{qj_n}/Y_{qj_n})(y_1 y_2 \cdots y_m)$.
Where each of $x_1,\ldots,x_n$, $y_1,\ldots,y_n$ is either a simple variable or a formula. As a formula includes more than one variables, n is equal to or larger than m. Then the rule (1) in Sec. 3.3.1 for the condition of P$\Rightarrow$Q is modified to;

(1') for every corresponding pair (x_i y_i), either the condition of Table 2 is satisfied or the implicative condition (x_i)$\Rightarrow$(Y_i) holds depending on both x_i and y_i being the simple variables or the inner formulas.

To realize this algorithm, the control structure of the deductive procedure is modified so that the TIC is recursively applied to the nest form of representation.

8. CONCLUSION

The outline of the first version of the KAUS(Knowledge Acquisition and Utilization System) has been presented This system is designed to bridge over the gap between man and computer and to aid man in his intelligent works. In this system, information is processed by being transformed among four different forms; external representation, program, databases and knowledge representation. The mappings between different forms and the deductive procedure to the representation of knowledge have been designed.

An experimental system has been implemented, which realized some important basic procedures as has been described in Sec.2 through Sec.6. The procedure to evaluate the credibility of information and the expansion of the framework are not realized yet but is in planning stage.

REFERENCES

[1] Bobrow, D.C.(chaired) "A Panel on Knowledge Representation "In Proc. 5th IJCAI,1977, pp983-992.
[2] Chang,C.L. $ Lee, R.C.T. "Symbol Logic and Mechancal Theorem Proving" Academic press, 1973.
[3] Codd,E.F. "Recent Investigarions in Relational Data Base System", Proc. IFIP Congress, Aug. 1974.
[4] Date C.J. "An Introduction to Database Systems", Addison-Wesley, 1975.
[5] Emden Van "Programming with Resolution Logic", Macgine Intelligence 8, pp266-299.
[6] Feigenbaum E.A. "The Art of Artificial Intelligence, Themes and Case Studies of Knowledge Engineering, 5th IJCAI pp1014-1029.
[7] Green C. "The Use of Theorem Proving Techniques in Question Answering System", Proc. 23-rd Nat. Conf. ACM, 1968.
[8] Huet G.P. "A Mechanization of Type Theory" 3rd IJCAI 1973. pp139-146.
[9] Kellog C.H. & Klahr P. "Adding a Deductive Capability to a Data Management System", Second VLDB Sept. 1976.
[10] Kowalski R.A. "Predicate Logic as Programming Language", Proc. IFIP-74. pp569-574.
[11] Krivine J. "introduction to Axiomatic Set Theory", D.Reidel pub. Co. 1971.
[12] Ohsuga S. "Semantic Information Processing in Man-Machine Systems", Proc. 1977 IEEE Conference on Decision & Control. pp1351-1358, Dec. 1977.
[13] Ohsuga S. "Towards Intelligent Interactive Systems", Proc. the IFIP W.G.5.2 Workshop Ssilac on Methodorogy of Interaction (to apper) North-Holland Pub. Co. 1979.
[14] Robinson J.A. "A Machine Oriented Logic Based on the Resolution Principle", JACM Vol. 12 Jan. 1965.
[15] Shapiro S.C. "Path-Based and Node-Based Inference in Semantic Networks", TINLAP-2. Jan. 1978 pp219-225.
[16] Smith J.M. and Chng P.Y.T. "Optimizing the Performance of a Relational Algebra Databases Interface", CAMA vol.18,No.10, 1975, pp568-579.

Scientific Information Systems in Japan
H. Inose (ed.)

USE OF HANDWRITING ACTION IN CONSTRUCTION OF MODELS

Mamoru HOSAKA and Fumihiko KIMURA***

** Institute of Space and Aeronautical Science, University of Tokyo, Komaba, Meguro-ku, Tokyo*
*** Faculty of Engineering, University of Tokyo, Hongo, Bunkyo-ku, Tokyo*

This paper treats uses of handwriting action of drawing as direct input to a computer for constructing in it the internal description of an object which has the same meaning to the machine as the external drawing has to man. Various aspects and functions of drawings and their elements are analyzed to fully utilize the information on the drawing. A method of recognition of handwriting actions and their locations is described, which is insensitive to variations of symbol shapes. The various uses of recognized symbols and patterns are explained. To show the effectiveness of our method, two nontrivial examples are selected. One is general geometrical model building and processing, the other is drawing description of turning operation planning.

1. INTRODUCTION

Two dimensional expressions of information, such as figures, graphics, mathematical expressions, charts and diagrams and etc., are widely used in our daily works. Because it is easier for us to express our integrated concepts in such forms concisely and also to make their meanings understandable to other people than to use language for describing them. If we could use these expressions directly as input to the computer, our informatic activities will be very much helped by the computer. In order to make this feasible, we have to solve various problems involved. To grasp the problems clearly, we had decided to treat engineering drawings. Because they contain various types of two dimensional expressions, structures and meanings of which cannot easily be described by usual computer languages. And if our objective described below can be attained, its effects on industries will be large.

The main objective of this study is to develop a method of using handwriting actions of making engineering drawings and their records on a sheet of paper as input data and commands to the computer for constructing internal models of objects to be designed, which will be used in future engineering data bases of CAD and CAM. The most essential task in data base construction is consistent description of a world which the data base is to represent. When differences in expressions of description for man and for the machine are great, tasks of making internal description become difficult. For example, engineering drawings cannot be understood by the machine even if their precise copies could have been stored in it. Accordingly man has to describe the meanings conveyed by the drawings so as to be processed by the machine. This is again very difficult and cumbersome tasks requiring both professional knowledge and excellent programming techniques.

As importance of the model building of three dimensional objects has been recognized for constructing integrated CAD/CAM systems, the research works for it are being carried on at several places in the world. Our research work is one of the most comprehensive and advanced ones, and its software package called GEOMAP has been transferred to several places in Japan and abroad. Our current and future study is to add technological information to the models by writing necessary data on engineering drawings, and to make the models as data elements of engineering data bases. So far as we know, the other places are contemplating to proceed almost in the same direction, but they do not seem to have attained the level we had reached in many respects.

As design activity is a kind of problem solving process with its goal and constraints being given, designer's creativity is highly required to attain optimally its goal. Though the activity involves many routine tasks, such as calculation, drawing, referencing etc., drawing processes are used throughout to make designer's own ideas concrete and to proceed his designing further and to communicate with other people. The final main product in the designing is the engineering drawings with their associate infomation, which contain the detailed specification of the object to be manufactured, and are the principal media to hold and convey the correct design information. Thus the engineering design without the drawing can hardly be considered so long as man is engaged in it.

The engineering drawing of an object can be looked upon as a model of the object, which is understood only by man who knows the general rule of drawings and can form a clear mental picture of the object. Even if a machine could recognize their patterns on the sheet, it would not be able to understand the object the drawing represents. We cannot expect that the machine could read and understand the drawing just as man can do. If we want to make the machine understand the drawing, we have to construct in the machine a model of the object which gives the same meanings to the machine as the drawing does to the man. If this construction is realized, the machine can further process the model and extract the necessary information for subsequent production preparing processes, just as man can do the same tasks with the drawings.

In our CAD system, it is not our aim to make a precise drawing with the so-called graphic language, but to construct an internal model of the object, from which the precise drawing can be made when required. We want to communicate our intention of designing to the machine by using the simplified drawing of the object, for it is cumbersome and error-prone for us to use programming languages to dercribe the internal model of the three dimensional object. Drawing is, so to speak, a designers' language with which they can express their conceptual models on a sheet of paper and define their shapes, dimensions, functions, constraints, materials and manufacturing methods. Though there are general rules and standards of the engineering drawing, they are not sufficient for understanding or model construction by the machine and frequently additional information is necessary for them. Accordingly we had studied an internal description method of drawing information and then, their use in constructing an internal model of the objects.

We first analyzed styles, structures, elements, symbols used in a drawing and then create a method of expressing constraints and construction sequences. We had to develop an on-line recognition method of symbols handwritten on the drawing sheet, which is used for communication with the machine. As the application fields of our method are considered to be wide, we select in this paper two examples of general and special applications to show the effectiveness of our approach. One is rather complicated general three-dimensional geometrical processing, and the other is generation of information for turning operation planning from freehand drawing input.

2. STRUCTURE OF DRAWINGS AND FUNCTIONS OF SYMBOLS

2.1 Problems to be Solved in the Use of Drawings

Description with a language is one-dimensional, but there are intrinsically two-dimensional descriptions such as engineering drawings, diagrams, tables, graphs, maps, flow-charts, sketches etc. We call them simply 'drawings' in this paper. They are very important styles of description for human use. They do not have precise grammar as computer languages have, but they have the general drawing rules, sometimes the symbol standerdization. And man can understand their meanings correctly not only with these rules but also with relation to their environments, associations and references to other symbols or descriptions. The drawings cannot become direct input data for the computer which does not have the ability of recognizing complex patterns, much less the knowledge necessary to understand the meaning of the drawings.

As the engineering drawing is made by a sequentially defining process of the object, it can be described by an appropriate command sequence. However, our objective is not to use command languages with parameters, because the command sequence can only be written by man who carefully observes and analyzes the drawings. Accordingly, we thought it better to use directly man's drawing actions as input to the machine. We use machine recognition of man's drawing actions and at the same time we supply necessary information which is also given by drawing or writing on the appropriate parts of the sheet or by referring to them. Then we construct the description of man's intention of the drawing with the accompanying information. With these data the machine can construct or process the internal model of the object. In order to realize the above stated tasks, the following studies have to be performed.

(1) Category of drawings and their semantics and syntax have to be analyzed. Methods of describing them in the machine have to be established.

(2) Auxiliary information which is necessary only for machine must be found. A method of its description and transmission has to be settled.

(3) Simplified drawing rules, which are easy to use and preserve features easily recognizable by man, have to be defined.

(4) An algorithm of recognizing handwriting symbols has to be devised, which is insensitive to variations of each symbol shape.

2.2 Category and Elements of Drawing

Drawing consists of unit information - symbols, character strings, graphic elements - distributed on a sheet. Groups of them represent certain meanings. Generally speaking, there are three types of representation of drawings: fixed form, non-fixed form and geometrical form.

In the fixed form representation, form of unit information and their classes are predetermined. Man writes in their entries the unit information of the specified class. Tables, arrays, records and simulated registers belong to this category.

In the non-fixed form representation, relations among records are important as well as records themselves. Relations are explicitly indicated on the drawing. Graphs, list structures, block diagrams, networks and flow charts belong to this category.

Description of drawings of the above stated categories by computer languages is not difficult for man, for he can easily recognize their meanings. But the reverse is not easy for him, therefore its description errors are very difficult to find. If machine recognition of symbols and their location is possible, the conversion of a drawing of these cateogories to the internal description is not difficult for the machine. Therefore direct use of the drawing is desirable.

Characteristics of geometrical representation is quite different from those of the above stated categories. By using graphical elements such as lines and curves, projected shapes of real objects are shown on the sheet. Auxiliary lines and attached symbols are written to define the elements or to make their meanings clear. The drawing rules and symbols used are different according to fields of engineering and applications, though their basic geometrical relations are held.

In the descriptive geometry, the model building and processing are performed graphically, therfore the drawing must be as accurate as possible. Whereas, if computer is to be used for them, drawings should be used only for input information. In this case, dimensional accuracy of the drawing is not necessarily required, but the defining information and the processing operations applied have to be written in the drawing so as to be easily observed by man and recognized by the machine.

In the engineering drawings, simplified or symbolic drawings are widely adopted. Various symbols are attached to graphical elements to show geometrical constraints, dimensions, materials to used and machinig properties. Most of these symbols can be recognized and used directly by the machine, but some ones have to be supplemented with auxiliary information, as they refer implicitly to other information which can easily be recognized. From these drawing data and supplied auxiliary data, the internal description of the drawing and the internal model of the object can be constructed. This construction sequence must be recorded in the simulated registers on the sheet, and they constitute the process progress record which is used by the machine in the processing.

2.3 Unit Information on Drawing

The unit information is categorized as follows.
(1) Names and values attached to graphical elements of an object, those written in entries of records, and names of a drawing sheet and its subdivisions and control figures. Sometimes an 'indicating arrow' symbol is written between the element and defining symbol of unit information.
(2) Mathematical expressions such as fraction, square root, subscript etc.
(3) Symbols to indicate certain operations, operands, relations and groups of elements, displacement directions and regions where certain features exist. These symbols or marks are written when their associated actions are to be taken. Therefore, some of their sizes are not fixed. As man can recognize two dimensional global features easily, the machine operations concerning to these features are greatly simplified with these information given to the machine by man. Similarly a certain region or a range is easily indicated by man.
(4) Control points. It is possible to connect a certain action to the touching of a corresponding control point of the sheet. An explanatory figure is used to indicate the control points. This is a function figure corresponding to function keys.
(5) Commentary information. Man can write on the sheet any comment information for easy understanding in any forms, styles or colours. This information does not affect the machine operations. Frequently in processing of three dimensional objects he has to draw shapes of component bodies to indicate their existence whose dimensional values are given by the other means. Frames of records also belong to the same category.

2.4 Preparation for Drawing Input and Functions of the System Routines

As it is difficult to use statical drawings as computer input, we use an on-line recognition method in which man's handwriting actions are measured with a graphic tablet and an accompanying pen. The recognition algorithm is described in the next section. A sheet of paper on the tablet is the working area for drawings or instructions. The system routine first prepares the following control information and connects them to specific application programs: sheet name, division of sheet area, category of a drawing in it, its name, record name and format definition, function figures and menu definition, description of new symbols to be recognized. The system software includes the following routines: on-line recognition and action routines which are indicated with special symbols, such as searching of an object pointed to by an arrow symbol, or objects encircled by a loop, deletion of drawings and changing of defined base lines etc.

When man manually constructs and processes an object, he observes and understands what he did already and he can proceed his drawing. Similarly when machine processing is required during the interactive drawing, the response from the machine must be written by man in the entry of the process progress record whose uses must be established beforehand. Observing this record and his drawing, he can continue his subsequent tasks. When new class of problem is to be solved with the drawing method, man has to define methods of communication by drawing and a form of sequential progress record. And he must prepare to convert input data to its internal description by use of available system routines and finally his application programs. These processes constitute the methods of the meaning recognition of man's drawing actions by the machine. Its complexity depends on applications. Examples are shown in sections 4 and 5.

3. SYMBOL RECOGNITION PROCESS

The recognition of symbols and their locations by measuring man's writing actions is the most important function in our input system. As explained in the previous section, the numbers of the types of symbols used on a drawing are rather large, their sizes are not uniform and their shapes cannot be strictly controlled. The recognition algorithm must be insensitive to these kinds of variations.

At present, we have to use a conventional graphic tablet to get data of the pen trace of handwriting, though we are now developing new types of the measuring device. The tablet is not made for this purpose, but for getting values of isolated points. We had to modify the pen by removing a vertical play in the pen, and attaching LED's and buttons on the pen shaft for the communication between a user and the computer. There are still disturbing effects in the measured signal, which become prominent when sizes of written symbols are small. They were ascribed to the measuring principle of the tablet and human behaviour in handwriting. We analyzed the experimental data and made the recognition algorithm also insensitive to these effects.

Our symbol recognition process consists of the main and its back-up step. The main recognition process comprises the initial filtering and local processing of measured data, the rough feature extraction, and the decision making. The back-up step includes the refining and correction of the previously extracted features and it is used when the previous decision making step cannot give a definit answer.

The filtering process is adopted to reduce the amount of raw data as well as to find regions of big or abrupt change in the direction of the curve. At first, the midpoint of the two consecutive measured points is assumed to lie on the original curve. A virtual square frame of a fixed size is put with its center on an assumed point and the nearest point outside of the frame is taken as the next sampled point on the curve. Then the center of the frame is moved to this new point and the process proceeds in the same way. The directions of chords between adjacent points are calculated roughly along the curve and, from their information, the starting and ending spurious transient regions and intermediate corners and small loops can be determined.

At the other sampled points, the point values are revised by taking the average of the consecutive four measured points. Then the coordinate values and chord directions are listed as functions of the arc length of the curve, which is approximated by the summation of each chord length. After this process finished, a rectangular area which covers the symbol is divided into 3x3 regions. From the data the following rough features of the symbol can easily be extracted: the number of pen-ups, the number and the regions of extremum coordinate values, the directions and the locations of the starting and ending strokes, and the types and the locations of the intersecting points of the strokes. From these rough features, more than eighty percent of the symbols used can be determined.

For the undetermined ones, the back-up routine is called. In it, the more precise calculation of chord directions and lengths are performed, and the stroke feature descriptions are written along the abscissa of arc length, such as almost constant direction, right or left rotation, invisible stroke, cusp or other singular point, and unspecifialbe region. After this is compelte, unspecifiable and small heterogenous regions other than singular points are determined or changed according to their adjacent regions or features. Using this stroke feature descriptions the undetermined symbol is recognized. In the still undecided case, the stroke feature descriptions are revised by consulting to the record of mistakable cases.

Usually symbols in a drawing for design are rather clearly written and separated each other, so that this recognition process works very well and error rate is negligible. The first half of the recognition program can be used for analyzing a user-defined new symbol in order to construct possible feature descriptions. from these, the prestored reference data for new symbol can be prepared. In fig.1, an example of a set of original data, and in fig.2, some of the symbols we use in drawings are shown.

The pen of the tablet is used not only for writing the machine recognizable symbols, but also for writing appropriate commentary information at suitable locations of the sheet, which is very useful for the user's memory and understanding. These commentaries may consist of handwriting figures, hatchings, comments or reference symbols etc. The pen is also used for pointing out information written on the sheet, making relational connection between written information blocks, separating areas on the sheet, redefining the base lines of the tablet surface. This recognition software consists of FORTRAN subroutines and programs of the microcomputer.

4. GENERAL THREE DIMENSIONAL GEOMETRICAL PROCESSING SYSTEM

4.1 General

For generation and processing of geometrical three dimensional bodies by computer we developed a comprehensive subroutine package called GEOMAP (GEOmetrical Modelling And Processing). Its functions are explained below.

(i) Generation of models of basic geometrical components

Internal models of several basic three dimensional bodies, surfaces and their coordinate systems are generated. They are prisms and pyramids with arbitrary polygonal bases, cylinders and cones, cylinders with rounded edge polygons, and bodies of revolution. External or internal names of generated bories are registered automatically in the Body Name Register (BNR) for easy access to them. This register has a stack structure and is maintained automatically.

(ii) Movement and position setting of models.

Any number of coordinate systems is set up in a space. Relative movement of bodies or coordinate systes can be specified. Their relative positions can be detected. Names of coordinate systems are registered in the Coordinate Name Register (CNR).

(iii) Generation and detection of interferences

Interferences between solid bodies are generated by set operations such as union, intersection, difference and inversion. With these operations, data structures concerned are reconfigurated and, as a consequence, a new object or objects can be generated or old one eliminated. Entries in the BNR are changed according to the results of these operations. Also, there are commands to test the possibility of interferences, and a command to perform the affine transformation.

(iv) Relations imposed among models

Hierarchical or network relations can be established between or among independent models or components of a model. Names of the relations can be registered in the CNR.

(v) Generation of two dimensional structure

By projecting three dimensional objects on a plane, structures of the two dimensional objects are obtained. These structures can be used for the output display or recognition of three dimensional bodies. For perspective projection and display, a simulated camera is set up in the CNR. By using this camera with parameters of a position, a direction and an angle of view, the pertinent data structure is established with or without eliminated hidden line indication.

(vi) Filing and temporary storage

The intermediate state of the work area, where models are constructed, can be moved to temporary storage to retain the state. With this process, alternative design can be tried without going back to the initial state. During handwriting input, generated data and control information of the input paper sheet are also temporarily stored with the name of the input sheet. When the model building is completed, the model can be stored in the parts storage file by being converted into the compact form. The command sequence used for the construction can also be stored for re-use or analysis.

(vii) Output display

As complete information on three dimensional bodies is kept, various types of external representation, such as various projection drawings, developed surface forms, sectional views, numerical values of geometrical elements and their topological relations, are possible and their display instructions for the terminal unit can be generated by specifying their external representations.

(viii) Other utilities

Utility commands for management of the work area, the BNR, the CNR, and data structures of models are provided. Also, the system has the control software of the microcomputer which is located near the terminal.

4.2 Drawing Input

We tried to connect this program package with our handwriting input method. In GEOMAP, objects, whose internal models are constructed, are registered in the BNR, through which the data of the object bodies can be accessed. Similarly the CNR are used to register the coordinate systems which are referenced during geometrical operations. These registers are simulated on the sheet by the user's drawing. During handwriting input, entries in the simulated BNR and CNR are maintained by the user. He can access the entity by writing an arrow symbol at the corresponding entry in the simulated registers.

The user writes on the sheet of paper on the tablet the following information: a name of the sheet, vertical and horizontal lines which divide the sheet area, base lines of the stack registers. Three of the divided regions of the sheet are used for three view projection drawing. Their coordinate directions have to be defined at the beginning by writing the arrows. The fourth region can be used for writing the registers and the diagram indicating relations or constraints among components of the model.

When a perspective display of objects whose names are in the BNR or in the constraint diagram is required, a coordinate framework of a camera must be established, and the objects in the BNR or in the diagram have to be pointed to. The style of the display such as hidden line elimination or simple rendering etc. is specified. Then a display on the storage tube is obtained and can be hardcopied. When the same output format as handwriting input is specified, the hard copy of the output display is used as the input sheet on the tablet for the subsequent processes.

4.3 Design Action

(i) Generation of the elementary bodies

The user writes a name of an object to be used in the subsequent operation together with its category symbol in the top entry of the current BNR. The specified body is generated in the position where its body coordinates coincide with the space coordinates. The user must sketch its projection shape and write the neccessary dimension values or identifiers in the location near the corresponding edge or the point to be defined in the drawing of this body. The identifiers and their values must be written in the DNR. When the predefined writing order is not used, the pertinent function point in the function figure attached to the tablet must be touched by the pen. In fig.3, an example of a function figure is shown. A new coordinate framework is defined in the same way and its name is written in the entry of the CNR by the user.

(ii) Placement or movement of bodies

The defined elementary body has to be moved to the place where it is processed by the synthesis operations. Displacement values and rotation values around the axes are defined separately or the final position of the coordinate framework of the body is specified. Movement of the synthesized body is treated in the same way, but the object name and the reference coordinate framework name must be specified in the BNR, the CNR or the constraint diagram by writing pointing arrows.

(iii) Synthesis of a body

Those bodies registered in the current BNR can be subjected to unary operations such as drilling, rounding and cutting, and binary operations such as set uinon, difference and intersection. They are specified by writing the pointing arrows with the operation symbols at their roots. As a result of the execution of the binary operation, two bodies are converted usually into one body, then the stack register is adjusted automatically. The new name of the new object may or may not be written in the corresponding entry of the simulated BNR. When more than one body or no body is generated, their number and shapes are displayed, and the user has to decide which one or ones he uses in the succeeding processes, and he has to register them in the BNR. In any case the user must write the new BNR to the right of the old one when the contents of the BNR are updated. This makes the progress record of the designing.

4.4 Example

An element of a mechanical link, whose shape is shown in fig.5, is considered to be made from a square block named 'BA' by removing appropriate blocks whose boundaries are defined in the orthogonal projections shown in fig.4. Rounding operations are performed at each corner of the removed object.

In fig.4(a) and (b), the coordinate system and the initial shape of the object 'BA' are defined. Then the boundaries of the removing parts are defined and removing operations corresponding to symbols S_1, S_2, S_3 and S_4 are performed. These processes are specified by the stack register operations. Boundaries of S_1 and S_2 are defined in the y-z plane only, S_4 is in the z-x plane only, whereas S_3 is defined in the y-z plane and z-x plane. Drilling size is defined in the y-z plane. In fig.4(b) a clockwise rotation of 45 degrees around the specified point and a translation are defined. In fig.4(c) the cutting shape of S_5 is defined in the y-z plane only and S_6 in the x-y plane. A square block 'BL' is defined and generated in the stack register. Its corners are rounded and this shape is removed from the object 'BA'. Then the internal model of the object is constructed, and its external representations are made and shown in fig.5.

5. DRAWING INPUT FOR TURNING OPERATION PLANNING

There are a couple of NC programming systems for turning operation planning. As they use, of course, one dimensional descriptions, we want to connect our drawing input method to any one of them to evaluate our approach. The input information of these system is considered to consist of the following five classes.

(1) Shape description of the part to be made and related machining information.
(2) Clamping data which determine the interference regions.
(3) Works to be applied to the elements and the tool movement data.
(4) Detailed data for each work.
(5) Data of the machine and the material, global data of the blank and the part, machining conditions.

Data belonging to classes (1), (2) and (3) should be written or drawn near the corresponding elements or at the approximate locations on a sheet of paper. The operation sequence of the works can be described by the suitable table form. A work name and its associated chuck name data should be registered in its entry. The form of the chuck should be drawn on the sheet to show the entry forbidden regions and to define the chuck location. And the cutter motion is indicated by writing its locus of movement. From these descriptions, our input system can produce output which can be used as input information of the existing NC software system. Though division of machining regions and exchange of clamps can be automatically made with these programming systems, it is possible for man to determine them by inspecting the drawing.

We show our method by an example. Fig.6 shows the table to define the following: the part name, the machine data, the material, the blank data, the part specification data in which the direction of contour drawing, global tolerance and roughness, oversize value of cutting, clearance of the cutter etc. are included. Fig.7 shows a work data table in which the work operation name and their detailed information are included. In fig.8 a shape drawing is shown. The coordinate system is defined first. X-axis is the center axis, and Y-axis coincides with a radial direction of the object. Contour elements of the shape such as face, taper, arc or longitudinal, are drawn from the left or the right, one by one with its dimensions, tolerance, fit and finish symbols. Its corners are defined with round or chamfer by writing C or R followed by its value. Thus the external and internal contours are defined together with their machining conditions. The sequence of work operations is to be determined by writing in the drawing the locus of boundary of tool movement along the corresponding boundary of shape elements. It is better to draw the shape of the chuck on the drawing to fix its location and to show the entry forbidden region determined by it, whose name and dimensions are obtained from the given chuck data file. The work name with the chuck are written in the work sequence table on the same sheet. From these input data, the computer can produce necessary input information for the existing software system. In a certain system, work sequence and chuck exchange can be automatically produced. But those information is rather easily produced by man inspecting the drawing. This may reduce the computer load or check both man and computer operation.

In our system it is easy to connect this drawing part to the GEOMAP which can generate a simulated work operation and display the object in two dimensional projection. This is shown in fig.9.

6. CONCLUSION

Using two dimensional representation of information is our familiar way of recording, by which we communicate what languages cannot adequately express to other people and also to ourselves to make our mental thought and pictures clear and concrete. That the drawing has not been used as computer input may be ascribed not only to the fact that there were no suitable pattern recognition techniques, but also to the fact that the drawing does not seem to have computer-oriented grammar. However, in engineering drawings there must be no ambiguity and the correct design information is carried with them. Therefore we thought that the engineering drawing can be used as computer input data and instructions.

We analyzed and categorized drawings and their elements to use them advantageously in description and input processes. We developed a simple and reliable symbol recognition technique which is combined with the drawing input. With our method cumbersome and mistakable conversion to a sequence of computer language description from the drawing can be avoided. Moreover man's capability of recognizing global patterns or relations can be utilized without giving him a special burden, and this fact can also make considerable reduction of a computer load in global pattern searching. Two nontrivial applications are described to show the effectiveness of our method. Therefore we can say that our approach and use of the pattern recognition technique in CAD and CAM are successful in experiments and promising for practical use.

REFERENCE

[1] M.Hosaka & F.Kimura, An interactive geometrical design system with handwriting input, Information Processing 77, pp.167-172, North Holland, 1977.

(a) Handwritten characters.

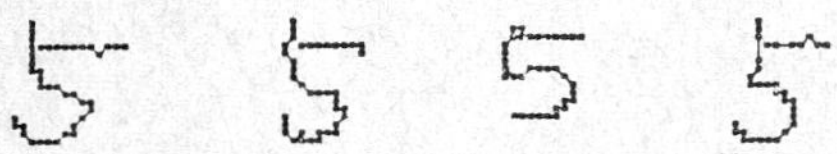

(b) Sampled raw data.

Fig.1. Example of admissible variation of a symbol.

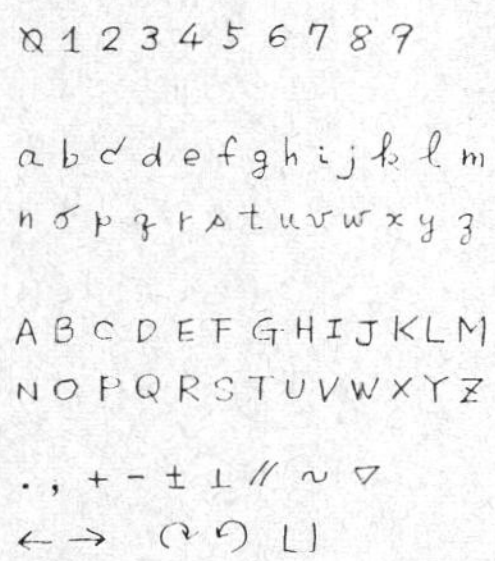

Fig.2. Examples of handwritten symbols.

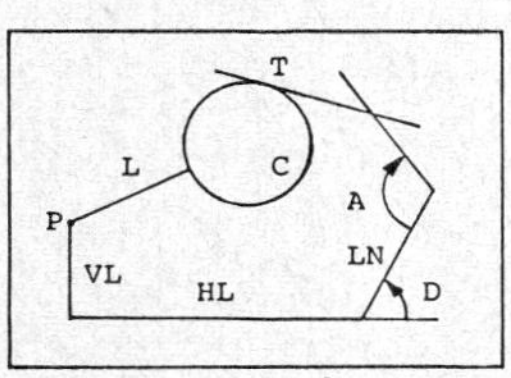

P : point
L : line
VL : vertical line
HL : horizontal line
C : circle
A : angle
D : direction
LN : length of a line segment
T : tangency

Fig.3. Function figure for a two-and-half dimension body.

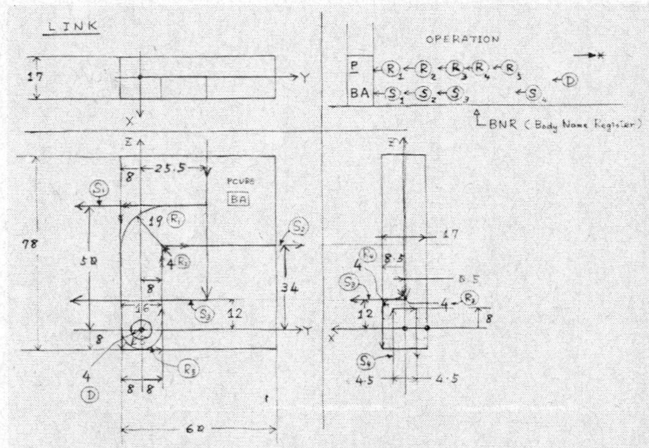

(a) Definition of the shape with dimensions.

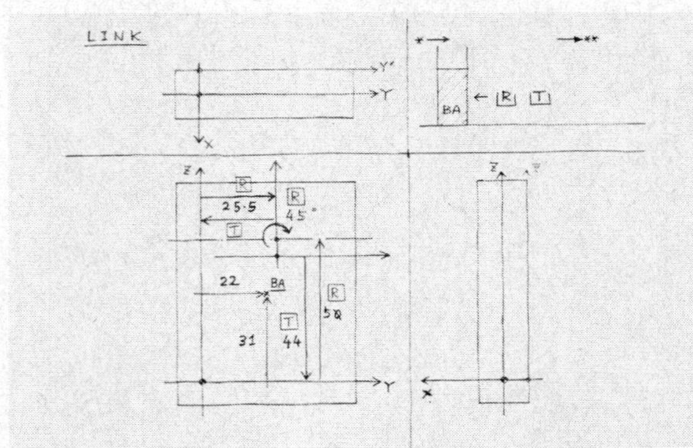

(b) Setting of coordinate systems.

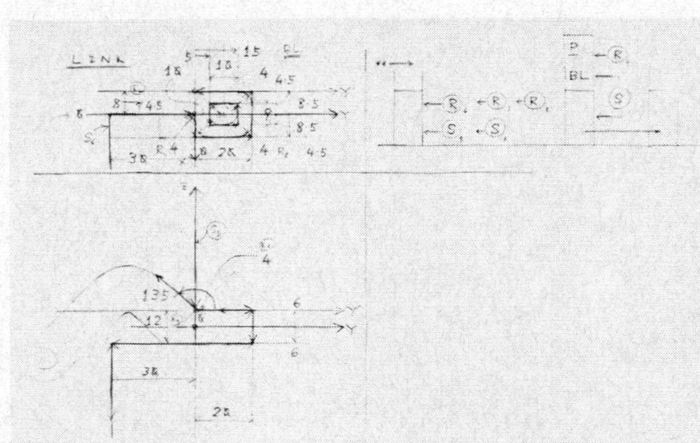

(c) Definition of the shape with dimensions.

Fig.4. Example of input drawings for GEOMAP.

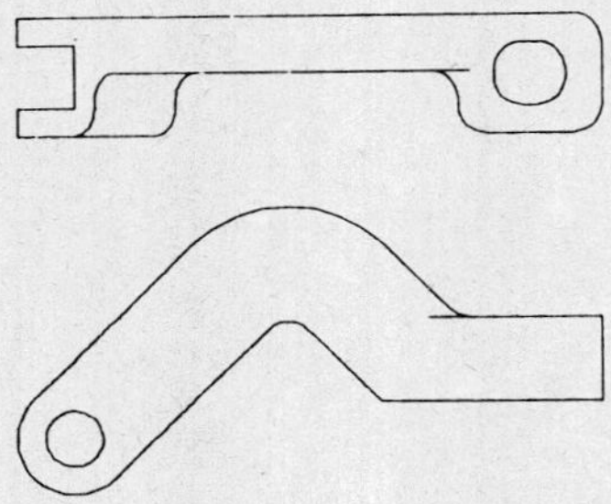

(a) Orthogonal projection.

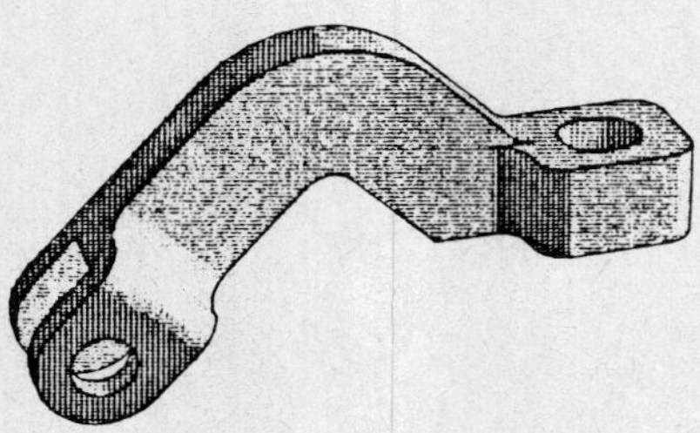

(b) Perspective projection with shading.

Fig.5. Output of the part defined in fig.4.

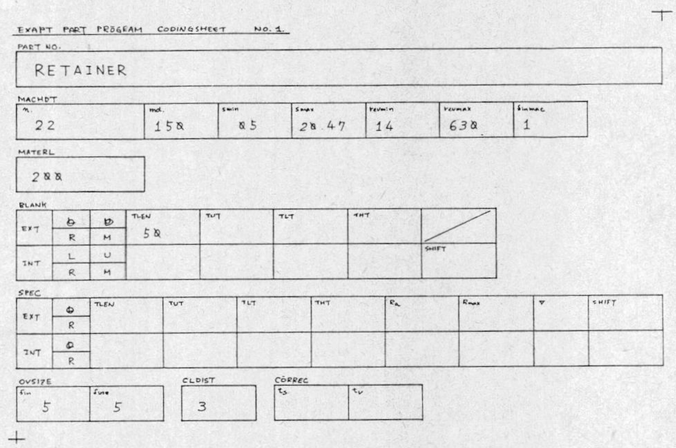

Fig.6. Input table for general data.

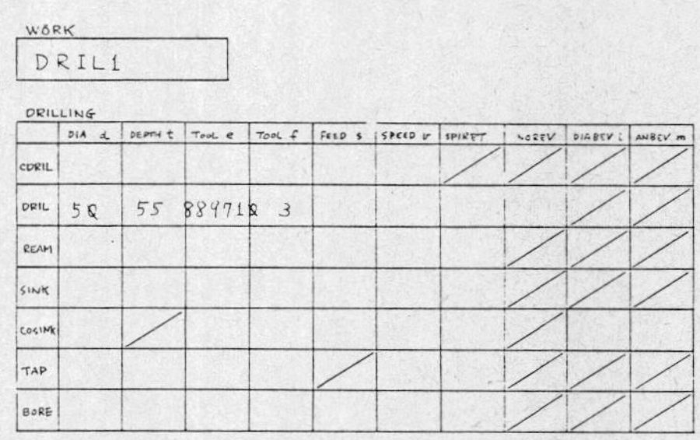

Fig.7. Input table for work data.

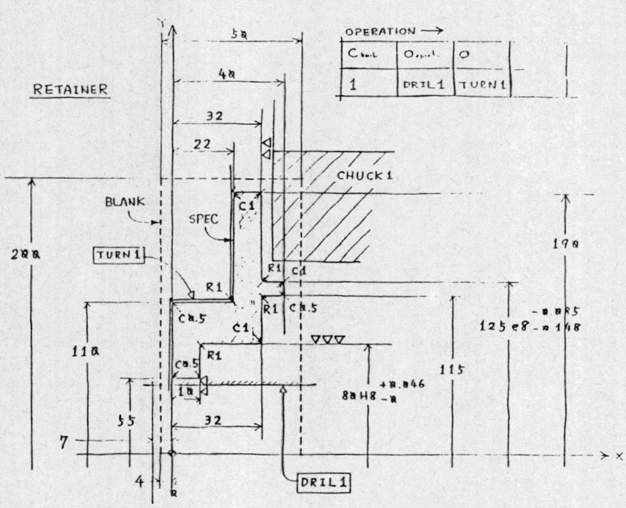

Fig.8. A part of input drawings for turning operation.

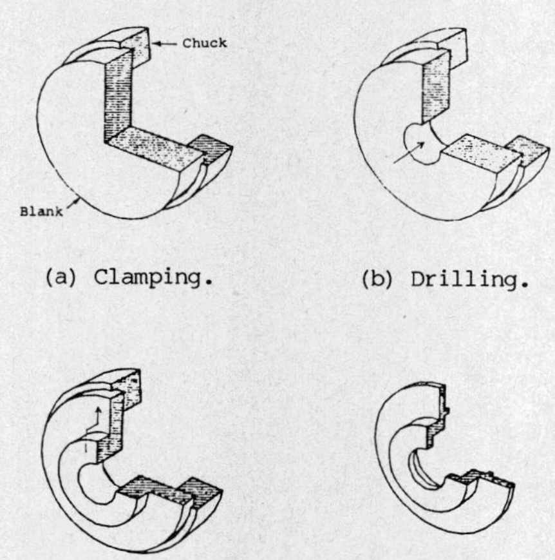

(a) Clamping. (b) Drilling.

(c) Turning. (d) Final product.

Fig.9. Simulation output of turning operation with GEOMAP.

Scientific Information Systems in Japan
H. Inose (ed.)

METHODOLOGIES FOR SCIENTIFIC INFORMATION INPUT SYSTEMS

Toshiyuki SAKAI
Faculty of Engineering, Kyoto University, Sakyo-ku, Kyoto 606

ABSTRACT:

General concepts and methodologies for scientific information input systems are described. We discuss classification of unknown input information, pattern understanding, artificial intelligence and robotics. Actual examples are taken in the field of speech and image: preprocessing (prefiltering) of speech information; degree of contribution of knowledge in speech understanding systems; model construction by man-machine interaction; and image interpretation based on both bottom-up and top-down methods. Through the discussions, we emphasize the importance of defining subworlds of information, knowledge utilization in pattern understanding, constructing models by means of man-machine interaction, and using the models effectively.

I. INTRODUCTION

Automatic or semi-automatic information input systems are very desirable not only for business area, but also for scientific area. By realizing such systems, constructing and updating scientific databases will become practically possible under a reasonable cost of information input. An input system should be designed from a total-information-system point of view, which includes input, storage, update, processing and retrieval.

In this paper, we present important methodologies which are concerned with techniques of pattern understanding and concepts of artificial intelligence.

Computer application is sometimes said to be limited only to routine works of imformation processing; such works in which updating of data is performed daily or weekly, and the programs used there remain the same almost permanently and the same programs are used in many places for the same purpose. What we mean by "scientific information" in research fields is very much different from that of business fields or from simple scientific calculation sevices. It includes not only numerical data but also various kinds of graphs, diagrams and photographs. Moreover, since the scientific information processing has so many varieties and spot-wise characteristics in the time domain, cost for programming should be smaller compared with that for the routine works.

We stress here that an information input system aims at the database into which information acquired is organized and from which information is retrieved for the purpose of its processing and distribution. If the input system needs too much human assistance during the input stage, it will be troublesome, errorful and expensive. Fig.1 shows blocks of pattern recognition and pattern understanding systems. The latter includes knowledge representation, model and learning control blocks that interact with models and procedures.

In the following section, information filter is first examined. The idea of information filter is preclassification of processing into subworlds depending on the information sources. Since it serves for the processing system to feed more uniform information by preprocessing the mixed information, it leads to better results of the processing and larger reduction rates of information transmitted. For the purpose of economical storage of the input information, the information filter also works as an ON-OFF switch that rejects the input with no new information. In Fig.1, the term "model" should be interpreted widely; it refers not only to pattern recognition techniques but also to artificial intelligence concepts. It may be mathematical formulas, stored reference patterns, dictionary (tables), rules of symbol strings, selection rules, generation rules or priority rules of procedures, world descriptions, or representations of knowledge for the task. It should be noticed that the block diagram shown in Fig.1 includes the following four subsystems.

1. "objects and world → measurement → model → preprocessing"
   In this subsystem, we discuss the necessity of knowing subworlds of information source. The classifier is included in the block of preprocessing.
2. "feature extraction ↔ knowledge & model → decision"
   In this subsystem, we discuss quantitatively the degree of contribution of various knowledge to speech understanding.
3. "human ↔ monitor ↔ model & knowledge ↔ feature extraction; information filter"
   In this subsystem, we discuss model construction by man-machine interaction, mainly guided by machine. Also time-varying images are traced by machine and the results can be used to make the ON-OFF switch which controls transmission intermittently.
4. "knowledge & model ↔ learning ↔ feature extraction → decision"
   In this subsystem, we discuss a method which utilizes both of the bottom-up and top-down techniques for the segmentation of outdoor scenes. Knowledge is embedded in the block of model and learning as

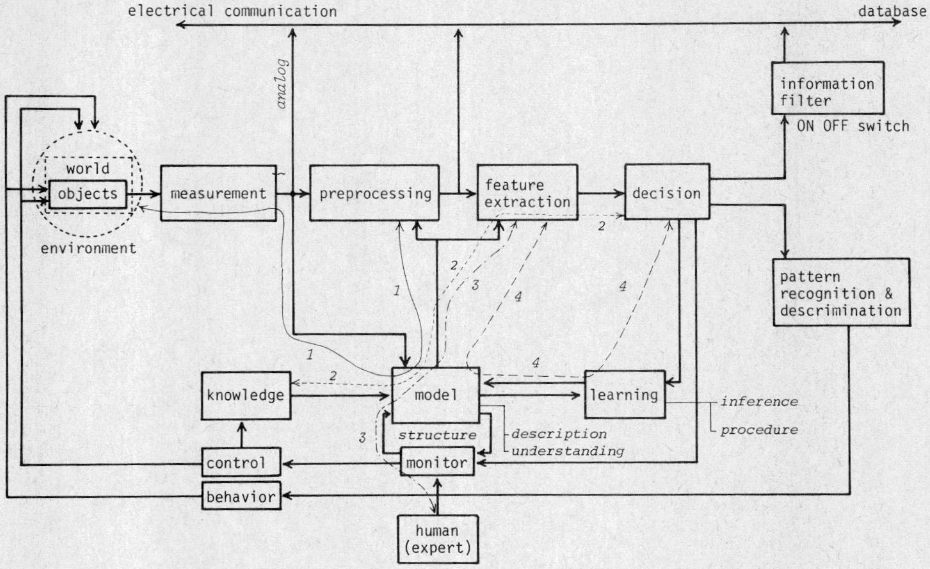

Fig.1 Schematic diagram of pattern recognition, pattern understanding and robotics

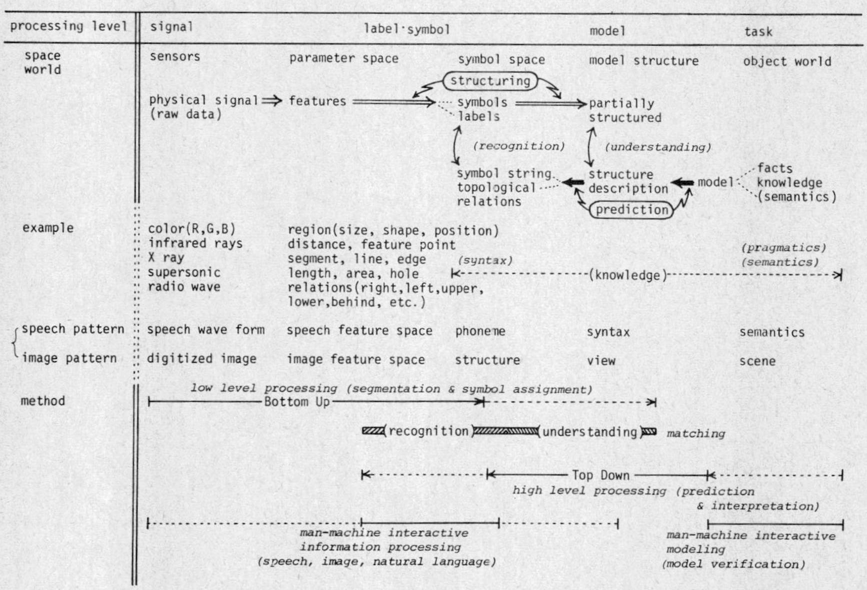

Fig.2 Hierarchical structure of pattern understanding system and suitable methodologies

the semantic network and procedure selection.

Fig.2 shows a hierarchical view of pattern understanding system and suitable application areas of those methodologies such as bottom-up method, top-down method, and man-machine communication in the system. In this paper, we consider a general-purpose system of scientific information input, rather than a special purpose one. When a method is too closely task dependent, it is adhoc. In Fig.2, top-down methods in the task level and the parametric expression level are indicated as dotted lines. Bottom-up method and top-down method have overlapped in the hierarchy. Bottom-up method just under the model level does not generally exist except a special case. Usually, bottom-up methods are universal in trasforming from signal to symbol. Most important features of having a model are:

(i) Compact and efficient descriptions of object worlds
(ii) Prediction and guidance in the processing
(iii) Applications of subprocesses of various levels under the consistent control by any model.

The input system of scientific information has: (i) to be simple for users who are not familiar with computer science, (ii) to be small in size in both hardware and software, (iii) to be easy to develop its programs, and (iv) to provide machine-guided communication subsystem to construct a model for interesting objects and tasks.

2. NECESSITY OF KNOWING SUBWORLDS OF INFORMATION SOURCE (PRECLASSIFICATION AND PREFILTERING)

Scientific information is processed and stored in the database in order to be retrieved later mainly by man-machine interactions. This concept is closely related to a system of "robotics". Robotics is the science in which sender and receiver or object and actor are considered in the same world with mutual interaction. Fig.1 shows the system of pattern recognition, pattern understanding, artificial intelligence and robotics according to the order of actual human communication behavior. It also includes various paths of electrical communication: analog information transmission without processing, encoded information transmission and intermittent information transmission through an intelligent ON-OFF switch. In this section, we discuss the effect of preclassifying sources of speech information according to the structure of human articulatory organs (subworlds of the information source).

We usually process human voice as being produced from one kind of source not only in electrical comminication but also in speech recognition. Human voice is very different from other voices of animals and from sounds of mechanical apparatus. In speech recognition, the variance in human voices due to sex has been conventionally handled as one of global factors of speech sounds. However, we have examined speech sounds from the point of world structure of information source, and have tested how the recognition rate depends on with or without a preclassifier corresponding to the subworlds.

Sample speech sounds are grouped into 6 categories according to age and sex as shown in Table 1. Japanese 5 vowels were uttered 3 times per one person, and in total 1800 utterances (5 vowels x 3 times x 6 groups x 20 persons) were collected. Various experimental results were obtained using our speech recognition system LITHAN [1]. We analyzed them by the multivariate analysis method. Table 2 shows the summarized classification rate and vowel recognition rate.

Table 1. Average pitch frequencies (in Hz)

samples: Japanese 5 vowels
each group: 20 persons

sex	age (years old)		
	11	~20	>40
male	269	146	139
female	268	246	205

We can extract the following remarks. The acoustic organs may be divided into 3 subworlds depending on pitch frequency of vocal cord and size of vocal tracts: children (male and female), male (adult), and female (adult). When we preclassify the input speech sound into these subworlds, we can expect higher uniformity compared with other classification such as "male and female", "without classification" and "original 6 groups".

The difference in the recognition rate itself is not large in each case, but this is because the experiments are for 5 Japanese vowels. Large differences will be observed when we do the experiments for consonants and for more unknown speakers. The knowledge of the input information source and the preclassification according to the subworlds of the input help much in the succeeding processing.

3. DEGREE OF CONTRIBUTION OF VARIOUS KNOWLEDGE TO UNDERSTANDING OF INPUT INFORMATION

Speech understanding has been one of the major topics in the research field of pattern recognition in the last decade. Many papers have been reported, but much has not been discussed on the degree of contribution of knowledge to the recognition rate in various level of hierarchy of the understanding system. In this section, we discuss these levels of knowledge (lexical, syntactic and semantics), and their contribution to the recognition rate of a speech understanding system LITHAN [2]. The tasks employed in these experiments are 4 kinds:

i) Digits: 10 words; Only acoustic analysis is employed for the recognition of isolated spoken digits.
ii) Arithmetic expression: 24 words; Syntactic rules for arithmetic expressions are used in addition to the acoustic analysis. The restriction are that the numbers are less than ten thousands and up to two operators with or without one parenthesis appear in one sentence.
iii) Calendar: 30 words; Pragmatic information concerning date, week and month are included.
iv) Computer network: 101 words; The task deals with Japanese simple sentences which are operational commands and queries of the status of a computer network. All the knowledge of lexical, syntactic, semantic and pragmatic levels are employed in this task. The vocabulary size of this task is 101; 21 predicates (mainly verbs), 60 nouns, 10 postpositions, 5 adjectives, and 5 interrog-

Table 2. Subworld classification and vowel recognition rate

class	recognition rate (%)			
	classification I	classification II	vowel recog. I (1800 utterances)	vowel recog. II (450 utterances)
child, male(adult) female(adult)	80.7	79.3	94.4	91.1
3 classes	class instructed (100)		96.4	92.2
male(11), male(20) male(40), female(11) female(20), female(40)	63.8	56.4	94.1	88.0
6 classes	class instructed (100)		98.1	89.6
male, female	72.4	70.7	92.5	90.0
2 classes	class instructed (100)		94.7	89.8
no classification 1 class	—	—	92.6	88.9
male(adult) female&child	94.6	93.8	94.6	90.7
2 classes	class instructed (100)		95.7	90.4

Classification I, recognition I:
120 speakers [1800 utterances]
All utterances are used for making classifier and used for recognition

Classification II, recognition II:
120 speakers [1800 utterances]
30 speakers [450 utterances: unknown for classifier] used for recognition
90 speakers used for making classifier

atives. In this task, the pragmatic information is the interconnection relationship among computers and their I/O devices.

Notice that sentences in the two tasks, "Arithmetic expression" and "Calendar" are not, strictly speaking, "sentences" because they contain no predicate.

The experimental results are shown in Table 3 [2].

Table 3. Four tasks and recognition rates vs. knowledge utilized

task		digit	arithmetic expression	calendar	computer network
vocabulary size		10	24	30	101
no. of speakers		20	5	12	10
no. of test sentences		/	80	120	200
no. of words in test sentences		1500	560	647	1983
acoustics	sentence		20	8	/
	word	97%	78	54	/
acoustics syntax	sentence		71	40	58
	word		94	83	91
acoustics syntax pragmatics	sentence		/	70	64
	word		/	90	93

The row "acoustics" gives the recognition results by the LITHAN system using only the acoustic-phonemic source of knowledge, and so forth. From this table, we can note the followings:

(1) The contribution of linguistic information to the recognition rate is apparently observed. The contribution is closely related to the content of a task.

(2) It shows quantitatively that automatic recognition of continuous speech is more difficult than that of speech uttered in isolation. This comes mainly from longer phonemic strings and more complex coarticulation.

In the task of computer network, it is very difficult to have data for the case of using only acoustic knowledge.

4. MODEL CONSTRUCTION BY MAN-MACHINE INTERACTION

We have examined model construction and automatic tracing for the images that are moving or changing. For analyzing moving pictures, we adopted the following three methods to organize the tracing system: model construction, man-machine interaction, and trace using the context. A model is useful to extract the object from a large quantity of picture information. The model that is symbolically described is used at the signal level to extract the interesting features and/or changes of the pattern. As well as the static information, the model should include the dynamic information to follow the change of object patterns. For the system which aims to deal with various types of objects, it is essential to construct, reserve, and retrieve the models and apply them to the image patterns.

A model must be described in terms of useful parameters to extract the correct pattern. Man-machine interaction is effective for discovery of these parameters, and for checking the validity of the model.

Since we can expect the position and properties of the pattern to be continuous between consecutive frames, the extraction of the current pattern can be done on the basis of the results of the previously analyzed frames.

The followings are four steps in tracing the pattern, and detecting and/or measuring interesting changes or features. Fig.3 shows a schematic of the system. Phases A through D correspond to these four steps. For "Model Construction(A)", an operator describes an image pattern in terms of its components (such as region, line or point) and their associated properties and relationships. Next the operator matches the image pattern in the first frame of a sequence with the model already constructed. We call this phase "Initial Matching (B)". For "Measurement Specification (C)" phase, the operator specifies the features to be measured and/or the changes to be detected by tracing the pattern. In "Trace(D)" phase, the system extracts the present image pattern, based on the results of the previous frame, using the context between the two consecutive frames. The system also measures or detects the specified features and/or changes. Finally the system terminates the trace if the condition of the termination is satisfied.

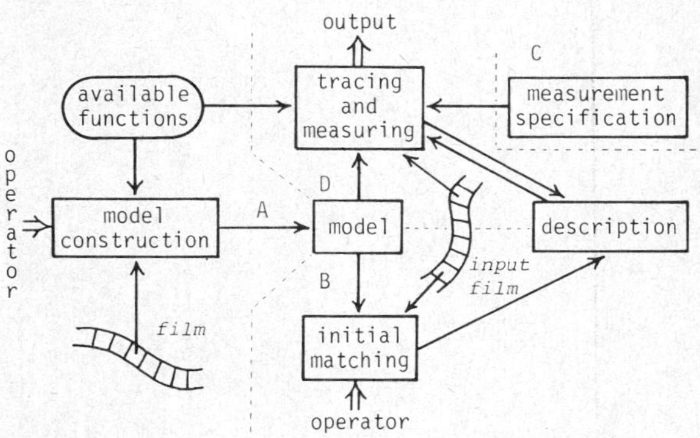

Fig.3 Scemantic of model construction system

At an earlier stage of our experiment, we had examined a model construction method based on primitive components [3]. But the description based on primitive components was not hierarchical one, and it turned out difficult to utilize effectively the context in a moving picture. The new version for the description of patterns is based on the hierarchical tree structures of points, lines, regions and object. For these nodes, the expression of corresponding properties and relations which exist among nodes, are associated as the results of man-machine interaction guided by the computer. It has been known that a hierarchical description of pattern is useful to describe patterns and control the extraction in a single static image, but we have shown that it is also useful to detect changes in a moving picture analysis.

We have examined time-varying images on a role of 16-mm monochrome film. They are 100-200 consecutive images of Xenopus laevis, echinus, ascidian and etc. Fig.4 shows the result of extraction of nucleus (white) of echinus. Fig.5 shows the plot of compactness and area of nucleus as a function of time.

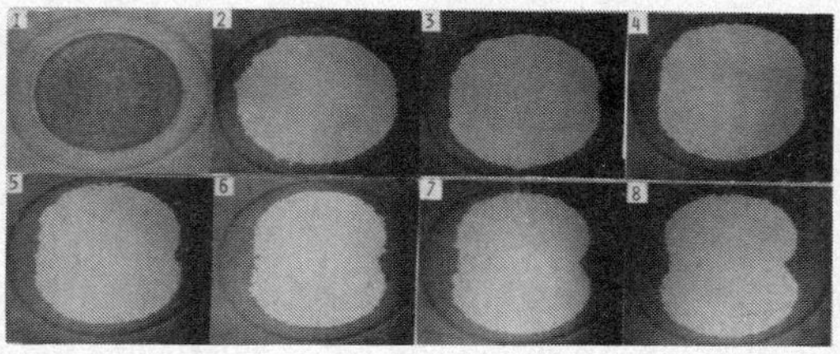

Fig.4 Extraction of nucleus of echinus(white). Upper-left is the original image in the first frame.

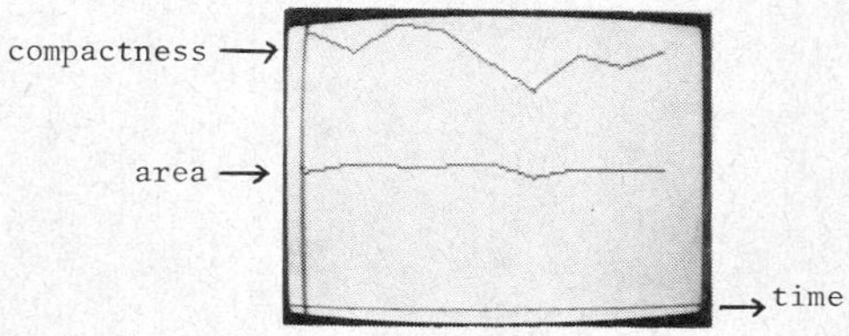

Fig.5 Plot of compactness and area of nucleus of echinus as the function of time.

5. BOTTOM-UP AND TOP-DOWN METHODS FOR IMAGE UNDERSTANDING [4][5]

In this section, we describe a scene analysis system which can perform semantic segmentation of a scene including buildings, sky, trees, road, etc. An input color image is first segmented into a structured symbolic description using color data. Here the structured symbolic description means that the image is represented by patches, boundary lines, vertices, holes and their topological relations. A patch is an elementary region almost homogeneous in color.

Semantic interpretation is performed on this description. Top-down control and bottom-up control are combined in the framework of semantic region growing which is to merge patches according to their interpretation. Knowledge is represented as two sets of rules; one for the bottom-up process and the other for the top-down process. They are organized into a form of semantic network as shown in Fig.6. Each node of the network is called a knowledge block in our model. It holds knowledge about a matter in the world; for instance, object "sky", material "concrete", property "blue", relation "linear-boundary", etc. The system is general, but in this section we explain the method applied to outdoor-scene analysis.

Fig.7-a is an original digitized color scene,and Fig. 7-b is the result of segmenting it into a set of coherent patches according to the color data. We employed the recursive thresholding technique for the segmentation [4]. It uses histograms of color features to determine the cut-off value of the thresholding operation for region extraction(feature extraction in Fig.1).

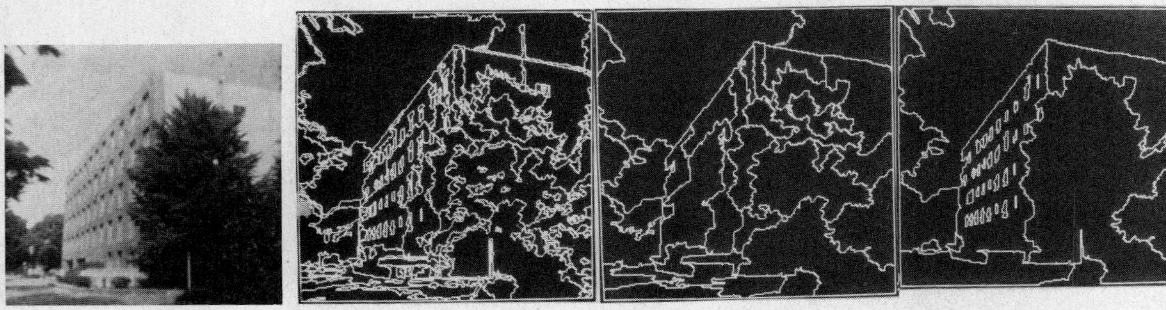

Fig.7-a
Digitized color scene

Fig.7-b
Result of preliminary segmentation

Fig.7-c
Plan image

Fig.7-d
Result of semantic segmentation

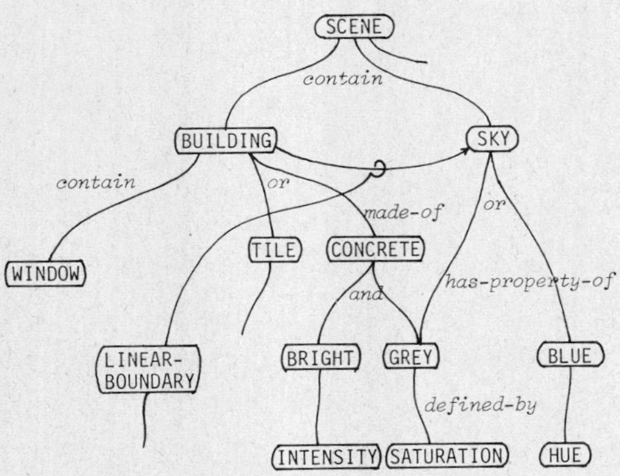

Fig. 6 Semantic network for knowledge representation

It is an important problem where to locate the interface between bottom-up process and top-down process. As rough interpretation of the scene, we make a plan by the bottom-up process. The patches with large area are selected as key-patches from the segmented image. A plan image is generated by merging all small patches with the key-patches (Fig.7-c). The top-down process can interface with the bottom-up process through this plan. The plan has a role to guide the top-down process; that is, which knowledge should be applied to what part of the scene. The purpose of generating the plan image is to make the relations between key-patches clearer. Labels are assigned to each patch in the plan image. (This corresponds to transformation from signal to symbol in Fig.1). The correctness value of each assigned label is first evaluated using properties of each patch. Then they are updated using their relations. Finally we get a set of labels and associated degree of correctness for each patch in the plan image, for instance (sky=0.6, tree=0.2, road=0.1, building=0.1). The top-down analysis is performed in a framework of region growing. But the analysis process is not simple iteration of labeling and merging operations. A description of the input scene is built as the result of the top-down analysis. The rules used in the top-down process are production rules. Each production rule works toward the construction of the scene description based on the segmented image and the plan. Fig.7-d shows the semantic segmentation of Fig.7-a which is generated from the scene description finally obtained by the top-down process.

ACKNOWLEDGEMENT

The author wishes to express his thanks to the staffs who have devolved research works presented in this paper. Concerning to speech processing, Dr. Sei-ichi Nakagawa contributed very much with students who have worked for their thesis at the university. In the research field of image processing, he is very much indebted to Dr. Takeo Kanade who is actually leading the studies. Mr. Yu-ich Ohta and Mr. Yasuo Ariki contributed in the research of outdoor scene analysis and moving images, respectively. Last, but not least, many thanks should go to Miss Akiko Ichiyama for her efficient and hearty supports for our research.

REFERENCES

[1] T. Sakai and S. Nakagawa, "A Speech Understanding System of Simple Japanese Sentences in a Task Domain", Trans. of IECEJ, Vol.E 60, No.1, p.13, (1977).
[2] T. Sakai and S. Nakagawa, "Speech Understanding System -LITHAN- and Some Applications", Proceedings of the third IJCPR, p.621, (1976).
[3] Y. Ariki, T. Kanade and T. Sakai, "An Interactive Image Modeling and Tracing System for Moving Pictures", Proceedings of the 4th IJCPR, pp.681-685, (NOV. 1978).
[4] Y. Ohta, T. Kanade and T. Sakai, "An Analysis System for Scenes Containing Objects with Substructures", Proceedings of the 4th IJCPR, pp.752-754, (NOV. 1978).
[5] Y. Ohta, T. Kanade and T. Sakai, "A Production System for Region Analysis", Proceedings of the 6th IJCAI, pp.684-686, (AUG. 1979).

Scientific Information Systems in Japan
H. Inose (ed.)

INTERACTIVE GRAPHIC DATA ENTRY SYSTEM

Tohru MOTO-OKA, Hidehiko TANAKA and Tadashi SAITO
Faculty of Engineering, The University of Tokyo, Hongo, Bunkyo-ku, Tokyo

Entries of a great deal of graphic data are necessary in many systems including a computer aided design system. One of the main bottle necks in practical use of CAD systems is the difficulties of data entries. The interactive graphic data entry system for logic design is investigated in this project and seems to be applicable to similar graphic entry systems with minor changes. Two methods are applicable to solve the data entry problem. One method is minimizing the quantity of the entry data by using data which have been input previously as many as possible, and another method is establishing a man-machine interface adapted to graphic data entry. In this system, the utilization of data base management system which is based on the relational data model and multiple input media such as tablet/display system, graphic data description language and data manipulation language are attempted and are evaluated.

1 Introduction

With the improvement of performance in digital computers and the expansion of their application fields, the quantities of data entry are increasing very rapidly and varieties of data which should be dealt in recent systems are extending to many data types such as texts, graphics, images, vocals and signals.

In order to expand the application fields of digital computers and realize the information systems which are natural and easy to use for users, it is necessary to accomplish the man-machine interface, in which users are able to deal with multiple information representations in natural manners for them.

In this research project, a interactive graphic data entry system is investigated as one example of above-mentioned data entry systems.

Graphic data are abstract data comming after numeric data and character data. Graphic data have different characteristics from image data because graphic data such as graphs, circuit diagrams and logic deagrams are expressed easily by languages, but image data are a set of image element data.

Automatic curve followers had been investigated as the input system of drawing but are still research objects , because it is not a practical system from the view points of reliability, speed and cost. In the graphic data entry system which is used in an industry, the co-ordinates are specified by digitizers and connections among co-ordinates are usually given by functional keys.

The system which is developed by this project is considered as the descendant of the above mensioned system basically, is the system utilizing positively the characteristic of graphic data such as abstract data. Distinguished characteristics of this system are efficient entry of a great deal of data, generalized input system and storage to the relational data base.

2 Data Entry System for Logic Design

The logic diagrams are used often in logic design of a logic system such as a digital computer and become input data for a physical implementation design system, because designers can understand it easily.

In order to input logic diagrams into a computer, input language describing a logic diagram or input system digitizing a drawing of a logic diagram have been used. The mis-manipulating in translation to Language expressions or long turn around time necessary for drawing make these system inconvenient.

Designers require interactive logic diagram entry systems which are used as well as notebooks. As a result of recent development of LSI technology, designers of LSI could not work without computer aided design system as well as designers of logic systems which consist of LSI's. The establishment of convenient data entry system used in CAD system of LSI are desired strongly.

It is desired for input data to be managed by the data base managements system in an information system. Quantity of input data necessary to be entered newly could be decreased by using old input data repeatedly. For example, input data format for an application program can be organized by combination of old input data and new input data sometimes.

Retrieving and combining old design drawings seems to be one of the main job steps in mechanical design. Each design step in a design automation system of logic devices which is shown in Figure 1, is a comparatively simple job dealing with a great deal of data as well as in many other design systems. The output data of each design step are used as the main input data of the next design step repeatedly. Therefore, in the consistent design system in which input and output data are managed by DBMS, manual I/O operations are not needed except input data in the initial design step, necessary manual correction and additional external data entry.

The most natural data model for logic diagram seems to be a network model and many design system use a network data model for the representation of logic diagram. However, a network model is not the most suitable for many applications in common. In this project, a relational model is tried from the viewpoints of flexibility and understanding of designers.

3 Constitution and Characteristics of the System

The block diagram of this system is shown in Figure 2. The intelligent terminal using a Z-8O micro-computer is the interface equipment between a designer and this system. The logic diagram description language ELIS is used in the communication between the intelligent terminal and the design system installed in a host computer. A SEQUEL-like language is used as the data base manipulation language in a host computer system. At present system, the time sharing system in Computer Centre of the University of Tokyo is used as a host computer and the relational data base management system TRDS which was developed in this project, is installed in the Centre.

The data base management system will be transfered on a distributed file management system which is installed in the Network Oriented Operating System NOS of the experimental laboratory computer network TECNET in future.

The characteristics of this system are as follows:

(1) Graphic data entry from a tablet as well as the input descriptions in the logic diagram description language ELIS and in the data base manipulation language SEQUEL from a keyboard are used as means of entry and editing of a logic diagram. These three media can be used to adapt to the purpose and edit the same data base.

(2) The input data from a tablet are indications of the kinds of modules and their names and indications of names of connecting nets and their connecting relations. The kinds of modules are added and changed easily by users. By registering a stored logic diagram as a new module, repeated input of the same circuit are simplified.

(3) The logic diagrams entered from a tablet are displayed in forms of graphics or ELIS language on a graphic display device in realtime, in order to verify the input. Editing data are entered from a tablet as well as through the ELIS descriptions.

(4) The ELIS language has module structures corresponding to graphic structures and is designed for describing logic diagrams. ELIS language is a input language having simple and clear

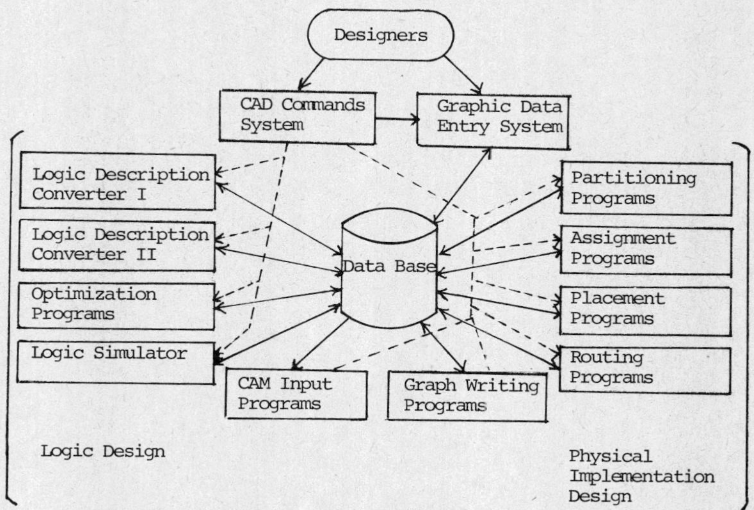

Figure 1. Computer Aided Design System for Logic Devices

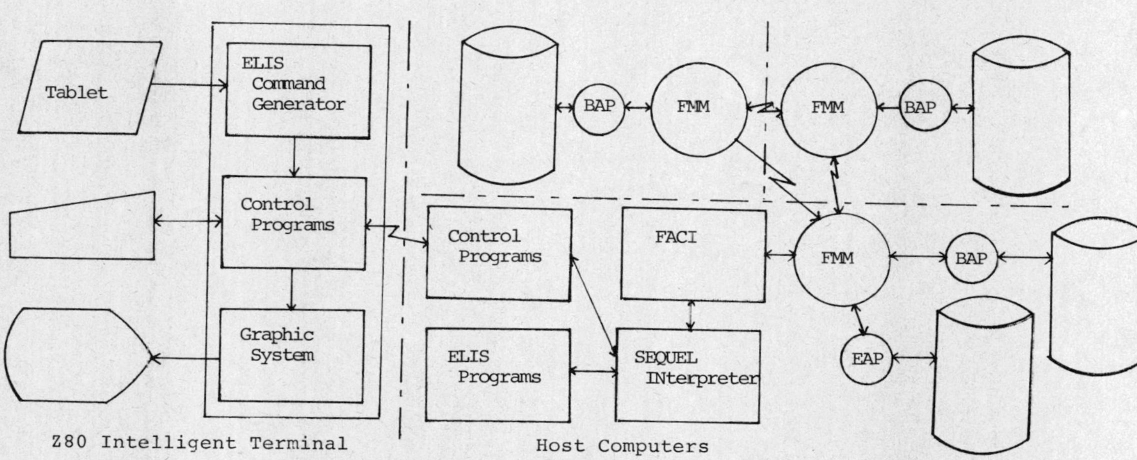

Figure 2. Interactive Graphic Data Entry System

description capability and editing capability. The length of ELIS description, however, is 1,5 or 2 times longer than graphic input from a tablet if they are compared in a basis of numbers of pen touches.
(5) When the ELIS language is used in communication between an intelligent terminal and a host computer, response times of TSS are short enough to graphic entry, because number of bits in communication data are small relatively and communication frequency can be decreased.
(6) The basic part of a logic diagram can be expressed by an element table and a pin table in a relational data model. The element table includes data concerning to logic circuit elements. The pin table expresses connecting nets. The some part of necessary data are stored in these two tables and the other part in new tables such as a net tables and a line tables. It was confirmed that design data were managed freely and easily by relational data base management system.
(7) In this system, the data manipulation language SEQUEL is used from a terminal in order to edit data base directly and also is used in a form of subroutine in FORTRAN programs. This latter means is very useful in inter-process communication between the data base and application programs for design. Input, output processing in ELIS language seems to be one example using this means.
(8) The performance such as retrieval speed is considered as a weak point of a relational data base management system in general. It was examined that the technique generating image contributed to increase the retrieval speed. If this technique is used as occasional demand in each application program for design, the difficulties in performance will be overcome.
(9) The necessity of a distributed data base is not strong in this interactive graphic data entry system. However, it will be necessary in future from the view point of the flexibility of a total design system. In this system, distributed file system is installed in host computers PPS-1 and OKITAC 43OOC which are menbers of the computer network TECNET having interprocess communication facility.

4 Data Representation of Logic Diagrams

Logic circuit diagrams which are refered in design, test and maintenance include many design data in logic devices. The logic circuit diagram is a typical graph used in design automation system of logic devices such as digital computers and are included in this diagram almost all important data in design of logic devices.

One example of the relational model corresponding to this diagram is shown in following sections. The element of logic circuits diagram are usually expressed as shown in an example of Figure 3. Many informations on the connection of elements are included in logic circuits diagram, too. These informations are expressed by following six tables in relational data models.

DIA-POSITION		B4	
LOGIC	CLOCK	A	
E-NAME		LOP1F	
IMP-POSITION		2017	
PACKAGE-CODE	IC-POSITION	L30	24
IC-CODE	ICE#	1285	0
CIRCUIT		4NAND	

Figure 3. Element Representation in Logic Diagrams

ELEMENT TABLE (E-NAME, LOGIC, CLOCK, IC-CODE, ICE#, PACKAGE-CODE, IC-POSITION, IMP-POSITION, CIRCUIT, [DIAGRAM#], DIA-POSITION)
PIN TABLE (P-NAME, E- NAME, ET#, NET-NAME, NEGATE,IO)
NET-TABLE (NET-NAME, NICKNAME, LINE, WIRING)
REFERENCE TABLE (NICKNAME, DIAGRAM#, DIA-POSITION)
LINE TABLE (POINT-1, POINT-2, NETNAME)
POINT TABLE (POINT, X-POSITION, Y-POSITION, P-NAME)

An element of Figure 4 corresponds to a tupple of a ELEMENT TABLE. A PIN TABLE includes data of input and output pin as of elements. The data of nets are stored in a NET TABLE. Nicknames of nets are stored in a REFERENCE TABLE. Segments of nets and end points of segments are stored in a LINE TABLE and a POINT TABLE respectively. These two tables are necessary for graphic representation of logic circuits diagram. If these connecting lines are generated automatically and rapidly, these tables are not necessary to be stored in a data base.

5 Relational Model Data Base Management System:TRDS

The relational model data base management system which is used in this project are experimentally developed and installed by FORTRAN at Computer Centre in the University of Tokyo. The SEQUEL of TRDS can be used from a TSS terminal. The logic circuit diagram are handled by SEQUEL in TRDS as shown in Figure 4.

In order to treat this data base through a host language, following interface subroutines are installed.

BIND (<program variable name>, <program variable address>)
SEQUEL (<cursor name>, <any SEQUEL statement>)
FETCH (<cursor name>)

In order to generate or eliminate images on a table, following statements are used.

CREATE IMAGE <table name> (<column name>)
DROP IMAGE <table name> (<column name>)

In order to decrease the information retrieval time, it is desirable to make images for expected many reference conditions. On the other hand, many images cause to increase the update time of a table, because images should be changed at update. In an application program such as a data entry routine in which table updating occur often, optimization of number of images is needed. For example, the number of disk accesses which was 9743 times without an image, decrease to 1641 times by optimization of images, in a job of data entry. Further increase in number of images cause to increase the number of disk accesses.

6 Editing System of Logic Information (ELIS)

ELIS system was developed as editing system of logic design information initially. In this system, ELIS commands are used as internal representation mainly. ELIS commands consist of circuit description commands, editing commands of circuit information, circuit display commands, circuit print out commands and SEQUEL call commands. The example of 4 bits adder described in ELIS Commands is shown in Figure 5 (a) and they are stored in the form of Figure 5 (b) and (c) in the data base.

7 Logic Circuit Diagram Entry from a Tablet

Direct data entry from hand-written logic circuit diagram seems to be more convenient to designers than data entry using languages such as ELIS commands.

```
SELECT NETNAME,ID FROM PIN ;
NETNAME   ID

N1        I
N3        O
N2        I
N4        O
N4        I
N5        O
N3        I
N6        I
N6        O
N5        [illegible]
N7        [illegible]
N8        O
N8        I
N9        O
***** SEQUEL PLEASE *****
@ SELECT COUNT(*) FROM PIN ;


     21
***** SEQUEL PLEASE *****
@
 SELECT MAX(ICE) FROM ELEMENT ;
  ICE

   4
***** SEQUEL PLEASE *****
@


 SELECT PNAME,ENAME,ET FROM PIN
@
  WHERE  NETNAME='N1' OR NETNAME='N2')
@
          AND ID='I'       ;
PNAME ENAME        ET

   1  A            1
   2  A            2
   4  B            1
   5  B            2
   7  C            1
  16  F            1
  17  F            2
***** SEQUEL PLEASE *****
@

  SELECT * FROM ELEMENT WHERE ENAME IN
@
          SELECT ENAME FROM PIN WHERE NETNAME='N8' ;;
EN        ICCODE    LOGIC        ICE

F         SN7400-2 NAND          2
G         SN7400-3 NAND          3
***** SEQUEL PLEASE *****
@
```

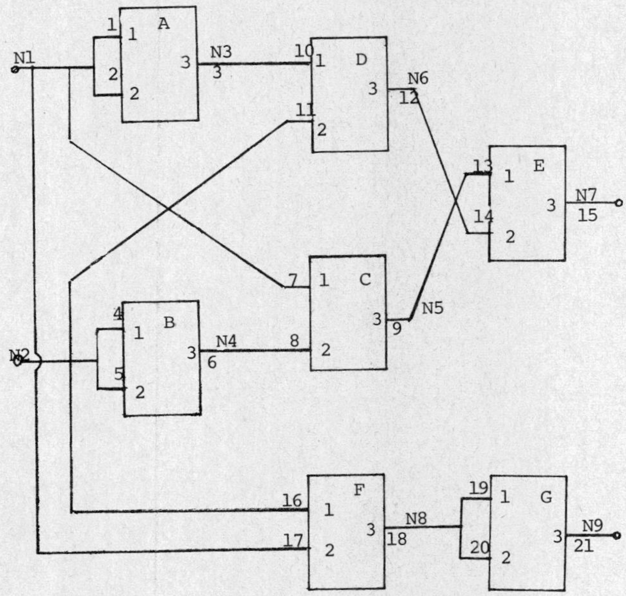


```
***** SEQUEL PLEASE *****
@SELECT * FROM ELEMENT ;
ENAME     ICCODE    LOGIC        ICE

A         SN7400-1 NAND          1
B         SN7400-1 NAND          2
C         SN7400-1 NAND          3
D         SN7400-1 NAND          4
E         SN7400-2 NAND          1
F         SN7400-2 NAND          2
G         SN7400-3 NAND          3
***** SEQUEL PLEASE *****
@SELECT * FROM PIN        ;
PNAME ENAME        ET NETNAME    ID

    1  A           1  N1         I
    2  A           2  N1         I
    3  A           3  N3         O
    4  B           1  N2         I
    5  B           2  N2         I
    6  B           3  N4         O
    7  C           1  N1         I
    8  C           2  N4         I
    9  C           [illegible]  N5         O
   10  D           [illegible]  N3         I
   11  D           [illegible]  N6         I
   12  D           3  N6         O
   13  E           1  N5         I
   14  E           2  N6         I
   15  E           3  N7         O
   16  F           1  N2         I
   17  F           2  N1         I
   18  F           3  N8         O
   19  G           1  N8         I
   20  G           2  N8         I
   21  G           3  N9         O
***** SEQUEL PLEASE *****
@
```

Figure 4. Examples of Logic Diagram, its Relational Model and SEQUEL sentences

```
FIL(FN=ADDER4)
MM(AND:3,1,0,A,1)
MM(OR:3,1,0,0,0)
MM(NOT:3,1,0,I,1)
MM(HA)
 TM(I:ITX;ITY/O:OTC;OTS)
 CM(OR1:OR,2.1/A1:AND,2,3/NT1:NOT,3.2/A2:AND,4.1)
 NT(W1:ITX-OR1.I001-A1.I001/W2:ITY-A1.I002-OR1.I002)
 NT(W3:A1.0001-NT1,I001-OTC/W4:OR1.0001-A2.I002)
 NT(W5:NT1.0001-A2.I001/W6;A2.0001-OTS)
MM(FA)
 TM(I:IX1;IX2;IX3/O:OX1;OX2)
 CM(HA1:HA,2.2/HA2:HA,3.1/OR:OR,4.2)
 NT(IX1:IX1-HA1.ITX/IX2:IX2-HA1.ITY/IX3:IX3-HA2.ITY)
 NT(S1:HA1.OTS-HA2,ITX/C1:HA1.OTC-OR.I001)
 NT(C2:HA2.OTC-OR.I002/OX2;HA2,OTS-OX2/OX1:OR.0001-OX1)
MM(ADD)
 TM(I:IX0;IY0;IX1;IY1;IX2;IY2;IX3;IY3)
 TM(O:01;02;03;04;OCR)
 CM(HA0:HA,2.1/FA1:FA,3.2/FA2:FA,4.3/FA3:FA,5.4)
 NT(IX0:IX0-HA0.ITX/IY0:IY0-HA0,ITY/IX1:IX1-FA1.IX2)
 NT(IY1:IY1-FA1.IX3/IX2:IX2-FA2.IX2/IY2:IY2-FA2.IX3)
 NT(IX3:IX3-FA3.IX2/IY3:IY3-FA3.IX3/C1:HA0.OTC-FA1.IX1)
 NT(C2:FA1.OX1-FA2.IX1/C3:FA2.OX1-FA3.IX1/01:HA0.OTS-01)
 NT(02:FA1.OX2-02/03:FA2.OX2-03/04:FA3.OX2-04)
 NT(OCR:FA3.OX1-OCR)
```

(a) ELIS Commands

FNM	MFNC	MNAME	STD	FNC
ADDER4	AND			&AND
ADDER4	OR			&OR
ADDER4	NOT			&INV
ADDER4	HA			&T
ADDER4	HA	OR1		OR
ADDER4	HA	A1		AND
ADDER4	HA	NT1		NOT
ADDER4	HA	A2		AND
ADDER4	FA			&T
ADDER4	FA	HA1		HA
ADDER4	FA	HA2		HA
ADDER4	FA	OR		OR
ADDER4	ADD			&T
ADDER4	ADD	HA0		HA
ADDER4	ADD	FA1		FA
ADDER4	ADD	FA2		FA
ADDER4	ADD	FA3		FA

(b) Element Table

MFNC	MNAME	INAME	TNO	NET	IO
AND		I001	1		I
AND		I002	2		I
AND		I003	3		I
AND		0001	4		0
OR		I001	1		I
OR		I002	2		I
OR		I003	3		I
OR		0001	4		0
NOT		I001	1		I
NOT		I002	2		I
NOT		I003	3		I
NOT		0001	4		0
HA		ITX	1	W1	I
HA	OR1	I001	1	w1	I
HA	A1	I001	1	w1	I
HA		ITY	2	w2	I
HA	A1	I002	2	w2	I
HA	OR1	I002	2	w2	I
HA	A1	0001	4	w3	0
HA	NT1	I001	1	w3	I
HA		OTC	3	w3	0
HA	OR1	0001	4	w4	0
HA	A2	I002	2	W4	I
HA	NT1	0001	4	w5	0
HA	A2	I001	1	w5	I
HA	A2	0001	4	w6	0
HA		OTS	4	w6	0
FA		IX1	1	IX1	I
FA	HA1	ITX	1	IX1	I
FA		IX2	2	IX2	I
FA	HA1	ITY	2	IX2	I
FA		IX3	3	IX3	I
FA	HA2	ITY	2	IX3	I
FA	HA1	OTS	4	S1	0
FA	HA2	ITX	1	S1	I
FA	HA1	OTC	3	C1	0
FA	OR	I001	1	C1	I
FA	HA2	OTC	3	C2	0
FA	OR	I002	2	C2	I
FA	HA2	OTS	4	OX2	0
FA		OX2	5	OX2	0
FA	OR	0001	4	OX1	0
FA		OX1	4	OX1	0

MFNC	MNAME	INAME	TNO	NET	IO
ADD		IX0	1	IX0	I
ADD	HA0	ITX	1	IX0	I
ADD		IY0	2	IY0	I
ADD	HA0	ITY	2	IY0	I
ADD		IX1	3	IX1	I
ADD	FA1	IX2	2	IX1	I
ADD		IY1	4	IY1	I
ADD	FA1	IX3	3	IY1	I
ADD		IX2	5	IX2	I
ADD	FA2	IX2	2	IX2	I
ADD		IY2	6	IY2	I
ADD	FA2	IX3	3	IY2	I
ADD		IX3	7	IX3	I
ADD	FA3	IX2	2	IX3	I
ADD		IY3	8	IY3	I
ADD	FA3	IX3	3	IY3	I
ADD	HA0	OTC	3	C1	0
ADD	FA1	IX1	1	C1	I
ADD	FA1	OX1	4	C2	0
ADD	FA2	IX1	1	C2	I
ADD	FA2	OX1	4	C3	0
ADD	FA3	IX1	1	C3	I
ADD	HA0	OTS	4	01	0
ADD		01	9	01	0
ADD	FA1	OX2	5	02	0
ADD		02	10	02	0
ADD	FA2	OX2	5	03	0
ADD		03	11	03	0
ADD	FA3	OX2	5	04	0
ADD		04	12	04	0
ADD	FA3	OX1	4	OCR	0
ADD		OCR	13	OCR	0

(c) Pin Table

Figure 5. ELIS Commands and Corresponding Tables of a Four Bits Adder

Tablet layout is shown in Figure 6. Through touching a hand-written circuit diagram, a symbol selection sheet and a control sheet on a tablet by a stylus, the data of the circuit diagram are transfered into the description of ELIS commands. In order to input a same circuit diagram, a number of stylus- touches in tablet entry are approximately a half of a number of key-touches in a ELIS entry. If the names of elements and nets are generated automatically by system, a number of stylus touches becomes 20% of a number of key-touches in some examples.

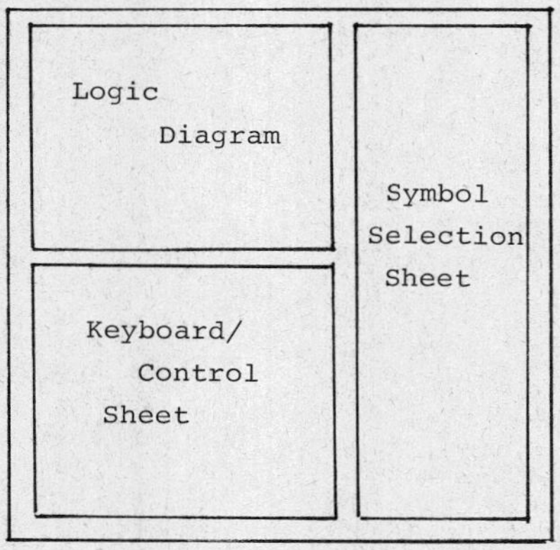

Figure 6 Tublet layout

8 An Intelligent Terminal

In an initial system, co-ordinates of a tablet were sent to a host computer and ELIS commands of input data were sent back for confirmation. The rapid data entry becomes impossible sometimes because the response time becomes long for a length of communication data and a busy condition in a host computer.

In order to overcome these demerits, the multiple media intelligent terminal dedicated to logic design was constructed with Z 80 microcomputer, a tablet, a graphic display, a typewritter/keyboard and disketts. The design concepts and distinguished feature of this terminal are as follows:
(1) The conversion program from tablet input to ELIS commands is installed in Z-80. The length of communication data and communication frequency between the terminal and a host computer are decreased, so that response time of a system becomes reasonable. Also, it is used as stand-alone system by storing data in disketts. Basic facilities of graphic output are installed in Z-80 and data transmission for graphic display are decreased.
(2) Editing and verifying of graphic input are usually done by using a tablet and a graphic display. ELIS commands or SEQUEL statements are sometimes convenient for these purposes. It is very convenient to be able to use these different means from a same terminal easily.
(3) Changing a module set indicated by a tablet is possible by replacing with other symbol selection sheets which are stored in a diskett. A graphic input can be registered as a new module in a symbol selection sheet.

9 Distributed Data Base

In order to realize a distributed data base, a distributed file system was installed in TECNET. A user can access any file of the computer network without consciousness of a host computer which stores the file. The distributed file sytem is composed by File Main Managers(FMM), Basic Access Processes(BAP), Extended Access Processes(EAP) and File Access Command Interpreters(FACI) as shown in Figure 2.
Each host has one FMM which manage all files of the host and communicate with FMM of the other host. FMM facilitates following four functions:
(1) Seize and release of a file
(2) Generate and Eleminate of a file
(3) Open and Close of a file
(4) Directory Management
A BAP is a process which is installed for each file organization and communicate with a user process and access to a file actually according to the demand. A BAP performs basic functions such as READ and WRITE only. On the other hand, the functions of an EAP will be extended to COPY, COMPARE of files and RETRIEVE and UPDATE of data. A FACI is a set of reentrant routines for file accesses.

10 Conclusion

This system is a graphic data entry system using positively the distinctive feature that graph is information easy to be abstracted . The intelligent terminals and data base management system which are developed through this project are applicable to similar application fields with minor changes.

Acknowledgment

The authors wish to thank Takashi Matsushima, Noriyoshi Ito and Toshio Misaka for the contribution to the development of this work during their master thesis.

REFERENCE

(1) T. Moto-Oka et.al: Logic Design System in Japan Proc. of 12th Design Automation Conference 241-250 (Jun. 1975)

(2) D. D. Chamberlin —SEQUEL: A Structrued English Query Language— Proc. ACM-SIGMOD Workshop on Data Description, Access and Control 249-264 (May 1974)

(3) M. M. Astrahan —Implementation of a Structured English Query Language— Comm. ACM Vol.18, No.10 580-588 (Oct. 1975)

(4) H. Tanaka and T. Moto-Oka: Network Oriented Operating System Proc. of PACNET Symposium 171-179 (Aug. 175)

(5) J. A. Smith —Non-Gridded Graphic Input— Proc. of 13th Design Automation Conference 392-398 (1996)

Scientific Information Systems in Japan
H. Inose (ed.)

INPUT PROCESSING SYSTEMS FOR IMAGE DATA

Saburo TSUJI

Faculty of Engineering Science, Osaka University, Toyonaka, Osaka 560

Scientific experiments have provided us very useful and important information with a variety of image data such as microscopic or radiographic images. Effective utilization of these data needs scientific information systems which find important portions of the data, analyze them, and organize compact and well structured informaion on the images at their input stages. Briefly introducing research activities of Group A on the input processing systems of the image data, this paper describes how knowledge on images, both given *a priori* by users and acquired through analysis, is utilized to generate and update models of objects, which are very useful for efficient and accurate analysis of moving images, video or film records of long sequences of images observed in experiments of medicine and biology. Another essential feature of scientific information systems, the versatility to accomodate to a wide variety of needs of sceintists, has been obtained by developing an interactive subsystem, in which a user can easily communicate with a computer to teach procedures for analysis or flexible structural models of objects.

1. INTRODUCTION

Scientific experiments have provided us very useful and important information with a variety of image data such as microscopic and radiographic images in biology, medicine, chemistry, or engineering. Effective utilization of these data, however, needed (or needs) much human labor and time for analyzing them. Recent remarkable progress of science and technology has brought rapid increases both in a number of these images and in complexity of analysis, making it inefficeint, or impossible to examine these data by the human labor.

Consequently, it is highly desirable to develope integrated systems which handle a variety of images as scientific data with effective aids by computer. Since the number of images taken is so great, still growing, and since each image contains a large amount of information contents, we should select important portions of the data, analyze them, and organize compact, well-structured information on the images at input stages of the scientific information systems, otherwise costs for transferring, storing, and accessing these data are too expensive. Thus, six research teams in Group A, which is responsible for studying input processing systems, have devoted their efforts to study input systems of handling and processing of halftone and multi-spectral pictures. Common strategies and key ideas, I believe, for the researches are as follows.

(1) The target of the researches has not been directed at constructing a special, *ad hoc*, system for a very limited class of images but at finding general principles in designing man-machine systems to handle, store, and analyze image data.

(2) Since the system must be versatile to accomodate to a wide variety of needs of scientists, it is highly desirable for a scientist, even not familiar with computer programming, can easily communicate with the system in an interactive mode.

(3) Because of the large amount of information contents, the system searches a set of images for those or portions of those worthy to examine, then analyzes and converts them into a compact, well-structured information for easy, quick, less expensive storage and access.

(4) Knowledge on images and objects, both given *a priori* by users and acquired through analysis, is utilized to generate and update models of objects, which are very useful for efficient and accurate analysis of image data

Briefly introducing activities of research teams of Group A, which have studied on image processing of gray and color pictures, this paper describes how the above ideas were applied to analysis of a long sequence of images, a video or film record of moving images.

2. RESEARCH ACTIVITIES OF GROUP A ON PROCESSING OF IMAGE DATA

Following six research teams in Group A have studied on input processing of image data.

A1: Professor Tosiyuki Sakai, Kyoto University
A3: Professor Morio Onoe, Univeristy of Tokyo
A10: Professor Mikio Takagi, University of Tokyo
A11: Professor Chiyoshi Yoshimoto, Hokkaido University
A12: Professor Saburo Tsuji, Osaka University
A13: Professor Yoichi Iidaka, University of Tokyo

If a computer keeps to observe a scientific experiment, for example a living cell in a microscopic image, for a long period, it accepts an enormous amount of information, and one wastes much cost for transferring and storing these data, unless the computer automatically selects important, or interesting portions of the data for the scientists. Professor Sakai's research team, A1, has studied a flexible computer system which processes such type of time varying scenes, with an aim of developing an on-line intelligent monitoring of scientific experiments. Guided by models of objects, which were given by a user in an interactive mode, the system efficiently tracks variations of the objects in the scene and obtains a structural description of the time-varying objects.

Although a number of research groups in Japan have studied image processing, most of them independently examined the effectiveness of their methods by doing experiments on different picture data. Professor Onoe's team, A3, has developed a picture data base of a wide variety of input pictures, and distributed it among major research groups in Japan, by which one can compare procedures for analysis and transmission of pictures developed by different laboratories, as well as performances of input and output equipments of pictures. A standard format of raw picture data, proposed by this team, was already approved by a technical comittee on image processing of Information Processing Society, and has been widely used in Japan. A3 has also studied various data compression methods for storage and transmission of picture data.

In order to automate analysis and inspection of microscopic images in biology and medicine, Professor Takagi's team, A9, has developed an on-line analyzer of microscopic images which controls X-Y movement of a stage, focusing, and wave length of light for illumination of a microscope. The system automatically searches the specimen for interesting objects, for example candidates for white blood corpuscles, and examines their images in detail to classify them. Another interesting study by this team is to track a great number of moving granules in microscopic images to analyze their motions.

Research activities of Professor Yoshimoto's team, A11, cover a wide area of processing of medical images; transmission of medical images on a telephone line from a remote site to a computer center, an intelligent display system of medical images, a high resolution two dimensional detector for RI images, and computer analysis and a data base of Xray images of stomach.

A wide variety of processes in biology and medicine have been recorded on video tape or cine film as moving images. In order to effectively utilize these records, Professor Tsuji's team, A12, has developed a plan-guided analyzer, an inexpensive yet powerful system of analyzing a long sequence of images. The performances of the system have been evaluated by applying it to two practical examples; motion analysis of fishes in a video record and measurement of spatio-temporal patterns of a heart wall in a cine-angiogram, a record of noisy, low contrasted Xray images of a heart on cine-film.

Professor Iidaka's team, A13, has studied detection and analysis of complex diffraction patterns in very weak Xray images to study complicated molecule structures. Image data from a newly developed, position sensitive, two dimensional Xray detector are analyzed by an on-line computer, which then controls the Xray diffraction equipment in such a way that only significant data for the analysis are acquired.

3. ANALYSIS OF TIME VARYING IMAGES

A wide variety of processses observed in experiments of medicine, biology, or engineering are recorded as time varying images on video tape or cine film for later analysis. Three research teams, A1, A8, and A12, of Group A have studied on analysis of these images, long sequences of pictures called frames. Their research activities have covered three typical examples of motion analysis or understanding of time varying images; (1) tracking of particle-like objects, (2) measuring parameters of spatio-temporal patterns of specified parts of an object, and (3) interpretation of a sequence of images into a structural description on components in the scene and their temporal changes.

Because of a large amount of information contents of a long sequence of images, too much computing time is spent, if whole areas of consecutive frames are examined in detail. Analysis of time varying images, however, does not require all information in the records, but only small fractions are utilized; for example, we analyze only small subareas covering moving objects for tracking, or abruptly changing frames are more carefully examined than almost still ones for motion pattern analysis. Motion analyzers developed by this project have been designed so as to save computation by utilizing knowledge, both given *a priori* by designers or users and acquired knowledge through analysis by themselves, about the input scenes. An example is a plan-guided analyzer of moving images developed by the author's team. *A priori* knowledge which contains general properties such that moving objects are dark and blob-like is useful to design efficient procedures. However it lacks quantitative information about a particular record, such as locations, velocities, brightness levels, and shapes of the objects. The analyzer, therefore, examines a small number of low reolution pictures of the record to collect such information and make a coarse model, a sketch, of the process, by which it can estimate which frame and what regions in it are worthy to be examined, or what type of feature extractor is most effective at each point in the regions. Now, the system makes a plan, specifying (1) frames to be sampled, (2) regions to be examined, and (3) effective feature extractors to be applied. Since more accurate knowledge is acquired as a result of this plan-guided analysis of a frame, the coarse model is replaced with a fine and more reliable one, describing exact properties of the objects in the current frame, which is then used to make a more effective plan of sampling and analyzing next pictures. By iterating this process, planning, analyzing and updating, the system constructs a final model of the process, containing all information required.

As mentioned before, flexibility of the system to various needs of scientists is highly desirable. An interesting and important idea, proposed by Professor Sakai's team, is that each scientist constructs a model of the input scene with a computer, in an interactive mode, by giving the structure of the scene based on his *a priori* knowledge about the objects, and also by specifying suitable portions in a sample input frame to acquire and evaluate each feature value of the components in the model. After matching of the first input frame to the model, the user interactively specifies measuring parameters, such as areas of components or a common boundary length of two specified regions. Then, the system automatically analyzes the consecutive frames and records the parameters measured. The structural model, updated by the result of analysis of each frame, guides the analysis of the coming frame by specifying analysis regions and their feature values, and is very effective for saving of computation.

3.1 Motion Analysis of Granules in Microscopic Images

Movements of a number of particle-like objects are observed in many scientific experiments in biology, medicine, physics or engineering. Manual tracking of these objects, however, is very tedious or impossible, especially when a great number of frames are examined. Professor Takagi's team has studied a system which automatically tracks such particle-like objects in microscopic images [1]. An example of its input pictures is a sequence of photographs of an internal secretion cell of pancreas, which is living in a cultivation chamber mounted on a microscope. Motion analysis of dark granules, carrying insuline to the cell memebrane, in the cell is very important for a study on the mechanism of producing hormone.

Input pictures taken at every second are converted into a sequence of 2280 by 2250 digital pictures. Because of a large number of granules in the scene, as shown in Fig.1, an efficient motion analyzer has been developed such that it first detects only moving granules by a simple, coarse analysis, then small areas covering these objects are examined to precisely measure their movement vectors.

In order to handle picture data in a limited size of memory in a minicomputer, each picture is compressed into 9 to 1 and divided into 20 subregions with 128 by 128 pixels, overlapping each other by a width of 20 pixels. Fig.2 shows a time varying image of Subregion (2,3) sampled at t=1,2,...,6.

Analysis of temporal changes in grey values between two frames is very useful to detect moving objects. Fig.3 shows a process of analyzing two images containing simple-shaped objects. Subtracting image data of (a) from those of (b) at every point in the subregion, we obtain a temporal difference picture (c), in which clusters of pixels showing significant changes are extracted as (d) by thresholding. Then, their centroids (e) are calculated. Since all objects are darker than their background, we can evaluate an approximate displacement of a particle as a vector from a centroid of a positive valued cluster to that of a negative valued one, if both clusters belong to the particle. Therefore, each pair of positive and negative valued clusters located closely each other are examined to test whether they are interpretable as a forward and a backward part of a particle or not. Fig.3 (e) shows results of verification of pairs in (d) and calculation of approximate displacement vectors of the particles.

Fig.1 Digital picture of granules in a pancreatic cell divided into 20 subregions.

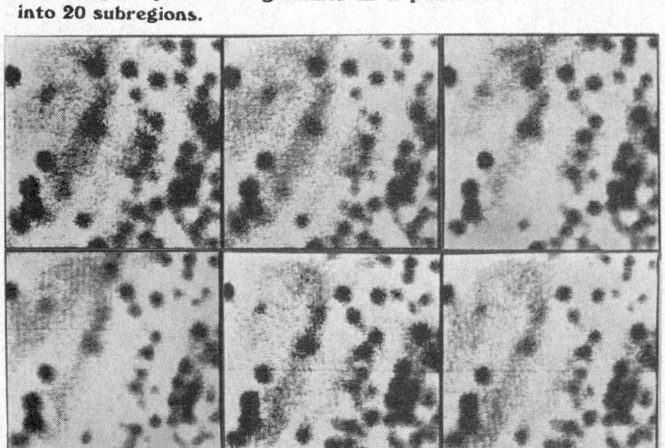

Fig.2 Time varying image in Subregion (2,3)

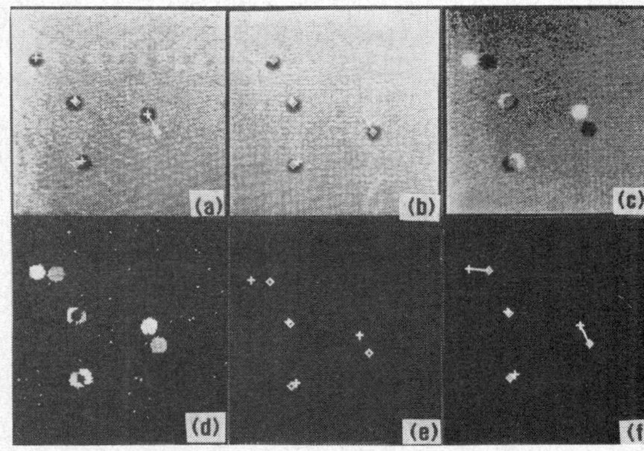

Fig.3 Detection process of displacement vectors of particles.

(a) (b) sampled frames, (c) difference picture of (a) from (b), (d) extracted changing portions, (e) their centroids, and their dispacement vectors.

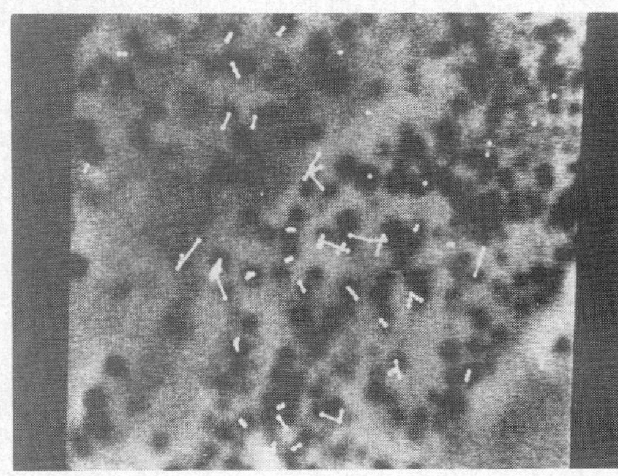

Fig.4 Results of tracking of particles in Subregions (2,2),(3,2) (2,3) and (3,3).

Next step is to precisely measure the displacement vectors by applying SSDA (Sequential Similarity Detection Algorithm) and an adaptive thesholding method to small areas, neighborhoods of the pairs of clusters.

Iterating this process to consecutive frames, the system obtains a list of displacement vectors. Finally, the movement of each particle is tracked by examining connectivities of these displacement vectors. Fig.4 displays a result of tracking of moving granules in 4 subregions (2,2), (3,2), (2,3), and (3,3).

3.2 Motion Analysis of Fishes in Video Records

Another example of motion analysis of particle-like objects has been studied by author's group as an application of the plan-guided analyzer of moving images [2]. The problem is to detect and track fishes in a sequence of pictures recorded on video tape to obtain a meaningful measure of their movements such as linear velocities.

The problem is difficult because (1) efficient procedures should be developed to analyze a large number of frames, (2) a reliable method should be developed to detect boundaries of objects blurred by their rather quick movements, (3) an intelligent method is needed to interpret images of occluding and occluded objects.

In order to solve the above-mentioned problems, following sophisticated techniques are used:

(1) Temporal difference of gray values (difference between consecutive frames) are used to separate moving objects from the background in addition to spatial difference of gray values to detect the blurred moving objects reliably.
(2) The results of previous frames are used to guide the boundary detection process of next frame so that efficient processing of moving images is possible.
(3) Difficult or uncertain objects are analyzed or deduced using models from previous frames.
(4) Misinterpreted or unknown parts in the previous frames are reanalyzed using the results of later frames where those parts could be found.

The input of the system is a video recorded image of fishes swimming in a pool taken by an overhead TV camera. The data are first sampled in a coarse mode, which divides an entire picture into 128 by 128 points, at a rate of 2 frames/second to know coarse information about the process. Analyzing a set of sampled pictures by a modified temporal difference method, the system detects approximate locations and velocities of the moving objects and makes a plan of the next fine data acquisition and analysis. Then, the video tape is rewound and replayed for fine sampling. At this time, a window (or windows) of a quarter area of the entire frame is sampled in a fine mode and digitized to a 128 by 128 picture, where its location and frame number are specified by the plan. Fig.5 shows typical input pictures sampled in the fine mode.

The system first detects approximate boundaries of moving objects by a temporal difference method. Next, a suitable threshold value for detecting a more precise boundary of each objects is evaluated by examining a grey value distribution in a neighborhood of its approximate boundary. Fig.6 shows boundaries of objects in Fig.5 (a)

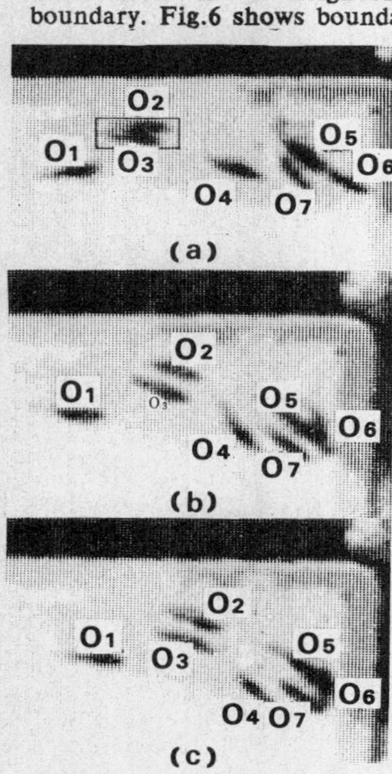

Fig.5 Input pictures sampled in a fine mode at (a) 1st frame, (b) 4th frame, and (c) 5th frame.

extracted by a boundary detector using the threshold values. Then a model of the first frame is generated; each object in the model is described in terms of its area, thickness ratio, location, thereshold value for detection, and chain coded boundary. Since there are small differences in gray values between two objects, it is very difficult to discriminate whether an object is occluded or not. Therefore, if objects are occluded in the first frame, they are interpreted as a single object.

The objects in the consecutive frames are efficiently tracked with aids of the scene model iteratively updated by the results of analysis of each frame. The analyzer generates a prediction map, as shown in Fig.7, from the last model and velocities of the objects evaluated from the last two models. Each object in the current frame is easily found by examining a limited area, a neighborhood of its location predicted, if its image does not overlap others. The prediction map is also used to predict whether an object is occluded or not, and which objects are overlapping which others. A sophisticated coloring method, which also utilizes the models of the objects, interprets the overlapping image into individual objects. Fig.8 shows an example of the interpretation.

When all the frames have been processed, then those objects misinterpreted, such as object O2 and O3 in the first frame, are reanalyzed in a backtracking process utilizing the models of the later frames where those objects are successfuly analyzed. The backtracking process uses the same procedure for the tracking process except that the frames are processed in the reverse order, and only the objects marked as misinterpreted are processed. Fig.9 displays a pass trajectory of the centroid of each object from the first frame to the tenth frame.

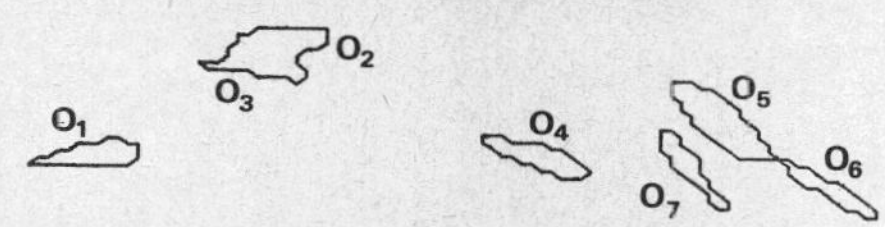

Fig.6 Boundaries of objects in Fig.5 (a).

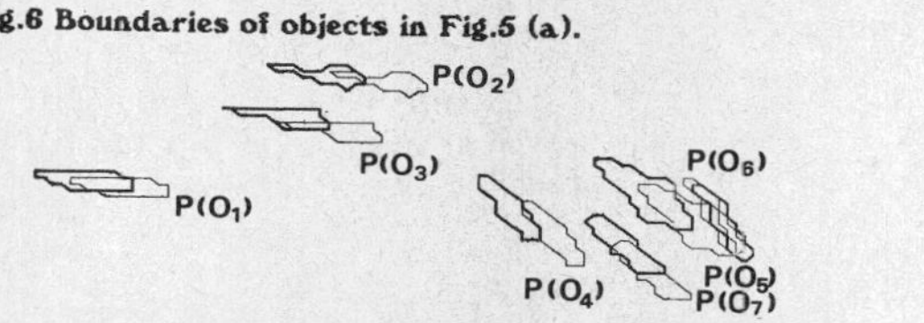

Fig.7 Prediction map generated from model of previous frame. Thick and thin lines show the model of 4th frame and prediction in 5th frame, respectively.

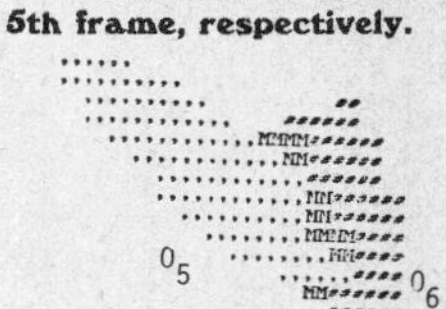

Fig.8 Interpretation of an overlapping image in Fig.5 (a). *(b)*

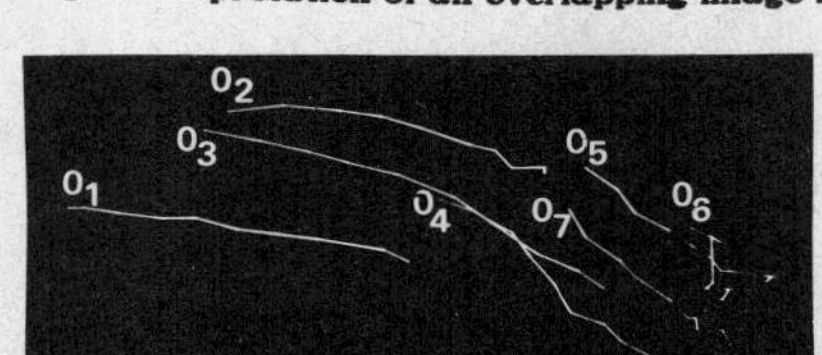

Fig.9 Pass trajectories of centroids of objects.

The system was implemented on a minicomputer HP2108, a 16 bit machine with 32 kword memory, three disks (50 Megabytes), two displays, and 256 bytes buffer memory to store image data of several consecutive frames. Although the computation time depends both on the number of objects in the scene and on the complexity of the picture, the figures obtained for the pictures shown in Fig.4, 11 seconds for the first frame and about 1 second/frame for the other frames, are satisfactory for the practical uses.

3.3 Analysis of Spatio-Temporal Pattern in Noisy Time Varying Image

In recent years, people have studied computer analysis of a cine-angiogram, a cine film record of a radiographic image of a beating heart in which Xray opaque dye is intermittently injected through a catheter. Their primary interest is in measuring temporal change in the volume of the heart; the computer detects the internal surface of the heart in each frame, and calculates the volume by utilizing a three dimensional model of a heart.

In this section, a more advanced analysis is described, which detects both internal and external surfaces of the heart, and measures the spatial and temporal change in thickness of the heart wall, which gives us important measure of heart diseases. Since cine-angiograms for this purpose contain much noise, and since an accurate detection of both surfaces is needed, usual simple level detecting methods are not applicable to this problem. Another important problem is computation efficiency. We need to examine a large number of high resolution pictures, therefore it is necessary for the analyzer to fully utilize its knowledge, both a priori and acquired knowledge, about the dynamic image so as to extract correct information from the record at an extremely low computation cost.

The author's team has developed a plan-guided analyzer of which features are as follows: (1) detection of edge points in the first frame is guided by a plan which specifies regions to be examined and promising feature extractors to be used, (2) smooth boundaries of the wall are obtained by an efficinet search method, and (3) another plan is utilized to analyze consecutive frames; it selects next frame to be sampled and guides the analyzer so as to examine it in a much more efficient manner than the first frame [3].

Input of the system is a sequence of Xray images of a left ventricular chamber recorded on 35mm cine film at a rate of 60 frames/sec. A frame is converted into a 256 by 256 6bit digital picture. An example is shown in Fig.10. Because of the low contrast of gray values between both sides of each surface and much noise, a level detecting method cannot reliably find a heart wall. In order to reliably detect edge points in a noisy picture, edge features should be extracted from averaged gray values in some larger local areas. Since the heart wall is a narrow region, a group of paired rectangular areas of different directions shown in Fig.11 are used as the smoothing areas. A pair of local areas, of which absolute difference of average gray values is largest in the group, is selected at every point in the picture. Each selected pair gives reliable features of the edge at that point. This method, however, requires too much time. If differences of averaged gray values in a pair of 10 by 1 areas are computed and compared for 12 different directions at every point in a 256 by 256 picture, then a minicomputer HP-2108A spends about one hour for the computation.

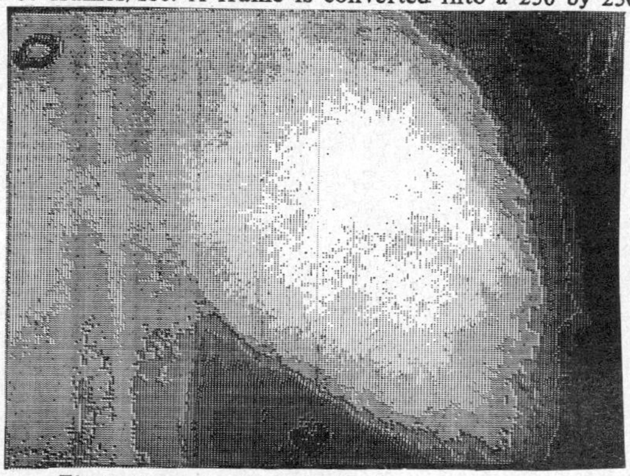

Fig.10 Example of input picture.

We, therefore, have developed a plan-guided edge detection method to shorten the computation time. The system first compresses a 256 by 256 original picture into a 64 by 64 picture by local averaging. Applying a simple 3 by 3 gradient operator to every point in the compressed picture and thresholding it, the system decides analysis regions, regions to be examined further, and an approximate direction of an edge at each point in them. A plan of the next edge detection is described as a map of the analysis region specifying edge directions in them, as shown in Fig.12. The analyzer utilizes this map to efficiently find features; an edge feature at each point in the analysis regions is detected by a pair of rectanglar areas perpendicular to the direction. Since the analysis regions occupy 1/3 of the input picture and only one selected edge direction from 12 variations is examined at each point, the computation for edge finding by this method is 1/36 of that without the plan. The result of applying the plan-guided edge detecting method to Fig.10 is shown in Fig.13.

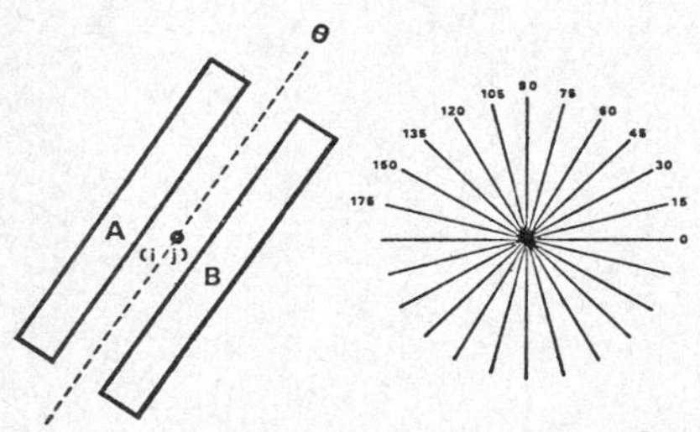

Fig.11 Smoothing areas for edge detection and their directions.

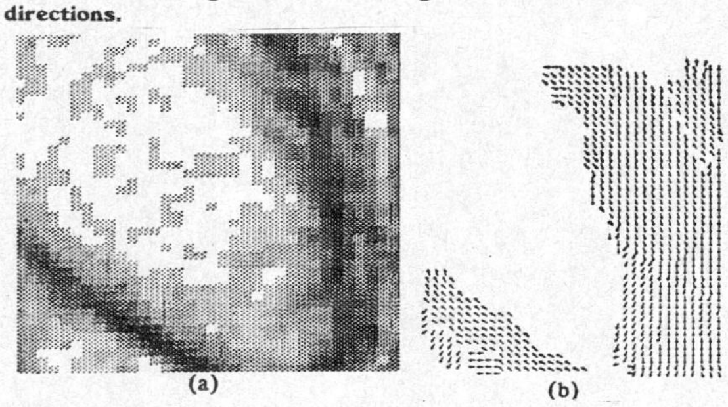

Fig.12 (a) Compressed picture of Fig.10 and (b) analysis regions specifying directions of edges.

Fig.13 Result of plan-guided edge detection

Next phase to the edge detection is to recover the outer and inner boundaries of the heart wall in the analysis regions. Since the edge picture still contains considerable noise, and since precise detection of the boundaries is needed, usual curve followers are not applicable. Optimization methods are known as useful to find a curve of specified properties in a noisy picture, however they require too much computing time, especially when a gobal shape property of a curve is specified. A modified method has been studied to avoid this difficulty. Since the curve to be detected is smooth, we can approximate it with a sequence of linear segments. We embed the smoothness over a wide range of a sequence of linear segments connecting points $x_1,x_2,\ldots,x_n$ into a figure of merit $C(x_1,x_2,\ldots,x_n)$ as

$$C(x_1,x_2,\ldots,x_n) = \sum_{i=1}^{n-1} E(x_i,x_{i+1}) + k_1 \sum_{i=2}^{n-1} |\theta(x_{i-1},x_i)-\theta(x_i,x_{i+1})|$$

where $E(x_m,x_n)$ is a mean of edge strength along the segment connecting x_m and x_n , $\theta(x_m,x_n)$ is the inclination angle of the segment, and k_1 is a constant. We apply a heuristic search to find a solution of maximizing the figure of merit, and dark lines in Fig.13 are results of applying the above-mentioned procedure to Fig.10.

In order to reduce computation for the consecutive frames, we effectively utilize the knowledge obtained from the result of the previous analysis. First of all, narrow analysis regions with a width of 7 pixels are selected on the both sides of the detected boundaries. We can save the computation by examining only these regions. Next, the system selects frames worthy examining from a great number of frames in the record. Because of intermittent movement of the heart, the analyzer skips over consecutive frames in which little changes from the previously analyzed one are observed. Sums of a gray value change at each point in the both regions from that of the previously analyzed one are computed for a coming input frame, and if at least one of the sums is greater than a threshold, then the frame is analyzed, else it is skipped. Thus, the system analyzes about 1/3 of frames in a record as a mean, and only 1/16 at the end of the contraction or expansion period, when the wall shows very slow movement.

Edge detection and search for the boundaries in the sampled frame is guided by the result of the previously analyzed frame. The analyzer measures an edge component parallel to the nearest boundary at every point in the analysis regions. Then it searches each analysis region for a boundary by the modified optimization method.

In order to measure dynamic behaviors of the heart wall, we must find correspondence of each portion of the wall between the analyzed frames. At present, a human operator specifies several small areas, of which patterns seem useful for finding the correspondence. The patterns are matched to those along the boundary in the next analyzed frame, by a simple correlation technique. Fig.14 shows the correspondence of each portion of the wall between selected frames, giving us information about the thickness variation.

The computation time by the minicomputer HP-2108A is about 6 minutes for the analysis of 1st frame, 14 seconds for analyzing a next sampled frame, and less than 1 second for deciding whether an input frame is skipped or not.

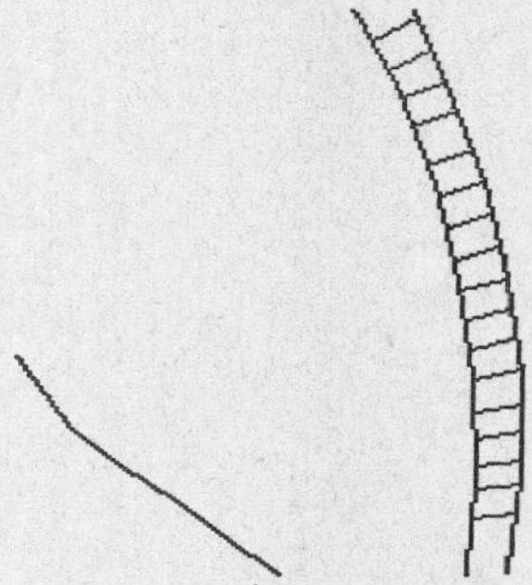

(a)

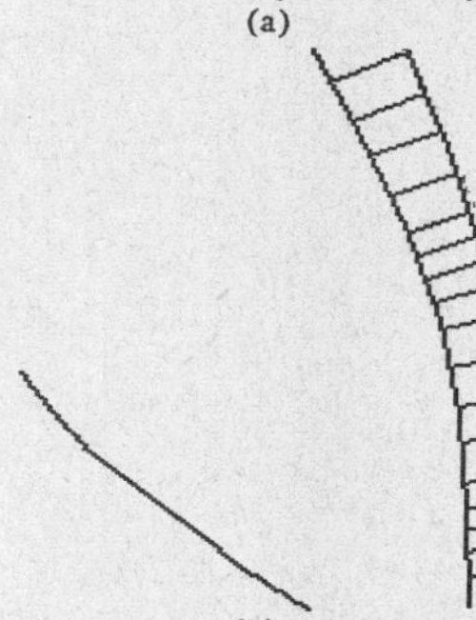

(b)

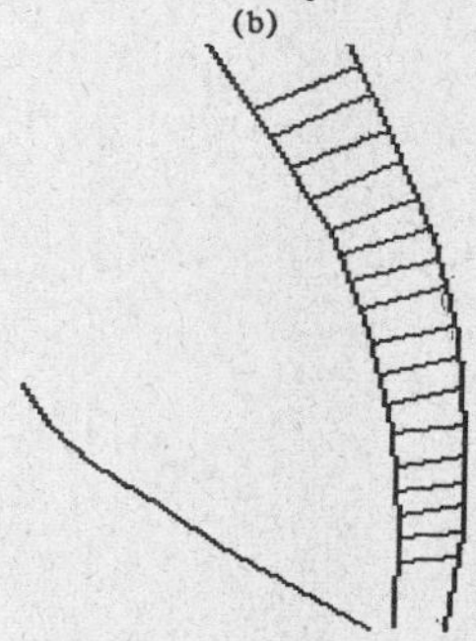

(c)

Fig.14 Thickness of heart wall in (a) 10th, (b) 26th, and (c) 34th frame.

3.5 Interactive Image Modeling and Tracing System for Time Varying Pictures

With an aim of developing an on-line intelligent monitoring of scientific experiments, Professor Sakai's team has studied a versatile and efficient detection system of significant or meaningful changes in a wide variety of time varying images observed in biological and medical experiments. The most important feature, flexibility to various needs of scientists, is obtained by developing a sophisticated man-machine system, in which a scientist, even not familiar with computer programming, easily interacts with a computer to construct and verify a model of the input scene and to specify important parameters to be measured. Since the system utilizes the model to predict objects in the next frame, we can considerably save computing time, as the examples described in the previous sections.

A user and a computer analyze a time varying image in the following 4 phases. Fig.15 shows a schematic of the system.

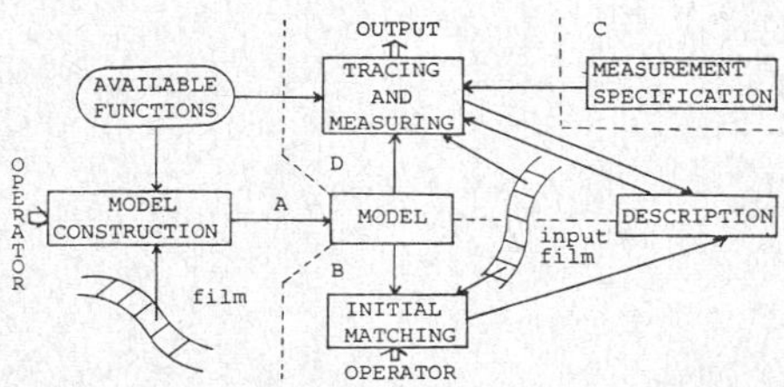

Fig.15 Schematic diagram of the system.

Model Construction (A) : A user interactively describes an image pattern in terms of its components, such as area or line, and their associate properties and relationships. In other words, this phase is to give *a priori* knowledge of the user about the scene to the computer.

Initial Matching (B) : An image pattern in the first frame of the sequence is matched, with an aid of the user, to the model before tracing.

Measurement Specification (C) : The user specifies the parameters to be measured and/or the changes to be detected.

Trace (D) : the system iteratively extracts the objects in the consecutive frames, guided by the model of the previous frame; it effectively utilizes the context between two consecutive frames. The system also measures or detects the specified parameters and/or changes. Finally, the system terminates the trace when conditions of the termination are satisfied.

A hierarchical model, of which structure is shown in Fig.16, is used as a representation of an object, an image pattern, in the scene. The model consists of Primitive Components such as regions or lines, which are characterized by their two types of features, point feature and component feature, and the relationships between them. An example of the objects in the time varying images is a dissociated embryonic cell of Xenopus laevis, of which psudopodium shows a circular movement around its black-dyed nucleus as shown in Fig.17. We define two regions of its nucleus and psudopodium as PCs. Gray values and gradients are used to describe features of pixels in PCs, and an area and a length of the contour of each region are selected as the component features. Adjacency and a length of the common boundary are defined as the interrelational descriptive properties between two PCs. Fig.18 shows a model of Xenopus leavis shown in Fig.17.

The user describes a structure of the model by answering a series of questions asked by the computer. At any time, in the course of constructing the model, he (or she) can suspend the description process and locate efficient features and/or relationships to extract the object; for example, he specifies samples of primitive components and non-interesting areas (background) with a cursor of a color display. Fig.19 shows the sample regions for the object in Fig.18.

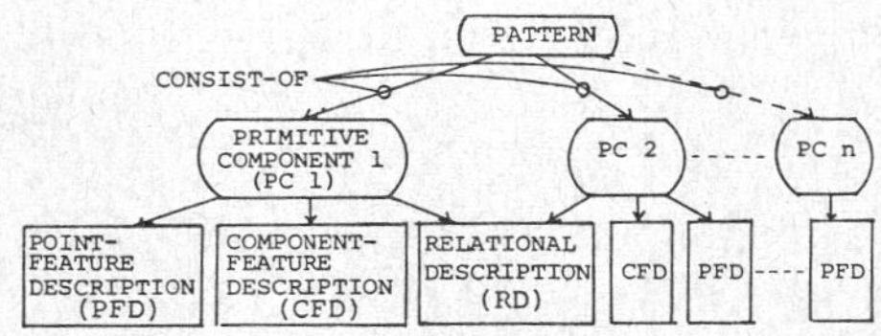

Fig.16 Model of an object.

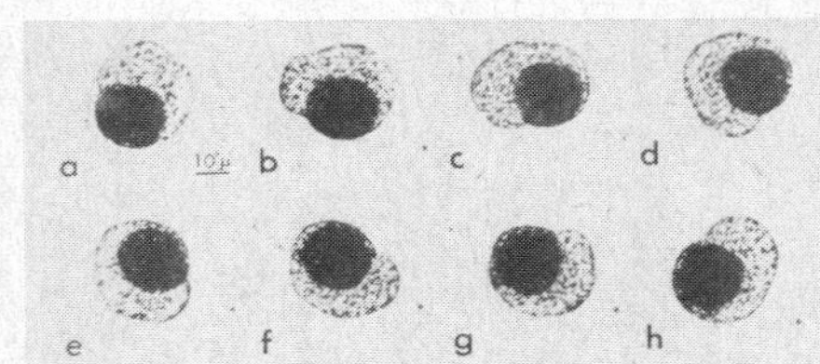

Fig.17 A time sequence of pictures of dissociated embryonic cell of Xenopus laevis.

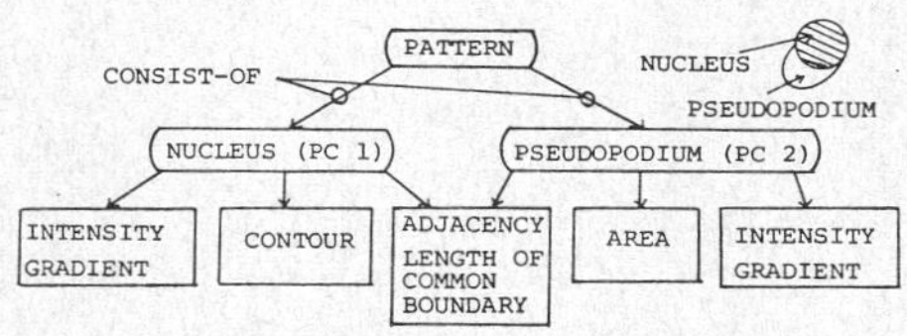

Fig.18 Model of the object in Fig.17.

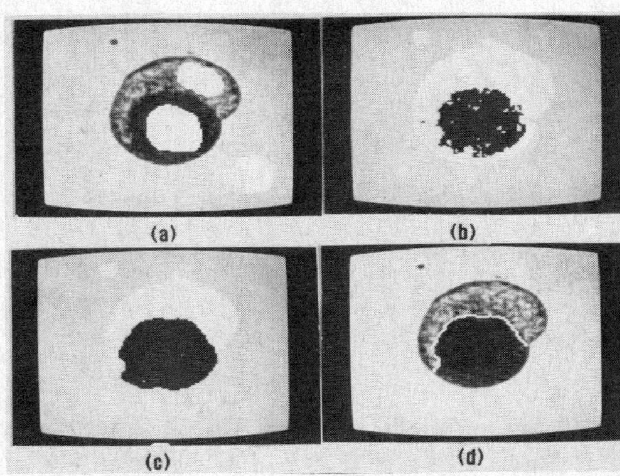

Fig.19 (a) Sample regions (displayed as bright regions) for evaluating feature values for primitive components, (b) result of classified picture elements, (c) extracted regions of nucleus and pseudopodium, and (d) common boundary of neucleus and psudopodium.

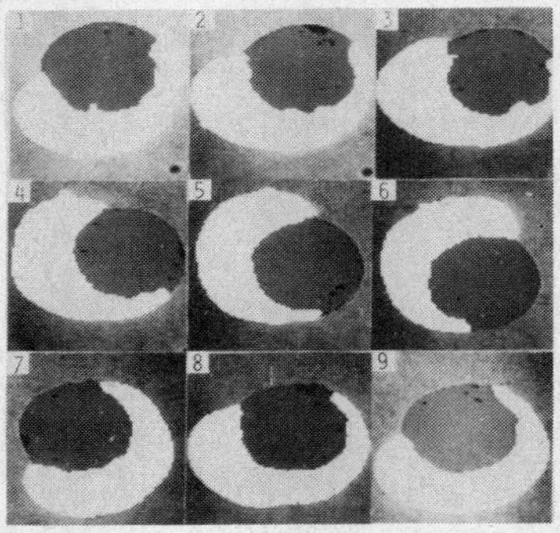

Fig.20 Extraction of Xenopus leavis in 8 consecutive frames.

Neucleus and psudopodium are displayed as dark and white regions, respectively.

Length of the common boundary of nucleus and psudopodium.

Distance between the centers of gravity of neucleus and psudopodium.

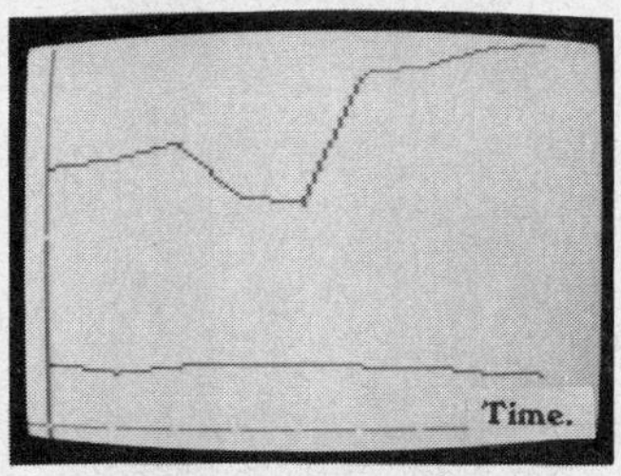

Fig.21 Plot of measured attributes of Xenopus leavis along time axis.

One of the interesting features of the system is that the user can verify the effectiveness of the model by observing the result of matching the model to a sample frame. The system first generates a decision rule in a feature space defined by the point feature description in the model, based on the given samples in PCs, regarding the PCs as classes. Next, it applies the rule to the sampled frame, and classifies the picture elements as shown in Fig.19 (b). Executing commands descrined in the model for image processing such as removal of noises, the system extracts regions that satisfy the feature values and relationships in the model. If the result shown on the color display is not satisfactory, the user iteratively modifies the model until each region is accurately extracted. Fig.19 (c) illustrates the regions extracted by utilizing the model shown in Fig.18.

Before tracing, the object in the first frame is matched to the model already constructed by a similar process described above. Next, the user specifies parameters to be measured and terminating conditions of the analysis, as well. Fig.19 (d) shows that a length of the common boundary of the nucleus and the psudopodium is specified as a measuring parameter.

The system efficiently analyzes the coming frames under an assumption of the differences between two consecutive frames are very small. Fig.20 shows how the nucleus and the psudopodium are successfully detected in 8 consecutive frames, and Fig.21 displays temporal changes of the parameters measured. The versality of the system has been demonstrated in the analysis of various types of moving patterns in biological experiments, such as moving images of an echinus and an ascidian.

4. CONCLUSION

Reseach activities of Group A on procssing image data observed in various types of scientific experiments are described. Because of the large amount of information contents in image data, it is necessary to develope versatile and efficient systems to extract important information for scientists. The system should be intelligent to effectively utilize both *a priori* knowledge of the user and acquired knowledge by itsself about the objects, and several examples developed for analysis of time varying image data show this idea is very useful.

ACKNOWLEDGEMENT

The author wishes to express his hearty thanks to all members of Group A for their kind discussions and advices. He is grateful especially to Professor Sakai of Kyoto University, the leader of Group A, and Professor Takagi of University of Tokyo for their kindness of providing photographs in this paper.

REFERENCES

1] M. Takagi and K. Sakaue, The analysis of moving granules in a pancreatic cell by digital moving image processing, *Proc. 4th Intern. Joint Conf. Pattern Recogintion*, pp. 735-739, 1978.

2] M. Yachida, M. Asada and S. Tsuji, Automatic motion analysis system of moving objects from the records of natural processes, *ibid.*, pp. 726-730, 1978.

3] M. Yachida, M. Ikeda and S. Tsuji, Plan-Guided analysis of noisy dynamic images, *Proc. 6th Intern. Joint Conf. Artificial Intelligence*, pp. 978-981, 1979.

4] Y. Ariki, T. Kanade and T. Sakai, An interactive modeling and tracing system for moving pictures, *Proc. 4th Intern. Joint Conf. Pattern Recognition*, pp. 681-685, 1978.

Scientific Information Systems in Japan
H. Inose (ed.)

FLATS, A MACHINE FOR NUMERICAL, SYMBOLIC AND ASSOCIATIVE COMPUTING

Eiichi GOTO,**, Tetsuo IDA*, Kei HIRAKI**, Masayuki SUZUKI** and Nobuyuki INADA**

** Institute of Physical and Chemical Research, 2-1 Hirosawa, Wako-shi, Saitama 351*
*** Department of Information Science, Faculty of Science, University of Tokyo, 7-3-1 Hongo, Bunkyo-ku, Tokyo 113*

Functional aspects of a machine called FLATS, presently (June 1979) in the design stage, are described. FLATS aims to efficiently run both numerical and algebraic programs. Overflow free and variable precision arithmetic, table look-up computation, and associative computation based on single-hit content addressed tables are introduced for advanced numerical, algebraic and symbolic computing. Hashing hardware, tag mechanism and hardware list processing are used to realize these features.

1. INTRODUCTION

Numerical computation and symbolic formula manipulation are the major practices in scientific computation. While numerical computation has a long history, computer-aided algebra or large scale formula manipulation has become practical only recently.

In this paper, we describe the architecture of FLATS, a machine which aims to efficiently run scientific computations. We consider Fortran (F) and Lisp (L) as major scientific programming languages for reasons:

(i) Scientific numerical programs are written mostly in Fortran.
(ii) Lisp is the major host language used for symbolic and algebraic systems such as REDUCE[1] and MACSYMA[2]. Moreover, Lisp has been used for writing a very large number of programs in the AI area.

Hence, high efficiency in running programs written in 'F' and 'L' is considered a minimum requirement, which is symbolized in the first two letters of 'FLATS'. 'A' in FLATS stands for associative capabilities; 'T' and 'S' stand for Table, Set and String which are the data types used in FLATS.

Many efforts have been made to increase the speed of computations by introducing parallelism[3,4,5], in which we include the pipe line architecture as a specific implementation (of parallelism). Vectorized operations on floating and fixed point numbers of fixed bit length are recognized as a class computations suited for highly parallel machines. It should be noted, however, that highly parallel algorithms are not known for or not applicable to certain computations. For example, take the Euclidian gcd algorithm. The gcd of two integers r_0 and r_1, say, is obtained by successively computing the remainder sequence $r_0, r_1, r_2, r_3 \ldots (r_i$ is the remainder of r_{i-2} divided by $r_{i-1})$ until it reaches the first zero remainder $r_n = 0$ with r_{n-1} giving the gcd.

Parallel schemes faster than the Euchidian algorithm are not known. Hence, for gcd, a machine equipped with a single sophisticated high speed remainder unit would do better than parallel machines with hundreds of slower units. List processing, arbitrary precision integer arithmeter, and variable precision floating point arithmetic are other examples of operations which are not suited for contemporary highly parallel machines. The objective of FLATS is to efficiently run those operations which are not suited for highly parallel machines.

FLATS is basically a single instruction stream machine and parallelism shall be pursued only to a limited extent such as advance control, parallel hashing and parallel interpolation but it shall not incorporate parallel processor arrays.

FLATS may hence be regarded complementary to the highly parallel machine approach.

The numerical support to algebraic manipulation systems is indispensable since the process of algebraic manipulation requires exact calculation of intermediate (and final) numerical coefficients as opposed to approximations done by conventional fixed precision floating arithmetic. Furthermore, the results of the algebraic computation must often be given to numerical computation systems.

Functional requirement for FLATS is therefore to incorporate the different characteristics of the two languages (F and L), e.g. compile-time vs. run-time data type checks, and static vs. dynamic storage allocation, into a single architecture, without impairing the efficiency of either one.

Besides the advanced numerical features, associative capabilities have been found to be very powerful[6] in efficiently carrying out basic operations in formula manipulation.

We first give in section 2 functional specifications of FLATS from users' point of view, and in section 3 consider the implementation from architectural point of view.

2. FLATS FROM USERS' POINT OF VIEW

2.1 Overflow free and variable precision computing

We observe that most contemporary computer systems neither provide adequate means for handling (1) overflows, (2) precision higher than built-in standard precisions, e.g. double or quadruple, nor (3) variable precision arithmetic. The design and implementations of some important class of algorithms[7,8,9] are hampered by the lack of the systematic support of these three features.

To remedy such defects, FLATS is provided with following means:

- Integers are treated with automatic multiple-precision.
- Exponents and mantissas of floating numbers are expressed in integers as stated above, hence practically no overflow would occur.
- A variable precision scheme[10] is adopted, in which precision is specified by a status word called an ASW (Arithmetic Status Word), which is part of the PSW (Program Status Word).

These features are made available to users in an extended FORTRAN called BFORT (Bignum Fortran).

2.2 Tabulative computing

Before high speed computers came into extensive use, look-up of mathematical tables was an essential process of numerical calculation. This fact is in quite contrast to 'on-demand' computing prevalent today. To pursue further the speed-up of the present numerical computation, table look-up is becoming again a practical method, owing to the recent advances in memory fabrication technology. In FLATS, we provide following means for tabulative computing[11];

(i) Tabulative computation of numerical functions with a floating number argument such as x^{-1}, $\sqrt{x}$, e^x, sin x, log x and atan x.
(ii) Associative tabulation for (sparse) integer and/or symbolic arguments.

Linear (or quadratic) interpolation shall be used for 24 bit (single length) mantissa and cubic interpolation for 48 bit (double length) mantissa. Namely f(x) is to be computed by $f(x) = f(x_0) + f'(x_0)\Delta x + (1/2)f''(x_0)\Delta x^2 + (1/6)f'''(x_0)\Delta x^3$ where $x = x_0 + \Delta x$ and x_0 is the most significant 12 bit part of the 48 bit mantissa x and each term is assumed to be less than the previous term by a factor of 2^{-12}. The coefficients f, f', f'' and f''' are tabulated (in ROM or RAM). This cubic interpolation can be made within time of two stages of multiplications by introducing parallelism:
Stage 1. Read the coefficient table and the table of cubes of the 12 bit (in precision) Δx^3 and compute the 24 bit Δx^2 in parallel. (Table look up is assumed to be no slower than multiplication.)
Stage 2. Multiply and add in parallel.

For computing the 48 bit inverse $y = x^{-1}$, this scheme is faster than Newton iteration, which would consist of five stages of successive operations:
Stage 1. Read the 12 bit approximate inverse $y_0 = x^{-1}$ from table.
Stage 2,3. Two multiplications to get the 24 bit inverse $y_1 = (2-xy_0)y_0$.
Stage 4,5. Two multiplications to get the full 48 bit inverse $y = (2-xy_1)y_1$.

When a function, say g with (sparse) integer or symbolic arguments, is computed in tabulation mode, the value $v = g(i_1, \ldots, i_n)$ is tabulated in a content addressed table (CAT), using the tuple $(g, i_1, \ldots, i_n)$ as a search key in the CAT. In the subsequent computations of g with the same values of the arguments, the value v is fetched immediately from the CAT. Associative tabulation effectively and automatically speeds up recursions by avoiding repetitive computation. For example, the notoriously slow (exponential time) recursive computation of the Fibonacci numbers:

```
procedure Fib[n] = [n<1 -> 1;
                    T -> Fib[n-1] + Fib[n-2] ]
```

is turned into a fast algorithm (linear time for the first evaluation and O(1) time for the subsequent evaluations).

2.3 Associative capabilities and built-in data types

Starting from the pioneering associative language LEAP[12], a variety of associative language and architectures for the associative processing have been proposed and implemented.

We consider an AMT (Address mapping table) and an h-op hash table (Fig. 1) as the two basic entities for building associative data structures[13]. We define an AMT as a mapping from bit patterns of fixed sizes (actually 32 bits/64 bits) into addresses (consecutive integers). An AMT may be regarded as a single-hit associative memory. A multi-hit associative scheme such as found in LEAP is to be dealt with by software, making use of arrays, linked-lists and AMT's. Table 1 gives the built-in data types in FLATS. The data types STR, INT, FLOAT, VECT (one dimensional array) are the same as in most other languages except that INT and FLOAT are of arbitrarily high precision as stated in 2.1. PAIR is the same as in Standard Lisp[14].

The data types prefixed by H denote those types obtained by hashing (h-op). h-type objects are guaranteed to be unique by virtue of hashing. The Lisp concept of 'intern' corresponds to that of h-op with the following differences; that h-op does not allow remob[15] and that h-op can apply to all data types in FLATS, whereas intern op in Lisp 1.5 is applicable only to atomic symbols. h-type objects must be read-only so as for the hashing search mechanism to work properly. Hence, HSTR objects are the same as atoms of Lisp; i.e. every different string of characters is represented as a unique pointer to a table called oblist in Lisp 1.5, or to name tables in assemblers and compilers. HINT and HPAIR objects are multi-precision integers and hashed lists, respectively. Equality check of two long integers of type HINT or two lists of type HPAIR is reduced to that of two pointers. HPAIR and HINT can be made by the shared linked list method as shown in Fig. 1.

Insertion, deletion and membership test of an AMT entry correspond to the same operations on a set. A 'set' created by h-op (represented by an AMT) is given[6,16] and shown in Fig. 1. In FLATS the hashed AMT (HAMT) is a synonym to 'set'.

A CAT consists of an AMT and RAM (random access memory, i.e. hardware term for VECT sharing the same addresses). A CAT can be hashed similarly to an AMT and the resultant data type is called HCAT.

These data types, especially the hashed types, have been found to be very useful in polynomial manipulation. For example, take a polynomial $P = 3AX^2 + 4BY^3$, where AX^2 and BY^3 are term identifiers; and 3 and 4 are coefficients. There are many (non-unique) ways to represent a term identifier in list forms owing to the commutative nature of multiplication. AX^2, for example, is represented in a list form ((A 1) (X 2)) and ((X 2) (A 1)) corresponding to AX^2 and X^2A respectively. The HCAT representation {A:1, X:2} is unique (AX^2 and X^2A are represented by a pointer to a unique HCAT structure). In this case, P is represented uniquely by nested HCAT's as
P = {{A:1,X:2}:3, {B:1,Y:3}:4}.

A FLATS FORTRAN program for multiplying polynomials in the HCAT normal form can be written in 9 lines:

```
      FUNCTION CATMULT (P,Q)
      C = MKCAT(0)
      SWEEP 1, P, TP=KEY, FP=VAL
      SWEEP 1, Q, TQ=KEY, FQ=VAL
      T = H(CATADD(TP, TQ))
    1 C(T) = C(T) + FP*FQ
      CATMULT = H(C)
      RETURN
      END
```

Here SWEEP is similar to DO in FORTRAN; SWEEP scans all the occurrences of keys in the specified CAT's P and Q. Function H corresponds to *h*-op above mentioned. This program would give the best known time complexity for multiplying multi-variate polynomials[6].

3. FLATS FROM ARCHITECTS' POINT OF VIEW

3.1 Tagged architecture

A practical solution for efficient detection of the built-in data types given in 2.3 is a tagged architecture.[10,17] Table 2 gives tags to be used in FLATS.

Take numerical computation for example. Arithmetic operations on small numbers tagged as "sint" (short integer) or "sfloat" (short floating number) are to be made with standard hardware at high speed, while those on tagged "big" numbers are to be trapped and handled by microprogramming. Since "sint" or "sfloat" numbers are likely to appear with high probability in most programs, the slow down factor due to the "big" numbers is very small.

Bit-loss, a common objection against tagbit(s), can be remedied by using a specific exponent value as a 'trapping tag'. In the case of 7-bit exponent $-64 \le p \le 63$ FLOAT representation, $p = -64, \pm 63, \pm 62$ are used as the tags to denote data types other than "sfloat" with the mantissa part having different meanings given in table 2. The effective bit loss is only $\log_2(123/128) = -0.06$ bits in this scheme.

3.2 Hashing Hardware

AMT's can be implemented either by using CAM (content addressable memory) chips or by hashing. In a previous paper[18], we presented a hardware hashing scheme which makes use of parallel read-out mechanism of memory. A parallel hashing hardward consists of multiple RAM banks, hash address sequence generators and a hashing control unit. A single comparator is attached to every RAM bank (cf. Fig. 2) in the parallel hashing hardware, whereas in CAM chip such as Intel 3104 a comparator is attached to every memory cell. Even at a chip level, a CAM requires several times more gates than a RAM chip. Hence total number of gates required for realization of CAM is greater by at least one order of magnitude than that for the hash table, with no significant speed gain over parallel hashing: The items in the hash table are searched in average O(1) memory cycle (a quantity independent of the number of items entered in the hash table), in comparison with strict O(1) used to search items entered in CAM. Since we are interested in the overall efficiency in our intended applications, it is more advantageous to invest the resource on enlargement of RAM memory than on still expensive CAM chips. Moreover, *h*-op of complex types such as AMT, CAT and VECT cannot be performed with CAM's.

Figure 2 shows the schematic diagram of the parallel bank hashing hardware of FLATS. Hashing is essentially the cycle (hash probe cycle) of (i) hash address generations, (ii) memory accesses, (iii) key comparisons and (iv) judgement of termination based on the comparisons. The hash tables, (i.e. AMT's and CAT's and an *h*-op table), are taken in main memory consisting of J banks. In a simple parallel hashing scheme with J banks, a single hash address generator and J comparators are used together with some auxiliary hash sequence control circuits. To handle key deletion, detectors of reserved words for denoting empty and deleted states are provided in each bank.

Assuming that the operations (i) - (iv) are performed in a single memory cycle, the performance of parallel hashing is given in number of memory cycles used until the completion of the probe cycles. Figure 3 shows that the average number of memory cycles in a successful search (finding a key in the table) in the worst case[19]. We presume the worst average performance in the performance evaluation when key deletion and insertion are repeated alternately. It shows that the basic hash operations can be performed in a time comparable to an indirect addressing operations, e.g. for $J = 16$ and the load factor of a hash table $\alpha = 0.9$. Note that the hashing hardware handles a fixed length key at a time. For handling variable length strings, arbitrarily long integers and lists, a shared linked list above mentioned is used with the parallel hashing hardware.

3.3 Hardware support for list processing

The hardware mechanisms essential to the speed-up of Lisp are following;

(1) run time data type checking,
(2) hardware stack,
(3) high memory bandwidth
(4) segmented memory space with automatic boundary checking for arrays etc., and
(5) instructions for Lisp primitives, such as car, cdr, cons and atom.

Note that the tagged architecture given in 3.1 is a solution to point (1).

As for point (2), we incorporate in FLATS both general (global) register and stack frame machine architecture. The basic op codes consist of four 8 bit fields as (op, r_1, r_2, r_3), where r_1, r_2, r_3 are register addresses each specifying either one of 128 global registers or one of 128 registers on the current stack frame. While the global registers are used to hold global entities, the stack registers are used for local entries and for (recursive) subroutine linkage.

In contrast to the two register address operations, r_1: = op(r_1, r_2) notably used in IBM 360/370, three register address operations r_3: = op(r_1, r_2) are used in FLATS. The three register scheme would greatly reduce the number of register to register transfer operations. For example, while no register to register transfers are needed in the three register scheme for the following codes

```
a:=add(x,y);
b:=sub(x,y);
c:=mul(x,y);
d:=div(x,y); ,
```

50% of the corresponding codes would be register to register transfers in the two register scheme as in

```
a:=x; a;=add(a,y);
b:=x; b:=sub(b,y);
c:=x; c:=mul(c,y);
d:=x; d:=div(d,y);.
```

The unit of memory banking used in hashing is actually taken to be the bit width of a single list cell (64 bits). Expansion of memory to increase the degree of parallelism (i.e. to increase J) will speed up hashing operations as shown in Fig. 3; this does not imply the requirement of higher memory bandwidth since the comparators are provided in the memory units.

Need for segmentation is clear since the main memory is conceptually divided into several segments; AMT's, CAT's, arrays, list areas, program space and so on.

Fundamental Lisp primitives given above are executed in a single memory cycle. In fact, when run-time data type check is made by hardware, car and cdr instructions are equivalent to simple load operations. Complex Lisp primitives such as member and reverse are microprogrammed.

Furthermore, by virtue of the hashing hardware, operations on property lists and the oblist which rely on linear search in most Lisp systems are performed more efficiently in FLATS.

3.4 Hardware garbage collector

Dynamic storage allocation of variable length segments of memory in FLATS requires a compactifying garbage collection scheme. The most time-consuming process in compactifying garbage collection schemes[20,21] is pointer adjustment. FLATS is to incorporate the pointer adjustment hardware given in Fig. 5.

The pointer q in Q register is to be adjusted to be $q' = q - \Delta q$, where Δq is the total number of inactive cells in the memory block between base q_0 and q. A bit table of 2^n width is set up in the memory local to the hardware. The bit table is marked by a garbage collector: '1' in the m -th bit position in column ω corresponds to the inactive cell at $q_0 + \omega 2^n + m$ in the main memory. Therefore Δq is obtained by counting 1's between the base q_0 and q. The column $(q - q_0)/2^n$ of the offset table contains the cumulative bit count Σ. The adjustment can be done in a main memory cycle when fast memory for the offset table and the bit table is used.

4. CONCLUDING REMARKS

At a users' level, FLATS is a machine both for numerical processing and for formula manipulation. Instead of devising yet another language, we provide the facilities in the framework of existing languages, Fortran and Lisp (REDUCE is to be used for the user interface language). New compilers for Fortran and Lisp are under development, which incorporate the facilities for associative processing and overflow free and variable precision arithmetic, at the same time to accept programs written in Lisp and Fortran, which have been accumulated in libraries of scientific application packages. The underlying techniques include hardware hashing, tag mechanism, and hardware list processing (including hardware garbage collection). Concurrent garbage collection is not considered, since FLATS is not intended to be used for real time applications. FLAT is to be used as a back-end processor to a conventional computer system; hence the average performance, i.e. throughput, is a more important factor than responsiveness. For the same reason, hashing is best suited for the associative processing discussed.

The associative capabilities of FLATS would also be suited for some operations for data base management.

Implementation of unified memory management of secondary and main memories using virtual tapes[22] is under consideration.

Substantial amount of software for FLATS has already been written and has been running under a FLATS simulator. Architectural design of FLATS is an outcome of the experiences with the development of the software system HLISP[6,16] REDUCE is now running under HLISP system and are used for various applications. A version of BFORT mentioned before which translates the extended Fortran into HLISP has already been running.

REFERENCES

[1] Hearn, A.C. REDUCE2 user's manual, Utah Computational Physics UCP-19 (March 1973)
[2] The MATHLAB Group, MACSYMA reference manual, Laboratory for computer science, MIT (1974)
[3] Watson, W.J. The TI ASC - A Highly Modular and Flexible Super Computer Architecture, AFIPS 1972 FJCC, AFIPS Press

[4] Russell, R.M. The CRAY-1 Computer System, CACM, Vol. 21, No. 1 (1978)

[5] Barnes, G.H., Brown, R.M., Kato, M., Kuck, D.J., Slotnick, D.L. and Stokes, R.A. The ILLIAC IV Computer, IEEE Trans. Computers, Vol. C-47, No. 8 (1968)

[6] Goto, E. and Kanada, Y. Hashing Lemmas on Time Complexity with Application to Formula Manipulation, Proc. ACM-SYMSAC, (1976)

[7] Takahasi, H. and Mori, M. "Double Exponential Formulas for Numerical Integration", Pub. Research Institute for Math. Science, Kyoto University, 9 (1974), 721-741

[8] Grau, A.A. "Algorithm 256, Modified Graeffe Method", Collected Algorithms from ACM, 256-P1-0

[9] Brent, R.P. "Fast Multiple-Precision Evaluation of Elementary Functions", JACM, Vol. 23, No. 2, (April 1976), 242-251

[10] Ida, T., and Goto, E. Overflow Free and Variable Precision Computing in FLATS, Journal of Information Processing Vol. 1, No. 3 (1978)

[11] Goto, E. and Terashima, M. MTAC - mathematical Tabulative Automatic Computing, Technical Report of Information Science, 77-03, Faculty of Science, University of Tokyo, (1977)

[12] Feldman, J.A. and Rovner, P.D. An Algol-Based Associative Languages, CACM, Vol. 12, No. 8 (1969)

[13] Goto, E., Sassa, M. and kanada, Y. Algorithms and Programming with CAM's (Content Addressable Associative Memories) Technical Report of Information Science Department, 78-04, Faculty of Science, University of Tokyo, (1978) submitted for publication

[14] Marti, J.B., Hearn, A.C., Griss, M.L., and Griss, C. Standard Lisp Report, UUCS-78-101, University of Utah, 1978

[15] McCarthy, J., et. al., Lisp 1.5 Programmers Manual, MIT Press, Cambridge, Mass., 1962

[16] Sassa, M. and Goto, E. A Hashing Method for Fast Set Operations, Information Processing Letters, Vol. 5, No. 2, (1976)

[17] Feustel, E.A. On the advantage of tagged architecture, IEEE Trans. Computers, Vol. C-21, No. 9 (1972)

[18] Ida, T. and Goto, E., Performance of a Parallel Hash Hardware with Key Deletion, Proc. IFIP Congress 77, North-Holland (1977)

[19] Ida, T. and Goto, E. Analysis of Parallel Hashing Algorithms with Key Deletion, Journal of Information Processing, Vol. 1, No. 1 (1978)

[20] Wegbreit, B. A Generalized Compactifying Garbage Collector. The Computer Journal, Vol. 15, No. 3, (1972)

[21] Terashima, M. and Goto, E. Genetic Order and Compactifying Garbage Collectors, Information Processing Letters, Vol. 7, No. 1

[22] Sassa, M. and Goto, E. "V-tape", a Virtual Memory Oriented Data Type, and its Resource Requirements, Information Processing Letters, Vol. 6, No. 2, (1977)

[23] This paper is the updated version of the paper by the same authors with same title given at ACM IEEE Symposium on Computer Architecture, May 1979, Philadelphia, PA.

main types	examples	hashed types	examples
ATOM			
STRing	3STOM	HSTR	3HTOM, 'TOM', TOM (three optional representations)
INT	-0137	HINT	-137
	+137		137
FLOAT	3.1415	HFLOAT	3.1415
PAIR		HPAIR	
DLIST	(A.B)	HDLIST	(A;B)
LIST	((A 1) (B 2))	HLIST	((A,1), (B,2))
FUNC	Fib		
VECT	[3 5 6 7]	HVEC	[3,5,6,7]
AMT	{A B}	HAMT	{A,B}
CAT	{A:1 B:2}	HCAT	{A:1,B:2}

Table 1. Built-in data types in FLATS

value of exponent, p	hardware data type	meaning of a mantissa
63	address	a pointer to some data structure
62	character	upto three characters
61 ~ -61	"sfloat", exponent of short floating number	mantissa of short floating number
-62	"sint", short integer	value of short integer
-63	software use	used as a trap code which may denote software (and microprogram) identified data types, such as "big" integer and "big" floating numbers
-64	reserved bit patterns	reserved values denoting nil, empty, deleted etc.

Table 2. Hardware tags used in FLATS

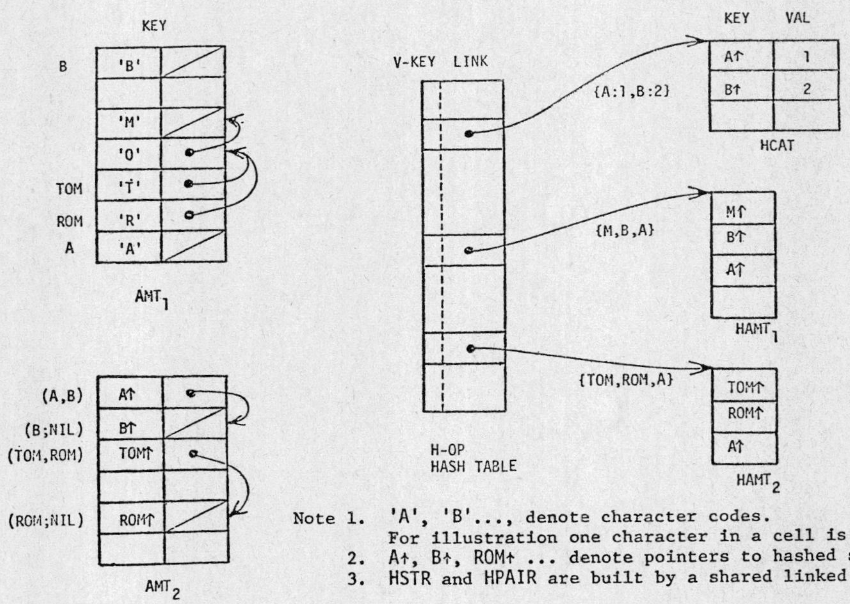

Note 1. 'A', 'B'..., denote character codes.
For illustration one character in a cell is stored.
2. A↑, B↑, ROM↑ ... denote pointers to hashed strings.
3. HSTR and HPAIR are built by a shared linked list.

Fig. 1 Realization of Built-in Data Types

Fig. 2 Parallel Hashing Hardware

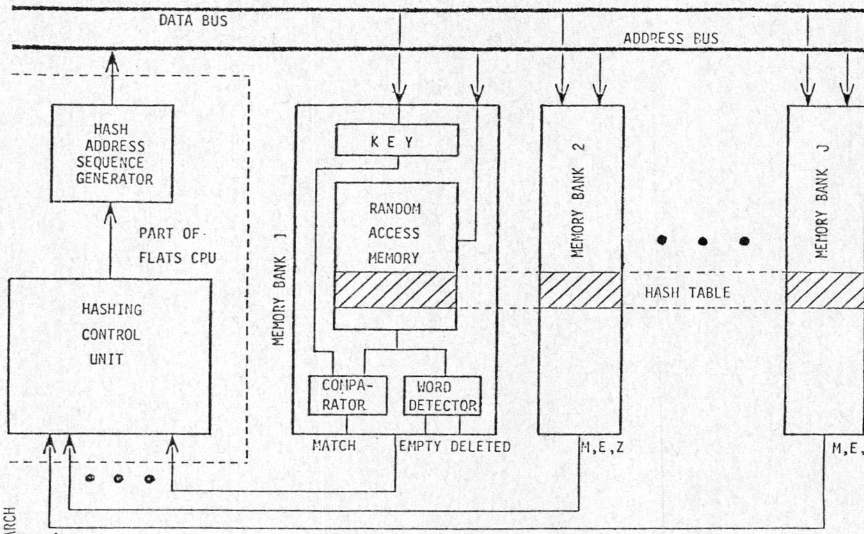

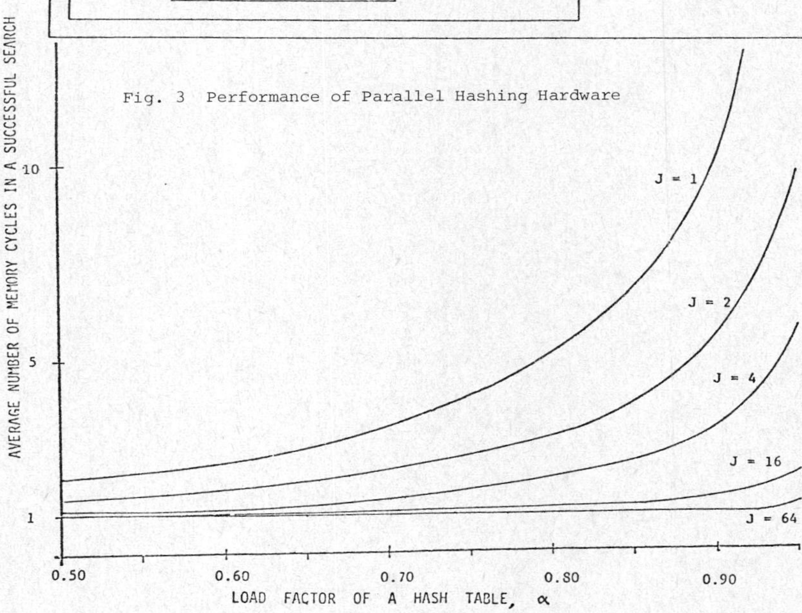

Fig. 3 Performance of Parallel Hashing Hardware

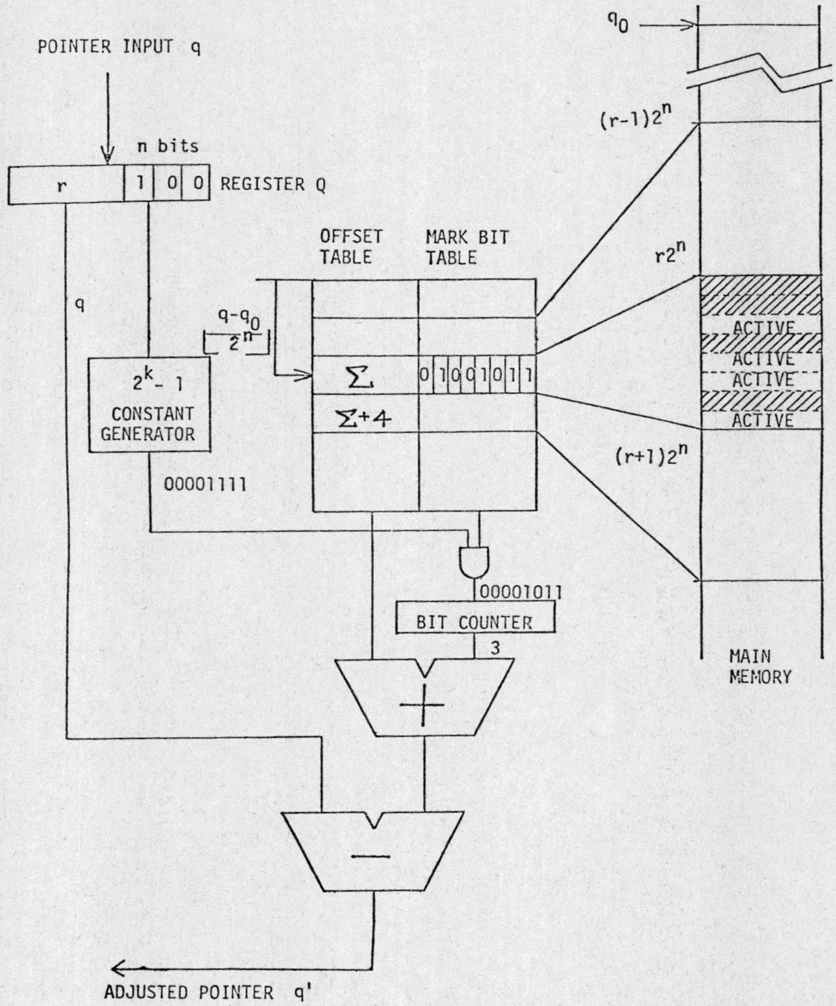

Fig. 4 Hardware for Pointer Adjustment

Scientific Information Systems in Japan
H. Inose (ed.)

STATISTICAL DATA PROCESSING SYSTEMS

Masashi OKAMOTO, Chooichiro ASANO** and Toru NAKAGAWA*

**Faculty of Engineering Science, Osaka University, Toyonaka, Osaka*
*** Research Institute of Fundamental Information Science, Kyushu University, Hakozaki, Higashi-ku, Fukuoka*
**** Faculty of Science, University of Tokyo, Hongo, Bunkyo-ku, Tokyo*

As a subgroup of the A-Group "Input/Output" which dealt with the input-output aspect of scientific information systems, we determined our objective of developing two statistical data processing systems, SALS and NISAN. As for SALS system, the first version and second version were finished and made public in April 1978 and May 1979, respectively, at the Computer Centre of the University of Tokyo and the latter is now being transferred to several other university computer centres in Japan. The designing and developing of NISAN system was started in 1977 after a basic survey on statistical program packages throughout the world as well as on Japanese user's needs to such packages. The preliminary version of the system was installed at the Computer Centre of Kyushu University in January 1979.

1. INTRODUCTION

Research Project A-4 "Research of statistical program packages" to which the authors belonged started its activities in 1976 as a subgroup of the A-Group "Input/Output" in the Special Research Project "Formation process of information systems and organization of scientific information", supported by Ministry of Education, Japan.

Our objectives were two-fold as follows:
(i) to develop two statistical data processing systems including computer programs of various statistical procedures, especially for multivariate analysis and least-squares nonlinear fitting for natural or behavioral sciences, and
(ii) to develop statistical methodologies useful in data analysis in natural or behavioral sciences.
Since this paper is concerned mainly with the first objective, the second shall be touched upon only briefly in this section because it serves as an important background of the first.

The reaearch team was divided into four subgroups corresponding to the following four subject fields: SPP(statistical program package), natural science, behavioral science, and methodology. The first two subgroups were organized primarily for the objective (i), while the other two worked for the objective (ii). What we did in this research project will be summarized below.

SPP subgroup consisting of 7 investigators led by Asano started its research activity by conducting in 1976 an extensive survey on statistical program packages throughout the world. Asano and two other investigators traveled in Europe and the U.S.A. in order to collect and review program packages GLIM, GENSTAT, RGSP, CLUSTAN 1A etc. in Europe and BMD·BMDP, SPSS, PSTAT, OSIRIS etc. in the U.S.A. [1]. The subgroup also carried out a survey on Japanese user's needs to statistical program packages. In the second year, 1977, the subgroup set out into the essential stage of the research, designing and developing NISAN (New Interactive Statistical ANalysis) system, which will be described in Section 3 at full length.

Natural-science subgroup consisted of 4 investigators and several other natural-scientists and was directed by Nakagawa. It first held a symposium "Data Analysis for Natural Science" in November 1976 to review the current situations of least-squares method and to search user's needs for a computer program system [9]. Then, through ardent collaboration of the members, the subgroup developed SALS (Statistical Analysis with Least Squares fitting) system which is a first all-round system in Japan for such a purpose. The system was initially intended to be used for data analysis in chemical and physical sciences but may be useful in many other branches of sciences, natural or behavioral. The detail will be shown in the following section.

Behavioral-science subgroup (5 investigators, led by Asano) held 7 symposia or workshops in 1976-79 at Osaka U., Hokkaido U., Kyushu U., Hiroshima U., Okayama U. and the Institute of Statistical Mathematics in order to discuss modeling and data analysis in various fields of behavioral sciences. Some of the topics treated were optimal scaling method [24], sequential selection procedures [3] and constellation graph [25].

Methodology subgroup (7 investigators, led by Okamoto) held 13 symposia or workshops in three years at Osaka U., Tokyo U., Kyushu U., Hiroshima U., Tsukuba U. and Okayama U. in order to discuss mainly methodologies in multivariate statistical analysis. Topics treated therein include Monte Carlo integration [19], quantification method [20], ridge regression [6], factor analysis [22], model fitting [5] and latent class analysis [21].

2. SALS SYSTEM

2.1 Needs and Problems of Data Fitting

In various fields of natural science and applications, statistical data analyses like least-squares fitting are required to deal with elaborate theoretical models which are worked out for each problem. Thus, a great number of people have made much efforts to write least-squares programs for solving their own problems.

However, they often meet difficulties in getting

reliable solutions, in selecting the best model, and in extracting statistical information. Such difficulties are mostly originated in

(a) near singularity of normal equations,
(b) unstability in the nonlinear iterative refinement,
(c) contamination of unreliable observations, and
(d) poor guideline to select the best model.

Since all these difficulties are mainly concerned with numerical analysis and statistics, they should be solved in a way universal to various fields of application. Thus, it is desirable to develop a general-purpose program of least-squares which is well organized for wide variety of applications and thoroughly examined from both numerical and statistical points of view [7].

2.2 Development of SALS System

In April 1975, a year earlier than the start of the present Special Research Project, Nakagawa and Kunio Tanabe, Institute of Statistical Mathematics, proposed to build a new program system for least-squares fitting for the above mentioned purpose. About fifteen volunteer scientists working in the fields of mathematics, statistics, computer science, physics, and chemistry and belonging to various universities and research institutes around Tokyo gathered to form SALS Group led by Nakagawa. This new project was soon supported by the Computer Centre of the University of Tokyo (UTCC) as a development project of its application program library. Then, as the present Special Research Project started in 1976, SALS Group joined A-4 subgroup to carry the mission of building a data processing system mainly for natural science.

Basic specifications of the SALS system were discussed by the whole SALS Group, and settled by a four-membered working group in 1975 [8]. Then, Nakagawa and Yoshio Oyanagi, now at Tsukuba University, started to design and code the system in 1976, and had to struggle for full three years to finish the SALS system. SALS-MINI system was built in 1977 as a pilot to the standard SALS system. The first preliminary version of the SALS standard system was finished in April 1978 and its full revised version (SALS Version 2) was finished and made open to public use in May 1979. A large number of SALS Group members collaborated to improve the design and to debug the system. Among others, Hayato Togawa, now at Nihon University, was in charge of linear least-squares algorithms and Tetsuzo Ito, Institute of Phycical and Chemical Research, tested the effectiveness of robust estimation methods. Discussions and encouragements by other subgroup members of A-4 group were also helpful [9, 10]. In actual programming, a flowcharting method FORTFLOW [11] was introduced. The FORTFLOW flowcharts, which keep one-to-one correspondence with Fortran statements, are helpful to rationalize the initial coding, to record the whole history of program revisions, and to maintain the up-to-date source programs, at the same time.

2.3 Structure of SALS System

SALS is a full system including least-squares fitting algorithms and statistical treatments [12, 13]. It is written in Fortran JIS 7000 and constructed as a single system composed of about 130 subroutines with neary 10000 steps. Figure 1 illustrates the calling structure of SALS; each block in Fig. 1 is composed

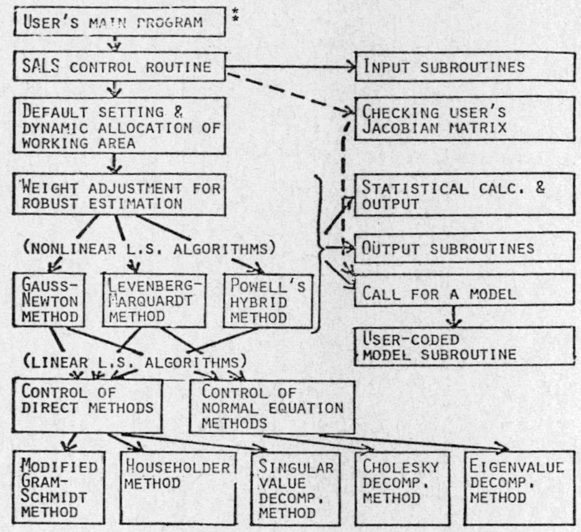

Fig. 1 Structure of SALS System

of one to twenty subroutines. This hierarchical structure makes the system easy to understand and maintain.

The user has to supply a main program just for calling SALS system and also a subroutine to calculate the theoretical model to be fitted. It is desirable but not always necessary to calculate the Jacobian matrix (first derivatives) of the model in the user subroutine. There is an option to check the correctness of the user-coded Jacobian matrix. If desired, the user may also supply his own input/output routines suitable to his problem. The distinction of the roles of system routines and those of user-coded routines makes SALS flexible and applicable to a wide variety of fields.

2.4 Numerical Algorithms for Least-Squares

2.4.1. Basic Formulation

The least-squares problem is symbolically expressed as

$$\mathbf{y} \overset{\sim}{=} \mathbf{f}(\mathbf{x}) \quad \text{with uncertainty } \boldsymbol{\sigma}, \tag{1}$$

where $\mathbf{y}$, $\mathbf{f}$, and $\boldsymbol{\sigma}$ stand for the n-vectors of observed values, model functions, and observation uncertainties, respectively, while $\mathbf{x}$ for the m-vector of the parameters. By expanding $\mathbf{f}(\mathbf{x})$ into Taylor series around the initial guess of parameters $\hat{\mathbf{x}}^{(0)}$, Eq. (1) is reduced in the first-order approximation into

$$\Delta\mathbf{y} \overset{\sim}{=} \mathbb{A}\Delta\mathbf{x} \quad \text{with uncertainty } \boldsymbol{\sigma}, \tag{2}$$

where

$$\Delta\mathbf{y} = \mathbf{y} - \mathbf{f}(\hat{\mathbf{x}}^{(0)}), \quad \Delta\mathbf{x} = \mathbf{x} - \hat{\mathbf{x}}^{(0)}, \quad \text{and}$$

$$A_{ij} = \partial f_i / \partial x_j \ : \text{Jacobian coefficients.} \tag{3}$$

The linear (or linearized) least-squares problem, Eq. (2), is traditionally solved through the normal equation

$$(\mathbb{A}^T\mathbb{W}\mathbb{A})\Delta\mathbf{x} = (\mathbb{A}^T\mathbb{W}\Delta\mathbf{y}). \tag{4}$$

Setting up Eq. (4), however, has been recently found to lose accuracy especially in ill-conditioned cases. New methods are based on the QR decomposition of the weighted Jacobian matrix $\mathbb{A}'$ as

$$\mathbb{W}^{1/2}\mathbb{A} = \mathbb{A}' = \mathbb{QR}, \tag{5}$$

where $\mathbb{Q}$ is an orthogonal matrix and $\mathbb{R}$ an upper triangular matrix, and then give the least-squares solution with higher accuracy as

$$\Delta\hat{\mathbf{x}} = (\mathbb{A}^T\mathbb{W}\mathbb{A})^{-1}(\mathbb{A}^T\mathbb{W}\Delta\mathbf{y}) \tag{6a}$$

$$= \mathbb{R}^{-1}\mathbb{Q}^T\mathbb{W}^{1/2}\Delta\mathbf{y}. \tag{6b}$$

Hence in SALS system these new methods without setting up the normal equation are introduced as the default procedures.

In case that the model $\mathbf{f}(\mathbf{x})$ is nonlinear with respect to the parameters $\mathbf{x}$, they need to be refined by using the correction term $\Delta\hat{\mathbf{x}}$. The simple Gauss-Newton refinement

$$\hat{\mathbf{x}}^{(1)} = \hat{\mathbf{x}}^{(0)} + \Delta\hat{\mathbf{x}}, \tag{7}$$

however, sometimes fails to reduce the residual sum of squares S. Hence, various techniques are introduced to SALS for the purpose of reliable convergence of the parameters.

2.4.2 Linear Least-Squares Algorithms

For the linear least-squares calculation, the following five selected algorithms are available in SALS:
(a) Modified Gram-Schmidt orthogonalization method,
(b) Householder orthogonal transformation method,
(c) Singular-value decomposition method,
(d) Cholesky decomposition method, and
(e) Eigenvalue decomposition method (or diagonalization method).

Algorithms (a)-(c) obtain the solutions directly from the Jacobian matrix without setting up the normal equation, while (d) and (e) solve the normal equations. The formers are to be used whenever possible, because they are more accurate than the latters especially in ill-conditioned cases and yet as fast as the latters. On the other hand, (d) and (e) may be used to save the memory for large scale problems.

All the five methods contain pivoting algorithms and treatments of rank deficiency. Especially, (c) and (e) give the minimum-norm solutions in rank deficient cases.

2.4.3 Nonlinear Least-Squares Algorithms

Two algorithms are now available for the nonlinear least-squares calculation:
(a) Gauss-Newton method with a damping option, and
(b) Levenberg-Marquardt method with Fletcher's modification.

Gauss-Newton method is most efficient in case of weak nonlinearity and/or good guess of initial parameters. If the residual sum of squares S does not decrease for the refined parameter values of Eq. (7), the damping option reduces the parameter corrections $\Delta\hat{\mathbf{x}}$ into halves stepwise until getting a smaller S. The criteria on convergence and termination of the iteration are set after a number of tests, and also can be tuned by the user's input.

Levenberg-Marquardt method (b) artificially adds some extra terms $\lambda\mathbb{D}$ to the diagonal elements of the normal equation matrix in order to avoid the increase in S:

$$(\mathbb{A}^T\mathbb{W}\mathbb{A} + \lambda\mathbb{D})\Delta\mathbf{x} = \mathbb{A}^T\mathbb{W}\Delta\mathbf{y}, \tag{8}$$

where $\mathbb{D}$ is taken to be either a unit matrix $\mathbb{I}$, a diagonalized matrix diag $\{\mathbb{A}^T\mathbb{W}\mathbb{A}\}$, or $\mathbb{I}$ + diag$\{\mathbb{A}^T\mathbb{W}\mathbb{A}\}$. Large λ value with $\mathbb{D} = \mathbb{I}$ tends to the steepest descent direction; with a proper adjustment of λ after Fletcher's algorithm, the iterative refinement is expected to converge.

Furthermore, a third method of nonlinear least-squares algorithm,
(c) Powell's hybrid method,
is under way of development. This method does not require the calculation of the Jacobian matrix in the user-coded subroutine, and has been found useful in strongly nonlinear cases of some test problems.

2.5. Statistical Features

2.5.1 Weighted Least-Squares Fitting

Weighted least-squares estimation is a basic statistical method in the present system. The weight w_i of the observed datum is taken as

$$w_i = \text{const}/\sigma_i^2 = \sigma_0^2/\sigma_i^2, \tag{9}$$

where σ_i is the observation uncertainty and σ_0 the standard error of observation.

Some variable parameters which are found difficult to determine from the present data alone may be either fixed to any input values or loosely bound around them with the specified uncertainties.

2.5.2 Robust Estimation

A drawback of the ordinary least-squares estimation is its weakness against the contamination of unexpected error in observation; even one outlier due to a mistake in input can destroy the whole fitting. It is desirable to be resistant against such outliers by relying only on the other normal data. This is the idea of robust estimation.
In addition to the ordinary least-squares estimation, SALS provides the following methods for the sake of robust estimation:
(a) least-squares method with a cut-off option,
(b) Huber's M-estimation method, and
(c) Tukey's biweight method.

The default option in SALS is (c), which proceeds in the following steps:
i) By use of the initial guess of parameters, the residuals v_i are calculated, and a scale of the residuals is evaluated by

$$s = \text{median}\ \{|v_i/\sigma_i|\}. \tag{10}$$

ii) Weights of the observed data are readjusted according to their residuals as

$$w_i^{\text{eff}} = \begin{cases} w_i\cdot[1 - (v_i/\sigma_i sc)^2]^2 & \text{if } |v_i/\sigma_i| < sc, \\ 0 & \text{if } |v_i/\sigma_i| \geq sc, \end{cases} \tag{11}$$

where the constant c is to take a value around 10.
iii) By use of the effective weights w_i^{eff} in place of w_i, (nonlinear) weighted least-squares method is applied to get refined parameter values.
iv) Steps i) through iii) are repeated in cycles until the convergence of the effective weights or until the allotted limit number of cycles.
These procedures have been found robust and useful for diagnostical fitting even for nonlinear problems.

2.5.3 Statistical Information and Diagnostics

The standard deviation $\hat{\sigma}_0^{\,2}$ is calculated from the residuals. As a safer guess of the standard error σ_0, however, $\hat{\sigma}_0' = \max\{\hat{\sigma}_0, \sigma_0\}$ is also output. Error matrix $\Sigma_{\hat{x}}$ of the parameters is calculated as

$$\Sigma_{\hat{x}} = \hat{\sigma}_0'^{\,2}(A^T W A)^{-1}. \qquad (12)$$

The uncertainties of parameters,

$$\sigma_{\hat{x}j} = (\Sigma_{\hat{x}})_{jj}^{\;1/2}, \qquad (13)$$

and the correlation matrix are also output. Furthermore, uncertainties in the calculated values $\hat{y}_i = f_i(\hat{x})$ are evaluated as

$$\sigma_{\hat{y}i} = (A \Sigma_{\hat{x}} A^T)_{ii}^{\;1/2}. \qquad (14)$$

Any of three kinds of graphs of residuals may be output on the line printer to check the goodness of fit:
(a) histogram of residuals,
(b) plot of residuals against any variable in the input data,
(c) normal probability plot of residuals.

2.5.4 Model Selection with AIC

For the purpose of selecting the most plausible model among several trial ones, Akaike's information criterion AIC may be useful. The general definition,

$$\text{AIC} = -2 \log_e(\text{maximum likelihood}) + 2m \qquad (15)$$

is reduced for the normally distributed error to

$$(\text{AIC})_m = n \log_e S + 2m, \qquad (16)$$

where n and m are the numbers of the observations and parameters, respectively, and S is the residual sum of squares. The model which gives the minimum value of AIC is selected as the most plausible one.

2.6. Control of SALS

2.6.1 Example of SALS Input Job

An example of a SALS input job deck is illustrated in Fig. 2. The job consists of a main program simply calling the SALS system, a user-coded subroutine for theoretical model calculation, and input data of controls, initial guess of parameters, and observed values. As this example shows, the user has to supply only the minimum essentials for using SALS.

2.6.2 Block Structure of Input

About 30 commands are defined to control the SALS system. Each of them has its registered name and

```
//SALS:JOB
//LSALS                                                 ..... Access to SALS library.
//FG   (S,O=O,D)
      CALL SALSI ( 4000)                                  } User's main program.
      STOP
      END
      SUBROUTINE MODELF(N,M,NC,X,Q,FO,F,R)                } User's model subroutine.
      DIMENSION X(N),Q(M),FO(M),F(M),R(M)
      DO 10 I=1,M
      F(I)=X(1)+X(2)*Q(I)**X(3)                            f_i(x) = x_1 + x_2.q_i^x_3
   10 R(I)=FO(I)-F(I)
      RETURN
      END
/*
PROBLEM    1.          5.     3.      4.      4. .... Control of algorithms,
PARAMETER 3.           1.                             statistics, and outputs.
 X(1)       25.                       1.0           } Input of parameters.
 X(2)       10.                       0.1
 X(3)       -0.5                      0.01
DATA       20.                                        } Input of observations.
 DATA  1    2.00       31.50
 DATA  2    3.00       30.06
 DATA  3    4.00       28.74

 DATA 14   35.00       26.10
 DATA 15   40.00       25.32
ENDSALS                                          ..... End of SALS input.
//END
```

Fig. 2. Example of SALS Input Job Deck.

six-or-less control data, and is read in a unified format of (2A4, A2, 6F10.0).

In the input, these commands form the blocks of problem specification, of method, of parameters, and of observations, in this order. If multiple blocks of the same kind are involved in a problem, the data fitting is carried out for all combinations of the blocks. The command analyzer is designed flexible enough so as to require only four commands (PROBLEM, PARAMETER, DATA, and ENDSALS) as the essentials, while leaving all others omissible in ordinary runs as shown in the example in Fig. 2.

PROBLEM command is read at the beginning and specifies the controls of algorithms, statistical treatments, and outputs. PARAMETER and DATA commands lead the inputs of parameters and observations, respectively. ENDSALS command terminates the input to SALS system.

2.6.3 Hierarchy in Default Setting

In order to control various treatments in SALS in proper standard ways and also to allow the user to tune the controls flexibly, a hierarchical structure is introduced in setting the controls and their defaults. Default options are set with reference to the upper-rank control data when a command is omitted in the input, when zero or blank is read as a control datum of a command, or when an improper value is read as a control datum.

Six control data of the PROBLEM command are at the top of the hierarchy to control algorithms, statistical treatments, and outputs. They are:
LMODEL specifies the type of user's model subroutine,
LALGO selects a nonlinear or linear least-squares algorithm,
LCYCLE sets the limit of iteration cycles in the nonlinear algorithms,
LWADJO chooses a least-squares or robust estimation method,
LOUTO controls outputs of final and intermediate results,
LSTATO specifies statistical outputs.

2.7. Implementation and Use

SALS system has been developed on a HITAC 8800/8700 computer at UTCC, which is operated under an OS7 monitor in a virtual memory system. It has been transferred successfully to about ten computer centers in Japan, including Osaka U., Kyushu U., Nagoya U., Tohoku U., Tsukuba U., and Tokyo I.T., where machines such as FACOM M-200, NEAC ACOS 900, HITAC M-170, and FACOM 230/75 are in operation.

The source program with detailed comment cards are open to users, and User's Manual in two volumes [12, 13] are available. Some introductory reviews [14, 23] and conference reports [15~17] were published and an original paper is to be published [18]. Lecture Sessions on SALS Usage were held four times since 1978 at UTCC.

For these five months since the opening of SALS Version 2 in May 1979, the system was accessed 2670 times in total by users of UTCC. More than 25 users executed SALS over 10 times each. Chemistry, pharmacology, and medicine are the major fields of users at present, and the problems solved seem to be relatively small and simple. Extensive uses of SALS system for larger and more sophisticated problems are also growing in various fields such as chemistry and physics.

3. NISAN SYSTEM

3.1. Introduction

With the advance of electronic computer and the systems, many kinds of data-base have been developed, as well as small ones of observational data at hand of experimental researchers. Thus the whole extent of statistical analysis is eagerly desired for the exploratory and confirmatory situations of statistical data analysis. Furthermore, through such a process, statisticians sometimes have been forced to spend a lot of time for data screening and preliminary inspection on raw data. For this matter, it may be required for statistical software system to exert all possible effects to emancipate statisticians in charge from such cumbersome labor as carrying out such collection, arrangment and analysis of the data. In view of recent development of mathematical methodology of statistics, new and useful methods should be included in data analysis and must be newly developed as our research subjects if any is not developed yet.

In response to such demands for statistical data analysis, a New Interactive Statistical Analysis (NISAN) program package has been planned by a research group of Japanese statisticians and information scientists with the consensus of academic societies and with the Grants of Research Project of the Ministry of Education of Japan, since 1972. At the start of planning NISAN, the research group held several symposia to discuss properties and performances on data-handling and statistical analysis with many users in every field of universities. Thus the group took many proposals of them into consideration.

NISAN system is being built upon FACOM M-190 OS-IV/F4, being equivalent to IBM system, OS/370-MVT/TSO. The minimum equipment for a NISAN user is a conversational teletypewriter of CRT to access NISAN system through the public telephone lines. By a graphic display terminal connected with the system, graphic representations in the system will be also possible.

3.2. System Construction

The computer language used in NISAN system is basically and mostly high transportable FORTRAN IV to maintain machine independence of the system, but a few assembly is temporarily used, because the research group is now concentrating to a conversational function with the most interest.

The test version consists of five independent modules, i.e. NISAN monitor, STAT, DATA, DOC and HELP modules, and the system files consist of a data file for analysis and a document file for records. DATA is a data-base handling module, mainly taking care of the data file. DOC controls the document file. Without ever considering the file structures, STAT module can easily access any file through DATA and DOC modules. HELP module has the complete inquiry function to give answers and instructions to any user's question on interpretation or utilization of the system. There are three kinds of message levels. Even if the user is a beginner, he may easily access the system with the aid of the instruction from HELP module if necessary.

In the face of constructing NISAN system, the input data in every program of statistical analysis, written in FORTRAN IV, have to be stored on a data file, being independently of the respective programs. The data file, called NISAN standard file, consists of a file directory and a data part as a self-descriptive system file. The logical structure of data part is a set of records being called cases. Each case consists of a set of items referred by variable names like in FORTRAN. Group names of variables are also assigned as key names. The data on NISAN file are stored with a character type, because of the efficient use of file utility programs provided by the system.

The system is designed so that a command as an interface between the system and its user has the same capability as any command of TSS. A series of programs for non-statistical analysis is developed for decoding a command in the system. These programs are all subroutine subprograms written only in FORTRAN IV, too. The main functions of NISAN monitor are (i) to control the states of system, (ii) to decode the command syntax and (iii) to control FORTRAN subroutine subprograms and the files. A group of FORTRAN subgroutine subprograms takes charge of the essential part of the conduct of analysis,and NISAN monitor takes charge of effectively integrating the interactive system.

Thus NISAN research group has made special efforts for a set of convenient data handling commands, as well as those of statistical analysis, and actually such a data management function is very important for a statistical program package. NISAN system was firstly and successfully tested between Kyushu University and Osaka University in January 1979, and is still being rigged and decorated.

3.3. Statistical Principle

As for statistical methodology, the most emphasized functions are as follows; 1) data investigation, 2) graphic representation, 3) generalized methods of optimal scaling, 4) various methods of cluster analysis and multidimensional scaling, and 5) studies on successive process of statistical inference.

In applying statistical methods in practice, it is very important to know features and properties of input data in advance, i.e. data investigation before statistical analysis. Since most of the current statistical program packages have not enough data investigation function, we studied and strengthened such a function, especially examination of multivariate normality, transformation of variables, check of outliers etc.

By inspecting graphical representation at every stage of analysis, we often obtain important insights, effective suggestions and interpretation not only on the input data but also on the way of analysis and the results. NISAN system can give us such graphical representations any time at users' request.

For categorical input data, the NISAN system includes generalized methods of optimum quantification, newly-developed by us, as well as ordinary nonparametric methods. We can choose any of them in accordance with the presence or absence of external criterion and/or order relations among item-categories.

Methods of cluster analysis and multidimensional scaling give us some of exploratory methodology in data analysis. Many of them are included together with our newly-developed ones and are applicable with relation to the quantity and properties of input data and to the hierarchical or nonhierarchical structure of clustering.

Methodologies of testimating, testipredicting and testitesting are included in view of the unspecified mathematical model of data, e.g. pooling methodology is available in cases of ANOVA and MANOVA with the resultant estimations. Facing the statistical data analysis, we can obtain statistical wisdom on the accuracies and precisions for estimates, powers of test for various alternative hypotheses and so on, by applying a simulation function in NISAN system. Thus, as a whole process of statistical inference, we can examine the performances and strategies, and can finally proceed with the optimal procedure of statistical analysis.

3.4. Emphasized Functions for Methodology in NISAN

3.4.1 Data investigation

In order to investigate features and properties of input data before applying a method of statistical analysis, NISAN system provides two kinds of examinations, for basic and simple data inspection and for further investigations on mathematical assumptions. The former is related to misinput, missing values, extraordinary data etc. The latter concerns the type of distribution, transformation of variables, trend, strata etc. Especially, since many statistical methods require the normality of distribution, the system includes tests of skewness and kurtosis, Geary test, Gap test, Shapiro-Wilks test, Shapiro-Francis test and D'Agstino test for univariate normality, and Mardia test for multivariate normality, for the convenience to obtain the global feature, asymptotic property and easy algorithm of computation for the frequency distribution. For a non-normal distribution, the system provides two kinds of normal-transformations of variables, i.e. variance-stabilizing transformations and power-transformations. Transformations of the first kind are square-root method, angular method, log-transformation, log log-transformation, logit-transformation, probit-transformation, rankit-transformation etc. Naturally after transformation, if necessary, the performance of transformed distribution is examined by the normality tests mentioned above. The examination for outliers may depend largely on the identification of population, the feature of distribution type, the choice of variables etc. Actually it may be rather complicated for us to define and treat such outliers, especially in multivariate cases. Therefore the system includes the omnibus methods, i.e. FUNOP and FUNOM proposed by J.W. Tukey for single variate and the influence method proposed by F.R. Hampel for multivariate.

3.4.2 Graphic representations

Inspecting the feature of observations gives us some important insights, effective suggestions and interpretation not only on the input data but also on the way of analysis and the results. Generally speaking, such trials are descriptive and rather informal but still they are fruitful. NISAN system contains various kinds of graphic representation as follows; histogram, hanging rootgram, stemleaf graph, scatter diagram, radar chart, face graph, body graph*, constellation graph*, tree graph*, linked vector graph*, biplot graph, statistical ellipse, non-linear mapping graph, dendrogram and so on, where the symbol * means new graphs proposed by NISAN group. Setting each variable or composite variable to each portion defined on the graph, we can get figures as quickly as we like.

These graphs are selected according to the aim of analysis, the field of application, and the character of data. Naturally it will be expected that graphic representations and analytical methods complement mutually on a sound statistical data analysis. In particular, our graphic representations are suitable for multidimensional observations to grasp visually the structures as a whole, and are simultaneously helpful to investigate characteristics of individuals. Some of them are demonstrated below. Body graph may represent the global status of subjects more effectively than the face graph like in case of investigation of physique, physical strength and other latent structures. Then the graph depends on observations for body weight, chest circumference, sitting height and so on. Tree graph may represent multi-items or multi-traits by branches, roots, fruits, leaves, the height and shape of a tree, and thus it may help to recognize globally characteristics and profiles of the subjects. It seems to us that the tree graph is particularly superior to other graphs in grasping the global interrelationship among items, latent traits and time-dependent changes, too. The physical characteristic of constellation graph is to show data in the upper half circle. Specifically, after transforming the original data $\{x_{j\alpha}\}$ to standardized values $\{\xi_{j\alpha}\}$ for location and scale, $j = 1,2, \ldots, p$ and $\alpha = 1,2, \ldots, n$, the graph is drawn by a set of complex numbers $\{P_\alpha\}$, where $P_\alpha = \sum_j w_j \exp(i\xi_{j\alpha})$, $i = \sqrt{-1}$ and $\sum_j w_j = 1$. Further details are given in Wakimoto et al. [25].

3.4.3 Extensive methods of optimal scaling

Methods of optimal scaling, or quantification, are classified into two types. Those of the first type are for the cases with external criterion or explained variable, using information related to the qualitative attributes of each subject, and may analize the contribution or influence of each attribute to the explained variable. Methods of this type include those proposed by Fisher(1948), Johnson(1950), Hayashi (1952), Bradley, Katti and Coons(1962), Nishisato and Arri(1975). Tanaka and Asano[24] recently generalized them. Methods of the latter type are based only on the relationships within data and are applied to represent relatively the data set in an appropriate multidimensional space and occasionally to classify the data. Those proposed by Guttman(1941, 1968), Hayashi(1952, 1954, 1974), Okamoto and Endo (1973) belong to the latter type.

Thus in the NISAN system, a variety of optimal scaling methods widely used in various fields are going to be available systematically. A feature of the system will be to give those methods generalized for unordered or ordered itemcategories, to give an evaluation for reliability of the optimal scores, if necessary, and also to complete detailed inference about the contributions of factor items on the explained variables, by the use of asymptotic theories. As a matter of course, since the system has full functions of preliminary analysis and editing data and graphical representations, it may be very convenient and reliable for us to search the optimal model and to interpret results of the analysis.

3.4.4 Systematical techniques for cluster analysis and multidimensional scaling

Techniques of cluster analysis are nowadays popular for data analysis. Although there are available a lot of them, NISAN system includes systematically those which are defined objectively with clear algorithm and are frequently referred to with wide adaptability for various kinds of data. The techniques may be generally classified into three categories. (i) Hierarchical types with agglomerative type and divisive type, including Automatic Interaction Detector type, (0,1) Association Analysis and their modified types due to us. (ii) Nonhierarchical types with k-means method, Iterative Self-Organizing Data Analysis method and the modified for outliers. (iii) Other types with Fuzzy clustering and two-way clustering. Naturally these techniques are interactively connected each other and also connected to the next step to analyze. It may be the advantage of the present system for statisticians to evluate the criteria of measures of similarity or distance, of numbers of clustering, and of the optimization. Also it may be worth to say that a number of sensitivity analysis is available on the basis of simulations by the use of NISAN random numbers. As for the methods of multidimensional scaling included, the techniques are based on MDA-OR and MDA-UO, proposed by Hayashi(1974). They also give us effective methods for exploratory data analysis.

3.5 Studies on Processes of Statistical Inference

3.5.1 General principles

In order to specify adequate mathematical model(s) for the data, statisticians may apply their systematic process of statistical methods to their data in accordance with their inference logic. From the viewpoints of such uses of statistical methods in a package, statisticians tend to examine objectively the performance of a total system of statistical inference on the validity, reliability, sensibility, efficiency etc. The NISAN system enables statisticians to evaluate their inference process. For instance, pooling methodology may be applied. Keeping away from awfully complicated formulation and the numerical computation for a multiple-stage process, the system allows to obtain numerical evaluation on the basis of simulating the process by NISAN random numbers implemented. These random numbers were generated physically and transformed to 2.4 million physical random figures. They were assured of good properties by fifteen tests for randomness (Goto and Asano, 1978). According to such studies on inference process, statisticians may obtain several important comments for the analysis as follows; (i) improvement on bias, precision, or the combined risk for parameter estimation, (ii) improvement or adjustment for the power of tests, (iii) investigations for various associations among statistical methods, (iv) choice of a serial process of statistical methods and a mathematical model, fitting better to the given data, and (v) estimation of the additional number of samples.

3.5.2 Multistage testing process

A process called "a multistage testing process" here is a compound process which is objectively constructed with several statistical tests and corresponding preliminary treatments following the general principle of successive statistical inference. It is not only an independent process but also a part of general statistical inference system, which may have the following functions: data investigation, learning by simulation, processing of statistical inference, retrospective reconsideration, rebuilding mathematical models, etc. These subgeneral-systems can be built in the NISAN system with commands and/or related operations of NISAN from TSS-terminal.

3.5.3 System constructions

A multistage testing process to be partly built in a general inference system may be presented with an abstract tree structure. On each node of the tree, some statistical methods to strengthen the efficiency of tests (e.g. pooling of statistics, scaling of data, or adjustment of location etc.) (U_i) would be set pairing with corresponding test (T_i). In this tree, each test (T_i) on a node (N_i) should be acted as a preliminary test for other tests on following nodes. Each leaf of the tree would present a compound hypothesis defined by the route from a root to the leaf. If we consider the testing process as a tree, we may be able to constitute a tree-structured statistical machine on a computer with plexus, in which every node should be changed with any box organized by a treatment (U_i) and test (T_i) if necessary.

3.5.3.1 An interpretation of multistage testing process

For a given set of preliminary treatments (U_i) and tests (T_i), there may be several ways to construct multistage testing process, under situation of data analysis and accumulated knowledge of investigators. Assignments of ordinary sizes of test (a_i) to each test should be changed with objective hypothesis and requirements for testing made by investigators. We shall call a tree constituted with (U_i, T_i) except (a_i) as a skelton tree of a testing process (STTP). Any testing after testing process (TTP) can be constituted from a STTP and a set of constants (a_i), and this will be refered to as TTP. Any TTP is a function defined on some unspecified models (V_j), and takes some leaves of the testing tree as its value. An evaluation of performances of TTP can be made as a comparison between it and a standard test (T_0) or other TTP.

3.5.3.2 POSTTS and POSTTA

Two sub-systems POSTTS and POSTTA are presented in NISAN. POSTTS is a tool for designing TTP and evaluating, and POSTTA for analysis of actual data. Activation of POSTTS or POSTTA is made by a command POSTTS or POSTTA from a TSS-terminal. These commands possess some further functions presented by key-word parameters(k.p.) or sub-commands(s.c.), and other informations during a processing would be handled interactively by forms of system-lead.

Table 1 Functions of POSTTA and POSTTS

MNEMONIC	FUNCTIONs	NOTE *	*
FI	data file for in-put	k	A
POP	define populations by v.name	k/s	A
STRA	define strata	k/s	A
RESP	define response-variables by v.name	k/s	A
TTD	design of a TT-process	k/s	A/S
TTS	design of a competitor	k/s	S
TSIZE	size of tests for TTD	k/s	A/S
TSIZES	size of tests for TTS	k/s	S
SAMPL	conditions of sampling	k/s	A/S
SIZE	sizes of samples	k/s	S
CYCL	major cycle for simulation	k/s	S
RFI	define file of random number	k	S
RPOP	define distribution of samples	k/s	S
RUN	start of commands	s	A/S
END	end of running	s	A/S

* k: key-word-parameter s: sub-command
A: POSTTA S: POSTTS

A simple illustrative example is given by notational forms;

```
LOGON  ... activate a TSS
NISAN  ... activate NISAN
ready  ... signal from the TSS as 'ready'
POSTTS RFI(RNUMF),RPOP(NORMAL(0.0,1.0),NORMAL(0.0,1.0)),
       CYCL(1000)          .........  activate POSTTS
TTD    (T1,U1,a1,(T2,U2,a2,T3,U3,a3)) ... design a TTP
                                      ... start
RUN
RPOP   (NORMAL(0.0,1.0),NORMAL(0.0,2.0)) ... change RPOP
RUN
  :
  :
END   .... simulation to be ended
```

where each T_i, U_i and a_i in TTD correspond to a test, preliminary treatment and size of test.

3.6. Conclusion

NISAN system was planned in 1976 by a statisticians' group in Japan and the first version was completed early 1979. The attempts, principle and methodology, of the system are considerably different from those of the ordinary statistical program packages available currently.

REFERENCES (* means "written in Japanese")

[1] Ch. Asano et al., "An Investigation Report on Statistical Program Packages in Europe", Report of B-1 Section, Special Research Project "Formation Process of Information System and Organization of Scientific Information". (1978). *

[2] -----, "The Statistical Principle and Methodology in NISAN system", Proc. 3rd Symp. COMPSTAT, Leiden, 431-437 (1978).

[3] -----, "A Generalized Sequential Selection Procedure with Play-the-Winner Sampling and Drop-the-Loser Decision Rule in a Finite Population", ICCS, (1978).

[4] -----, "System Construction of New Interactive Statistical Analysis (NISAN)", to appear in Proc. 42nd Session, Int'l Stat. Inst., Manila, (1979).

[5] N. Inagaki, "Two Errors in Statistical Model Fitting", Ann. Inst. Statist. Math., 29(2), 131-152 (1977).

[6] N. Kawai and M. Okamoto, "A Generalization of the Ridge Function Theorem", Math. Japon., 24(2), 175-178 (1979).

[7] T. Nakagawa and K. Kuchitsu, "Data Fitting --- Its Ideal and Reality", J. Spectrosc. Soc. Japan 24, 109-126, 165-187 (1975). *

[8] T. Nakagawa, "Development of a Standard Program for Least Squares. (1), (2) Histrory and Goal of the Development of SALS System", UTCC News 8(5), 68-72 ; (6), 89-95 (1976). *

[9] -----, Ed., "Proceedings of Symposium on Data Analysis for Natural Science", SALS Group, Tokyo, (1977). *

[10] ----- and Y. Oyanagi, "Development of a Standard Program for Least Squares. (5) An Overview of SALS System", UTCC News 11(5), 63-72 (1979).

[11] -----, "Introduction to Fortran Programming for Scientists and Engineers", Tokyo Kagaku Dojin, Tokyo (1978). *

[12] ----- and Y. Oyanagi, "Program System for Statistical Analysis with Least-Squares Fitting : SALS (Version 2). User's Manual, Part 1. Basic Usage", UTCC, Tokyo (1979). *

[13] -----, Y. Oyanagi and H. Togawa, "ibid, Part 2. Controls and Algorithms", UTCC, Tokyo (1979). *

[14] -----, "New Approaches to Data Fitting", Bunseki (J. Soc. Anal. Chem. Japan). 1979, 230 (1979). *

[15] -----, "Program System SALS for Statistical Analysis with Least-Squares Fitting", 46th Meeting, Japan Statist. Soc., Sendai, July (1978). *

[16] ----- et al., "Program System SALS , I-III", 20th Meeting, Japan Inf. Proc. Soc., Tokyo, (1979). *

[17] -----. Y. Oyanagi and H. Ohiwa, "Program System SALS for Nonlinear Least-Squares Fitting", 10th Int'l Symp. Math. Program., Montreal, (1979).

[18] ----- and Y. Oyanagi, "Program System SALS for Nonlinear Least-Squares Fitting : System Design", to be published.

[19] M. Okamoto, "Asymptotic Normality in Monte Carlo Integration", Math. Comp., 30(3), 831-837 (1976).

[20] ----- et al., "Optimality of Multidimensional Representation in Hayashi's Fourth Mathod of Quantification", J. Japan Statist. Soc., 8(2) 63-69 (1978).

[21] ----- et al., "Stability of the Solution in Latent Class Analysis", 11th SDA Symp., Osaka, Jan. (1978). *

[22] -----, Model Specification and Estimation in Factor Analysis", 13th SDA Symp., Osaka, Jan. (1979). *

[23] Y. Oyanagi, "Robust Estimation and Data Analysis", Butsuri (Bull. Phys. Soc. Japan) 34(10), 884-891 (1979). *

[24] Y. Tanaka and Ch. Asano, "A Generalized Method of Optimal Scaling for Partially Ordered Categories", Proc. 3rd Symp. COMPSTAT, (1978).

[25] K. Wakimoto et al., "Constellation Graphical Method for Representing Multidimensional Data", Ann. Inst. Statist. Math., 30(1), 79-104 (1978).

Scientific Information Systems in Japan
H. Inose (ed.)

FOUR YEARS OF TOOL-IR: THE RECEPTION OF AN ON-LINE INFORMATION SERVICE BY JAPANESE USERS

Takeo YAMAMOTO, Masamitsu NEGISHI**, Mamoru USHIMARU*, Hidehiro ISHIZUKA*** and Shizuo FUJIWARA***

** Computer Centre, Univ. of Tokyo, Yayoi, Bunkyo-Ku, Tokyo 113*
*** Research Center for Library and Information Science, Univ. of Tokyo, Hongo, Bunkyo-Ku, Tokyo 113*
**** National Institute of Japanese Literature, Shinagawa, Tokyo 142*

TOOL-IR (University of **To**kyo **O**n-**L**ine **I**nformation **R**etrieval) is the first on-line bibliographic information service of practical scale in Japan. It was started in July 1975 for the academic community of Japan. The service and the software are described briefly. The users' reaction towards the system is analyzed through the usage statistics and user questionnaire.

1. INTRODUCTION

In late 1969, some 10 chemists and computer scientists including three of the present authors formed a study group for on-line bibliographic information systems. Funded by the Japanese Ministry of Education, Science and Culture, the group proceeded to design and develop an experimental system, TSIR-1[1]–[2], for on-line storage and retrieval of bibliographic data. The system was implemented on HITAC 5020 DAT TSS. Although the system was based on linear search of natural language data and could not be used for large-scale data bases, some of the concepts developed in the system, especially the sharing of query files among the users, have been used in the following project.

In the spring of 1973, a new team was formed at the University of Tokyo for the design and development of a more practical system, TOOL-IR (University of **To**kyo **O**n-**L**ine **I**nformation **R**etrieval). The first version of the software was completed in August 1974 on HITAC 8800/8700 OS7 of the Computer Centre, Univ. of Tokyo. A test data base, initially consisting of half a year (26 weekly issues) of Chemical Abstracts Condensates (CACon), was made available to members of ten monitoring laboratories, two each from the faculties of science, engineering, agriculture, pharmacology and medicine of the University. The results of the test use by those monitoring users as well as by the members of the development team were evaluated at the development team meeting and were used for improving the software.

In July 1975, the service was made available to all the potential users of the Computer Centre, i.e. the academic community of Japan. The initial data base was one and a half years of CACon. In August 1975, the Crystallographic Data Centre (Cambridge University) Numerical and Bibliographic Data Files (XDC data base) were made on-line. In February 1976, INSPEC Computer and Control (INSPEC C) data base was added. The current status of the system, as of December 1979, is shown in Table 1. The data bases occupy approximately 2000 MB of disk memory.

Although many batch retrieval and some experimental on-line retrieval systems had been developed for bibliographic data bases, no other service had existed in Japan before 1975 which

(1) had the data base updated regularly by incorporating externally generated computer-readable data files, and
(2) was open to users external to the organization controlling the data base.

In the above sense, TOOL-IR was the first on-line bibliographic information service of practical scale in Japan.

In the present paper, the development of the software will briefly be described. Statistics on the usage of the service will then be analyzed. The user questionnaire will briefly be described. Statistics on the usage of on-line information services in the United States and in Europe have been extensively documented[3]. However, little data has been compiled on this subject in Japan.

2. DEVELOPMENT OF SOFTWARE

2.1 Design Targets

In developing the software, the typical user was thought to have the following characteristics:

(1) He is an end-user of the system. Little help from the information specialist or other intermediary was expected.
(2) He has an adequate *passive* knowledge of technical English in his field. However, his *active* knowledge may not be enough to enumerate alternative expressions or grammatical variations of a word. His typing skill may be minimal.
(3) Terminals are mostly low-speed (300 Baud), typewriting terminals, although other terminals are sometimes used.

From the above, some guidelines for user-interface design were derived:

(1) Make each command simple and short. Index words, not phrases.
(2) Let the user use all the information already input. The system should, therefore, remember all the reference sets formed during the session. Formation of stored

Table 1. Current Status of TOOL-IR Services

	DATA BASE	FIELD COVERED	DATA CONTAINED	PERIOD (# ITEMS) ONLINE	UPDATING FREQUENCY	COMMAND SYNTAX
Externally generated data bases	CA Condensates	Chemistry & related areas	Bibliography Keyword phrase	1977.1--1979.6 ($1.05X10^6$)	------	//CAS volno
	CASearch	Chemistry & related areas	Bibliography Keyword phrase Chemical substance index entries General subject index entries	1979.7-- ($1.5X10^5$)	Biweekly	//CAS //CAS NEW
	Crystallographic Data Base	Crystallography (Organic, non-polymer)	Bibliography Numerical	1935-- ($2.5X10^4$)	2--4 times/yr.	//XDC
	INSPEC-1 C	Computer & control	Bibliography Keywords Abstract text	1978.1-- ($6X10^4$)	Monthly	//IEE year
	COMPENDEX	Engineering	Bibliography Keywords Abstract text	Current month ($7X10^3$)	Monthly	//ENGINEER
	CA Ecology & Environment	Ecology & Environmental chemistry	Bibliography Keyword phrase Abstract text Chemical substance index entries General subject index entries	Current month ($4X10^3$)	Biweekly	//ECOLOGY
Thesauri	Molecule	Chemical substance	Chemical name Chemical formula CA Registry No.	(80,000 names)	------	//MOLECULE
	CODEN	Coden--Journal name conversion		(14,000 journals)	------	//CODEN
Local data bases	Library Books	Foreign book acquisition at Univ. Libr.	Simplified bibliographic	1976--1978 ($1.5X10^4$)	------	//RF
	Personal	Users own data	NMR bibliography	1965--1974 ($2.2X10^4$)	------	
			Solvent extraction bibliogr.	1945-- ($6X10^3$)	------	

query and its usage should be made simple.

(3) Tolerate user's input error as much as possible. This means through checking and recovery of user errors by the user-interface program.

(4) Let the user choose the detailedness of system messages. In TOOL-IR, commands SHUTUP and TALK, which may be used in the inturrupt mode, may be used to control the system message style.

2.2 Special Features of TOOL-IR

The following are some of the special features of the software:

(1) Left- as well as right-truncation index search is possible. For keywords longer than 8 characters, the last 8 characters are indexed together with the first 8 characters, only in reverse order. If there is a left-truncation search request in the query, the keyword is reversed and matched with the index.

(2) Automatic fragmentation of chemical names. simple algorithm which, however, prevents excessive fragmentation, is used to fragment both the database record and the query. The procedure allows for variations in segmenting chemical names and absorbs differences in nomenclature.

(3) Automatic transfer of molecule number (CA registry number) set to a reference number(bibliographic citation) set.

In Fig. 1 and Fig. 2, an example usage of the system is illustrated. In Fig. 1, //MOLECULE command was used to obtain a set of 16 CA registry numbers for BHC and related compounds. In Fig. 2, a keyword search through the bibliographic data was done. The result was merged with the result of transferring the registry number set into a reference number set, giving a complete document set dealing with BHC.

3. USAGE

3.1 Geographic Distribution

TOOL-IR has been widely used in Japan. Between July 1975 and September 1979, there were 1,241 users(user-id's) who accessed the CACon and CASearch data bases. They came from 26 prefectures (out of the total 45) in Japan, belonging to 90 campuses of 76 institutions (universities, colleges, research institutes or technical high schools). The number of userid's in each prefecture is shown in Fig. 3.

The distribution of the users is, as will be expected, heavily centered around Tokyo. However, there are several remote centers of activity including Osaka, Aichi(Nagoya), Kyoto, Hyogo(Himeji), Miyagi(Sendai) and Hiroshima. The distribution of the users, especially in the initial period, is influenced by the presence of one or two key users in a city. They help other users in their area by organizing informal workshops, distributing service-related documents, assisting in profile formulation and writing introductory articles in local publications such as school library newsletters.

The main problem with the remote usage of the system is the high long-distance telephone charge in Japan. Most of the users outside the Kanto area are sooner or later forced to use the batch service of TOOL-IR. The start of the N-1 network service is eagerly awaited by many users of TOOL-IR.

```
//MOLECULE

 WELCOME TO TOOL-IR CAS-REGISTRY NUMBER FILE SEARCHING!
   THIS IS VERSION-4/C, RELEASED ON 1979.6.22.

 INPUT FORMAT   DATA BASE
  //MOLECULE    FREQUENT 80000 COMPOUNDS.

 SUBCOMMANDS:COMPOUND(C),DISPLAY(D),AND,OR,DIF,DUMP,
             QUSE,QSAVE,KEEP,ENTER,REMIND,
             FRAGMENT,KILL.

 NEW FEATURES OF PROGRAM VERSION-4:
   1.  //SHUTUP   MODE GIVES CONCISE MESSAGES.
   2.  'DISPLAY M.B' IS REFINED.
   3.  //TYPE  COMMAND IS AVAILABLE. (SEE CENTER-NEWS VOL11,NO10,P34.)

 ************************************
 WHEN TOOL-IR DATA-BASE IS OFF-LINE FOR UPDATING, MOLECULE SYSTEM
 WILL BE OFF-LINE, ALSO.
 ************************************

KEYWORD INDEX FILE IS ON-LINE.
LINEAR DATA FILE IS ON-LINE.
SYSTEM QUERY LIBRARY IS ON-LINE.

TYPE IN COMMAND
@COMPOUND HEXACHLOROCYCLOHEXANE C6 H6 CL6

COMPOUND HEXACHLOROCYCLOHEXANE C6 H6 CL6
RETRIEVAL KEYS:HEXACHLO/CYCLO/HEXANE/C6/H6/CL6/OROLHCAX
K='HEXACHLO'
K='CYCLO   '
K='HEXANE  '
K='C6      '
K='H6      '
K='CL6     '
K='OROLHCAX'
COMPOUND-NAME SET #  1 CREATED
CONTAINS    16 NAMES

TYPE IN COMMAND
@KEEP MOLWORK

KEEP MOLWORK
COMPOUND-NAME SET #  1 KEPT, DTFNAME='MOLWORK'

TYPE IN COMMAND
@D M.B

D M.B
DISPLAY COMPOUND-NAME SET #  1
SET CONSISTS OF      16 NAMES
NAME & MOLFORM DATA OF LAST     16 WILL BE DISPLAYED
NM:NAME  NC:NAME CODE  MF:MOLFORM

RGSTRY#=000058-89-9
NM:Cyclohexane
   , 1,2,3,4,5,6-hexachloro-
   , (1.alpha.,2.alpha.,3.beta.,4.alpha.,5.alpha.,6.beta.)-
NC:0000100003NT1
MF:C6H6Cl6
          SYNONYM#  1
NM:Cyclohexane
   , 1,2,3,4,5,6-hexachloro-
   , (1.alpha.,2.alpha.,3.beta.,4.alpha.,5.alpha.,6.beta.)-
NC:0000100003NT1

RGSTRY#=000319-84-6
NM:Cyclohexane
   , 1,2,3,4,5,6-hexachloro-
   , (1.alpha.,2.alpha.,3.beta.,4.alpha.,5.be'
   '0000100003NT1
     ''6Cl6
```

Fig. 1. Example of using TOOL-IR. MOLECULE command.

Underlined parts are user input.

```
//CAS

 WELCOME TO TOOL-IR CASEARCH SERVICE.
  THIS IS VERSION-15/1, RELEASED ON 1980.1.23.

 INPUT FORMAT   DATA BASE                 REFS.
   //CAS NEW    CASEARCH VOL.91,25-26     18163
   //CAS        CASEARCH VOL.91,1--26    221961
   //CAS 90     CA CON   VOL.90,1--26    214553
   //CAS 89     CA CON   VOL.89,1--26    225616
   //CAS 88     CA CON   VOL.88,1--26    202558
   //CAS 87     CA CON   VOL.87,1--26    210532
   //CAS 86     CA CON   VOL.86,1--26    199309

 SUBCOMMANDS:SEARCH(S),DISPLAY(D),AND,OR,DIF,LIMIT,
             QUSE,QSAVE,KEEP,ENTER,REMIND,
             EXTRACT,DUMP,KILL

****************************************
 PLEASE USE '//EIR:BIR'
 FOR USING VERSION-15 PROGRAM.
****************************************
 NEW FEATURES OF PROGRAM VERSION 15:
  (1) LIMIT SUBCOMMAND ADDED.
      EX1---LIMIT 3,ALL:CORRELATION NMR
      EX2---LIMIT A:SAITO, NOBUO    (LAST DOCUMENT SET)
      EX3---LIM 7,005A-02:LOS ALAMOS SCI. LAB.
  (2) DISPLAY FORMATS IMPROVED.  NEW HEADERS ARE:
      D:DOCUMENT TYPE   L:LOCATION OF WORK
      U:COUNTRY SUBUNIT(STATE)   Z:ZIP CODE OR EQUIVALENT
      N:NATION   O:ORIGINAL LANGUAGE

 ****************************************
.NEXT TOOL-IR UPDATING:  FEB. 6, 1980.
 ****************************************
 ************************************************
 IF 'MPG OVERFLOW' OCCURS, INPUT THE FOLLOWING:
  //STP MPG=80
 AND TRY AGAIN.  FROM NEXT TIME ON, SPECIFY MPG
 IN //JOB COMMAND (//JOB 1234563001,MPG=80).
 ************************************************

KEYWORD INDEX FILE IS ON-LINE.
AUTHOR INDEX FILE IS ON-LINE.
LINEAR DATA FILE IS ON-LINE.
SYSTEM QUERY LIBRARY IS ON-LINE.
```

Welcome message (suppressed in SHUTUP mode except for the updating date announcement)

```
TYPE IN COMMAND
@S BHC,DIELDRIN

S BHC,DIELDRIN
K='BHC     '
K='DIELDRIN'
DOCUMENT SET #  1 CREATED
CONTAINS    49 DOCUMENTS
```

Search records on BHC or dieldrin

```
TYPE IN COMMAND
@ENTER MOLWORK

ENTER MOLWORK
RGSTRY#=000058-89-9
RGSTRY#=000319-84-6
RGSTRY#=000319-85-7
RGSTRY#=000319-86-8
RGSTRY#=000608-73-1
RGSTRY#=008064-48-0
RGSTRY#=008065-09-6
RGSTRY#=008066-27-1
COMPOUND-NAME SET CONSISTING OF    16 NAMES
ENTERED FROM 'MOLWORK'
DOCUMENT SET #  2 CREATED
CONTAINS    270 DOCUMENTS
```

Transfer the result of //MOLECULE search into the bibliographic search

```
TYPE IN COMMAND
@OR 1,2

OR 1,2
DOCUMENT SET #  3 CREATED
CONTAINS   298 DOCUMENTS
```

Merge the above results

```
TYPE IN COMMAND
@S METABOLISM

S METABOLISM
K='METABOLI'
DOCUMENT SET #  4 CREATED
CONTAINS 10753 DOCUMENTS
```

Search records on metabolism

```
TYPE IN COMMAND
@AND 3,4

AND 3,4
DOCUMENT SET #  5 CREATED
CONTAINS    29 DOCUMENTS
```

Find records on both BHC and metabolism

```
TYPE IN COMMAND
@D N.3
```

Display the result

```
D N.3
DISPLAY DOCUMENT SET #  5
SET CONSISTS OF     29 DOCUMENTS
BIBLIOGRAPHIC DATA OF LAST     3 WILL BE DISPLAYED
------------------------------------------------------------------------------
(00001) V/I/N:91/23/187624A   C:ZABKA6   D:J   V:24   P:245-9   Y:77
A:KORTE, F./PARLAR, H.   L:INST. OEKOL. CHEM.
  /GES. STRAHLEN- UND UMWELTFORSCH. M.B.H./MUNICH   N:FED. REP. GER.
J:Z. ANGEW. BAEDER- KLIMAHEILKD.   S:CA004004
T:FORMATION OF CHEMICAL ENVIRONMENTAL POLLUTANTS BY NATURAL EFFECTS   O:GER
K:DIELDRIN METAB EXCRETION
------------------------------------------------------------------------------
(00002) V/I/N:91/25/205378E   C:HCKHDV   D:J   P:12-15   Y:79
A:PEKING INSTITUTE OF ZOOLOGY   L:PEKING   N:PEOP. R. CHINA
J:HUAN CHING K'O HSUEH   S:CA004004
T:ACCUMULATION OF MERCURY, BHC AND DDT IN CARP (CYPRINUS CARPIO L.)   O:CH
K:CARP BHC DDT MERCURY METAB/CYPRINUS BHC DDT MERCURY METAB
------------------------------------------------------------------------------
(00003) V/I/N:91/25/207090D   C:AJBSAM   D:J   [illegible]   Y:79
A:HERITAGE, A. D./MACRAE, I. C.   L:DEP [illegible] ST. LUCIA
[illegible]4067   N:AUSTRALIA
       [illegible]. BIOL. SCI.   S[illegible]
```

Fig. 2. Example of using TOOL-IR. CAS command.

3.2 Change with Time

Fig. 4 shows the number of jobs which searched either one of the three major data bases in TOOL-IR; CAS(CACon or CASearch), XDC and INSPEC C, within each month. Fig. 5 shows the number of active userid's corresponding to the above jobs. Table 2 shows the yearly summary of the number of jobs and the number of active users for each of the above three data bases. From the figures and also from Table 2, it may be seen that the usage has increased approximately linearly with time.

In Fig. 6, the seasonal factor obtained by the EPA method of time series analysis, using the multiplicative model, for the monthly number of TOOL-IR(CAS, XDC and INSPEC C) jobs is shown together with the corresponding factors for the total number of batch jobs and of TSS jobs processed at the Computer Centre. The seasonal factor of the batch jobs is characterized by a low start in April, when the Japanese school year begins. Except for a small dip in August (the summer vacation starts at the end of July and continues through August) the seasonal factor rises until February, when the factor is almost 2.5 times that of April. The factor for March is in between those for February and April.

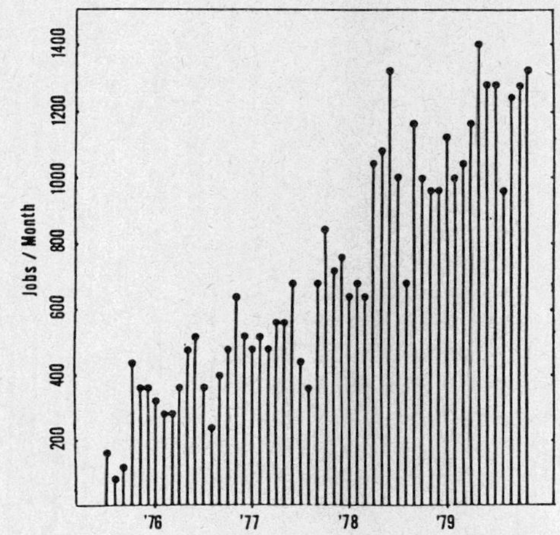

Fig. 4. Number of jobs which searched the three major data bases in TOOL-IR.

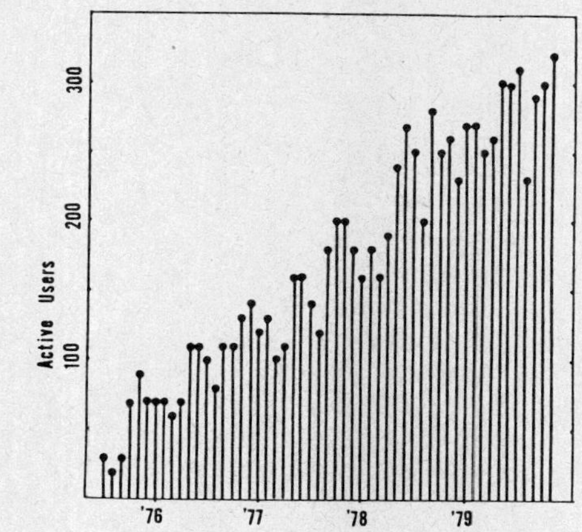

Fig. 5. Number of userid's whose jobs accessed the three major data bases in TOOL-IR during the month.

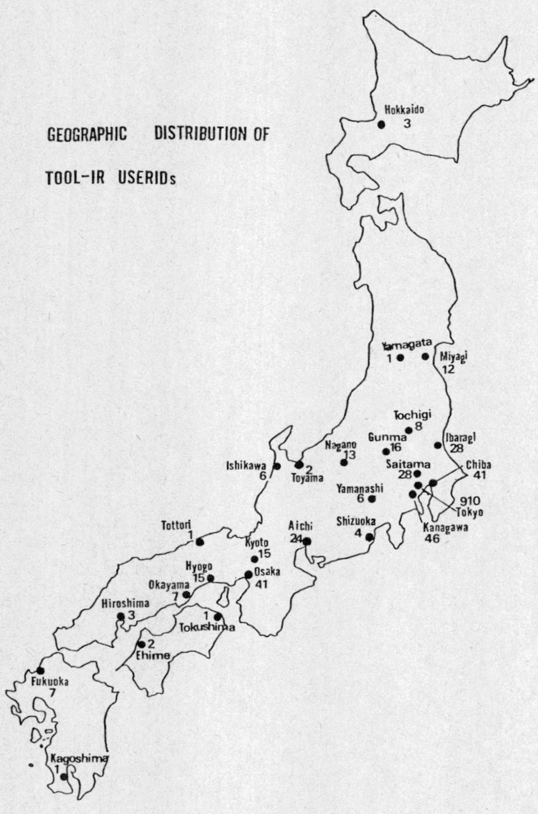

Fig. 3. Geographic distribution of 1,241 TOOL-IR userid's whose jobs accessed CAS data bases during July 1975--Sept. 1979.

Table 2. Yearly summary of TOOL-IR usage.

Year \ Data Base	CAS Jobs	CAS Users	XDC Jobs	XDC Users	INSPEC C Jobs	INSPEC C Users
1975-76	2,378 (9 Mo.)	169	80 (2 Mo.)	17	----	----
1976-77	4,755	300	667	91	277 (7 Mo.)	36
1977-78	6,496	386	724	130	536	69
1978-79	10,966	572	797	144	877	109
1979	9,398 (9 Mo.)	576	521 (9 Mo.)	115	455 (9 Mo.)	105

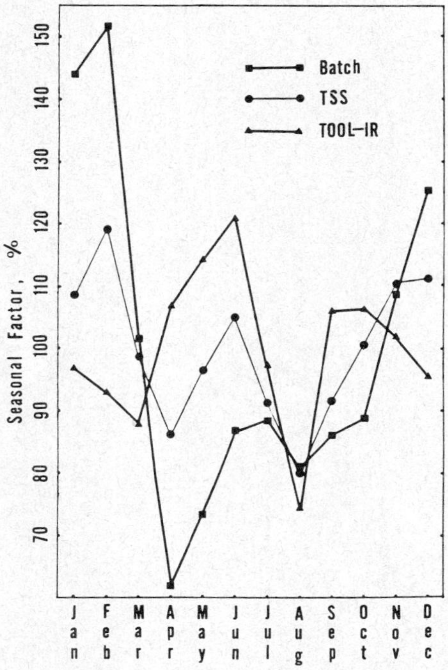

Fig. 6. Seasonal factors of usage at the University of Tokyo Computer Centre.

EPA method, multiplicative model.

The graph for the batch jobs seems to show that a typical batch user starts slowly each year and develops his program (most likely a number-cruncher) gradually, culminating in production runs at the very end of the school calendar.

The seasonal factor for TOOL-IR jobs is completely different from the above. The factor in April is higher than that in February or in March, suggesting that TOOL-IR users come to use the system as soon as the new school-year starts. In some laboratories, the new graduate students are given instruction in using the system by teachers or older students. It is not clear how much of the increase in spring and early summer is attributable to those activities; it may also reflect the fact that a bibliographic search is needed before starting a research project, which in many cases occurs early in the school-year. In any case, the seasonal factor increases until June and then decreases sharply in July and August.

In the autumn, TOOL-IR usage recovers up to its April level. However, during the winter months when the center is generally crowded, the seasonal factor keeps decreasing (although the actual number of jobs may not, due to the increasing non-seasonal trend factor). It may mean that the TOOL-IR users are not as desperate or as highly motivated for calculation as the batch users; or it may be that by the end of the school-year there is not so much bibliographic search to be done.

For TSS, the seasonal factor almost always falls between the curves for the batch and the TOOL-IR usage. It suggests that the behavior of an ordinary TSS user is in between the traditional batch user and a completely new class of users for Japan, namely, parametric users of an interactive software.

4. Users' Evaluation

Two user studies by questionnaire were conducted in December 1977 and August 1979. The same set of questions were sent to 328 and 650 past users of the system, and 228 and 227 of them responded, giving response ratios of 69.5% and 34.9%, respectively.

In the questionnaire, the user's motives for using TOOL-IR, his general assessment of the system, its strong and weak points, effect of the use on his research, and his former experience with other information systems or services were asked. Most of the questions were either single- or multiple-choice types.

In spite of the difference in user population and the response ratios, the results of the two questionnaires were remarkably similar. For example, in Table 3 where the users' choices as the major (single choice) and the minor (multiple choice) motives for using TOOL-IR are listed, the first row under each entry represents the response for the 1977 questionnaire and the second, the response for the 1979 study. It can be seen that the system is used for rather immediate needs; users are not looking for references outside their field of research but in most cases using the system for daily research and publication needs. In the table, differences between the 1977 and the 1979 figures are mostly insignificant. However, it may be said that, as there are more users, less are motivated by curiosity or novelty (H,K) and more by urgent needs in their fields of research (A,B).

Table 3. Motives for using TOOL-IR.

Entry	Major motive*	Minor motive**	Total	Year of study
A. Looking for papers relevant to the ongoing research	110(48.2%)	74(32.5%)	184(80.7%)	1977
	113(49.8%)	89(39.2%)	202(89.0%)	1979
B. Keeping up with progress in one's field of research	27(11.8%)	104(45.6%)	131(57.5%)	1977
	32(14.1%)	112(49.3%)	144(63.4%)	1979
C. Looking for reference in writing a paper	6(2.6%)	108(47.4%)	114(50.0%)	1977
	8(3.5%)	99(43.6%)	107(47.1%)	1979
D. Looking for hints in starting a research	9(3.9%)	68(29.8%)	77(43.8%)	1977
	7(3.1%)	66(29.1%)	73(32.2%)	1979
E. Searching references for other members of the group	14(6.1%)	46(20.2%)	60(26.3%)	1977
	8(3.5%)	41(18.1%)	49(21.6%)	1979
F. Searching other authors' work	1(0.4%)	52(22.8%)	53(23.2%)	1977
	2(0.9%)	61(26.9%)	63(27.8%)	1979
G. Used in writing a review article	5(2.2%)	38(16.7%)	43(18.9%)	1977
	1(0.4%)	42(18.5%)	43(18.9%)	1979
H. Wanted to know about information retrieval	5(2.2%)	38(16.7%)	43(18.9%)	1977
	4(1.8%)	29(12.8%)	33(14.5%)	1979
I. Looking for information on other fields of research	3(1.3%)	32(14.0%)	35(15.4%)	1977
	2(0.9%)	35(15.4%)	37(16.3%)	1979
J. Used in refereeing scientific paper	-	15(6.6%)	15(6.6%)	1977
	-	13(5.7%)	13(5.7%)	1979
K. Checking whether one's own paper is abstructed	3(1.3%)	10(4.4%)	13(5.7%)	1977
	-	4(1.8%)	4(1.8%)	1979

*single choice **multiple choice

Table 4 shows the users' general assessment of TOOL-IR. Because this study is not an independent study conducted by a third party, the results of Table 4 can not be taken too literally. However, it may be seen that the results of the two studies are similar, the new study being slightly more favorable.

Table 4. General assessment of TOOL-IR

Entry	1977 study	1979 study
Excellent	54(23.7%)	56(24.1%)
Fair	134(58.8%)	147(64.8%)
Ordinary	33(14.5%)	23(10.1%)
Poor	6(2.6%)	--
No answer	1(0.4%)	1(0.4%)

Questions were asked about users' assessment of the technical sides of the system such as the recall and the noise ratios. However, perhaps because most users had never used an information retrieval system (84.6% in 1977, 80.6% in 1979), there were no clear-cut answers. The major complaint of the users is that the depth of data (3 to 4 years of CAS data bases) was not enough and that TSS jobs could not be started due to system congestion.

A more detailed analysis of the user studies will be published elsewhere.

5. Conclusion

Doubts have often been expressed about the Japanese user's ability and willingness to use an interactive system. It seems safe to say now that, on the basis of the results of our study and also from other evidence such as an exponential increase in the number of TSS terminals at the Computer Centre, the Japanese user is capable and willing as long as there is an attractive resource to be used.

TOOL-IR has been designed and run as an information service for Japanese academic end-users. The system seems to have been well received by the users; though it has yet to go through the crucial test of user charging (the licence and the hit charges have been covered by the Ministry of Education, Science and Culture aids in research), it seems to have attracted a stable and expanding body of users.

Future developments in Japanese language processing and computer networking in Japan will make information services in Japan, hopefully including TOOL-IR, more accessible and more useful to the Japanese users.

ACKNOWLEDGMENT

The authors wish to thank Dr. Norimasa Nogami (now at Tsukuba University) for his valuable contribution to the software in general and the chemical-name fragmentation process in particular.

REFERENCES

[1] T. Yamamoto, T. L. Kunii, S. Fujiwara and H. Takahasi, "An On-Line Scientific Information Retrieval System Based on a Natural Language Data Base", *Proc. 1st USA-Japan Comput. Conf., pp.* 632-637(1972).

[2] T. Yamamoto, "Information Retrieval in an Academic Environment: An Experimental System TSIR-I", in S. Fujiwara and H. B. Mark, *Eds.*, "Information Chemistry", University of Tokyo Press, Tokyo(1975).

[3] For example, "On-Line Information *(Proc. international On-Line Information Meeting)*", Learned Information (Europe) Ltd., Oxford (1977-).

Scientific Information Systems in Japan
H. Inose (ed.)

DATABASE SYSTEMS ARCHITECTURE

Tosiyasu L. KUNII, Hiroyuki KITAGAWA, Minoru HARADA, Nobuo OHBO and Takahisa KIMURA
Department of Information Science, Faculty of Science, The University of Tokyo

1. Introduction

Our researches of database systems have been forcusing on:

1. identifying well desciplined engineering approaches, and
2. extending the data types to include picture and sound data as well as characters.

This is the interium report of such activities in our laboratory. The recursive graph formalism we developed for system specification is successfully applied to specify database systems design. Graph theoretical approaches also have proven to be powerful to describe the extended data types and to process the extended data in a logically uniform way. These results are currently applied to realize office automation and design automation systems. Further extension of the researches are in the direction of evolutionary and adaptable database systems architecture.

We expect that these theoretical and engineering results also provide basic building blocks to build very large information systems to support fundamental social subsystems such as for management of national resources, energy policies and international policies.

2. APAD: Application-Adaptable Database System [KIT 79]

An architecture and design of a new database system named APAD(the APplication-Adaptable Database system) have been investigated. APAD can satisfy major business automation requirements by providing users with facilities for:
1) specifying business forms which usually are structured tables and each row/column of which may be nested (that is, it may have sub-rows/columns, sub-sub-rows/columns, and so on);
2) dynamically creating new links to combine forms on hand, attributes of which may originally be unrelated. (Significant part of business activities is spent in this process. Currently available database systems do not support this type of data creation.)

In conventional database systems, queries and application programs are used to specify database applications. In hierarchical and/or network database systems, the database structure is too much oriented to a specific application [KUN 78]. Therefore, the formulation of queries becomes too complex, if the database structure is not suitable for the application. In such cases, costly database restructuring is required. But this problem cannot completely be solved by one time database restructuring, since there can be very large varieties of database applications. In contrast, in relational database systems, the database structure is too primitive. Formulating queries on a collection of flat tables is similar to making programs in terms of a machine language. It is rather difficult for end-users to use a huge collection of primitive flat tables, and to take application-oriented data out of them.

In APAD, to explicity specify each database application, a virtual database is defined by an application administrator, and dynamically mapped onto the currently available real database. Virtual databases can handle structured tables and dynamic links. The logical structure of a virtual database is a network consisting of flat tables, structured tables, and m-to-n links between two tables. Fig. 1 shows the logical structure of a sample virtual database. The virtual database has the following three tables.

PERSON(NAME, ADDRESS, SEX)
LAND(ADDRESS,LOT_SIZE,PRICE,OWNER(NAME, ADDRESS))
MOBILE_HOUSE(TYPE,ID#,SIZE,DESIGNER)

Tables PERSON and MOBILE_HOUSE are flat tables, and LAND is a structured table. A structured table has a nested structure of attributes. Attribute OWNER of LAND has attributes NAME and ADDRESS as its sub-attributes. The virtual database has a link OWN between PERSON and LAND. A link represents an interrelationship between two tables.

The real database of APAD is composed of data modules. Each data module is logically represented in a flat table format. Representation of data modules in flat table formats has the following two advantages.
1) Application Independence: Representation in flat table formats is practically free from any predefined structure that is strongly oriented to a specific application of the data modules [KUN 77, KUN 79].
2) Machine Independence: Representation in flat table formats is free from any machine-dependent representation [COD 70, KUN 79].

Virtual databases are defined on top of the real database in DSE (the Database Specification by Example) [ZLO 75] descriptions. DSE is a user-friendly language. To achieve the user-friendliness, example-oriented description facililty of QBE (the Query By Example) is adopted as the starting basis. DSE features the capabilities of handling structured tables and links, whereas QBE can handle only flat tables as uses' views.

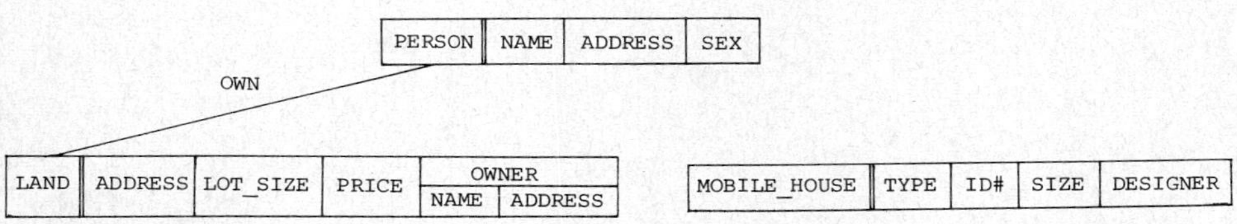

Fig.1 Sample Virtual Database

Suppose the following data modules are given.

PEOPLE(NAME,ADDRESS,AGE,SEX,SPOUSE)
LAND(ADDRESS,LOT_SIZE,PRICE,OWNER_NAME, OWNER_ADDRESS)
MOBILE_HOUSE(TYPE,ID#,SIZE,DESIGNER)

The DSE description which defines the virtual database on these data modules is given in Fig. 2. An example element is enclosed by a pair of quotation marks. The values of example elements have no meaning. Their role is to specify associations among table locations through value matching. The following formulas specify the tables in the virtual database.

PERSON(NAME,ADDRESS,SEX)
$\leftarrow\{(a,b,c)|\exists d\exists e(a,b,d,c,e)\in \text{PEOPLE}\}$
LAND(ADDRESS,LOT_SIZE,PRICE,OWNER(NAME, ADDRESS))
$\leftarrow\{(a,b,c,(d,e))|(a,b,c,d,e)\in \text{LAND}\wedge b>50\}$
MOBILE_HOUSE(TYPE,ID#,SIZE,DESIGNER)
$\leftarrow\{(a,b,c,d)|(a,b,c,d)\in \text{MOBILE\_HOUSE}\}$

PERSON is merely a projection of PEOPLE. LAND in the virtual database results from restricting data module LAND on condition that LOT_SIZE>50 and grouping attributes OWNER_NAME and OWNER_ADDRESS into attribute OWNER. MOBILE_HOUSE is just a data module. Various combinations of relational operations can be used to define tables in virtual databases. Link OWN represents an association between tuples in PERSON and tuples in LAND on condition that PERSON.NAME=LAND.OWNER.NAME and PERSON.ADDRESS=LAND.OWNER.ADDRESS.

SOURCE:

PEOPLE	NAME	ADDRESS	AGE	SEX	SPOUSE
	'TOM'	'211'		'MALE'	

TARGET:

PERSON	NAME	ADDRESS	SEX
	'TOM'	'211'	'MALE'

SOURCE:

LAND	ADDRESS	LOT_SIZE	PRICE	OWNER_NAME	OWNER_ADRESS
	'352'	'98'>50	'70'	'KENT'	'181'

TARGET:

LAND	ADDRESS	LOT_SIZE	PRICE	OWNER	
				NAME	ADDRESS
	'352'	'98'	'70'	'KENT'	'181'

SOURCE:

MOBILE_HOUSE	TYPE	ID#	SIZE	DESIGNER
	'A32'	'63'	'20'	'BROWN'

TARGET:

MOBILE_HOUSE	TYPE	ID#	SIZE	DESIGNER
	'A32'	'63'	'20'	'BROWN'

LINK: OWN

PERSON	NAME	ADDRESS	SEX
	'JOHN'	'453'	

LAND	ADDRESS	LOT_SIZE	PRICE	OWNER	
				NAME	ADDRESS
				'JOHN'	'453'

Fig.2 Sample DSE Description

It is a unique feature of DSE that information-carrying links [TSI 77] can dynamically be defined between two tables. Suppose a plan of placing mobile-houses on estates is made with the sample virtual database. After determining which mobile-house is to be built on each estate, an information-carrying link should be established between LAND and MOBILE_HOUSE in the virtual database. In APAD, the necessary work is to input the DSE description in Fig. 3 and to provide linking table HOUSE_PARK representing the association. The name of the information-carrying link is also HOUSE_PARK.

The overall architecture of APAD is shown in Fig. 4. Data modules are stored in the data module base. A DSE description for a virtual database is analyzed by DDA (the DSE Description Analyzer). DDA produces a virtual database definition specification, and stores it in the virtual database definition base. When a transaction which uses the virtual database is processed, real data values in relevant data modules have to be accessed. DDA also generates facilities used in the access to the data modules through driving the index generator, coupler generator, selector generator, and pointer array generator. By means of indices, tuples in a data module can directly be accessed. A coupler represents an association between two data modules. From a tuple in one data module, associated tuples in the other data module are directly accessed. A selector picks out a subset of the tuples in a data module. The use of selectors can narrow the scope of data search. A pointer array provides direct access paths to real data values from a table in a virtual database. Owing to a pointer array, requested data values can readily be accessed.

Once access facilities are generated for a virtual database, the transactions are efficiently processed. However, this efficiency has to be paid by the increase of the overhead of the system in generating and maintaining those application-dependent access facilities. In APAD, an application administrator has to select an appropriate adaptation made, when he defines his virtual database. The higher the adaptation mode becomes, access facilities which are the more oriented to the given application are generated, and the transactions are processed the more efficiently. On the other hand, the overhead of the system gradually increases as the adaptation mode becomes higher. For the sample virtual database shown in Fig. 1, access facilities such as COUPLE(PEOPLE,LAND,PEOPLE.NAME=LAND. OWNER_NAME∧PEOPLE.ADDRESS=LAND. OWNER_ADDRESS) and SELECT(LAND,LOT_SIZE>50) may be provided, depending on the adaptation mode given to the virtual database. Access facilities, thus generated, are stored in the index base, coupler base, selector base, and pointer array base. They are automatically shared among various virtual databases to prevent regenerating identical access facilities in the system, so long as the cost of the redundancy check is smaller than that of the regeneration. Transactions on virtual databases are processed by TP(the Transaction Processor). TP refers to the virtual database definition specification, and accesses to the data modules utilizing access facilities for the virtual database. All these processes are supervised by the system

LINK: HOUSE_PARK

LAND	ADDRESS	LOT_SIZE	PRICE	OWNER	
				NAME	ADDRESS
	'233'				

MOBILE_HOUSE	TYPE	ID#	SIZE	DESIGNER
		'7'		

USING:

HOUSE_PARK	ADDRESS	MOBILE_HOUSE_ID#
	'233'	'7'

Fig.3 Information-carrying Link

supervisor, and it carries out concurrency control and security check. Access facilities in the system are also maintained by the system supervisor, and they are properly updated in accordance with changes in data modules to preserve consistency of the system.

The principal features of APAD are summarized as follows.
1) A database application is explicitly specified as a virtual database, separately from the real database. In this sense, the application independence of the database is completely attained [KUN 77].
2) Virtual databases in APAD can fully support business data processing activities through their capabilities of directly handling structured tables and dynamic links.
3) Adaptive generation of access facilities makes it possible to process database transactions efficiently. Inefficiency sometimes becomes a mortal defect in practical use of database systems.
4) Effective resource sharing is largely improved through the two levels of sharing: the sharing of virtual databases among users and the sharing of data modules and access facilities among virtual databases.

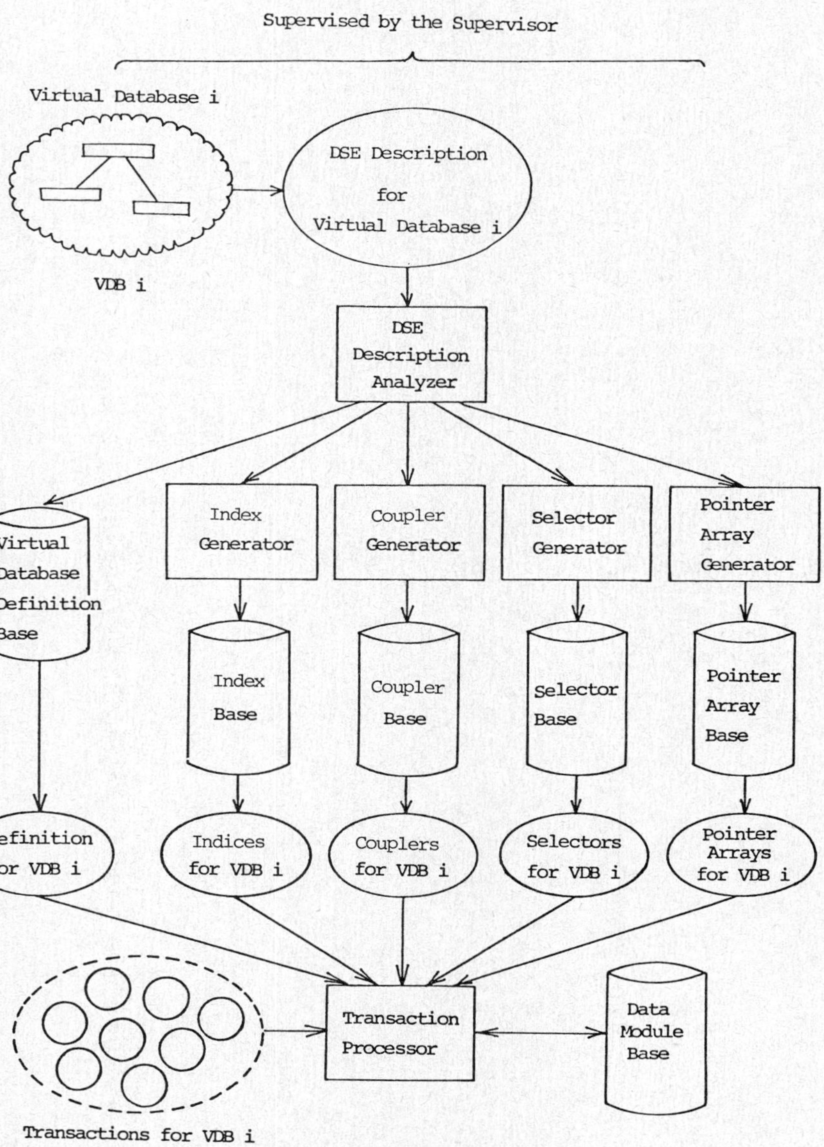

Fig.4 Architecture of APAD

5) Adequate selection of adaptation modes attains the cost-effectiveness in construction of virtual databases.

Implementation of the core of DDA is under way in Kunii Laboratory.

3. IDMS: Interactive Design Management System

In any industry, today's systems are very large in their scale and are required to be highly reliable to constantly perform very basic functions in the society. These systems are, for example, operating systems of very large computers, very large database management systems, chemical plant systems, medical information systems, and business automation systems. We have never experienced such large systems, constructed with so many parts and so many interactions among them.

In such a system design, it is unusual and also unrealistic to expect that the entire details of the system are designed from scratch. Most of the system are constructed efficiently by using a lot of already designed efficiently modules. Thus, it is required to share the design results of those modules. In addition, a design process of any system often contains similar or common processes with those of other systems. Those processes are, for example, the followings:

(1) to integrate or reduce design sheets in order to produce more detailed or more simplified design sheets.
(2) to check interface compatibility in integration processes.
(3) to check to see if designed modules completely satisfy the functional requirements for those modules.
(4) to analyze the performance of designed modules.
(5) to verify the consistency, unambiguity and/or logical completeness of designs.

All of these processes, if they are precisely specified, can be repeatedly applicable to wide range of designs of similar systems. Thus, if these processes can be shared by designers, it will save lots of their design efforts to achieve remarkable cost effectiveness. Hence, the importance of a systematic approach to share good design processes.

In this section, we propose an Interactive Design Management System IDMS which is a computer aided design support and management system. IDMS provides its users with the following facilities:

(1) For sharing design results,
  a. to specify a design of a system satisfying the requirements for the system.
  b. to correct or modify a design of a system.
  c. to retrieve various records of a system such as system name, designer's name and functional descriptions.
(2) For sharing design processes,
  a. to specify a design process by a given set of reserved operations to the stored design results.
  b. to registrate and manage thus specified design processes by assigning specific names and functional specifications to them.

All of these specifications of design results and design processes must be formal enough to process them by computers. To do so, we formalized design results and design processes as design schemas and design operations, respectively [HAR 78, HAR 79]. The design schema is a formal specification of design results and is represented as a set of recursive graphs (R−graphs). A recursive graph consists of eight graph components as shown in Fig. 5. The design operation is the formalized action taken to the design schemas by designers and is specified as a series of recursive graph operations (R−operations). The R−operations are classified into primitive operations which construct, modify or retrieve a graph dealing with one graph element at each opertion call, and macro operations which integrate, reduce, analyze or verify a graph dealing with it as a whole. A macro operation is equivalent to a sequence of primitive ones and reflect designer's macroscopic design view.

Notations and Meaning / Graph Component	Graphical Notation	Formula Notation	Meaning
Nodes and node semantic function	R1: 	$N_{R1}=\{1\}$ $ns_{R1}(1)=$buffer	There is a *buffer* system.
Subnode function	R2: 	$N_{R2}=\{1,2,3\}$ $sn_{R2}(1)=\{2,3\}$ $sn_{R2}(2)=sn_{R2}(3)=\phi$	*Buffer* has *sender* and *receiver* as its subsystems.
Port function	R3: 	$N_{R3}=\{1,2,3,4,5\}$ $pt_{R3}(2)=\{4\}$ $pt_{R3}(3)=\{5\}$ $pt_{R3}(1)=pt_{R3}(4)=pt_{R3}(5)$ $=\phi$	*Sender* and *receiver* have *syn* as their port.
Arcs, arc semantic function and arc function	R4: 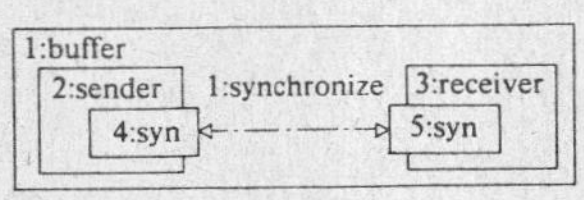	$A_{R4}=\{1\}$ $as_{R4}(1)=$synchronize $af_{R4}(1)=[4;5]$	There is a *synchronize* association directed from *syn* of *receiver* to *syn* of *sender*.
Subarc function	R5: 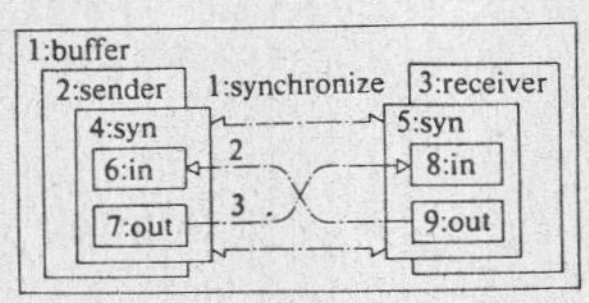	$A_{R5}=\{1,2,3\}$ $sa_{R5}(1)=\{2,3\}$ $sa_{R5}(2)=sa_{R5}(3)=\phi$ $af_{R5}(2)=[9;6]$ $af_{R5}(3)=[7;8]$	*Synchronization* has two *control flows* of opposite directions as its subassociations.

Fig. 5 Recursive Graph Components

System Types	Graphical Notation	Meaning
Processor	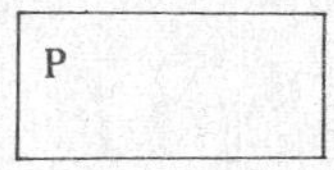	Processor P is a system which performs a single task, when activated; a processor must communicate with others through its ports.
Block		Block B is a system which performs a single task, when activated; a block can communicate with others directly without through ports.
Machine	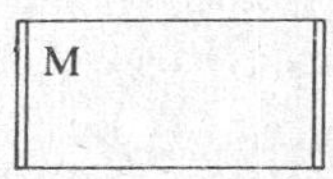	Machine M is a system which performs multiple tasks and consists of some processors performing those tasks and some data for communication among those processors.
Data		Data D is a system which holds some information.
Group		Group G is a virtual system which is only a named collection of systems having similar characteristics.

Association Types	Graphical Notations	Meaning
Control Flow	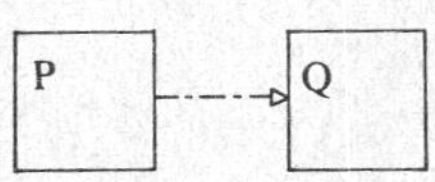	When processor P is terminated, then processor Q is activated.
Data Flow	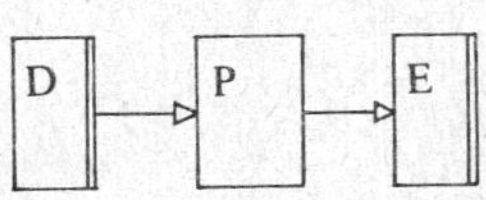	Processor P reads data D and writes data E.
Semantic Link	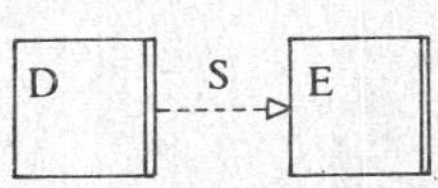	Data D is connected to data E under semantic association S.
Hierarchy Link	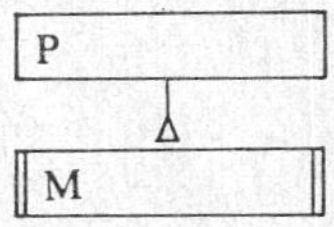	Processor P is implemented by using some processors composing(i.e., functionally abstracting) machine M.

Fig. 6 Reserved System Types and Association Types of RDSL-SP

R–graph provides basic building blocks and a framework for defining a machine checkable and/or verifiable design specification language called recursive design specification language RDSL. The syntax of RDSL consists of the set of allowable language elements called the alphabets, and the set of production rules to specify allowable relationships among those elements. The semantics of RDSL is given by assigning interpretations of language elements as well as relationships among them. In addition, because RDSL is a two dimensional language, shapes of nodes and line patterns of arcs must be defined depending on the system types and relationship types represented by nodes and arcs, respectively. These syntax and semantics of various RDSLs are defined depending on the conventions of the industries, such as software industry or chemical plants industry, in which designers are working. We have already defined RDSL for software products RDSL– as shown in Fig. 6.

IDMS consists of the following six subsystems: Interaction Supervisor IS, Display Routine Manager DRM, Grafic Display System GDS, Design Schema Machine DSM, RDSL Library RL and Database Management System DBMS. The design scheme specifying the architecture of the entire IDMS is represented in Fig. 7 in RDSL–SP. IS manages interaction commands implementing R–operations and supervises by those commands the interaction between IDMS and its user. The interaction with designers is performed through a graphic display terminal which is under control of GDS. DRM provides routines drawing or recognizing graph elements over graphic display terminals. DSM provides routines constructing, modifying or retrieving graph elements of R–graphs in a data base. DBMS stores and manages R–graphs in relevant storage forms. RL manages definitions of registered RDSLs and provides their syntax checkers and semantics analyzers. A submachine IDCL of a processor IS saves programs implementing macro recursive graph operations and made of DSM processor call statements. The programs saved in IDCL are

R6:

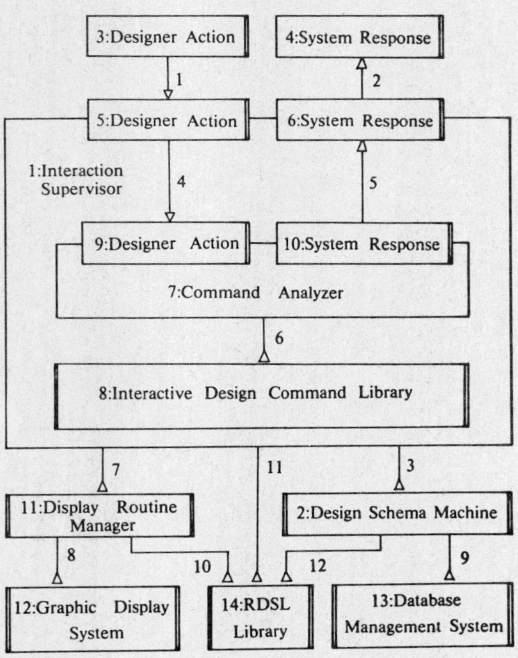

Fig. 7 An Integrated Design Schema Specifying the Entire IDMS

shared by designers as interactive design commands. Thus, the users of IDMS can share design results as well as design processes. This is the meaning of design process formalization.

The architecture of IDMS is designed to satisfy the following important requirements of independences:
(1) Machine independence from graphic display systems,
(2) Machine independence from database management systems,
(3) Application independence from recursive design specification languages.
The first two requirements mean that over any graphic display system and any database management system, IDMS can be installed easily with minimum modification of the system. The first is achieved by introducing DR between IS and GDS; the second by introducing DSM between IS and DBMS. The third requires IDMS to be able to process any RDSL if it is defined and registered in IDMS, and is achieved by managing concentratedly the language dependent parts of IDMS by RDSL Library.

4. Pattern Processing Tools

4.1 Minimal Spanning Tree Traversing Merge Method (MSTTM) [OHB 79]

A central problem in image analysis is that of segmentation, i.e., partitioning an image into disjoint regions that are homogeneous in some sense. For high contrast and noise free pictures, differenciation can be used successfully for edge detection and regions can be identified by an edge following algorithm. However, such a simple process fails for noisy pictures with fuzzy boundaries. This is also true when objects are defined by texture rather than brightness levels. Region detection came into the picture with the work of Muerle and Allen [MUE 68]. An important justification for working with regions rather than edges lies in the fact that natural objects are visually characterized more by regions, whereas man-made objects are often characterized by prominent edges. This section presents a general and computationally inexpensive algorithmic scheme of picture segmentation that falls in the region detection category.

The scheme of picture described in this section is non-semantical. The result is both data compression and organization. Algorithms of high computational complexity can be used afterward on this reduced description. A successful picture segmentation must take into account semantics. However, it is desirable to reduce the amount of data before the semantic phase by a general and computationally inexpensive step. This allows considerable simplification in the problem-dependent segmentation 'semantics' phase. This is especially important in processing large volumes of pictures stored in a pattern data base. Encoding of face information for the criminal investigation and extracting vegetation maps from aerial photographs are typical examples of the large scale picture processing.

In the scheme, a minimal spanning tree (MST) is used as a path of a sequential region grower. We believe that the key to the efficient picture processing lies in a system's capability to traverse the spatially adjacent graph while maintaining the structural organization of an image. A particular data structure which satisfies the criteria is an MST. Indeed, given an MST we can find hierarchical tree similar to the structural decomposition tree of Kirsh [KIR 70].

In the scheme, a digital picture is represented as an edge weighted graph in a matrix format. Then a set of edges at which mergeability is checked is generated by an MST constructing algorithm. A sequential region grower traverses edges in the MST in the proper order which is so generated that for each picture element weaker pairs must be checked first. Kruskal's MST constructing algorithm can select all edges at which the mergeability is checked and it can sort the edges in non-decreasing order of similarity measured locally. Although the idea that most similar pairs should be merged first is valuable, extra calculation for selecting candidate points is time-consuming. In most cases, the local evidence is very suggestive for the similarity measures. This allows considerable simplification in selecting candidate points for a region grower.

The region growing process in this subsection can conceptually be divided into two phases as follows:

1) Local phase in which a path of a region grower is determined locally. An MST is chosen as the path of a region grower.
2) Global phase in which heuristics (e.g., piecewise constant approximation and phagocyte heuristic) representing global information control a region grower at each point in the path.

This division allows significant simplification of the total region growing scheme. Indeed, in the present scheme, only (n-1) mergeability checks depending on the global information are required, where n is the number of pixels. The result of the segmentation by the MSTTM based on the piecewise constant approximation is shown in Fig. 8.

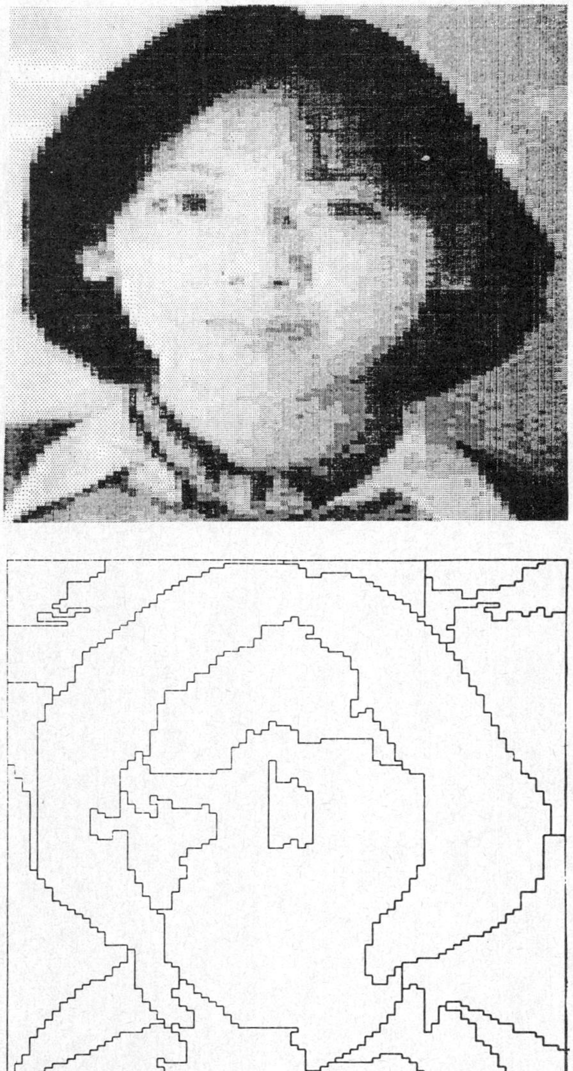

Fig. 8 An Example of Segmentation Based on the Simple Biregional Predicate Function

4.2 A Software System Based on a File-Coupling Structure

We have developed a software system which is used in a systematic approach to both design and verification of signal processors based on the linear predictive coding method. The aim of the system is to supply a software tool which is used to simulate experimental signal processors represented by a block-diagram notation. The system provides modules each of which performs the function of the box in the block-diagram. These modules are used to represent the block-diagram by coupling them within intermediate files. Contents of these files are used to evaluate performance of these modules. As a module is judged to have insufficient performances, this module is replaced independently with others.

To develop the software system, we have provided a set of routines each of which performs elementary mini-steps of linear predictive analysis and synthesis process. Routines are used in the following context:

```
module: program (INPUT,OUTPUT)
  {initialize}
    while not (eof(INPUT))
      read (INPUT)
        call {a routine}
      write (OUTPUT)
    end
  {house keeping}
end of module.
```

Note that only one routine is used in the module and two or more modules are linked through sharing files. Routines may consist of a sequence of basic routines, i.e., following routine can be appeared:

```
composite-routine: procedure
     {initialize}
     call procedure-1
     call procedure-2
       ......
     {house-keeping}
       end of composite-routine.
```

Provided that a sequence of modules is judged to have satisfactory performances, routines in these modules can be packed into one routine to reduce amounts of file-spaces used to link them. To make the conversion easy, we should agree some requirements to the programming style and languages employed to describe routines.

In order to clarify requirements to realize the system, we implemented a model system and designed a voice-separation system by using it for a case-study. The model system is written in a FORTRAN and partially an assembly language. Routines which back up the model system consist of subroutines each of which performs a small step of the linear predictive vocoder. Referring a heuristic procedure [MIL 73], we introduced smoothing procedures applied to both the excitation function and the area-function of the pseudo vocal tract. It is shown experimentally that the system separates a singing voice from a mixture of the voice and piano-sound. A phrase taken from "Der Linderbaum" is used for the experiment. It is difficult to implement such the filter by using traditional filtration schemes.

The motivation of the research was originally to clarify the sound understanding process of human beings. On the way of the investigation, a tool which can be used to define new processors and to evaluate their performances was required to make hypotheses about the sound understanding process. About a same time, a new theme on the "computer-ethnology" was brought. The requirement of the theme was to develop a software tool which can be used in the field of ethnomusicological research. Since ethnomusicologist's requirements to the signal processing technique are generally intuitive and not strict enough to realize the processor, the software system should have features to design a new processor and to evaluate it without much trouble. (The ultimate object should be a "block-diagram programming" to design a new processor for end-users.) Hence, we settled a goal to develop a software system which supports both design and verification of signal processors in a systematic way.

At the beginning, we implemented a model system to identify requirements to the system itself and its environments in order to achieve the goal. Until now, we have examined the model system and evaluated its performance. As the conclusion, we summarize the characteristics of the block diagram simulator to be achieved and tools to run the simulator effectively.

CHARACTERISTICS.

1. It can be used to evaluate performances of signal processors which are described by block-diagrams.(Except a time-factor.)
2. The system allows end-users to concentrate input and output data of boxes in the diagram.
3. The structure of any box is hidden from end-users. "Block-diagram programming" is allowed to end users.

TOOLS.

1. Design-tool for block-diagram design [HAR 78].
2. Monitor-system to analyze dynamic behavior of modules.
3. Description languages to describe modules and routines.
4. Back up libraries.
5. Report modules to analyze contents of files.

e.t.c.

REFERENCES

[KIT 79] Kitagawa, H. and Kunii, T. L., 'APAD: Application-Adaptable Database System -- Its Architecture and Design --,' Proc. of the 13th IBM Computer Science Symposium (1979, in press).

[COD 70] CODD, E. F., 'A Relational Model of Data for Large Shared Data Banks,' CACM, Vol. 13, No. 6, pp.377-387 (June 1970).

[KUN 77] Kunii, T. L. and Kunii, H. S., 'Design Criteria for Distributed Database Systems,' Proc. of the 3rd International Conference on VLDB, Tokyo, pp. 93-104 (Oct. 1977).

[KUN 78] Kunii, T. L., Browne, C. J. and Kunii, H. S., 'An Architecture for Evolutionary Database System Design,' Proc. of COMPSAC 78, Chicago, pp. 382-386 (Nov. 1978)

[KUN 79] Kunii, T. L. and Kunii, H. S., 'Architecture for Virtual Graphic Database System for Interactive Computer-Aided Design,' CAD Computer-Aided Design, Vol. 11, No. 3, pp. 132-135 (May 1979).

[ZLO 75] Zloof, M. M., 'Query by Example,' Proc. of NCC 1975, Anaheim, pp. 431-438 (May 1975).

[TSI 77] Tsichritzis, D. C. and Lochovsky, E. H., 'Data Base Management Systems,' Academic Press, (1977)

[OHB 79] Ohbo, N, Shimizu, K and Kunii, T. L., 'A Graph-Theoretical Approach to Region Detection,' Proc. of COMPSAC 79, Chicago (Nov. 1979, in press).

[HAR 78] Harada, M, Kunii, T. L. and Saito, M., 'RGT: The Recursive Graph Theory as a Theoretical Basis of a System Design Tool DESIGN-TOOL -- With an Application to Medical Information System Design,' Proc. of MEDIS '78, Osaka, pp. 503-507 (Oct. 1978).

[HAR 79] Harada, M. and Kunii, T. L., 'A Design Process Formalization,' Proc. of COMPSAC 79, Chicago (Nov. 1979, in press).

[MUE 68] Muerle, J. L. and Allen, D. C., 'Experimental Evaluation of Techniques for Automatic Segmentation of Objects in a Complex Scene,' Pictorial Pattern Recognition (Cheng, G. C. et al., Eds.) Thompson, Washington, pp. 3-13 (1968).

[HOR 74] Horowitz, S. L. and Pavlidis, T., 'Picture Segmentation by a Directed Split-and-Merge Procedure,' Proc. of 2nd International Joint Conference on Pattern Recognition, pp. 424-433 (Aug. 1974).

[WON 77] Wong, A. K. C. and Vogel, M. A., 'Resolution-Dependent Information Measures for Image Analysis,' IEEE Trans. on Systems, Man and Cybernetics, SMC-7, No. 1, pp. 49-60 (Jan. 1977).

[KIR 70] Kirsh, R. A., 'Computer Determination of the Constituent Structure of Biological Images,' Computers and Biomedical Research, Vol. 4, No. 3, pp. 315-328 (1970).

[MIL 73] Miller, N. J., 'Filtering of Singing Voice Signal from Noise by Synthesis,' Ph. D. Dissertation, University of Utah, Salt Lake City (Mar. 1973).

Scientific Information Systems in Japan
H. Inose (ed.)

RELIABILITY DESIGN CONSIDERATION IN INFORMATION NETWORK ENVIRONMENT

Tadao SAITO, Hiroshi INOSE, Kunihiro ASADA and Jun ADACHI
Faculty of Engineering, University of Tokyo

1. INTRODUCTION

Large scale information networks are always exposed to reliability and security problems which are caused by physical and human reasons. The exposures includes hardware and software failures, human carelessness as well as malicious damages and crimes. Various techniques have been developed to minimize the probability of happening of problems and to minimize the damage if it happened. Hardware errors at the center machine are rare today. Errors in telecommunication lines can essentially be eliminated using effective error control codes. However, the maintenance of remote intelligent terminals may cause problems because of the difficulty to quickly send maintenance personnel to distributed remote locations.

Software and application program errors are still the principal reasons of reliability problems of the information system. The answer to this problem is the complete program testing.

The problem of man-machine interface is also important to minimize errors. In an information network to access the data base, it is important to implement a query language to minimize the operation error.

The subject of our research group is to explore various means to improve the reliability of information mation networks. The subject included the remote hardware diagnostics for intelligent terminals, an intensive testing method for programs and the simplified query language for data base access.

An experimental network system has been constructed to demonstrate these results in the network environment. In the experimental system, the remote diagnostics system for intelligent terminals and simplified query language system have been implemented. The paper will describe the result of our research group, mainly on the experimental network.

2. EXPERIMENTAL MODEL OF INFORMATION NETWORK

The large scale information system consists of a number of data base centers each of which stores the specified data base informations to be used as the distributed data base (DB) and a number of host computers (HOST) which is used for various computation using the distributed data bases. The data base centers and the host computers are connected by way of a digital data network as is shown in Fig. 1. The terminals (T) to use the data bases and the host computers are connected to the host computers or the data base computers or connected directly to the digital data network.

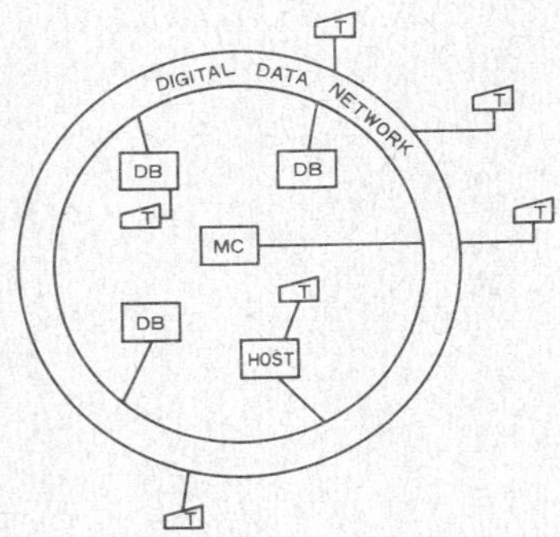

Fig. 1. A model of information network.

The information system should be constructed based upon a unified network architecture. To support the network architecture, the terminal system should also be an intelligent terminal implementing the network protocol. The maintenance of intelligence distributed in large area will cause a problem in the information network without automated maintenance function. For this purpose maintenance centers for distributed intelligence are to be provided in the network.

The computers at data base centers and host computers are to be maintained at the respective location by their own maintenance personnel. However switching node of the digital data networks and remote intelligent terminals will be operated without maintenance personnel. The intelligence at such a location should be constructed using a plurality of micro-computers. If a malfunction is found at such an intelligence, the maintenance center (MC) will send a diagnostic message to the intelligence through the network or through the host computers directly connected to the terminal. If the trouble is located, the micro-computer package having the trouble will be disconnected and the remaining package will continue the operation. After this emergency operation, the package having the trouble will be replaced at the next convenience.

As a model of high reliable information network, an experimental system having the feature of remote terminals automatic diagnostics and simplified data base query language has been constructed. Information networks are to be constructed to have data base systems and a maintenance center. Different from a centralized system, in a large scale information network environment the data bases and the maintenance center are constructed in different locations. As the smallest model to simulate such a situation, an experimental information network having two computers has been constructed. One computer (AICOM, C-5) is to simulate the data base center and the other (HITACHI Hitac-10) is to simulate the maintenance center. Fig. 2 shows the outline of the experimental information network.

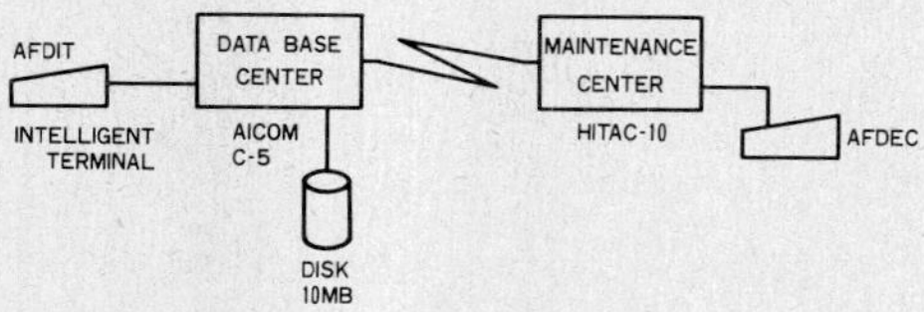

Fig. 2. Experimental network model.

Two mini-computers are connected by a high speed link. An intelligent terminal having automatic remote diagnostics function called AFDIT is connected to the data base machine. The terminal is to be diagnosed by the maintenance center which is indirectly connected to the terminal.

Since the reliability techniques for the communication network itself is already established, the experimental network simplified the structure of the communication network to be a single 9600 bps line.

Another intelligent terminal having automatic remote diagnostics function called AFDEC is also connected to the maintenance center.

The data base system is implemented in the computer of the data base center. The access to the data base is possible, through both of the intelligent terminals as well as CRT terminal connected to each of the minicomputer. The two intelligent terminals have the function to define user own macro language. This function can be used to convert data.

3. REMOTE DIAGNOSTICS FOR INTELLIGENT TERMINALS

3.1 Structure of the Intelligence

The terminals in the experimental system is composed of a plurality of micro-processors (CPU), a plurality of memories, a CRT display and a key board. These components are connected into a multi-processor system as is shown in Fig. 3.

The communication between a CPU and one of the memories or the I/O's is controlled by the CPU. In this sense the CPU is the master and the memory is the slave. Between the masters and the slaves, a group of switches (SW) is provided, which is controlled by the state register. The state register has 6 states 0 through 5. The connection between slaves and masters specified in each state is given in Table 1. In

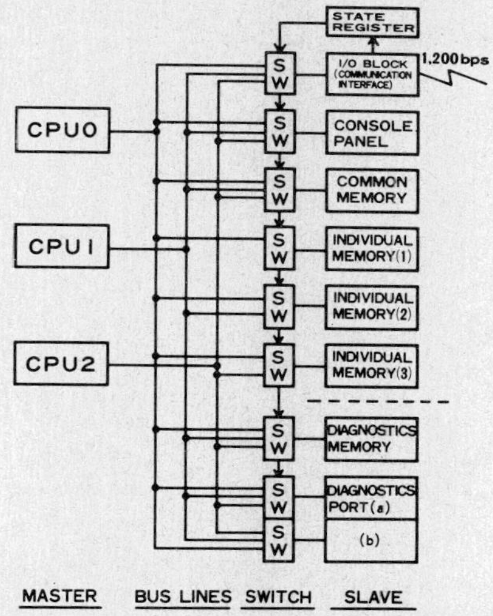

Fig. 3. Structure of an intelligence at the terminal.

Table 1. Assignment of CPU's to slaves.

STATE REGISTER	'0'	'1'	'2'	'3'	'4'	'5'
IO BLOCK	1	0	1	2	0	1
CONSOLE PANEL		0	1	2		
INDIVIDUAL MEMORY (1)	0					
INDIVIDUAL MEMORY (2)	1				0	1
INDIVIDUAL MEMORY (3)	2				2	0
COMMON MEMORY	012	0	1	2	02	01
DIAGNOSTICS MEMORY		0	1	2		
DIAGNOSTICS PORT (a)		0	1	2		

The meaning of the contents of the state register is as follows:

'0': normal state.
'1', '2', '3': diagnostics states at which CPU 0, 1, or 2 is the diagnostician respectively.
'4', '5': recovered state at which CPU 1 or 2 takes the place of CPU 0.

The value 0, 1 or 2 in the table means that the unit is assigned to the CPU 0, 1, or 2.

Table 1 the number of CPU connected to each slave is specified for each state. The content of the state register is loaded by the maintenance center.

State 0 is the normal state. In this state CPU 1 is used for communication control, CPU 2 is used for online data processing and CPU 0 is used for back-ground off-line processing. In the diagnostics state(states 1, 2, 3),one of the CPU is assigned to be the diagnostician. The diagnostic program is stored in the diagnostic memory, which is connected to the diagnostician CPU. To the diagnostician CPU the IO block, console panel and the diagnostic port (a) are also connected. The diagnostic port (b) is connected to the other CPU's, i.e., patient CPU's.

3.2 Structure of the Diagnostic Ports

Fig. 4 shows the structure of the diagnostics port which is used to interconnect the diagnostician CPU to the patient CPU. The port is used to transfer the test sequence and the results of the test sequence between the diagnostician and patient CPU's.

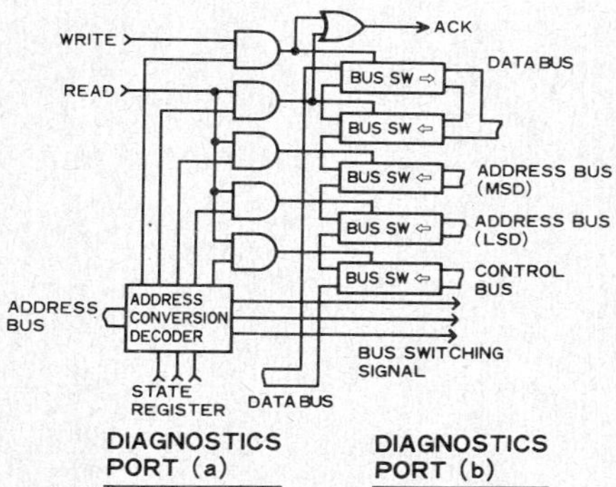

Fig. 4. Structure of the diagnostics port

The data bus of the port (a) can be connected to data bus, address bus or the control bus of the port (b). The connection can be controlled by the address conversion decoder which is connected to the address bus of the port (a).

In operation, the diagnostician CPU read the control bus of the patient CPU and determines whether the diagnostician CPU should read data or write data. After the completion of the exchange of data through the data bus of the patient CPU, an ACK signal is sent to the patient CPU, which notice that one machine cycle has been completed.

3.3 Test Procedure

The diagnostics of the terminal system is started by the command from the maintenance center. The command includes only two characters, the first of which is the diagnostic start code and the second is the state register data.

When the command is received at the terminal, a diagnostic interrupt signal is applied to each of the CPU's. By the interrupt, the contents of registers in each of CPU's are saved in the memory for the restart of the operation after the diagnostics. Then the second character is applied to the state register after which reset signal is applied to each of the CPU's.

The diagnostician CPU assigned by the state register observes the state of the patient CPU's. For an input request from the patient, a test data is applied. The patient CPU receives the test sequence as a data or a operation code. The state of the buses of the patient CPU's are observed by the diagnostician CPU through the diagnostics port.

Appling the same test sequence to two patient CPU's, if discrepancy is found between the results from two CPU's, it is interpreted as a simpton of failure. The simpton is sent to the maintenance center as the diagnostics report. The diagnostics report includes

i) Internal state of the registers of patient CPU's before the start of diagnostics.
ii) Initial value of the pseudo random number used as the test sequence.
iii) The external state of the patient CPU's at the detection of the discrepancy.
iv) The number of machine cycles tested before the detection of the discrepancy.

The maintenance center has the simulation program to simulate the operation of the CPU's of the intelligent terminal. When a diagnostics report is given, the maintenance center can identify the correct result of the test, by which the center determines which CPU has failure.

In some instance, the diagnostician CPU may have fault. To cope with this situation, the diagnostician CPU can be changed. If the result of diagnostics is different when the diagnostician CPU is changed, the third CPU is assigned as the diagnostician, then the diagnostic procedure is repeated. The final result is given based on voting principle. If the number of CPU in trouble is less than half of the total CPU, the result of the diagnostics is guaranteed to be correct.

3.4 Recovery Procedure

When the CPU having a failure is identified, the command from the maintenance center switches the content of the state register to remove the CPU having trouble from the normal operation. In this case since the result of operation before the diagnostics start is expected to be not correct, the intermediate results saved at the initiation of the diagnostics is invalid. Therefore, the operation is restarted from the initial condition.

If no failure is found by the diagnostics, the condition saved at the initiation of diagnostics is recovered and the operation is continued.

To identify the restart mode of operation, one particular bit in the common memory is assigned. This bit is set by the restart command from the maintenance center. Each CPU determines the restart procedure based on the bit.

4. DESIGN OF TEST SEQUENCE

4.1 Model for Test Sequence

The test sequence generated by diagnostician CPU is a pseudo-random sequence. If a fault exists in a patient CPU, when a test sequence is given, a test data may or may not effect to the state of the patient CPU. The effect may appear at the buses to be observed by the diagnostician CPU or may disappear by the effect of following test data before appearing at the buses. If the effect of a operation code exists in the patient CPU, the effect can be exposed to the buses by applying a test sequence which transfer the content of internal registers to the buses. Therefore the test sequence should be constructed by a predetermined length of random number code and the exposure sequence which functions to transfer the effect to the buses.

To design the test sequence, the length of random number code is to be specified. For this purpse, effectiveness of random number for diagnostics should be evaluated.

It is difficult to introduce various fault to LSI microprocessors. Therefore it was assumed that fault in LSI causes an errornous operation for a particular operation code. In other words, a CPU is assumed to not respond to a particular operation code when a fault occurs corresponding to that operation code. The effectiveness of the test sequence depends upon the operation code having fault.

The probability that the particular operation code appears in the random sequence is denoted by α. When the operation code appears, the internal state of the CPU is changed by the probability γ. The effect will disappear by the probability η/time unit. The effect may be exposed by the probability τ/time unit. η is called the disappearing rate and τ is called the exposure rate. The time unit is identical to the byte to count the length of random number sequence. The probability model is shown in Fig. 5.

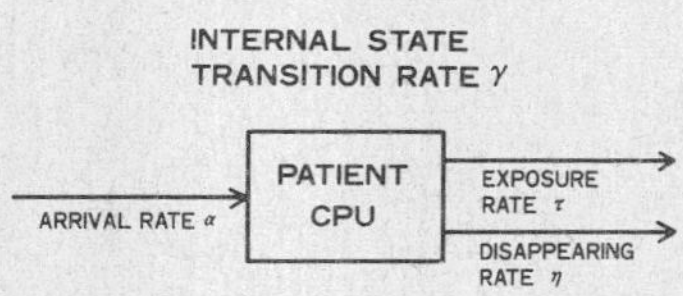

Fig. 5. Model for diagnostics

In the random test sequence, a particular operation code appears by the rate α. The effect of the code is caused by the rate $\alpha\cdot\gamma$. The effect may disappear or be exposed to the buses. The expected number of operation codes that the effect which has not been exposed to the buses remains in the CPU is denoted by q. Denoting the time by t, the following equation holds

$$\frac{dq}{dt} + (\eta + t)q = \gamma\cdot\alpha \tag{1}$$

The initial condition for the differential equation is $q = 0$ at $t = 0$.

Denoting the length of random sequence by S in bytes since the effect remaining at the end of S is exposed by the exposure sequence, average number of chances of observation of the effect caused for the operation code during one test sequence including the exposure sequence is given by

$$H(S) = \int_0^S \tau\cdot q(t)dt + q(S) \tag{2}$$

The effectiveness of the test sequence by for a particular operation code is given by

$$C(S) = \frac{H(S)}{\alpha S} = \frac{\gamma\tau}{\eta + \tau}\left\{1 + \frac{\eta}{\tau}\cdot\frac{1 - \exp\{-S(\eta + \tau)\}}{S(\eta + \tau)}\right\} \tag{3}$$

In Eq. (3) the value of $C(S)$ tend to γ for $S \rightarrow 0$. However, since the process is time discrete, $C(S)$ = at $S = 1$. Therefore the effectiveniss should by expressed by $C(S - 1)$. This is expressed by $e(S)$, i.e.,

$$e(S) = C(S - 1) \tag{4}$$

Denoting the length of test sequence including the exposure sequence by T and the length of random number sequence by S, the probability that the particular operation code is applied in a test sequence is given by

$$A(S) = \frac{\alpha S}{T} \tag{5}$$

Therefore, the probability that the effect of a particular operation code is observed for each test of one byte is given by

$$E(S) = A(S)\cdot e(S) \tag{6}$$

$E(S)$ may be interpreted as the average number of effective operation code for each test of one byte.

4.2 Evaluation of Parameter Values by Experiment

To obtain the optimum length of the test sequence, the values of parameters for the model must be known. For this purpose, an experiment using the intelligent terminal has been performed.

Fig. 6 shows the outline of the experimental set up. The set up is implemented in the software of the diagnostician CPU. The test sequence is generated by the test sequence generator. The test sequence is essentially pseudo random sequence, from which particular patterns corresponding to undefined codes or codes harmful to diagnostics are excluded. The sequence is applied to the fault pattern generator through which the test sequence is applied to the patient CPU's. The fault pattern generator has the function to convert the pattern (byte) corresponding to a particular operation code to the no operation (NOP) code. The converted pattern is applied to the CPU 2. To the CPU 1, original random number code is applied. The fault of the particular operation is thus simulated in the CPU 2. To avoid the possibility that the converted pattern is interpreted as an address part of the preceding codes, a predetermined number of NOP codes are applied to both of the CPU before the converted fault pattern is applied to the CPU 2. The number of applied fault is counted by the counter A. The counter L counts the total number of bytes generated by the test sequence generator. The responses exposed to the buses from two CPU's are compared by the comparator. The number of different exposures is counted by the counter D.

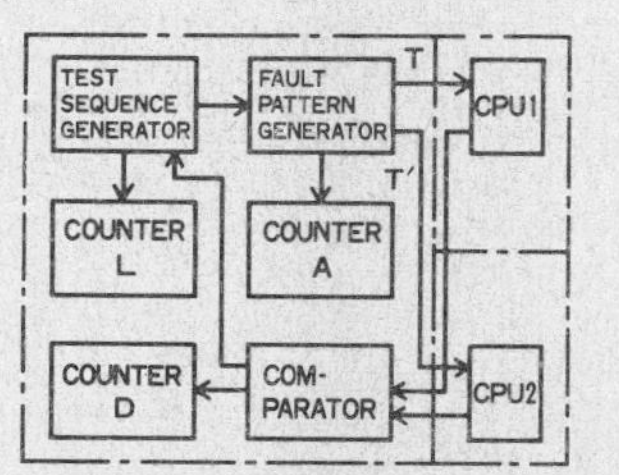

DIAGNOSTICIAN CPU PATIENT CPU

Fig. 6. Experimental set up to evaluate diagnostic parameters.

The experiment was repeated until the count in the counter A reaches to a predetermined value. In the experiment to be described, the value is selected to be 100. The data obtained by the experiment is

$$A_e = \frac{\text{content of the counter A}}{\text{content of the counter L}} \qquad (7)$$

$$e_e = \frac{\text{content of the counter D}}{\text{content of the counter A}} \qquad (8)$$

The parameters are obtained using Eqs. (3),(4) and (5).

The value of γ is obtained from Eq. (4) since $\lim_{S\to 1} e(S) = \gamma$. The value of α is obtained by Eq. (5). The parameters η and τ are obtained by Eq. (3).

Fig. 7 shows the experimental value of the effectiveness $e(S)$ of the test sequence for typical particular operation codes, i.e., ADD L, MOV B, L, CMP L for Intel 8080. Fig. 8 shows the distribution of the value of τ and η. The value of α is about 0.004 ($\fallingdotseq$ 1/256) and the value of γ has the distribution between 0.6 $\sim$ 1.0. The effect of the values of α and γ is smaller than the values of τ and η.

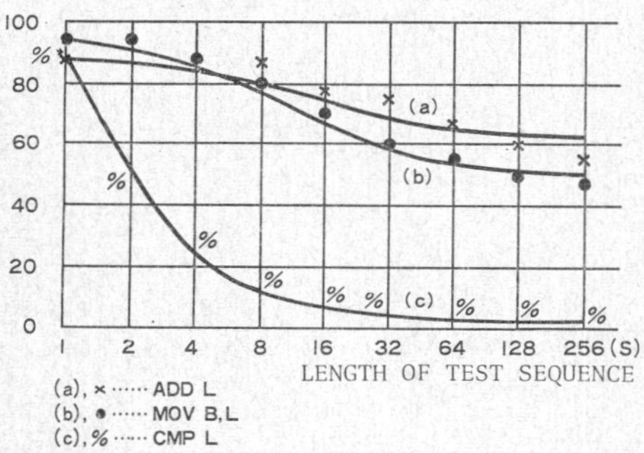

Fig. 7. Effectiveness of the test sequence for a particular operation code.

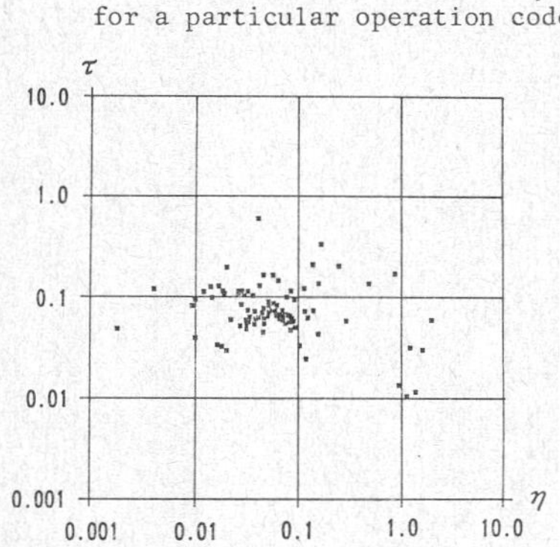

Fig. 8. Values of τ and η for various operation codes of INTEL 8080.

The probability $E(S)$ that the effect of a particular operation code is observed for each test of one byte is shown in Fig. 9 for various values of τ and η. Fig. 8 shows that for an operation code having the value of disappearing rate η is less than 0.1 the value of exposure rate τ is distributed around 0.1.

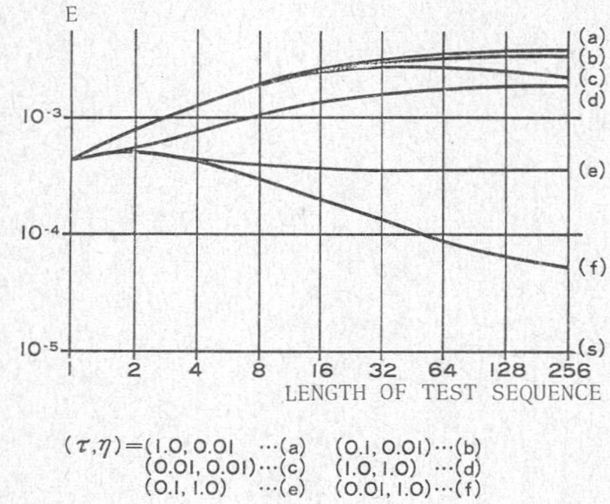

Fig. 9. Theoretical value of E for various values of τ and η, τ = 1.0, η = 0.01

Curves (a) and (b) of Fig. 9 show that for this type of operation code, the value of $E(S)$ is the larger as the length of sequence increases.

On the other hand for some operation codes, the disappearing rate is as high as 1.0 or more. For this type of operation code, the value of E sharply decreases as the length of test sequence increases, as is shown by curve (f) of Fig. 9. Operation codes of this type mainly includes the handling command for program status words (PSW).

4.3 Design of Test Sequence

If the length of test sequence is short, the value of E is small for the types of operations as are shown by the curves (a) and (b) of Fig. 9. However if the length of test sequence is long, the value of E is large for the type of operations as are shown by the curve (f) of Fig. 9. Since the latter operation codes are related to PSW, the test sequence that can compromise these two different tendency can by devised. In other words, if push PSW instruction to expose the content of PSW is mixed in the random test sequence by a fixed ratio, the effect to the PSW type instructions can be exposed even if the test sequence is fairly long. The test sequence thus designed is shown in Fig. 10. The optimum mix ratio is obtained from the curve (b) of Fig. 9. The distribution of τ and η when the mix ratio is 3 : 1 is shown in Fig.11.

Comparing Figs. 8 and 11, it is known that for all the operations having the value of η larger than 1.0 , the value of τ is larger than 0.1. This means that the improvement of more than 10 times is obtained. Using the test sequences as is shown Fig. 10, the value of E is tend to a constant as the length of test sequence increases. Therefore, the length of test sequence can be determined from the view point of testing time at the maintenance center irrespective of the value of E, i.e., the efficiency of diagnostics. The length of random number test sequence in the experimental system was 128 bytes.

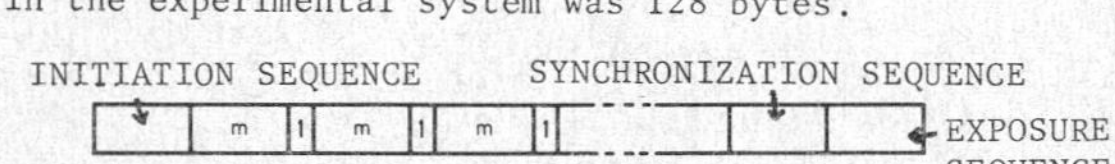

Fig. 10. Test sequence.

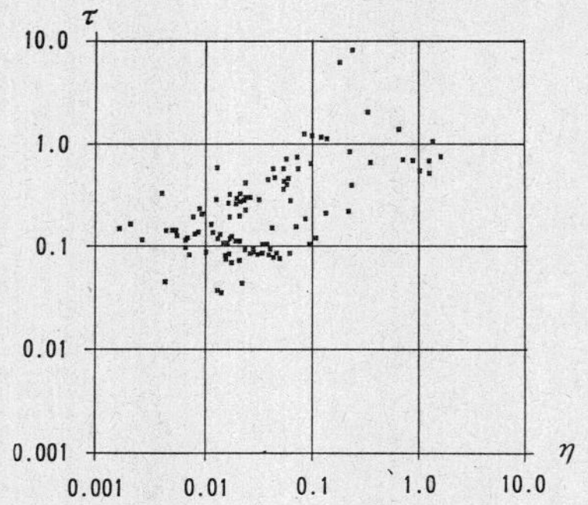

Fig. 11. Equivalent values of τ and η in improved test sequence.

The initiation sequence is to set all the registers to the "0" state. This is useful to give known initial state for all the registers. The length of the initiation sequence is 10 bytes. The synchronization sequence is continuous four no operation code (NOP). Since the maximum number of operand for one operation codes is 4 in Intel 8080, the following part of the test sequence is always interpreted as operation codes not as addresses or operand. The exposure sequence is to expose the content of registers for the buses. Four codes of push PSW, push B, push D, push H are used for this purpose.

Since it is difficult to generate physical fault in the actual microprocessors, experiment of diagnostics using actual microprocessor has not been performed, however, in the experiment using simulated fault, the fault has been detected 100 %.

5. DATA BASE USING SIMPLIFIED QUERY LANGUAGE

To improve the reliability of information systems, it is essential to reduce the possibility of human errors. In the system centered around a data base, the improvement of query language is essential. From this point of view, a data language having simplified query capability is proposed and implemented in the experimental system.

Its major feature is automatic join facility of tables in query processing. In a relational data base system, since the tables are subdivided into small pieces to satisfy the condition of the third normal form, it often becomes troublesome to obtain desired information by joining various tables. The automatic join capability implemented in the experimental system utilizes the deterministic relationships between attributes of the data base schema. Using the relationship the retrieval can be specified using only attribute names in user language. Fig. 12 shows an example of schema that consists of three tables. There can be various forms of query involving attributes MANAGER and EMPLOYEE. When the table name of each attribute is not specified, the possible forms of join relating these two attributes are,

1) to retrieve data by connecting these two attributes according to the relationship implied in table A.
2) to join table B and C based on the common attribute SALARY. etc.

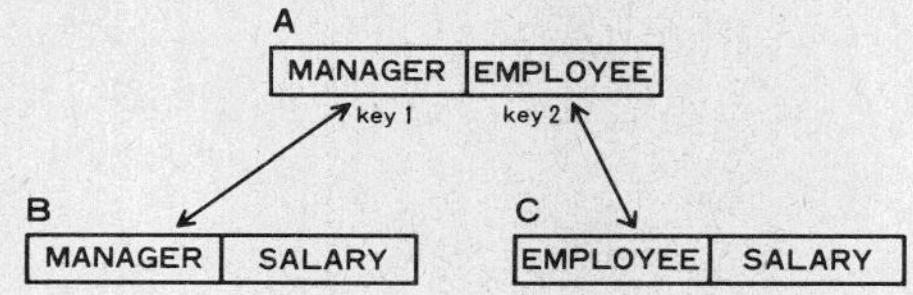

Fig. 12. An example of relation between tables.

The example 2) is the join based on the attribute which is determined by the specified attributes, that is, based on a lower-level attribute. On the other hand, the example 1) is the join based on the common attribute which determines all the specified attributes, that is, the higher-level attribute. The strategy used in the example 1) is reasonable in most cases when only the attribute names are specified without their table names. The reason is that the data base schema composed of the third normal form implies the semantic relationship between attributes derived from the functional dependency.

In the implemented data base system, if the query is according to the semantic structure implied in the data base schema, a user can issue a query by using only attribute names without specifying table name nor the strategy of table joins. The query processing strategy is as follows.

First, the possibility of automatic join is examined. The automatic join is possible if the highest-level attribute common to all the specified attributes is attained by searching higher-level attributes on which each specified attribute is dependent. Thus, a tree structured deterministic graph of attributes is obtained which has the highest-level attribute as the root. This is actually the partial graph of the data base schema. Then, from this graph, the sequence of tables and their joins is derived to relate the specified attribute each another.

After this analysis, the query processing procedure is generated which executes table joins, restrictions, projections and so on.

Fig. 13 shows another example of schema. Fig. 14 a) is a example of query written in the implemented query language which means to "List the names and hobbies of the students of the department of electron-

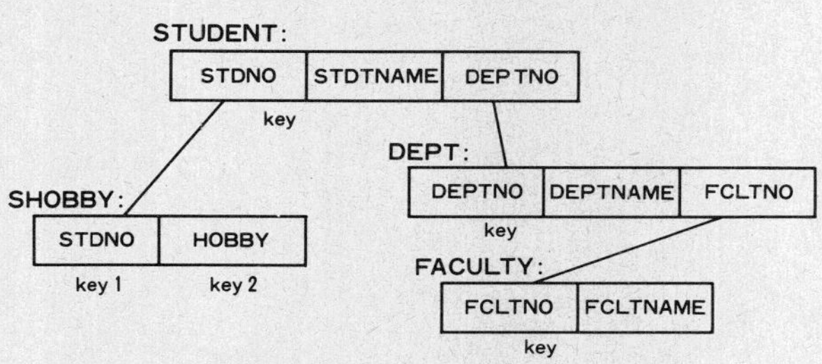

Fig. 13. An example of data base schema.

ics". This query involves three tables. In a conventional query language like SEQUEL in System R, the join is to be specified in the where clause as is shown in Fig. 14 b). In this case, the automatic join analyzer examines the possibility of join between three attributes, STDTNAME, HOBBY and DEPTNAME, and generates the query processing strategy to join three tables.

```
a)  SELECT  STDTNAME, HOBBY
    WHERE   DEPTNAME = 'ELECTRONICS';

b)  SELECT  STUDENT. STDTNAME, SHOBBY. HOBBY
    FROM    STUDENT, SHOBBY, DEPT
    WHERE   STUDENT. STDNO = SHOBBY. STDTNO
         &  STUDENT. DEPTNO = DEPT. DEPTNO
         &  DEPT. DEPTNAME = 'ELECTRONICS';
```

Fig. 14. An example of query.

As is mentioned in this example, the automatic join facility functions successfully in a query which depends on the semantic structure of the data base schema and reduces user's burden to describe a query. However, some times,a user wants to get information independently from the semantic structure of schema or the join analyzer may decide that the automatic join is impossible because of the ambiguity in searching the higher-level attributes. To cope with these cases, the query language also has the facility to describe an arbitrary form of table joins in a flat form.

The prototype data base system was implemented by using PL/I as a host language, and also has the facilities of data definition and manipulation. And this system uses a batch type system for query processing. However, since this automatic join facility is more useful and powerful in an interactive system, the experimental system is implemented as an interactive query processing system.

6. CONCLUSION

To improve the reliability of the information network, several methods are tried in our study group. That includes the systematic test sequence generation for software test, automatic diagnostics system for the remote intelligence and simplified query language for data base. The automatic diagnostics system and the simplified query language were implemented in the experimental network model.

The paper mainly described the automatic diagnostics system for remote intelligences. The reliability and security problem for the information network is the protection system from various unexpected accident or malicious attacks. If a new technology is developed, the protection mechanism for that should be devised from extensive view points. If a new protection system was developed,some other techniques to penetrate through the protection will always be found. In this sense the study for the reliability and security will be endless.

The study in our research group is oriented to more fundamental concept for reliability improvement. The techniques described in this report is believed to be a essential means to improve the maintenability of large scale information networks.

REFERENCES

[1] H. Inose, T. Saito, K. Asada, K. Hasuike and M. Kumazaki, "A Remote Diagnostics and Recovery System for Intelligent Terminals Including Microprocessors", Transaction of Institute of Electrical Engineering of Japan, Vol.99, No.10 (Oct., 1979)

[2] J. Adachi, T. Saito and H. Inose, "A Data Language Embedding Functional Dependency", Professional Groups DBMS of Information Processing Society of Japan, DBMS 11-2 (Jan., 1979)

[3] K. Asada, T. Saito and H. Inose, "A Systematic Method for Program Testing", Professional Group of Reliability of the Institute of Communication and Electronics Engineers of Japan, R 79-2 (May, 1979)

Scientific Information Systems in Japan
H. Inose (ed.)

REMOTE ACCESS AND RETRIEVAL SYSTEM FOR VISUAL PATTERN FEATURING MICROFICHE STORAGE AND FACSIMILE RECORDING

Hiroshi INOSE, Tadao SAITO and Hiroshi NAKAGAWA
Faculty of Engineering, University of Tokyo, Hongo, Bunkyo-ku, Tokyo

A data base that provides visual patterns in addition to alphanumeric information may play an important role for future information systems including automated library. Key factors for the organization of a data base system for visual patterns are the storage and on-line retrieval of large amount of visual patterns, the transmission of retrieved pattern information to remote terminals, and the recording of visual patterns at the remote terminals in an appropriate form of hard copy. The present paper proposes a system concept in which visual patterns are stored in microfiche, retrieved on-line, and through optical-to-electrical conversion, digital conversion and speed conversion, transmitted to remote terminals along with alphanumeric data, and recorded by facsimile receivers. An experimental system which has been constructed for the purpose of feasibility demonstration is also described. In addition to this, taking arabesque and floral patterns on roof tiles of eighth century Japan as examples, feature extraction techniques are described that help evaluate the keys for retrieval.

1. INTRODUCTION

In addition to data in alphanumeric form, diagrams, pictures, handwritten documents and other visual patterns will play an important role for data base services. Bibliographic information service may be enhanced by providing visual patterns along with the titles and abstracts. Data bases for various scientific disciplines, including chemistry, biology and archeology, require some sort of visual patterns to be stored and retrieved. An automated library may be organized in the near future that stores the original documents in the form of visual patterns, retrieves them upon request from remote terminals, and transmits back the requested patterns over communication lines.

Key factors for the organization of a data base system for visual patterns are pattern storage, retrieval and remote access. In connection with storage and retrieval, such pattern processing as redundancy reduction and feature extraction may also be required. The present paper describes the studies made on the organization of a data base system for visual patterns and its remote terminals, with particular reference to storage, retrieval and remote on-line access of visual patterns. An experimental data base system is also described that has been constructed for the purpose of demonstrating the feasibility of these studies. In addition, feature extraction techniques are described taking as examples, the arabesque and floral patterns on eighth century roof tiles in Japan.

2. SYSTEM DESIGN CONSIDERATIONS(3)

2.1 Storage Media

Due mainly to the fact that a visual pattern contains a great deal of information, selection of an appropriate recording media is one of the most important factors in designing a data base for visual patterns. Even if a sophisticated redundancy reduction technique is employed, a visual pattern still contains information in the order of several hundred kilobits. The memory for digitally storing one million patterns therefore is required to have the capacity of tens of thousand MegaByte. This means that one hundred large capacity disks are required for the purpose.

In other words, data bases for visual patterns call for a storage media with a higher storage density and lower cost. Although the use of video disks is foreseen, extensive development is required for this purpose, because a video disk is designed mainly, for television programs and its high cost of inputting is shared by a large number of reproductions. Therefore the use of an appropriate optical film recording may be considered as the most realistic approach at this moment. The use of roll film, microfiche and super microfiche may be possible. From the view points of required resolution, case of retrieval and of conversion to electrical signal, microfiche has been chosen as the storage media. A commercial microfiche sorter which stores more than 1000 sheets each containing 63 or 126 frames is now available on the market. The draw-backs of the use of microfiche are such that it is basically a read-only memory and requires photo-chemical process for writing, and that optical-to-electrical conversion is required for computer processing and transmission.

2.2 Signal Conversion

A television camera or a flying-spot-scanner may be used for optical-to-electrical conversion. Although a flying-spot-scanner has better conversion characteristics in terms of resolution and linearity, no flying-spot-scanner for reading out microfiche is yet available in local market. For this reason, a television camera is the only solution at this moment. The resolution of commercial television cameras, however, is not quite satisfactory for the purpose.

Speed conversion is required for the output signal of a television camera to be processed by a computer or to be transmitted over a transmission line. For this purpose, either an analog buffer such as a scan converter or a digital buffer such as a frame memory may be used. Through a preliminary experimental studies[4], the former which is cheaper, was found unsatisfactory, because it is less stable and flexible than the latter. When a frame memory is used, the output of a television camera is fed into an analog-to-digital converter and then the digitalized signal is stored in a frame memory. This permits the digitalized signal to be processed for format conversion and redundancy reduction before being transmitted over a transmission line.

2.3 Transmission

Format conversion depends upon the method of transmitting and recording the visual information at remote terminals. The constraint here is the cost of communication and of remote terminals. The use of switched telephone networks is inevitable when communication cost is taken into account. This limits the transmission rate to 2400 bps or less. The use of facsimile is preferable considering the permissible cost of a remote terminal that is required to have the capability of interactively communicating with the data base, receiving and recording alphanumeric data as well as visual patterns in the form of a hard copy.

Preliminary experimental studies[4] ruled out analog facsimile transmission which is cheaper, because of the inefficiency of switching over the use of a communication line from receiving analog facsimile signal to receiving digital alphanumeric information. The decision to use digital facsimile permits the transmission of alphanumeric data and visual pattern in an unified digital form.

As for the transmission control, some sort of data link control procedure is necessary to insure error-free transmission of information. For simplicity and for maintaining bit sequence transparency, the Extended Basic Mode of operation is preferable that provides efficient information transfer by synchronous data transmission and maintains bit sequence transparency by double DLE technique.

2.4 Remote Terminal

A remote terminal requires, as its input-output devices, a keyboard to interactively communicate with the data base, and a facsimile receiver with a digital-to-analog converter to record digitalized pattern information. In addition, judging from the difference between the transmission speed of a communication line and the recording speed of a commercial facsimile receiver, a speed converter is required. An interface is also required for communication control and for separating alphanumeric data and digitalized pattern information which arrive from the data base in an unified digital stream. Considering the cost constraint for a remote terminal, the use of an inexpensive buffer typically represented by a floppy disk is preferable for speed conversion. For the same reason, microprocessors are preferable for interfacing and for controlling remote terminal operations.

In addition to the above, a facsimile transmitter, an analog-to-digital converter and a speed converter may be provided at a remote terminal to enable visual pattern inputting, if required.

2.5 System Configuration

Based upon the above considerations, the configuration of the proposed system is drawn as shown in Fig. 1. The central equipment consists of a central processor unit (CPU), a data base (DB) for information retrieval, microfiche sorters with a television camera, a speed converter and an analog-to-digital converter (A/D); a communication control unit and other associated units. The communication control unit interfaces with a plurality of communication lines for the purpose of receiving and transmitting alphanumeric as well as pattern information in accordance with the Extended Basic Mode data link control procedure. For a system with larger scale, digitalized pattern information may be fed directly from the speed converter to the communication control unit to reduce overhead of the CPU, as shown by the broken line.

A remote terminal equipment consists of a keyboard terminal, a facsimile receiver with a speed converter and a digital-to-analog converter (D/A), and an interface unit. The interface unit separates pattern information from alphanumeric data and communicates with the communication control unit. Facsimile transmitters with associated analog-to-digital converters and speed converters, may be added to some terminals to permit pattern information inputting. The interface unit in this case has the function of combining pattern information with alphanumeric information to be transmitted in an unified digital stream to the central equipment.

3. EXPERIMENTAL SYSTEM

For the purpose of feasibility demonstration of the proposed system concept, an experimental model shown in Fig. 2 has been constructed. Due to the limitted fund available for feasibility studies, low-cost commercial units have been used for television camera, microfiche sorter, keyboard display and minicomputer, and the rest of equipments have been constructed in the authors' laboratory. Fig. 3 shows the software structure of the central equipment. The basic control includes the multiprocess monitor and works as the central part of the operating system. The communication control performs the data link control and controls message buffering. The microfiche control performs microfiche retrieval and controls transfer of pattern information to the speed converter. The speed converter control is to read out the pattern information from the speed converter to be fed to the computer. The input/output control administers the input and output to and from the disk, teletypewriter and other devices. The pattern data base retrieval system is an application program that uses above mentioned software modules to open and terminate a session between the remote terminal, to retrieve microfiche in response to a command from the remote terminal, to store the retrieved pattern information in a communication buffer within the disk to transmit the data to the remote terminal, and to perform such functions as the command error check. The above program amounted to 12 kilo word.

In the following sections describe the major design factors of the experimental system.

3.1 Pattern Retrieval and Conversion

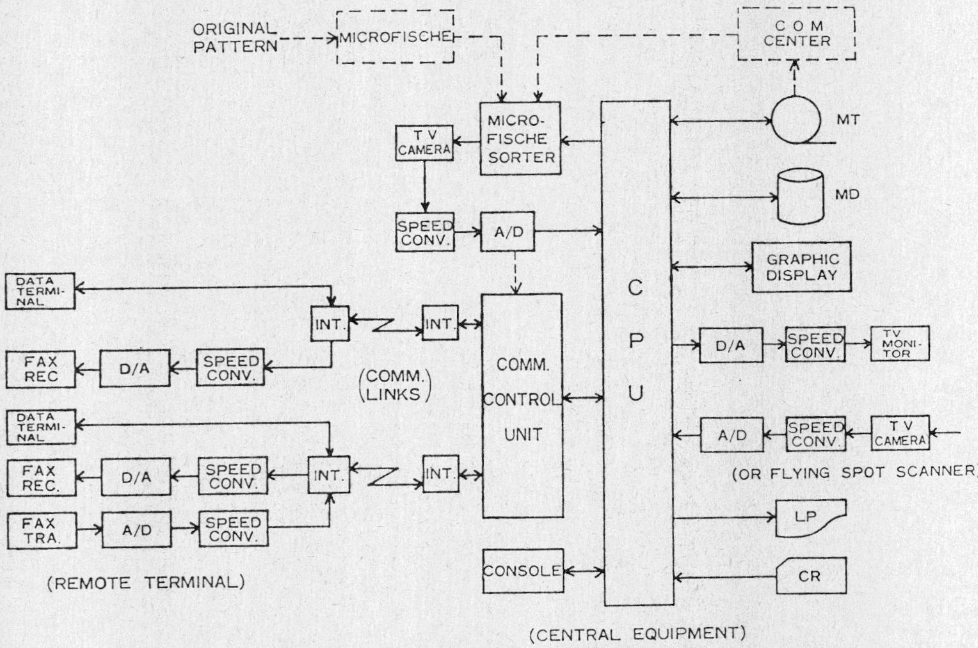

Fig. 1. Configuration of data base remote access system for visual information.

Fig. 5. External view of speed converter frame memory.

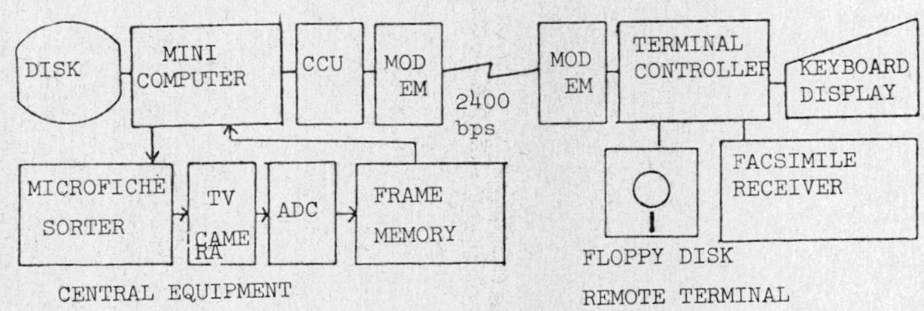

Fig. 2 Configuration of Experimental System.

Fig. 4. External view of microfiche sorter and picture input system.

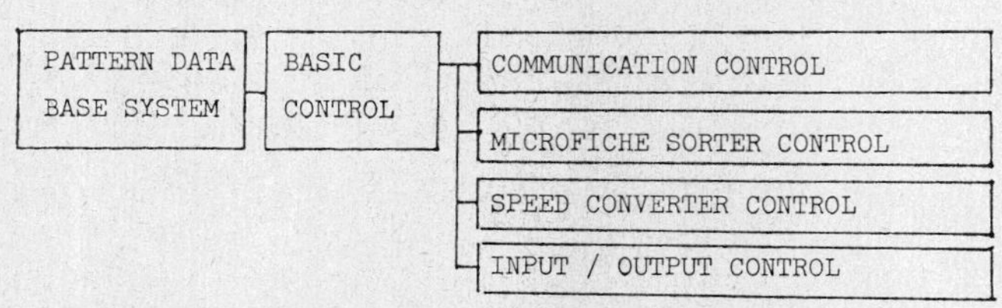

Fig.3 The Software Structure of Central Equipment.

The experimental system employs a commercial microfiche sorter that accommodated in a carrousel, up to 1000 sheets each containing 63 microfiche sorter by means of a minicomputer NEAC 3200/70, an interface has been constructed between them. A commercial television camera has also been added to the microfiche sorter as shown in Fig. 4. When retrieved by the computer, the particular microfiche is automatically placed over the television camera and the pattern is read out in the form of a standard television frame and then digitalized by an analog-to-digital converter with a sampling rate of 11 MHz.

Since the read-out speed by means of a television camera is 10^3 times faster than the transmission speed, a speed converter is necessary. The speed converter constructed for the experimental system is shown in Fig. 5. The converter has an IC memory with a capacity of 256 kiloByte and is capable of storing an entire television frame in digitalized form. A register is provided between the analog-to-digital converter and the IC memory, to store and then forward four digitalized picture elements at a time. The memory cycle time is therefore 360.8 nsec.

A commercial facsimile receiver used in the experimental system has a resolution of 3.78 lines per millimeter when recording an A-4 sheet in six minutes. It also can record the same sheet in three minutes or four minutes. By means of a microprocessor in the remote terminal, scale-up or scale-down recording has also been made possible. For this reason, three modes, in terms of the number of picture elements per frame, the number of bits per picture element and the interlace ratio, are provided as shown in Table 1.

Mode	No. of Picture Elements per Frame	No. of Bits per Picture Element	Interlace Ratio	Write-in Time from Television Camera
I	240 × 256	16	1 : 1	1/60 sec
II	480 × 512	8	2 : 1	1/30 sec
III	960 ×1024	2	4 : 1	1/15 sec

Table 1. Three modes of operation of speed conversion memory.

The speed converter provides a number of buffering benefits. For instance, it can be used for displaying on a television monitor, or for a work area for picture processing including bandwidth compression and feature extraction. The data transfer to the computer has been made possible in either a program mode or highspeed direct memory access (DMA). A Pattern transfer takes 3.05 sec in the former and 2.65 sec in the latter.

3.2 Information Transmission

The factors to be considered for transmitting information to remote terminals are such that pattern as well as alphanumeric information are being mixed, and that the amount of information to be transmitted is quite large even if bandwidth compression of pattern information is performed. The experimental system employs digital transmission of pattern information that permits combined transmission with alphanumeric data. Although this results in hardware complexity, the reasons for supporting this are: the flexibility to bandwidth compression and display techniques to be adopted, the expandability to larger and more sophisticated systems, and the compatibility to digital data networks.

The experimental system employs the data link control based upon the Extended Basic Mode procedure, because the bit sequence transparency is required for pattern information transmission and the error control capability is required especially when bandwidth compression is used. To permit the use of switched telephone network, half duplex synchronous transmission with the dat rate of 2400 bps is employed. The bit sequence transparency is maintained by the use of double DLE technique.

For the purpose of combined transmission of pattern information and alphanumeric data, the following transmission format is employed. As shown in Fig. 6, a header, named as a transmission header (TH) is included in each transmission block with the maximum length of 256 Byte. A transmission header provides information to distinguish whether the data block contains a keyboard input-output message, a pattern index, a pattern information, or a terminal control information. The header also includes a chain-flag that indicates the succeeding information block has the same attribute.

A data block when used to transmit pattern information, contains another header, named a block header (BH). A block header provides information on the block length (one Byte) and the block sequence number (one Byte). The block length is required when the length of a block is made variable by bandwidth compression. A block header is followed by the information that corresponds to one line of a pattern. By means of a plurality of data blocks, a complete pattern, which is called here as a pattern text is transmitted. A header, named as a text header (XH), is added to each pattern text. A text header indicates the text sequence number (two Bytes), the number of blocks in the text (two Bytes), the number of picture elements per line (two Bytes), the number of bits per picture element (one Byte), the bandwidth compression scheme (one Byte), the display scheme (one Byte), the pattern identity length (one Byte), and the pattern identity (n Bytes).

3.3 Remote Terminal

An experimental intelligent terminal named COMET (combined output microprocessor-oriented experimental terminal) has been constructed for feasibility demonstration. As shown in Fig. 7, the experimental terminal consists of a keyboard-display, a modem, a floppy disk, a facsimile receiver, a RAM, a ROM, a microprocessor with associated direct memory access (DMA) channel, all connected by way of a common bus through appropriate interfaces if necessary.

The facsimile receiver (PANA FAX 1000D) is a commercial drum scanning type. The facsimile output interface, associated with the facsimile receiver, detects the timing pulse generated by each rotation of the facsimile drum and feeds the pattern information line by line through its digital-to-analog converter. For

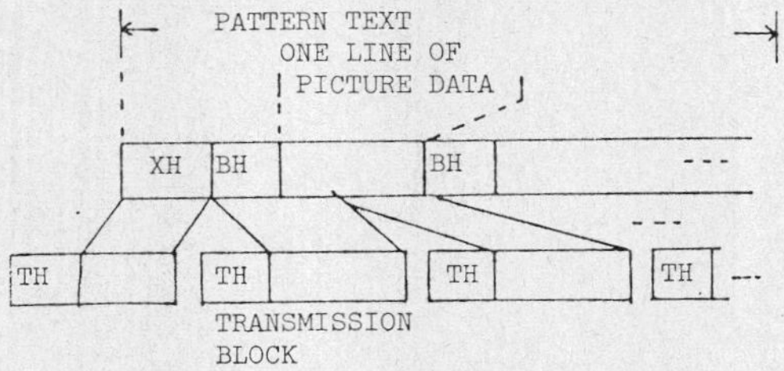

Fig. 6. Data transmission format.

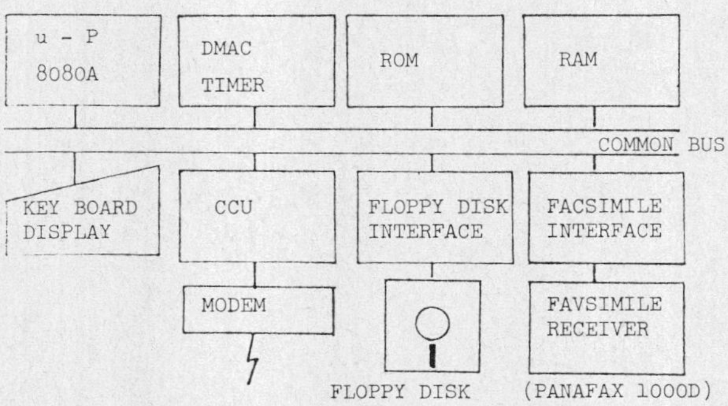

Fig. 7. Hardware configuration of COMET.

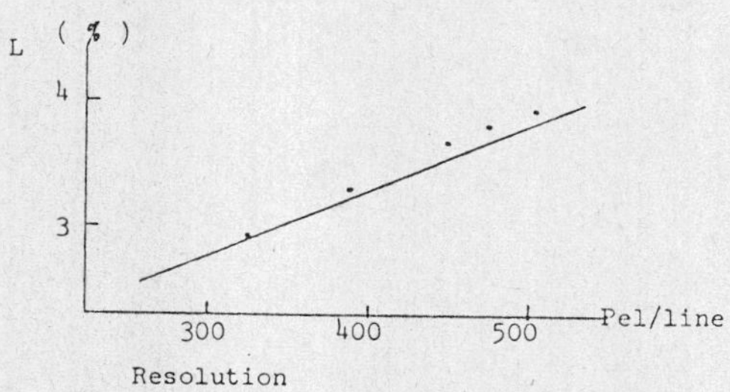

Fig. 10. Processor loading factor for pattern outputting.

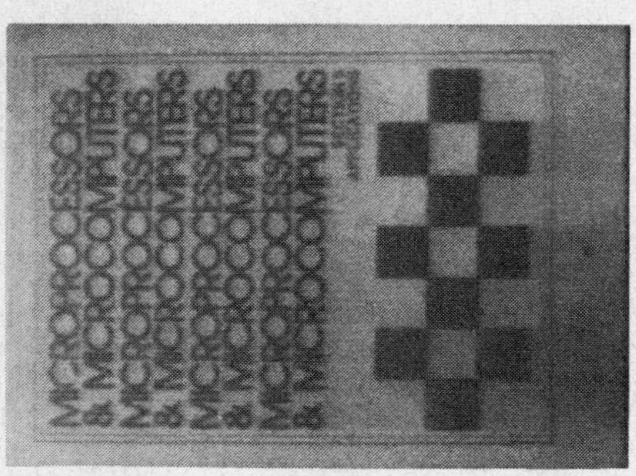

Fig. 11 (b). An example of pattern retrieval.

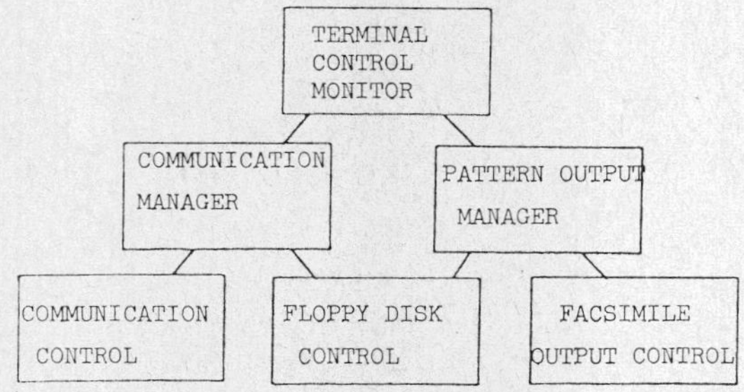

Fig. 8. Software structure of COMET.

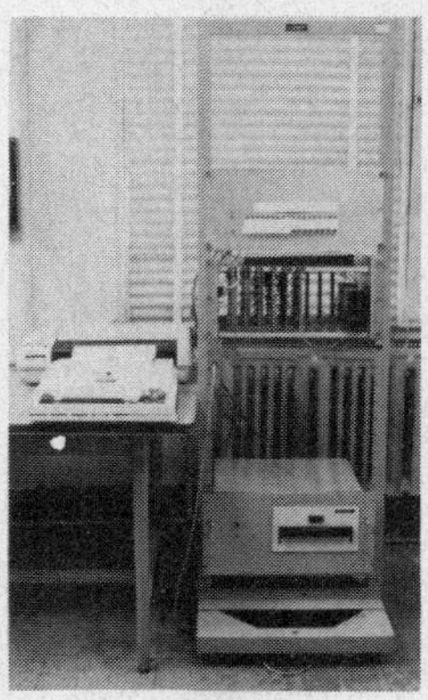

Fig. 9. External view of COMET.

Fig. 11 (a). Transmission test of a standard pattern.

the purpose of conversion between the transmission speed and the facsimile recording speed, the floppy disk is provided for temporary storage of pattern information. For outputting the pattern information from the floppy disk to the facsimile recorder, the direct memory access channel (DMAC) is used for the purpose of relieving the microprocessor from handling pattern information.

Software structure of COMET is shown in Fig. 8. The terminal control monitor controls the terminal operations as a whole. The communication control carries out the data link control based upon the Extended Basic Mode of operation. The floppy disk control performs the input and output to and from the floppy disk and permits high speed read-out by scheduling the movement of the disk head between tracks to the minimum possible extent. The facsimile output control feeds the pattern information in either of the three modes (see Table 1) to the facsimile recorder in synchronism with the rotation of the facsimile drum. The communication manager transmits the alphanumeric data from keyboard by adding a transmission header (TH). It also analyses the transmission header (TH) of a received transmission block and separates the alphanumeric data and the pattern information, and controls the outputting of them to the keyboard and the floppy disk respectively. The pattern output manager analyses the text header (XH), sets the control parameters in accordance with the mode of the pattern information, and control the outputting of pattern information from the floppy disk. The communication manager and the pattern output manager are under the interactive control through the keyboard input command. The amount of software that has been developed is about 12 kiloBytes. Fig. 9 shows the external view of COMET.

3.4 Experimental Results

Overall performance of the experimental system has been found satisfactory, although the use of low cost commercial equipment, which was inevitable due to the limitted fund available for feasibility studies, caused some problems. Major factors constituting the turn around time for providing a requested pattern information to the remote terminal are, the search time of the microfiche sorter, the data transfer time from the speed converter, the transmission time over a communication line and the recording time. The average search time is about 5 sec., the data transfer time is 3 sec., the transmission time is 2.5 minutes (for a pattern with 256 × 256 picture elements each with 16 levels), and the recording time is up to 3 minutes depending upon the mode of facsimile recorder.

Loading factor of the microprocessor at the remote terminal during reception of pattern information and during outputting of pattern information has been measured by counting idling loops of the terminal control monitor. When receiving the transmission blocks of 256 Bytes successively, the loading factor was measured as 8.14 %. Values of the loading factor when outputting were measured with respect to the number of picture elements per line as shown in Fig. 10. The measured values, shown in the figure by dots, conforms rather well with the calculation of program steps which is shown by a solid line. This indicates that the microprocessor is capable of receiving and outputting the pattern information at the same time if two floppy disks are provided, one receiving while the other outputting.

An example of the results of transmission test is shown in Fig. 11 (a). A standard pattern with 504 × 480 picture elements each of which is coded into 4 bits, is transmitted and recorded by the facsimile recorder. The results indicated that the transmission and recording provide output patterns with satisfactory quality. Over all system performance was tested by retrieving the microfiche by way of the keyboard of the remote terminal, and recording the requested pattern by the facsimile recorder as well as outputting alphanumeric data through the keyboard. One of the results of retrieval is shown in Fig. 11 (b). Although the experimental system worked as designed, the comparison between the facsimile outputs (a) and (b) indicates that precision of the optical system as well as resolution of the television camera have to be improved.

4. FEATURE EXTRACTION

Utilizing the pattern processing capability of the experimental system, feature extraction studies have been carried out with the intention of quantitatively evaluating and possibly extending the keys for pattern information retrieval.

Feature extraction techniques to be applied depend upon the pattern information to be dealt with. The specific patterns, which have been chosen for feature extraction, are the ancient patterns on roof tiles in eighth century Japan. Typical examples of the patterns are the floral patterns on round roof tiles and the arabesque patterns on flat roof tiles, examples of which are shown in the form of prints in Fig. 12 (a) and (b) respectively. In addition to these major patterns which have distinctive Greek and Roman origin, associated ornaments such as circles, dots and zig-zags are provided as are known from the figure. Because the exavated roof tiles are subject to aging and damage, and because the patterns are three dimensional rather than two dimentional, such pre-processing as noise elimination, line thining, interpolation, extrapolation and binary (black-and-white) representation are required when patterns on the prints are converted into electrical signal.[5]

In the case of round roof tiles, the first step for pattern extraction is the centering which is performed by taking cumulative distribution of black picture elements along the x and y axis. Then the percentage of black picture elements $R(r)$ along a circle of radius r is obtained. Because the domains of a pattern are separated by circles, measurement of $R(r)$ with respect to r shows peaks for specific values of r. This enables separation of domains. After the separation, the features of the zig-zag and the dots are extracted rather easily by counting the black picture elements along circles and by taking into account of their periodic property. As for the major domain which is a floral pattern, the patterns on various roof tiles are archeologically devided into four major categories, namely, simplepetal, single petal, double petal and superposed petal as shown in Fig.13. In addition, some patterns have intermediate petal between each petal. These features are extracted by the cumulative distribution of black picture elements along a circle of radius r. Fig. 13 shows the distribution, which is denoted by $D_r(\theta)$ with respect to

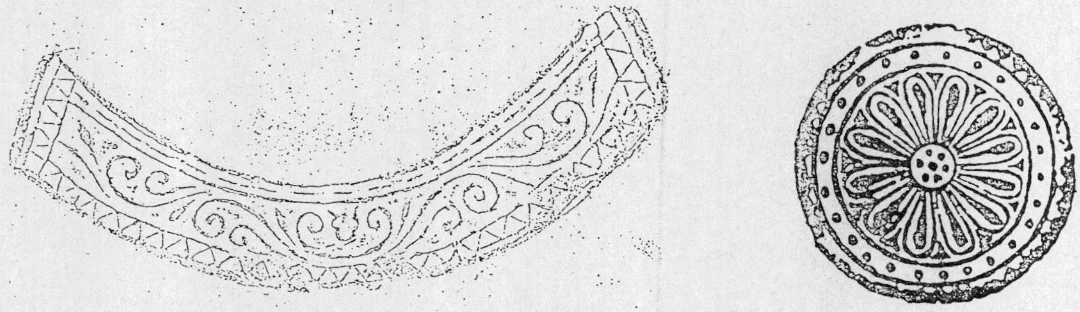

(b) Arabesque pattern

(a) Floral pattern

Fig. 12. Example of pattern on eighth century roof tiles.

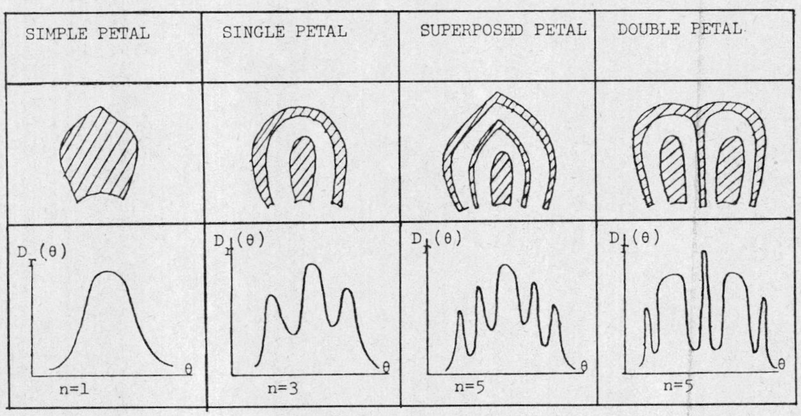

Fig. 13. Cumulative distribution of black picture element $D_r(\theta)$.

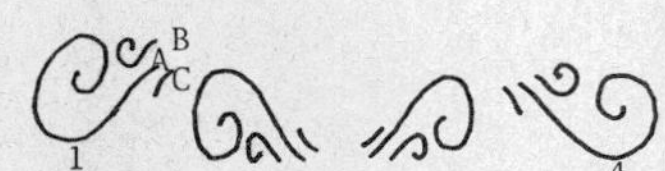

Fig. 14. An example of Arabesque with four subpatterns.

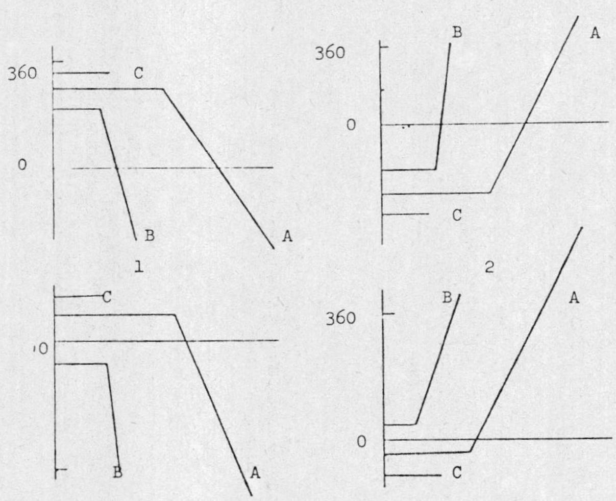

Fig. 15. Total curvature of three curves in each subpattern A B C.

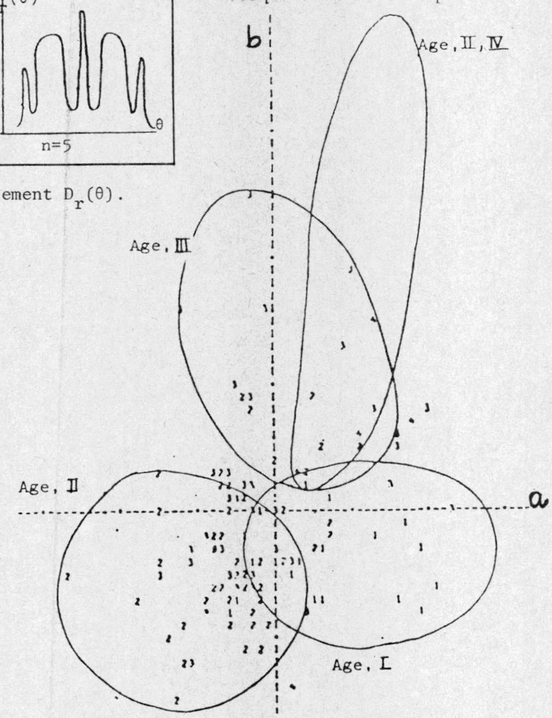

Fig. 16. The distribution of patterns in a-b plane.

the angle Θ. Denote the number of peaks of $D_r(\theta)$ by M and the number of petals by N, then $M = nN$. The values of n, which are distinctively related to each of the categories as shown in Fig. 13, are obtained from this relation. The value of M is obtained from the number that the curve $D_r(\theta)$ exceeds its average. The value of N, which is the period of the curve $D_r(\theta)$, is obtained by taking correlation between $D_r(\theta)$ and $D_r(\theta + 2\pi/j)$ where j is the estimated value of N, and choosing the value of j which maximizes the result of correlation. In some cases, the maximum appears for erroneous value of j due to disturbances. This is identified by the fact that N and n are integers.(6)

In the case of flat roof tiles, the major pattern is an arabesque. It generally consists of several sub-pattern that is further devided into several curves. As an example, Fig. 14 shows an arabesque with four sub-pattern each consisting of three curves, i.e., the major curve A, the first minor curve B and the second minor curve C. The total curvature which is the integration of curvature along a curve, was found to represent certain features of these curves. As examples, Fig. 15 shows the total curvatures of three curves in each sub-pattern. It may be observed that the total curvatures of three curves and their combination are alike in each sub-pattern and therefore represent the identity of a particular arabesque. Comparison of the total curvatures of arabesques from different roof tiles was found, to some extent, reveal the differences in the age of their manufacture.(6)

Another approach for the feature extraction of arabesques is curve fitting technique. Logarithmic spiral expressed as

$$r = ae^{b\theta} \tag{1}$$

was found convenient for the purpose of curve fitting because of its similarity and its simplicity for analytical treatment. For the purpose of estimating parameters a and b, the curvature function which is the curvature as a function of the curve length, is obtained by successively fitting arcs to five neighboring data points. Because the curvature function of the logarithmic spiral of Eq. 1 is given by

$$C(s) = \frac{1}{bs + a\sqrt{b^2 + 1}} \tag{2}$$

where s is the curve length, the error of measured value $C_i(S_i)$ is expressed as

$$E = \sum_i (C_i - \frac{1}{bS_i + a\sqrt{b^2 + 1}})^2 \tag{3}$$

The value of a and b are therefore obtained by minimizing the mean square error.

Taking the major curve of arabesques as the subject of consideration, parameters a and b have been extracted from 63 flat roof tiles whose age of manufacture has been known and categorized archeologically into four, i.e., Age I (708-721, AD), Age II (721-745, AD), Age III (745-756, AD) and Age IV (757-769, AD). The results which are plotted in Fig. 16, indicate that, in accordance with the values of a and b, the patterns can be grouped into four, i.e., Age I, Age II, Age III and Ages II and IV. Further studies taking into account such factors as longitudinal length, spiral angle and flatness, and performing component analysis, reveal that more detailed grouping is possible and that the results of analysis provide information of archeological value.(7)

5. CONCLUSION

Considering the fact that visual patterns when added to alphanumeric data greatly enhance the information to be provided, the system concept on data base systems for visual patterns proposed in this paper, may have wide area of applications. Feasibility studies by means of an experimental system indicates that the proposed system concept is physically and economically realizable. Feature extraction techniques taking ancient patterns as examples, reveal the possibility of quantitative evaluation as well as improvement of the keys for retrieval. For the implementation of operational systems, further developmental effort may be needed on the storage of larger amount of visual patterns, the optical-to-electrical conversion when optical storage media are used, the use of digital data network when it becomes available, etc.

ACKNOWLEDGEMENT

The authors wish to thank Messrs. K. Yanaka, T. Noro, T. Umadate, H. Kako and F. Sano for their collaboration in constructing the experimental system. They also wish to acknowledge the support of the Ministry of Education, Science and Culture through the Grant-in-Aid for Scientific Research, titled "Formation Process of Information Systems and Organization of Scientific Information".

REFERENCES

[1] T. Tomita, "Introducing the CAPTAIN system", Proc. Pacific Telecomm. Conf. 79 pp.4A-15~4A-23 (January, 1979)

[2] P. T. Kirnstein, "Facsimile Transmission in Message Processing System with Data Management", Proc. of 4th International Conf. or Computer Communication pp.717~726 (Sept., 1978)

[3] H. Inose, "Development of Data Base for Visual Patterns", Journal of Inst. Television Engineers, 31, 12, pp.933~939 (December, 1977)

[4] T. Tomiyama, et al., "An Image Data Base Access System by means of Facsimile", Proc. Pacific Telecomm. Conf., pp.3C-1~3C-6 (January, 1979)

[5] H. Inose, et al., "A Curves Recognition System -- to Apply the Arabesque Patterns on Ancient Rooftiles --", Transactions of Inst. Electronics and Comm. Engineers of Japan, Vol. J61-D No.6, pp.419~426 (June, 1978)

[6] H. Inose, et al., "Feature Extraction of Noisy Circular Periodical Pattern -- With Application to the Round Eaves —End Tile Pattern --", Trans. of Inst. Electronics and Comm. Engineers of Japan, Vol. J61-D No.9, pp.665~672 (Sept., 1978)

[7] H. Nakagawa, et al., "Feature Extraction of the Patterns on Ancient Roof Tiles", Technical Report of Inst. Electronics and Comm. Engineers of Japan, PRL 79-14 (June, 1979)

Scientific Information Systems in Japan
H. Inose (ed.)

LABORATORY COMPUTER NETWORK LABOLINK AND RESEARCH INFORMATION SYSTEM ARIS

Shuzo YAJIMA and Yahiko KAMBAYASHI

Department of Information Science, Kyoto University, Sakyo-ku, Kyoto 606

In the area of academic research, there are increasing demands for computer-based research support systems. In our project, we have concentrated on the research and development of laboratory-oriented computer network LABOLINK and research information system ARIS. One major objective of LABOLINK is to give a simple yet useful method to organize a system by which each user can utilize any combination of computers and terminals. The research information system ARIS has been developed to provide flexible research information handling capability such as bibliographic data. Some new ideas and our theoretical results are actually used to develop the systems LABOLINK and ARIS. This paper outlines the design principles and implementations of these systems and our theoretical results in both computer network and databases.

1. INTRODUCTION

In this paper design philosophies, features and technical details of an optically linked laboratory computer network LABOLINK and a research information system or a relational information system will be reported, which are the results of joint efforts of people in authors' laboratory in the Department of Information Science, Kyoto University, whose contributions are reflected in cited references.

Although one major objective of current computer systems is high-speed processing of large complicated computation problems, computer systems in future are expected to support a variety of communication and data storage/reorganization. Powerful database systems supported by computer networks seem to be one possible solution of organizing such systems. Especially in the area of academic research, there are increasing demands for such research support systems. Our project is motivated by the above fact, and we have concentrated on the research and development of laboratory-oriented computer network LABOLINK and research information system ARIS. Both systems employ several new ideas developed by us and the development of the systems helps to give motivations for our basic research as well. We also have collected over 2500 papers on database area, which are used for an experiment of high-level research information processing.

One major objective of LABOLINK [YAJIK7701] [YAJIK7711] is to give a simple yet useful method to organize a system by which each user can utilize any combination of computers and terminals. For this purpose, communication programs are written in extended BASIC and a microprocessor is used to switch and control computer terminals. Utilization of current technology and development of basic research are also objectives of LABOLINK implementation. LABLINK seems to be the first computer network which utilizes optical fiber cables which have many advantages over electric copper cables. Since we used only one cable in each direction, conversion between serial and parallel signals is required. We have developed a generalized serial-parallel converter called a Single-Line Multi-Connector (SLMC for short) which adaptively generates serial signals according to the speed of input parallel signals. For coding of serial signals, an efficient pulse-width modulation is used. We also have made research on the relation between the power spectrum of signals and the coding methods [YOSHY7605] [YOSHY7712]. For a systematic design of computer communication interfaces, a new automaton model called two-input/output-pair automaton is introduced [IWAMK7705] [IWAMK7809]. A design procedure using this model is actually applied to LABOLINK interface circuits. As a result of discussing this model a new automaton model for asynchronous circuits called a partially autonomous finite automaton is introduced [IWAMK7910]. This model also gives some interesting results from automata theoretic view points. Currently LABOLINK consists of PDP 11/40, MELCOM 70/35, HITAC 8350 (of the Department of Information Science), FACOM M-200 (of the Data Processing Center, Kyoto University) and a new microprogrammable flexible interface between PDP and MELCOM is currently under development [IWAMF7803] [INAGG7908]. A high resolution color graphics terminal for LABOLINK is also developed [HIRAK7710], which is used, for example, for developing data compression techniques for Kanji characters.

One of the motivations of developing information system ARIS (A Relational Information System) is to handle data for scientific research, since most currently available database systems are developed for business-oriented data and thus not suitable for such data. As an application of ARIS we have been developing a research information system based on the relational model. We have developed a language called RLB which is suitable for such research data processing. In order to improve system efficiency, a new file organization called RIS which is suitable for relational database operations is developed and used for files of ARIS. An improved partial matching algorithm is also developed, which is useful for handling incomplete data frequently appearing in scientific data. For research information systems, new index procedures are also developed, which are called Multi-KWIC Index and RS Index. By a combination of these index procedures and flexibility of relational characteristics of ARIS, a high level flexible research information system can be organized. We have collected over 2500 papers of database area, which are utilized to improve the system as well as to help our research activity. We also have obtained results in relational database theory, especially in design theory, dependency theory and view processing.

We have developed LABOLINK and ARIS which are motivated by requirements of future computer systems as well as requirements of improving our research environments. Some of theoretical results are actually used to improve the systems. Although there are still a lot of work to be done before achieving our objectives, we hope our contribution to computer network and database study will help to find ways for future computer usage.

2. LABORATORY-ORIENTED COMPUTER NETWORK LABOLINK

One major purpose of LABOLINK is to develop a user-oriented system which can be used for various research support. There are two kinds of major requirements which we feel must be satisfied by user oriented computer network. One of them is concerned with network operations and is summarized as follows [YAJIK7701] [YAJIK7711]:

(1) The user must be able to access the variety of computers/computer networks from any of the user terminals with different characteristics distributed in our laboratory. That is, any combinations of terminals and computers must be realized freely and simultaneously.

(2) Between remote computers and the central minicomputers in our laboratory, there must exist some communication lines supporting relatively high speed file transfer. It is also desirable to include data transmission capability between the remote computers via user-side minicomputers when required.

The other requirement is concerned with the design considerations.

(a) LABOLINK should be a simple, easy to develop system, comprehensible to the users, because the man-power to be engaged in the development is often restricted and the largest part of software development will be left to the users themselves.

(b) It should be provided with high expandability and flexibility for its possible modification in the future.

(c) A systematic interface design procedure is preferable.

As a result of such considerations, we have adopted the following basic design philosophies of LABOLINK:

(1) The basic system concepts is illustrated in Fig. 1. The user-side minicomputer, PDP 11/40 in our laboratory, not only offers its own computing power but also carries out communication control of the relatively high speed channels used to reach the remote computers/computer networks. As a terminal management subsystem, the microprocessor is introduced and controls the low speed channels for the user terminals [YAJIK7708]. It compensates for the specification differences of the various terminals and makes it possible for high-level languages such as FORTRAN and BASIC to control terminals in just the same way as assembly language does. This simple structure seems to satisfy the requirements mentioned above.

(2) Network supporting software is executed as one user program in each multiprogramming computers in LABOLINK, and such user processes can communicate with each other directly without a network control subsystem built into the oprating system. For the purpose of easy and powerful programming at the user-side minicomputer, a communication control facility is added to MU-BASIC (multi-user BASIC), so that the LABOLINK resources can be accessed simultaneously from the terminals. Thus modification of the operating system, which may require laborious work and lower performance of small or medium scale computers by the associated software overhead, has been avoided.

Furthermore, in LABOLINK several new experimental approaches have been adopted, which are considered likely to become powerful methods for network organization in the future.

(3) For a model of communication interfaces, a new, two-input/output-pair automaton was introduced, which is a finite automaton with a couple of input/output pairs [IWAMK7705] [IWAMK7809]. It enabled us to make a precise description of interfaces and to develop an automata theoretic interface design procedure. This procedure was implemented in the computer link between the HITAC 8350 and the PDP 11/40 and produced interfaces with several features such as minimum control signals and output memory realization. A partially autonomous finite automaton [IWAMK7910], another formal model originated in the LABOLINK development, can be regarded as a simplified version of a two-input/output-pair automaton. It enabled us to clarify several features of the partially autonomous property which means that (at most one) autonomous transition may be specified from a state and it occurs nondeterministically.

(4) Aimed at reliable and high speed communication, we have employed optical fiber cables as in-house communication lines of LABOLINK. If compared with conventional copper cables, they have many preferable characteristics, and will possibly become more economical in the future. We have installed them primarily to establish broad band communication channels between research rooms and to show a way of their practical use. By coupling the optical cables with a kind of universal digital multiplexer, called a single-line multi-connector (to be described later), several kinds of logical communication lines have been realized on an optical fiber cable. It has become possible for any network configuration to be embedded on previously established channels without concern about line capacity.

(5) The Single-Line Multi-Connentor (SLMC), developed in our laboratory, makes it possible to replace usual multi-wire cable in almost plug-in compatible fashion by a pair of optical fiber cables, one for each direction, or by a single copper wire. Two types of SLMCs have been actually developed and used for realizing both high speed communication lines for file transfer and low speed lines to terminals in LABOLINK.

Furthermore, a microprogrammable interface is being developed to connect two minicomputers PDP 11/40 and MELCOM 70/35 [IWAMF7803] [INAGG7908]. One advantage of this interface is that one computer can use the files of the other computer as if they were their own ones. Current system configuration of LABOLINK is shown in Fig. 2. The user of PDP 11/40 can freely use the file devices of MELCOM 70/35 as virtual devices without the modification of the operating system on PDP 11/40. For example, the 50MB disk of MELCOM 70/35 is virtually converted to a 20MB disk and two flexible disks for PDP 11/40 through this

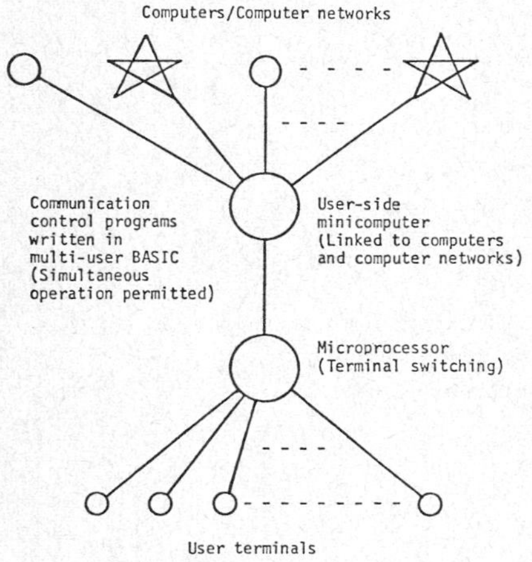

Fig. 1. LABOLINK concept.

interface. Two computers can communicate each other also as network host computers.

In many application areas of computer graphics, there is an increasing demand for much more high-level and high-performance graphic display device. We have designed and developed a new raster-scan computer graphic display device with random-scan functions having color specifying light pen facility which has the advantages of both types of random-scan and raster-scan graphic display devices [HIRAK7710]. Furthermore, we have introduced a software device simulator approach to realize it.

The main features of this device are following;
(1) Random-scan function: In spite of using a high resolution color TV monitor, it is possible to manipulate it as a random scan graphic display device including light pen functions through a computer,
(2) Color specifying light pen facility: Colors to be detected by a light pen can be specified,
(3) Device-simulator approach: The device-simulator simulates the display processor which is to be the main hardware in case of total hardware realization of it. Owing to this device-simulator, we have succeeded, although with the sacrifice of operation speed, in realizing the high-level graphic display with minimum hardware.

It is connected to LABOLINK in order to utilize it for our research and education. A graphic editing system GEST is developed on it and is used to make color slides. It is also utilized for research on data compression of Kanji character patterns digitized on the hexagonal mesh.

3. RESEARCH INFORMATION SYSTEM ARIS

We have been developing a relational database system called ARIS (A Relational Information System). The major objectives of ARIS development are as follows:
(a) To realize a flexible research information handling capability.
(b) To investigate the underlying problems of current relational database itself, such as database design, query languages, query processing techniques and efficient access path organization.

It is said that there has been a gap between (bibliographic) information retrieval systems and business-oriented database systems since each design technology has made progress independently and each of them has different goals. In a relational database system, since query types are not restricted, users' queries which are not even expected by the system designers can be accepted as well. In other words, users' query formulation facility is very flexible. On the other hand, in many bibliographic information retrieval systems, allowable query types are determined by the system designers and thus fixed. To handle research information including bibliographic data, the flexible query formulation facility is required.

Furthermore, unlike business-oriented data, to handle research information requires further facilities as follows:
(1) Unformatted textual data handling: Many kinds of variable length textual data shoud be handled in a research information system. The partial match retrieval facility is also essential in handling textual data.
(2) Incomplete information handling: The incomplete information problem (null value problem) occurs under the integration of several kinds of research information. The integration of commercial bibliographic data with different stored items or the integration of commercial bibliographic data and personal bibliographic data causes this problem.
(3) Flexible query modification facility: In a research information system such as bibliographic retrieval system, it may not be always the case that the user can faithfully express his desired queries by a query language just as in a business-oriented database system. Therefore, we need a flexible query modification facility, which supports for a

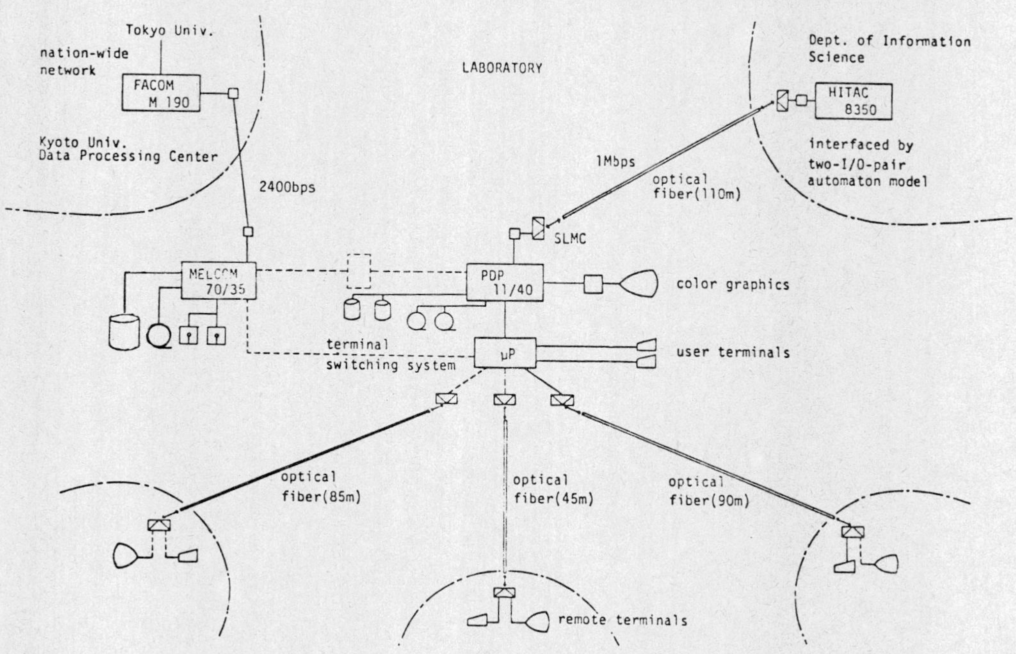

Fig. 2. Current system configuration of LABOLINK.

user to improve his queries based on intermediate results until he obtains the desired ones.
(4) Maintenance of high level integrity constraints: Especially for research information systems, integrity constraints, which cannot be expressed by dependencies, are also required to be maintained [KAMBT7710]. For example, in bibliographic data, many relationships between data items are not captured by functional dependencies [CODD7105F] that can be seen in business-oriented data. Typical relationships are many-to-many relationships such as the relationships between document ID's and references, or document ID's and keywords etc. Multivalued dependencies [FAGI7709] can capture some of these relationships, but are not sufficient [KAMBT7911].

The design principles of our system to handle these requirements are as follows [YAJIK8001]:
(a) Adoption of Codd's relational database model in order to allow flexible query patterns and access patterns of data,
(b) Improvement of conventional relational database systems by adding research information handling capability, and
(c) Providing a variety of output facilities and data analysis facilities, some primitive ones of which can be seen in the field of conventional bibliographic information retrieval systems.

The system configuration of ARIS is shown in Fig. 3. ARIS mainly consists of three parts: User Interface (UI), Data Language Subsystem (DLS) and Data Storage & Access Subsystem (DSAS).

UI offers several facilities to aid a user in formulating or modifying queries, such as catalog reference, command editor and switching among ARIS and IR packages.

DLS basically offers lexical analysis, parsing and logical optimization of queries written in ARIS query language RLB [KAMBT7710]. Two major characteristic facilities of DLS are as follows:
(1) to support a query language with simple binary relation handling capabilities, and
(2) to check user's queries for their semantic validity by several types of information.

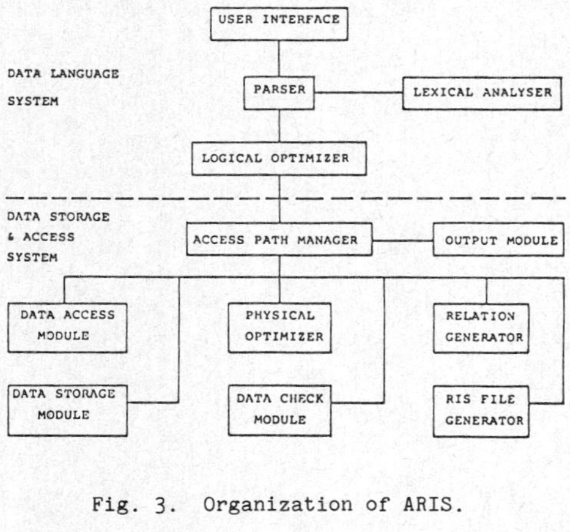

Fig. 3. Organization of ARIS.

(1) is motivated by the fact that conventional relational calculus based languages are not sufficient for end-users to formulate iterative/recursive join operations. Several operations for binary relations (defined as binary views) are introduced in RLB, based on relational calculus. This facility is useful to handle citation structure among documents or definitions of synonyms and related terms among keywords.

RLB allows partial match queries, for example, TITLE='*database*' (* denotes any string of characters). Several types of null values, meaning "unknown" and "inapplicable" values, are also allowed and used to formulate queries in order to handle incomplete information.

As for (2), although relational data languages offer flexible query formulation facility, some queries which are not semantically valid, may be sometimes allowed syntactically. Therefore, it is important to select semantically valid queries before accessing the database in order to increase the performance of the system. The logical optimizer of DLS offers improved connectivity test of queries, reduction of redundant queries and semantic validity check for relational operations such as restriction, join and projection.

DSAS consists of Access Path Manager(APM), Physical Optimizer(PO), RIS File Generator(RISGEN), Relation Generator(RELGEN), Data Check Module(DCM), Data Access Module(DAM), and Data Storage Module(DSM).

The access path organization used in DSAS is a combined index structure called Relational Inverted Structure (RIS for short) [TANAL7808]. The main objectives of RIS files are as follows:
(1) To support efficient processing of retrieval/update operations. This is to decrease the number of accesses to relation files(master files).
(2) To achieve as much as possible physical symmetry at the attribute, operation and management levels [TANAL7808]. This is motivated by the fact that queries on attributes having indexes are usually processed faster than those on attributes without indexes, and that most of known access structures are suitable only for restriction and join operations. Finally, the costs of retrieval/update operations are also desirable to be balanced.
(3) To allow easy access path maintenance [HAER7809].

A RIS file consists of a set of records qualified by domain values of joinable attributes. A record contains its corresponding domain value and a number of accessions lists. Each accession list corresponds to a relation and an attribute, and contains tuple identification codes(TID's), hashed values and other data. Fig. 4 illustrates the structure of a RIS file record. Fig. 5 shows the logical structure of a RIS file.

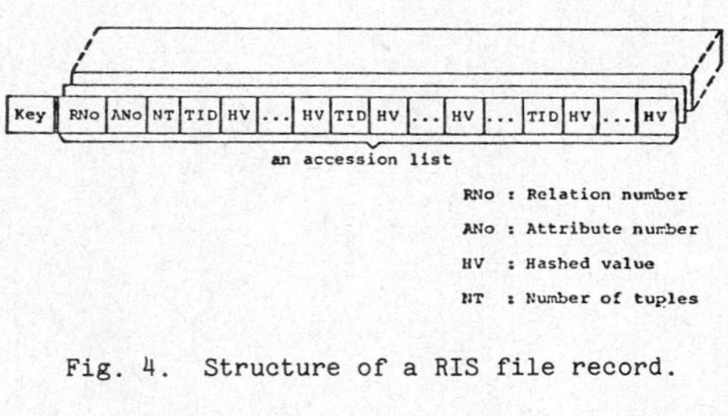

Fig. 4. Structure of a RIS file record.

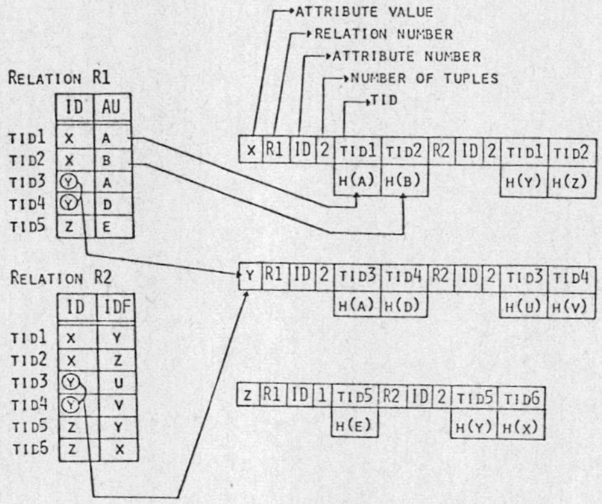

Fig. 5. Logical structure of a RIS file.

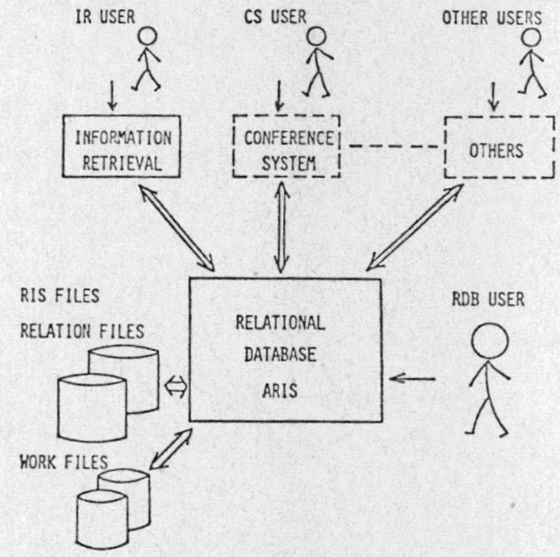

Fig. 6. Applications of ARIS.

The characteristics of RIS files can be summarized as below.

(1) Combined index structure: n-ary joins can be supported as well as binary joins. As a result of this, multi-relation queries can be processed efficiently as well as single relation queries.

(2) Pseudo-operations: A hash function is used to produce hashed values of attribute values. In pseudo-operations, these hashed values are compared instead of the attribute themselves in proper operations. The results obtained by pseudo-operations always contain those in proper operations. By using this hash method, pseudo-restriction, pseudo-joins, and pseudo-divisions can be supported. Furthermore, partial match can also be performed efficiently by the method.

(3) Use of abstracted characteristics of data [LEVIK7911]: Abstracted characteristics of data are defined as the characteristics abstracted from a set of tuples. They are classified into four classes: (a) time-independent global characteristics, e.g. functional dependency(FD), multivalued dependency(MVD) etc., (b) time-independent local characteristics, e.g. local FD, local MVD etc., (c) time-dependent global characteristics, e.g. time-dependent FD, bound number of associated attribute values, etc., and (d) time-dependent local characteristics, e.g. number of tuples of a domain value, number of hashed values etc.

Abstracted characteristics of data can be used to increase the efficiency of relational operations.

(4) Other apllications: View update operations and integrity constraints can be checked using RIS files only. In addition, operations concerning null values can be dealt with by incorporating null values into RIS files as special values.

The flexibility of access paths using pseudo operations in RIS files shows that the new unified file organization is not only suitable for basic operations but also essential in supporting an efficient and flexible data storage and access system.

As a typical application of ARIS, we have developed a bibliographic information retrieval system, whose overall system configuration is shown in Fig. 6. The system consists of a generalized relational database management system ARIS and information

```
     STON 7408

 1**1          2**1           3**1
STON 7408---|--STON 7503---|--BADA 7802
RELATIONAL  |              | INTEGRITY
DATA BASE M |              |
ANAGEMENT S |              |
YSTEMS ,INT |              |
EGRITY      |              |
            |              |   3**2
            |              |--BADAP7905
            |              | PERFORMANCE
            |              |  ANALYSIS ,
            |              | COST
            |    2**2      |   3**3          4**1
            |--STON 7505---|--HAMMM7509---|--HAMM 7601
            | INTEGRITY C  | RELATIONAL   |
            | ONSTRAINTS   | DATA BASE S  |
            | AND VIEWS ,  | YSTEM ,SEMA  |
            | IMPLEMENTAT  | NTIC INTEGR  |
            | ION          | ITY          |
            |    2**3      |              |   4**2
            |--HAMMM75 3**3|              |--ESWA 7610
            |              |              |  INTEGRATED
            |              |              |  DATA BASE ,
            |              |              |  ASPECTS
            |              |   3**4       |   4**3
            |              |--ESWAC7509---|--HAMM 76 4**1
            |              | SPECIFICATI  |
            |              | ONS ,SUBSYS  |
            |              | TEM          |
            |              |              |   4**4
            |              |              |--ESWA 76 4**2
            |              |              |   4**5
            |              |              |--WEBE 7701
            |              |              |  D-GRAPH MOD
            |              |              |  EL
            |              |   3**5           4**6
            |              |--FERNS7606------ESWA 76 4**2
            |              | SHARED DATA
            |              |  BASE ,ASPE
            |              | CTS
            |              |   3**6
            |              |--ESWA 76 4**2
            |              |   3**7
            |              |--WEBE 77 4**5
            |    2**4          3**8
            |--WEBE 7601------BADAP79 3**2
            | INTEGRITY C
            | ONSTRAINTS
            | ,MODEL
            |    2**5
            |--BADA 78 3**1
            |    2**6
            |--BADAP79 3**2
```

Fig. 7 RS-Index, Reference citation structure.

retrieval (IR) packages [YAJIK8001]. IR packages are classified into (1) output packages and (2) analyzing packages.

By output packages, retrieved results are printed in some special forms suitable for analysis by the assistance of human pattern recognition capabilities. The objectives of the output facilities are not only to display the retrieved set results in a comprehensive form, but also to provide useful information for a user to modify his query according to his intention. The output packages can generate RS index, improved KWIC index, MULTI-KWIC index and distribution tables etc. [KAMBY7906]. The RS Index program provides several ways for displaying a retrieved set of binary tuples in a form of tree-expansion [KAMBH7808], for example, a reference citation structure (see Fig. 7). The RS Index program provides the filtering facility by category and keyword etc., interactive editing facility and the usage together with the results of document clustering. In the KWIC list with an improved ordering, paper titles are ordered by a keyword, and by words before and after the keyword. By this ordering titles with the same subsentences are adjacently printed. In conventional KWIC lists, words before the keyword are not considered. As by-products, it is also possible to extract automatically a set of key phrases in titles. In the MULTI-KWIC list, titles with one common key phrase and one common keyword are adjacently printed (see Fig. 8). A distribution table is another way to represent a general binary relation. For example, instead of printing a two-column table (ID,YEAR), we can print like Fig. 9, where the values of YEAR are horizontally ordered and ID's for each year are printed under corresponding years. If horizontal values are values of YEAR, the tables are called year-distribution tables.

Analyzing packages facilitate to calculate the similarity between two attribute values according to the user determined formula. By the calculated similarities, clustering of values is carried out. The results generated by the packages are utilized for further analysis or are provided as data for the output packages.

For data to be used in the system, we have collected over 2500 papers of database area, most of which are published in 1970's. In order to simplify efficient usage of these data, MULTI-KWIC Index, RS Index and year distribution tables are also prepared.

In [KAMBN7908] and [KAMBN7911], we have shown efficient pattern matching algorithms by several kinds of preprocessing information. Some results in these papers are actually utilized in creating the hash values of RIS files.

4. RELATIONAL DATABASE THEORY

Our research in theoretical aspects of relational databases is mainly concerned with (1) logical database schema design and (2)view (defined relation) processing.

Normalizaton is a well-known technique to obtain a better logical database schema. Functional dependencies (FD's) and Multivalued dependencies (MVD's) play an essential role in normalization, which are important integrity constraints about attribute relationships. The advantages of normalization are to obtain a set of relation schemata in which less storage anomalies [DATE77] occur, and to store a relation in less redundant forms. To reduce the storage anomalies, it is

		COMMON KEY PHRASES	COMMON KEY WORDS
SCHN 7606	ODEL #A	RELATIONAL VIEW OF THE DATA INDEPENDENT ACCESSING M	INDEPENDENT
RISS 7712	#INDEPENDENT COMPONENTS OF	RELATIONS	INDEPENDENT
HELDS7503	#STORAGE STRUCTURES AND ACCESS METHODS IN THE	RELATIONAL DATA BASE MANAGEMENT SYSTEM INGRES	INGRES
HELDS7505	#INGRES - A	RELATIONAL DATA BASE SYSTEM	INGRES
WEBE 7601	#A SEMANTIC MODEL OF INTEGRITY CONSTRAINTS ON A	RELATIONAL DATA BASE	INTEGRITY
STON 7408	#HIGH LEVEL INTEGRITY ASSURANCE IN	RELATIONAL DATA BASE MANAGEMENT SYSTEMS.	INTEGRITY
MCLE 7609	ESSION OF SEMANTIC INTEGRITY SPECIFICATIONS IN A	RELATIONAL DATA BASE SYSTEM #HIGH LEVEL EXPR	INTEGRITY
HAMMM7509	#SEMANTIC INTEGRITY IN A	RELATIONAL DATA BASE SYSTEM	INTEGRITY
GRAV 7504	#INTEGRITY CONTROL IN A	RELATIONAL DATA DESCRIPTION LANGUAGE	INTEGRITY
LORIS7105	NS #A	RELATIONAL ACCESS METHOD FOR INTERACTIVE APPLICATIO	INTERACTIVE
CODDD7405	#INTERACTIVE SUPPORT FOR NON-PROGRAMMERS: THE	RELATIONAL AND NETWORK APPROACHES	INTERACTIVE
VALL 7503	ING OF DATA BASE RELATIONS: EXPERIMENTS WITH THE	RELATIONAL APPROACH #INTERACTIVE HANDL	INTERACTIVE
SMITC7505	#OPTIMIZING THE PERFORMANCE OF A	RELATIONAL ALGEBRA DATA BASE INTERFACE	INTERFACE
SMITC7510	#OPTIMIZING THE PERFORMANCE OF A	RELATIONAL ALGEBRA DATABASE INTERFACE	INTERFACE
BJORC7304	BJECT AND OPERATIONS #THE GAMMA ZERO N-ARY	RELATIONAL DATA BASE INTERFACE: SPECIFICATIONS OF O	INTERFACE
CODD 7408	#RECENT INVESTIGATIONS IN	RELATIONAL DATA BASE SYSTEMS	INVESTIGATIONS
NAVAM7505	#INVESTIGATIONS INTO THE APPLICATION OF THE	RELATIONAL MODEL TO DATA TRANSLATION	INVESTIGATIONS
CHAN 7802	#FINDING MISSING JOINS FOR INCOMPLETE QUERIES IN	RELATIONAL DATA BASES	JOINS
GOTL 7505	#COMPUTING JOINS OF	RELATIONS	JOINS
FADOF7505	#FINDING CANDIDATE KEYS FOR	RELATIONAL DATA BASES	KEYS
LUCCO7810	#CANDIDATE KEYS FOR	RELATIONS	KEYS
KAMBT7710	RELATIONAL DATA LANGUAGE WITH SIMPLIFIED BINARY	RELATION HANDLING CAPABILITY #A	LANGUAGE
KERSO7604	#A SYNTHETIC ENGLISH QUERY LANGUAGE FOR A	RELATIONAL ASSOCIATIVE PROCESSOR	LANGUAGE

Fig. 8 MULTI-KWIC Index.

INTEGRITY
INTEGRITY CONSTRAINTS
DATABASE INTEGRITY
IMPLEMENTATION OF INTEGRITY CONSTRAINTS

FREQ. 24

--1970--	--1971--	--1972--	--1973--	--1974--	--1975--	--1976--	--1977--	--1978--	--1979--
		WILK 7208		EVER 7404	STON 7503	DAVE 7605	DHALK7710	BADA 7802	
				FOSS 7404	STON 7505	FERNS7606	LORI 7703	CHAL 7808	
				STON 7408	ESWAC7509	GARDS7601	NIELR7708		
					GRAV 7504	HAMMM7610			
					HAMMM7509	INSEF7609			
					NASHM7508	LEE-S7610			
						MCLE 7609			
						MACH 7601			
						WEBE 7601			

Fig. 9. Year-distribution table.

necessary to obtain a set of relation schemata in which the necessity of update propagation is reduced and in which each relation can be updated as independently as possible. For this requirements, several problems have remained unsolved.

Our main results about logical database schema design are as follows [KAMBT7911R]:
(a) Minimization of the number of third normal form relations [KAMB7705] [KAMB7808] [KAMB7902] [KAMB7906] [KAMB7911R]: Bernstein's algorithm [BERN7612] is known as the best one so far, but it is shown that this algorithm may not always produce the optimum solution under the universal relation assumption [BEERB7809]. This universal relation assumption is that each relation in a database should be a projection of a universal relation consisting of all the attributes. Another problem of synthetic approach is pointed out by Arora and Carlson [ARORC7809]. That is, conventional synthetic approach does not guarantee that the join of the obtained relations is equal to the original universal relation. This problem is concerned with the lossless join property [AHO7710]. These problems are handled by our new synthetic approach.
(b) Properties of multivalued dependencies [TANAK7902] [TANAK7908]: For designing of fourth normal form relation schemata [FAGI7709], we need to know the properties of MVD's and EMVD's (embedded MVD's), which are MVD's defined on a proper subset of attributes of a relation. Several useful properties, such as interactions between MVD's and EMVD's or equivalence relationships between two sets of MVD's and EMVD's, are obtained. These properties are especially useful to examine whether or not an obtained set of relation schemata is equivalent to the original relation schema with MVD's. Furthermore, we also showed conditions for given MVD's to be preserved valid under update operations to designed relations. These results are useful to establish design criteria of relation schemata by using MVD's.
(c) Dependencies and data semantics [KAMBT7911S]: In order to investigate interface problems between a conceptual design, and a logical design, we discuss the disagreements between data dependencies (FD's and MVD's) and conceptual dependencies, a set representation problem (null valued problem) and an entity identification problem. The semantic analysis of data dependencies is provided to specify correct data dependencies. Especially, transitively specified MVD's are shown to often impose a semantically unnatural constraint. A sufficient condition for these type of MVD's to hold in a natural sense is provided. Based on these results, we introduce a mixed design approach in which conceptual and logical designs work to compensate each deficiency. Furthermore, some generalization of a relation data model is suggested to cope with the null value problem and the entity identification problem.

As for the processing of users' views, that is, virtual relations defined on stored relations, the following results are obtained: (d) Update viability conditions [TANAL7808] [LEVIK7902] [LEVIK7911]: Several conditions for updates through views to be reflected on stored relations are obtained. We classify available information for update viability checks, such as time independent and dependent information. According to each class of information, the conditions for update viability are obtained.
(e) Efficient view processing algorithm: In formulating queries for a relational database, it is convenient if the system automatically combines necessary relations. An efficient algorithm is obtained to perform necessary join operations automatically according to each user's query.

Some of these theoretical results are actually used to implement ARIS.

ACKNOWLEDGMENT

This research has been co-operated by the members of Yajima Laboratory whose contributions are greatly appreciated.

The following members have been engaged in the LABOLINK project: Associate Professor Susumu Yoshida, Dr. Kosaku Inagaki, Mr. Hiromi Hiraishi, Mr. Kazuo Iwama (Kyoto Sangyo University) and Mr. Hiroyuki Ogino.
The following members have been engaged in the ARIS project: Mr. Katsumi Tanaka (Kobe University), Mr. Osamu Konishi, Mr. Chung Le Viet and Mr. Narao Nakatsu (Aichi University of Education).
The authors are also grateful to other members in Yajima Laboratory for their continuing co-operation.

REFERENCES

[AHO7710]
A. H. Aho, C. Beeri and J. D. Ullman, "The Theory of Joins in Relational Databases," (Extended Abstruct), pp.107-113 (Oct. 1977).
[ARORC7809]
A. K. Arora and C. B. Carlson, "The Information Preserving Properties of Relationl Database Transformations," Proc. 4th International Conference on VLDB, pp.352-359 (Sep. 1978).
[BEERB7809]
C. Beeri, P. A. Bernstein and N. Goodman, "A Sophisticate's Introduction to Database Normalization Theory," Proc. 4th International Conference on VLDB, pp.113-124 (Sep. 1978).
[BEER7612]
P. A. Bernstein, "Synthesizing Third Normal Form Relations from Functional Dependencies," ACM TODS, Vol.1, No.4, pp.277-298 (Dec. 1976).
[CODD7105F]
E. F. Codd, "Further Normalization of te Data Base Relational Model," Proc. Couvant Computer Science Symposium 6, Data Base Systems, pp.33-64 (May 1971).
[DATE77]
C. J. Date, "An Introduction to Database Systems," 2nd Ed., Addison-Wesley, (1977).
[FAGI7709]
R. Fagin, "Multivalued Dependencies and a New Form for Relational Databases," ACM TODS, Vol.2, No.3, pp.262-278 (Sep. 1977).
[HAER7809]
T. Haerder, "Implementing a Generalized Access Path Structure for a Relational Database System," ACM TODS, Vol.3, No.3, pp.285-298 (Sep. 1978).

MAJOR PABLICATIONS

[HIRAK7710]
H. Hiraishi, K. Kawakubo and S. Yajima, "The design of a Microprogrammed Graphic Display Processor with Pipeline Organization," Proceedings of the 3rd EUROMICRO Symposium of Microprocessing and Microprogramming, pp.66-73 (Oct. 1977).
[INAGG7908]
K. Inagaki, K. Goto, S. Ishiguro, S. Yajima and K. Iwama, "A Microprogrammed Interface for Computer Communication and Its Design and Maintenance Aids" (in Japanese), Technical Report on Information Processing IEEJ, (Aug. 1979).
[IWAMK7705]
K. Iwama, Y. Kambayashi and S. Yajima, "Interface Design Procedure Based on Two-Input/Output-Pair Automata" (in Japanese), Research Institute for Mathematical Science Kokyuroku 296, Kyoto Univ., pp.168-197 (May 1977).
[IWAMF7803]
K. Iwama, H. Fujita, K. Goto, K. Inagaki and S. Yajima, "The Design of Microprogrammed Intelligent Interface for Computer Connection," Technical Report on Electronic Computers, IECEJ, EC77-75, (Mar. 1978).

[IWAMK7809]
K. Iwama, Y. Kambayashi and S. Yajima, "Computer Communication Interfaces Based on Two-Input/Output-Pair Automata and Their Implementation of the LABOLINK Network," Proceedings of the 4th International Conference on Computer Communication, pp.149-154 (Sep. 1978).

[IWAMK7910]
K. Iwama, Y. Kambayashi and S. Yajima, "Basic Properties of Partially Autonomous Finite Automata" (in Japanese), Trans. IECE Japan, Vol. J62-D, No.10, (Oct. 1979).

[KAMB7705]
Y. Kambayashi, "On the Decomposition of Normalized Relational Database Relations Utilizing the Minimization Method of Boolean Functions" (in Japanese), Research Institute for Mathematical Science Kokyuroku 296, Kyoto Univ., pp.105-128 (May 1977).

[KAMBT7710]
Y. Kambayashi, K. Tanaka and S. Yajima, "A Relational Data Language with Simplified Binary Relation Handling Capability," Proceedings of the 3rd International Conference of Very Large Data Bases, pp.338-350 (Oct. 1977).

[KAMBH7802]
Y. Kambayashi, T. Hayashi, Y. Tanaka and S. Yajima, "A Linear Storage Space Algorithm for a Reference Structure Index," Information Processing Letters, Vol. 7, No. 2, Feb., pp.66-71 (1978).

[KAMBN7803]
Y. Kambayashi, N. Nakatsu and S. Yajima, "Partial Matching Algorithms Suitable for Unsuccessful Matching Processing" (in Japanese), Research Institute for Mathematical Science Kokyuroku 332, pp.264-285 (Mar. 1978).

[KAMB7808]
Y. Kambayashi, "Equivalent Key Problem of the Relational Database Model," Lecture Notes in Computer Science 75, pp.165-191 (Aug. 1978).

[KAMB7902]
Y. Kambayashi, "Relational Database Design by a Synthetic Approach," Research Institute for Mathematical Science Kokyuroku 353, pp.11-19 (Feb. 1979).

[KAMBN7905]
Y. Kambayashi, N. Nakatsu and S. Yajima, "Hierarchical Pattern Maching Algorithms for Strings" (in Japanese), Trans. IECE Japan, Vol.J62-D, No. 5, pp.341-347 (May, 1979).

[KAMB7906]
Y. Kambayashi, "A New Synthetic Approach for Relational Database Design," Presented at AFIPS NCC (Jun 1979).

[KAMBY7906]
Y. Kambayashi, S. Yajima, O. Konishi and T. Hayashi, "New Indices for Bibliographic Data," AFIPS NCC, Vol.48, pp217-225 (Jun. 1979).

[KAMB7909]
Y. Kambayashi, "Logic Design of Programmable Logic Arrays," IEEE Trans. Computers, (Sep. 1979).

[KAMBN7911]
Y. Kambayashi, N. Nakatsu and S. Yajima, "Hierarchical String Pattern Matching using Dynamic Pattern Matching Machine," COMPSAC, (Nov. 1979).

[KAMBT7911S]
Y. Kambayshi, K. Tanaka and S. Yajima, "Semantic Aspects of Data Dependencies and Their Application to Relational Database Design," COMPSAC, (Nov. 1979).

[KAMBT7911R]
Y. Kambayashi, K. Tanaka and S. Yajima, "Relational Database Design," to appear in Proc. of the 13th IBM Computer Science Symposium, (Nov. 1979).

[LEVIK7902]
C. Le Viet, Y. Kambayashi, K. Tanaka and S. Yajima, "Query Processing in a Relational Database using Abstracted Characteristics of Data," Research Institute for Mathematical Science Kokyuroku 353, pp.20-29 (Feb. 1979).

[LEVIK7911]
C. Le Viet, Y. Kambayashi, K. Tanaka and S. Yajima, "Use of Abstracted Characteristics of Data in Relational Databases," COMPSAC, (Nov. 1979).

[NAKAK7902]
N. Nakatsu, Y. Kambayashi and S. Yajima, "Algorithms for Longest Common Subsequence Problem and Common Consecutive Subsequence Problem" (in Japanese), Research Insitute for Mathematical Science Kokyuroku 353, Kyoto Univ., pp.58-67 (Feb. 1979).

[TANAL7808]
K. Tanaka, C. Le Viet, Y. Kambayashi and S. Yajima, "A File Organization Suitable for Relational Database Operations," Lecture Notes in Computer Science 75, pp.193-224 (Aug. 1978).

[TANAK79020]
K. Tanaka, Y. Kambayashi and S. Yajima, "Organization of Quasi-Consecutive Retrieval Files," Information Systems, Vol. 4, No. 1, pp.23-33 (Feb. 1979).

[TANAK7902R]
K. Tanaka, Y. Kambayashi and S. Yajima, "Representability Problem for Relational Database Design by Multivalued Dependencies," Research Institute for Mathematical Science Kokyuroku 353, Kyoto Univ., pp.1-10 (Feb. 1979).

[TANAK7908]
K. Tanaka, Y. Kambayashi and S. Yajima, "Properties of Embedded Multivalued Dependencies in Relational Databases," Trans. IECE Japan, Vol. E-62, No. 8, (Aug. 1979).

[TOMIK7803]
N. Tomii, Y. Kambayashi and S. Yajima, "Algorithms for Plannarity Test of Multi-Level Graphs" (in Japanese), Research Institute for Mathematical Science Kokyuroku 322, Kyoto Univ., pp.230-263 (Mar. 1978).

[YAJIK7701]
S. Yajima, Y. Kambayashi, S. Yoshida and K. Iwama, "Optically Linked Laboratory Computer Network LABOLINK," Proceedings of the 10th Hawaii International Conference on System Sciences, pp.1-4 (Jan. 1977).

[YAJIK7708]
S. Yajima, Y. Kambayashi, H. Ogino and N. Nakatsu, "A Microprocessor-Based Terminal Switching System," Proceedings of the First Annual Rocky Mountain Symposium on Microprocessors, pp.302-310 (Aug. 1977).

[YAJIK7711]
S. Yajima, Y. Kambayashi, S. Yoshida and K. Iwama, "LABOLINK: An Optically Linked Laboratory Computer Network," IEEE Computer, Vol.10, No.11, pp.52-59 (Nov. 1977).

[YAJIK7811]
S. Yajima, Y. Kambayashi, O. Konishi, N. Nakatsu and T. Hayashi, "Some Techniques for Bibliographic Data Processing" (in Japanese), Information & Documentation, Vol.21, No.8, pp.607-621 (Nov. 1978).

[YAJIK8001]
S. Yajima, Y. Kambayashi, O. Konishi, K. Tanaka, C. Le Viet and T. Kato, "Bibliographic Information Processing Facilities for Relational Database System ARIS," to appear in Proc. of th 13th Hawaii International Conference on Systems Sciences, (Jan. 1980).

[YOSHY7605]
S. Yoshida and S. Yajima, "On the Relation between an Encoding Automaton and the Power Spectrum of Its Output Sequence," Trans. IECE Japan, Vol.E59, No.5, pp.1-7 (May 1976).

[YOSHY7712]
S. Yoshida and S. Yajima, "Reshaping of Power Spectra of Digital Coded Signals by Restricting Inputs to Encoding Automata," Trans. IECE Japan Vol.E60, No.12, pp.715-722 (Dec. 1977).

Scientific Information Systems in Japan
H. Inose (ed.)

A DATA MANAGEMENT SYSTEM FOR AN AGGREGATE OF MANY FILES WITH VARIOUS FORMATS -THE CASE OF ASTRONOMICAL DATA-

Yoichi TERASHITA

Computer Center, Kanazawa Institute of Technology, Kanazawa 921

In collecting, controlling, and disseminating machine-readable astronomical data, we are faced with the task of handling many data files which vary in size, in contents, and in format. These types of variability has been the major concern during the development of the Astronomical Data Services at the Computer Center, Kanazawa Institute of Technology. To circumvent the difficulty in dealing with the aggregate of such files, a specialized data management system has been developed. The main component of the system is a set of three secondary files. Placed at the highest level is the data directory, that controls the most basic pieces of information. At the next level come two versions ('human-readable' and 'coded') of the data description file, each element of which gives a detailed field specification for the corresponding data file. The data files themselves, which are at the lowest level, are then manipulated under the control of those secondary files.

1. INTRODUCTION

As in any scientific field, astronomy depends strongly on observational data, and, particularly in this field, there has been a long tradition of sharing such data among investigators. The reason for this tradition can be found in the fact that there are observational limitations due to the geographical reasons (observers in the northern hemisphere can observe only part of the southern sky, for instance), or those due to temporal reasons (a phenomenon stretching over a long period of time requires a succession of observations over several generations of observers). Accordingly, the results of observation are usually summarized in easy-to-read format, and then made available to investigators in many parts of the world, and to those in later generations.

Typically, such data are organized in the form of table, called 'catalog'. Entries in a catalog are individual astronomical objects, each of which is accompanied by a set of observational results and other pieces of information. In the past, the objects were stars and nebulae which can be observed optically. However, thanks to the recent advances in observational technology, more and more data for radio, infrared, and X-ray sources are coming out.

A great variety of catalogs have been published, depending on the field or the research interest of the observer (or, the compiler of the catalog, for that matter). This implies that a given object can be included in many catalogs, depending on what aspects of the object are in question. A catalog often consists of a main table and accompanying auxiliary tables that contain bibliographical references or other additional data. The number of objects included in a catalog is several hundreds in the case of small catalogs, and can be up to several hundreds of thousands in the case of large ones.

As an example, Fig.1 shows a portion of the General Catalog of Stellar Radial Velocities, produced at the Mount Wilson Observatory in 1953. This catalog lists, for each star, the sequence number (in this catalog), other identifications including the H.D. number, positional data, stellar magnitude, values of proper motion, spectral type, radial velocity, quality of measurement, number of plates measured, coded name of the observer, and note if any. Approximately 15,000 stars are included in the catalog. The H.D. identification is the sequence number assigned for the star in another catalog, called the Henry-Draper Catalog. The Henry-Draper Catalog itself holds basic quantities for each of about 270,000 stars, and is refered to by many other catalogs.

Although hundreds of such catalogs have been published and many of them have been widely used by astronomers, the compatibility among these catalogs are not always good. The accuracy for a given quantity, such as the position of the star, can vary from catalog to catalog, and there is no guarantee that the most up-to-date data are listed in a given catalog. Another problem is the ambiguity in the scheme for identifying the astronomical objects. Although common identification systems, such as H-D and DM numbering, are used in most catalogs, there has been no solid agreement as to which identification system(s) should be adopted while preparing new catalogs. In addition, the commonly used systems (such as those mentioned in the above) themselves are not necessarily complete, in the sense that many types of ambiguities appears when they are to be processed mechanically. The problem of identifying astronomical objects has been a great concern within the international astronomical community, especially among those dealing with machine-readable data (see, for instance, (1)).

Before the era of electronic data processing, these catalogs were, as a matter of fact, produced in print form. Since then, conversion of print-form catalogs into machine-readable form has been a great concern at many institutions which produce or compile astronomical catalogs. It can now be said that most of the important catalogs (those most often used) have been converted into machine-readable form, such as magnetic tapes. And the recent trend is that machine-readable form is the more common way of publishing astronomical catalogs than print form, especially when a large volume of data is involved. In the followings, astronomical catalogs in machine-readable form will be refered to as data files.

Since 1978, the Computer Center of Kanazawa Institute of Technology (KIT) has been engaged in astronomical data services(2). Our prime objective is to offer the scientific community in Japan a means for easy access to machine-readable astronomical data. The major diificulty in shaping up the services has been in the fact that we have to deal with data files with a great variaty in volume, in contents, and in format (currently, we have about 280 data files, with the total volume of about 400 megabytes, with the numbers still growing). In order to circumvent this type of difficulty, and to make possible efficient and reliable services, a specialized data management system has been developed, which is described in the followings.

2. CONSIDERATIONS IN THE IMPLEMENTATION OF ASTRONOMICAL DATA SERVICES

While print-form data have typically been disseminated through the publisher-library route, dissemination of machine-readable astronomical data files are in most cases depends on individual arrangements between the producer and the user. As long as the number of data files involved is small, such scheme does not pose much difficulties. However, when many different kinds of data files are to be dealt with, we face problems such as, how to obtain particular files, how recent data a file contains, or whether or not certain files are indeed available. Therefore, the basic requirement for an astronomical data center is to equip itself with a reliable control function for the data traffic.

It is equally important to realize that, unlike print-form data, machine-readable data are not so easy to handle, as one might think at first hand. For instance, in order to 'see' a portion of a magnetic tape, a certain amount of programming work is needed. If we wish to do more complicated processing of the file, the computer system at hand must have file-processing functions of certain levels, but unfortunately this is not always the case. In addition, programming works in such cases are sufficiently involved.

In view of the fact that recent years are a transient period toward the age of truly mechanized catalogs, there exists one group of users who are actively developing novel uses of the catalogs, and on the other hand there are those who simply regard them as just an alternative to conventional print-form catalogs. In the latter case, the technical complexity associated with the use of machine-readable data can be nothing but a nuisance.

From these considerations, the activities of an astronomical data center can be classified into three categories:

1) Dissemination of whole data files by way of machine-readable media such as magnetic tapes.
2) File processing jobs concerning the individual data files (or concerning combinations of several files). These include printing, editing, extraction, or statistical processing of the file (or files).
3) On-line data retrieval services.

Our emphases so far are on the first and second types of services. The second type of service is notably important in the user environment, since in astronomical research the common tendency is to extract a sizable chunk of data, rather than to retrieve a few data items, out of a given data file(s). The third type becomes important when we are to deal with bibliographical data or index type information.

General Catalogue of Radial Velocities 3

Cat. No.	Star		H.D. No.	1950 R.A.		1950 Decl.		Magn.	P.M.	Spec.	Vel.	Q	No. Pl.	Obs.	Notes
				h	m	°	′		″		km/sec				
101	NGC	40	826	0	10.2	+72	15	....	....	Pf	- 20.5	b	7	L	Em Pl. neb
102		228	829		10.2	+37	25	6.6	0.025	B3	- 9.0	b	9	3	IS -15.7 b *
103		232	877		10.5	-22	45	6.7	.018	sgG5	- 16	c	4	W	SB
104		234	874		10.6	+16	39	6.6	.090	G5	+ 9.6	b	4	D	
105	γ	Peg	886		10.7	+14	54	2.9	.010	B2	+ 4.1	a	106	8	IS 0 c 4 *
106	UX	Cas	.....		10.7	+63	11	12.0v	....	R2	- 11	d	2	W	SR 102
107	A	161A	895		10.8	+26	43	6.3	.046	F5	- 13.0	b	4	S	
108	A	161B	.....		10.8	+26	43	9.0	....	K0	- 17	c	5	D	
109		244	905		10.9	+40	46	5.7	.189	dA7	- 28.8	b	6	V	
110	75°	4	919		11.2	+75	45	7.6	.03	gM4	- 0.7	b	5	W	
111		254	936		11.3	+59	43	7.0	.040	K0	0	c	2	V	
112	43°	28	949		11.4	+44	14	8.2	....	dF4	+ 15	d	1	L	
113	18°	25	955		11.4	-17	49	7.2	....	B5n	- 33.6	b	8	L	
114		256	952		11.4	+32	56	6.1	.026	A0	+ 0.9	b	3	V	
115	75°	5	947		11.5	+75	45	7.9	.011	gG4	- 15.9	b	3	W	
116	4°	12	966		11.5	-04	11	7.5	.057	gG6	- 23	c	2	L	

Fig. 1. A sample catalog (General Catalogue of Radial Velocities, R.E Wilson, 1953).

3. THE DATA MANAGEMENT SYSTEM

Unlike print-form data, processing of machine-readable data requires computer-related procesures, such as programming and machine-operation. If we are to deal with a homogeneous set of data, such tasks would be rather straightforward. However, when the data consist of many files with various format, manual programming and operation can become fairly complicated, and we would be faced with a high possibility of mishandling. Accordingly, the main objective of the data management system has been to establish a unified (and automated) scheme for handling the data files, which should then eliminate the isolated efforts in programming and machine operation.

Fig.2 summarizes the overall system configuration. System components (files and program modules) are shown inside the dotted box. On top of it are indicated major types of operation which activate or maintain the system, and underneath the products from the system. System products currently available to users are the information bulletin (the data directory being as part of it), the documents describing each individual data files, copies of data files, and outputs from various types of file processing. An on-line retrieval subsystem for stellar identifications has been developed, which, however, is not presently available to the public. The solid arrows indicate data flows, while the dotted arrows indicate referencing.

The data files themselves are stored in thirteeen magnetic tapes (as of October, 1979), in multi-file, fixed-length-record, sequential format. Although it is obviously more convenient to store them in on-line disks, cost considerations do not seem to justify such a scheme, at least for the time being. The system is designed in such a way that there should be no operational differences if the data files are to be stored on disks.

As shown in the figure, the system has three types of secondary files for data management purposes. The first, and the most basic, is the data directory. The other two are what we may call the human-readable data description (HDD for short), and the coded data description (CDD). The data directory and the HDD are under the file management facility of CMS (a conversational programming system from IBM), and can be manipulated through CRT terminals. The contents and functions of these secondary files are discussed in the below.

3.1 Data Directory

This file holds the basic information on each of the astronomical data files. Fig.3 shows a portion of it. Its contents are, from left to right, the catalog (data file) number (-2, -3, ..., indicate subfiles), the name of the catalog, the record length, the number of records in the file, the availability of documentation, the acquisition date, the dataset name, the volume-serial number of the tape in which the file is stored, the file sequence

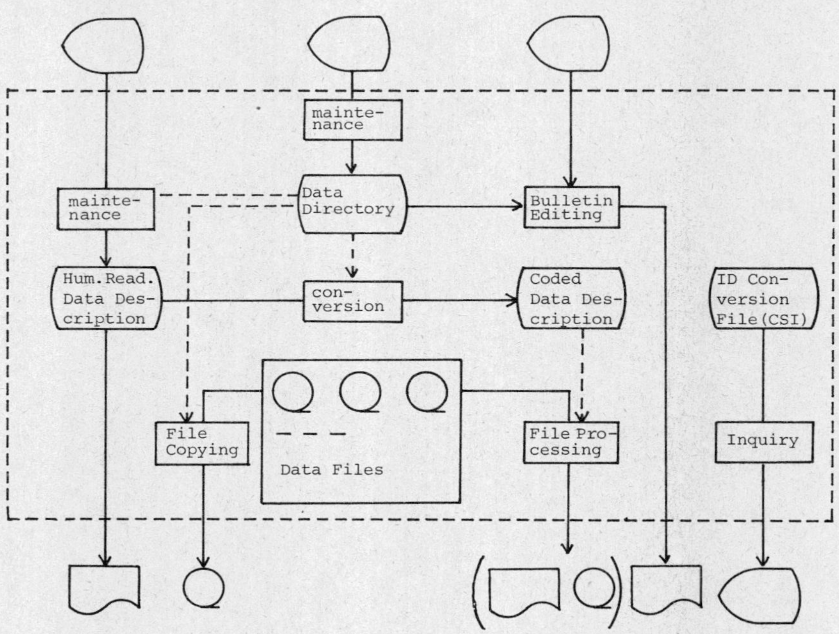

Fig. 2. Data management system.

CAT.NUM	CAT.NAME	R.LEN	R.NUM	AVAIL	DATE	DSN	VOL	SEQ	BSIZE	REMARK
3001	Henry Draper Catalog	49	225301	DE	04/25/77	HD	CDS460	22	15974	
3001 -2		69	46872	D	04/25/77	HDE	CDS460	23	15870	
3002	Pre. Cat. of Early Type Emission Stars	80	3216	D	01/13/78	EEMISS	CDS437	43	8000	Bertiau,-,1969
3003	Catalog of Rotational Velocities	21	3941	DE	04/25/77	CROTV	CDS460	24	15960	Uesugi,-,1970
3004	Bibl. of Stellar Radial Velocities	128	44135	D	04/25/77	BIBRADV	CDS460	25	15872	Abt,-,1972
3005	Faint OB Stars between Car. and Cen.	80	285		01/13/78	OBCARCEN	CDS437	24	8000	Lynga,1968
3006	Milky Way from Cen. to Norma: OB Stars	80	484		01/13/78	MWCENOB	CDS437	25	8000	Lynga,1964
3007	Survey of Faint OB Stars in Carina	80	454		01/13/78	OBCAR	CDS437	26	8000	Graham,-,1965
3008	Luminous Stars in S. Milky Way =>3043	83	5132	D	12/16/77	LSSMW	CDS443	25	7968	Stephenson,-,1971
3009	F8-G2 Stars in the N. Gal. Pole Region	80	1141		01/13/78	FGNGPR	CDS437	28	8000	Upgren,1963
3010	G5 & Later Stars in N. Gal. Pole Region	80	4045		01/13/78	GLNGPR	CDS437	29	8000	Upgren,1962
3011	F2 & Earlier Stars in S.A.28,54,106,107	80	498		01/13/78	FESA	CDS437	30	8000	Upgren,-,1969
3012	F5 & Later Stars in S.A.28,54,106,107	80	2181		01/13/78	FLSA	CDS437	31	8000	Upgren,-,1970
3013	Vyssotsky's Catalogs 1950.0	80	915		01/13/78	VYSS	CDS437	32	8000	1943-58
3014	F2 & Earlier Stars in N.Gal.Pole Region	80	626		01/13/78	FENGPR	CDS437	27	8000	Slettebak,-,1959
3015	Luminous Stars in the N. Milky Way	83	7389	D	12/16/77	LSNMW	CDS443	26	7968	HAMBURG,1959-65
3016	Orbital Elements of Spec. Bin. Systems	132	737	D	04/25/77	BATTEN	CDS461	2	15840	Batten,1967
3017	Early-type Stars with Emission Lines	80	10652	D	04/25/77	WACKLNG	CDS461	1	16000	Wackerling,1970
3018	MK Classification	102	21001	DE	04/25/77	MKCLS	CDS461	3	14484	Jaschek,-,1964
3018B	MK Classification (Single Values)	80	15168		04/25/77	MKCLSSIN	CDS461	6	16000	Jaschek
3019	MK Classification Extension	80	24456	DE	04/25/77	MKCLSE	CDS461	4	16000	Kennedy
3019 -2		80	264	DE	04/25/77	MKCLSEB	CDS461	5	16000	
3021	Gen. Cat. of Stellar Radial Velocity	80	15106	D	04/25/77	GCRADV	CDS461	13	16000	Wilson,1953
3022	Rotation of Evolving A & F Stars	80	599		01/13/78	ROTEAF	CDS437	34	8000	Danziger,-,1972

Fig. 3. Data directory.

number in the volume, the block size, and the additional information (usually, the producer or the producing institution). When new data are acquired or when there is a change in the storage media, the data directory is updated through CRT terminals.

The information bulletin (circulated among the interested scientists) contains the list of currently available data files, which is essentially a direct printout of the data directory (excluding the information on the storage media that is not relevant to the user). The bulletin itself is produced by means of a standard text processing system.

3.2 Human-readable Data Description (HDD)

This secondary file gives the descriptions of individual data files. As shown in Fig.4, the essential part of HDD is the field specification of a record belonging to a given data file. The printed version of HDD is used as the document that is sent to the user of the file. Since a considerable amount of key-in work is required to produce this file, an input editing routine has been developed. The field names, given at the right-most position in Fig.4, are used when we wish to manipulate a given data file, and comprise the basic component of the CDD, which is described in the below.

3.3 Coded Data Description (CDD)

This is a DDT (Data Description Table) library of the GIS file manipulation facility (an IBM software product). The DDT for each individual data file is obtained from the corresponding HDD of the file, with the use of a transformation routine. The routine scans the field definition part of the HDD (column positions and field names) to produce a DDT source text for the file. The source text is then compiled by the GIS language processor, and saved in the DDT library.

Fig.5 shows the DDT source text that corresponds to the HDD shown in Fig.4. The keyword FLD defines a field, which can be further partitioned by the keyword RDFN. Only second-level partioning is possible when GIS is used for a simple sequential file. Note that, in HDD, re-partitioning of the field is indicated by indentation. JUST=R indicates that the data item is a right-justified quantity, i.e., a number. Otherwise, JUST=L (character string) is assumed. The DATM line at the bottom gives the characteristics, as well as the storage information, of the file. These pieces of information is retrieved from the data directory when the HDD is created.

The CDD is used for file processing services, excluding simple file copying.

4. FILE DISSEMINATION

Dissemination of data files is the most basic function of the Astronomical Data Services at KIT. It is conceivably a simple task. For, what is required is to copy those data files ordered by a user, on to another tape(s), and sent it to him. In practice, however, manual works involved in such a task can be quite cumbersome. For example, when ten or so data files are ordered, the standard tape-copy routine (those supplied by the computer manufacturer) would require 30 to 40 control records. We would also have to compute the length of the tape reel needed for the data files being copied, since the blocking requirement varies from user to user. In order to simplify the procedure so that the possibility of erroneous operations can be minimized, a specialized tape-copy routine has been developed. The routine accepts as input the file numbers ('catalog numbers') and the corresponding block sizes, and searches through the data dictionary for other information on the data files under request. After the length of the output reel (therefore the number of tape cartridges) is computed, the routine generates a set of standard copy routines ready for execution. The instructions for the machine operator, the invoice to the requesting person (including the description of the tape contents), and the processing log, are produced at the same time. The HDD's for individual files are also produced for delivery.

GENERAL CATALOGUE OF STELLAR RADIAL VELOCITIES

Wilson, R.E.

Carnegie Institute of Washington, 1953

3021

RECORD LENGTH: 80 NUMBER OF RECORDS: 15106 FILE NAME: GCRADV

COL	DESIG	UNIT	REMARK	FIELD NAME
1-5	GCRV	---	Running Number	SEQ
			Other identifications	ID
6-10	GC	---	B.Boss General Catalogue Identification (1)	GC
11-16	HD	---	Henry Draper Catalogue Identification	HD
17-18	alp (hr)	h	Right ascension for the year 1950 (hour)	ALPHR
19-21	(min)	min	(minute)	ALPMIN
22-24	del (deg)	deg	Declination for the year 1950 (degree)	DELDEG
25-27	(')	'	(minute of arc)	DELMIN
28-30	m	1/10 magn	Magnitude (0 when unknown)	MAG
31	Rem	---	Remark for magnitude : blank : visual magnitude P : photographic magnitude V : variable	REM
32-35	mu	1/1000 '/yr	Proper motion (9999 when unknown)	MU
	Sp		Spectral type	SP
36			Luminosity class	SPL
37-38			Spectral sequence	SPS
39-41			Remark	SPR
42-46	RV	1/10 Km/s	Radial velocity	RV
47-51	l	1/100 deg	Galactic longitude	GL
52-56	b	1/100 deg	Galactic latitude	GB

Fig. 4. Human-readable data description.

5. FILE PROCESSING USING CDD

Under the current user environment of machine-readable astronomical data, it is often more important to use the entirety, or a sizable portion, of a data file, than to retrieve a few pieces of information. This involves such tasks as the reorganization of the file by selecting those records that satisfy certain conditions, or selecting only specified fields in the record, or creating a new file by joining several different files. In many of such processing jobs, the use of sorting routines or the utilization of work files is required. And usually it becomes very important to produce well-formatted printout.

The file processing jobs, such as these, can be done in principle by means of conventional programming languages (typically, PL/I or COBOL). In practice, however, programming works involving file processing can be quite cumbersome. A typical program would contain at least several tens of statement lines, most of them being devoted for reading and writing files, or for editing the printout. This is followed by a lengthy succession of debugging runs and validations. It should also be noted that computer systems at research institutions are not always set up for the convenience of the users whose jobs involve processing of large data files.

To facilitate processing of individual data files, we have been using the GIS facility, a file processing software. One advantage of using GIS is that the data definition part (DDT) and the procedure part in a program are segregated from each other so that, once the DDT for a file has been created, the processing of the file can be accomplished simply by using the predefined field names. Furthermore, the procedural language of GIS is considerably simplified in comparison with conventional programming languages. For instance, the SORT instruction (sorting a file) and the HOLD instruction (writing onto a work file) are among the very useful. In either case, various types of file definitions, which otherwise should be prepared by the programmer, are automatically created by the system. The language also has a report writing function, which automatically takes care of heading, field adjustment, pagination, and so on.

```
DDT
 FILE:NAME=GCRADV
*                      3021
*
 FLD:NAME=SEQ,      LENGTH=5
 FLD:NAME=ID,       LENGTH=11
                             RDFN:FIELDS=GC,5,HD,6
 FLD:NAME=ALPHR,    LENGTH=2, JUST=R
 FLD:NAME=ALPMIN,   LENGTH=3, JUST=R
 FLD:NAME=DELDEG,   LENGTH=3, JUST=R
 FLD:NAME=DELMIN,   LENGTH=3, JUST=R
 FLD:NAME=MAG,      LENGTH=3
 FLD:NAME=REM,      LENGTH=1
 FLD:NAME=MU,       LENGTH=4
 FLD:NAME=SP,       LENGTH=6
                             RDFN:FIELDS=SPL,1,SPS,2,SPR,3
 FLD:NAME=RV,       LENGTH=5, JUST=R
 FLD:NAME=GL,       LENGTH=5
 FLD:NAME=GB,       LENGTH=5
 FLD:NAME=Q,        LENGTH=1
 FLD:NAME=NRV,      LENGTH=3
 FLD:NAME=OBS,      LENGTH=3
 FLD:NAME=RMU,      LENGTH=1
 FLD:NAME=DM00,     LENGTH=6
 FLD:NAME=IV,       LENGTH=5
 FLD:NAME=DM01,     LENGTH=1
 FLD:NAME=QIV,      LENGTH=1
 FLD:NAME=NIV,      LENGTH=2
 FLD:NAME=DM02,     LENGTH=1
*
 SEGM:NAME=GCRADV, LEVEL=00, TYPE=RECORD
 DATM: DSORG=PS, CREATE=NO, CATLG=NO, VOLUME=SER=CDS461, LABEL=13,
£ RECFM=FB, BLKSIZE=16000, LRECL=80, DSNAME=GCRADV, UNIT=TAPE
END
```

Fig. 5. Coded data description (GIS DDT source records).

```
QUERY GCRADV
  LOCATE RECORD
  IF (MAG LE 30) AND (MAG NE 0)
  HOLD HOLDFL1 ID, ALPHR, ALPMIN, MAG, SP, RV
  EXHAUST RECORD

SORT HOLDFL1 MAG, ALPHR, ALPMIN

QUERY HOLDFL1
  LIST ID, ALPHR, ALPMIN, MAG, SP, RV
  EXHAUST RECORD

END PROCEDURE
```

Fig. 6. GIS query example.

An example of file processing using the GIS language is shown in Fig.6. The file under consideration is the one shown in Fig.4 (or 5). As described earlier, the CDD (DDT) of a particular data file has been created from the corresponding HDD through the transformation routine. The file name and the field names are conserved throughout the process. Therefore, in writing a GIS procedure, one only needs to use these names by looking at the HDD.

Such programming works are currently done by the staff at KIT. However, since it would be easy enough for a non-professional programmer to compose this type of program, and since a more detailed specification of a task would be possible that way, we are preparing an educational program aiming at such a scheme.

6. ID CONVERSION SUBSYSTEM

As mentioned earlier, a number of different naming systems are in use for identifying astronomical objects, depending on the field and the convenience of the investigator. It has also been mentioned that any of the identification systems is not free from the ambiguity of one type or another. If the catalogs are to be handled manually (therefore, the number of objects at hand being essentially small), the diversion (or the ambiguities) in the identification problem would not pose much difficulties, considering the fact that several of major identification systems are jointly used in most of the catalogs. However, when we wish to process the catalogs by means of computers, the situation is grave. This is especially so when the processing task involves more than one catalogs using different identification systems. This problem is compounded by the fact that, most of the identification systems in existence are based on coding rules which are sufficiently useful for manual processing, but rather ambiguous for machine processing.

A special kind of index file, called the Catalog of Stellar Identifications, have been developed by the Stellar Data Center, Strasbourg, France, for the purpose of cross-referencing stellar designations. It is essentially a collection of lists each of which contains a set of various identifications assigned to a star. About 440,000 stars are included in the index currently. Since, in most cases, the designation of a star in a given ID system is the sequence number in a particular catalog, CSI gives an important means when joint processing of several catalogs is under consideration.

It can be said that an effective use of the CSI is essential to further developments in astronomical data services, and our efforts are being devoted for that direction. At present, we are in the stage of having completed an on-line retrieval system for CSI. The system retrieves a whole set of designations for a star when a particular designation is given as input. Although such a system itself would be useful in many applications, it seems that a more elaborate design is required, in order to accomplish efficient execution of such file processing tasks as discussed in the above.

ACKNOWLEDGEMENTS

The idea of the Astronomical Data Services grew out of discussions made within the C-6 unit of the current project. The author is indebted to the colleagues of the unit for their constant encouragements and assistances throughout the development at Kanazawa. His gratitudes are also due to the colleagues at the Stellar Data Center, Strasbourg, France, who have shown lasting interests and offered every assistance in shaping up the present work.

REFERENCES

(1) C. Jaschek and G. A. Wilkins (eds.), 'Compilation, Critical Evaluation and Distribution of Stellar Data', D.Reidel, Dordrecht (1977).

(2) Y. Terashita and I. Fukuda, 'Astronomical Data Services at the Kanazawa Institute of Technology', Bull. Inform. Centre du Donnees Stellaires, 15, 93-97 (1978)

Scientific Information Systems in Japan
H. Inose (ed.)

USE OF BIOCHEMICAL DATA BASE IN RELATION WITH MOLECULAR EVOLUTION

Haruhiko NODA, Teruo YASUNAGA**, Takashi MIYATA**, Hirotsugu MATSUDA**, Masami HASEGAWA*** and Koji OHNISHI†*

** Department of Biophysics and Biochemistry, Faculty of Science, University of Tokyo, Hongo, Tokyo 113*
*** Department of Biology, Faculty of Science, Kyushu University, Fukuoka 812*
**** The Institute of Statistical Mathematics, Minami-Azabu, Minato-ku, Tokyo 106*
† Department of Biology, Faculty of Science, Niigata University, Niigata 950-21

Several methods to systematize and to organize biochemical data base in relation with molecular evolution are developed.

1. INTRODUCTION (H. Noda)

According to the Central Dogma of Molecular Biology, the total information of an individual of a species is carried through generations by means of semiconservative replication of DNA. At each generation, necessary information is transcribed into mRNA and then it is translated into proteins, which then perform prescribed functions such as structural proteins, enzymes and so on. In the cases of multicellular organisms, DNA of germ line cells, and that in somatic cells seems to differ in their fate and function, but all the processes proceed without apparent contradiction with the Central Dogma.

The mass of information contained in a single organism of a member of a species is called genome, and is assumed to be self-consistent, although the nature of consistence has much to be elucidated. Information content of genome for various species coexistent on the present earth are supposed to be mutually adjusted by the process of evolution, and thus form a consistent and cooperative ecological system. The total amount of information, operating among present day biosphere of the earth, is estimated to be around $10^{15} \sim 10^{18}$. Among such a large amount of information, present day human knowledge is a very small fraction. Primary structure of only 1081 proteins has been analized [1], but it is enormous when compared with our knowledge of 20 years ago. Our knowledge of primary structure of DNA and RNA is less, but is increasing at an enormous speed because of the recent improvement of the method of sequence analysis. In order to have a general view of the biosphere, penetrating many levels, through the level of information content of DNA, of molecular organization of macromolecules, of organization of cells, of organization of multicellular organisms, of society within a species and of the interspecies relationship, it is necessary to organize the total knowledge. Such organization should be the object of biology in general, but has never been considered as a practical aim. Due to the recent progress of analytical methods of amino acid sequence in proteins and nucleotide sequence in nucleic acids, a new field of the study of molecular evolution is established, and basic knowledge of genetic information in the form of the primary structure of the nucleic acids and proteins is systematized on the basis of the phylogenetic tree traditionally established and confirmed by the knowledge of molecular evolution. Although the fraction of systematized data is very small, it is already possible to imagine and foresee the future of biology unified and organized as is explained above.

We have tried, in the present project, to proceed as many steps as possible in the direction of finding the best way to systematize and organize our present knowledge.

2. METHOD FOR OBTAINING AN OPTIMUM ALIGNMENT OF HOMOLOGOUS SEQUENCE (T.Yasunaga, T.Miyata & H.Matsuda)

The alignment of nucleotide or amino acid sequences is the first step for the study of molecular evolution. So far, this problem has been studied by several authors [2-6]. Needleman and Wunsch[2] have reported the method for calculating the maximum matching which is defined as the largest number of amino acids or nucleotides of one sequence that can be matched with those of another sequence while allowing for all possible deletions. Sankoff[3] has improved their method from the view point of computational economy, and elaborated an algorithm to find the maximum match under the deletion/insertion constraint where the number of deletions introduced is arbitrarily restricted. If the Sankoff's latter algorithm is used, the alignment of two sequences which satisfies the maximum match under a certain deletion/insertion constraint could be obtained in principle. In practice, however, it is impractical for long sequences, because a large number of storage of computer is required. Indeed, if the lengths of two sequences are m and n, respectively, and q deletions are permitted, q+1 matrices, whose dimensions are m×n, are required. In contrast, Needleman and Wunsch's and Sankoff's former algorithm may be a realistic tool for aligning longer sequences. Here, only one matrix (V matrix) is required. The alignment is obtained as a path in V matrix. However, because there are usually a great number of alignments or paths each of which satisfies the maximum match, we should choose an optimum path among them. In this section, we present a method which is designed to select an optimum alignment efficiently.

2.1 Sankoff's V Matrix and a Path in the Matrix

Let $a_1,\ldots,a_m$ and $b_1,\ldots,b_n$ be two nucleotide or amino acid sequences of lengths m and n, respectively. Following Sankoff, a matrix V where V(i,j) is the highest value possible for a path to (i,j) is defined as follows: For $i=1,\ldots,m; j=1,\ldots,n$,

$$V(0,j) = V(i,0) = 0, \quad (1)$$

$$V(i,j) = \max_{\substack{h \le i-1 \\ k \le j-1}} V(h,k) + f(a_i,b_j) \quad (2)$$

where, for nucleotide sequences,

$$f(a_i,b_j) = 1 \qquad \text{if } a_i = b_j, \tag{3}$$

$$f(a_i,b_j) = 0 \qquad \text{if } a_i \neq b_j. \tag{4}$$

In the case of amino acid sequences, there are many different ways to define $f(a_i,b_j)$ in addition to that defined by Eqs. (3) and (4); for example, as a score matrix of Dayhoff. When $f(a_i,b_j)$ is defined by (3) and (4), the value $V(i,j)$ represents the number of identical pairs in an alignment of two partial sequences $a_1,\ldots,a_i$ and $b_1,\ldots,b_j$. Then the maximum match, $v = \max_{i\leq m, j\leq n} V(i,j)$, means the largest number of identical pairs in an alignment.

Once V matrix is given, an alignment which attain maximum match is defined as a path in V matrix. A path is the sequence of the cells in V matrix which is constructed in the following way:
First find (α_1,β_1) such that

$$\alpha_1 = m \text{ or } \beta_1 = n$$

and $$V(\alpha_1,\beta_1) = \max\{\max_{j\leq n} V(m,j), \max_{i\leq m} V(i,n)\}. \tag{5}$$

Next find (α_2,β_2) such that

$$\alpha_2 = \alpha_1-1,\ \beta_2 \leq \beta_1-1 \text{ or } \beta_2 = \beta_1-1,\ \alpha_2 \leq \alpha_1-1$$

and $$V(\alpha_2,\beta_2) = \max\{\max_{j\leq\beta_1-1} V(\alpha_1-1,j), \max_{i\leq\alpha_1-1} V(i,\beta_1-1)\}. \tag{6}$$

In general, if (α_K,β_K) is determined, then $(\alpha_{K+1},\beta_{K+1})$ is defined as

$$\alpha_{K+1}=\alpha_K-1,\ \beta_{K+1}\leq\beta_K-1 \text{ or } \beta_{K+1}=\beta_K-1,\ \alpha_{K+1}\leq\alpha_K-1$$

and $$V(\alpha_{K+1},\beta_{K+1}) = \max\{\max_{j\leq\beta_K-1} V(\alpha_K-1,j), \max_{i\leq\alpha_K-1} V(i,\beta_K-1)\}. \tag{7}$$

If this procedures are continued until $\alpha_L=1$ or β_L-1, then a path $(\alpha_1,\beta_1),\ldots,(\alpha_K,\beta_K),\ldots,(\alpha_L,\beta_L)$ is obtained. Here, L is defined as a path length, which is the number of the sites in an alignment where a base (or amino acid) in one sequence corresponds to a base (or amino acid) in the other sequence (not to a deletion).

It should be noted that (α_K,β_K) satisfying (5),(6) or (7) does not necessarily be determined uniquely, so that there are many paths which satisfy the condition of maximum match. The number of the paths depends on the sequences to be aligned and on their lengths. They increase rapidly as the lengths of the sequences become longer. For example, when N-terminal portions of amino acid sequences(100 amino acids) of A genes in phages ϕX174 and G4 are compared, the number of possible paths(alignment) is over 10^5. Therefore, it is necessary to choose the optimum one(s) to actually get the alignment(s). Now, we consider the path, the length of which is longer than any other paths, as the optimum path. In estimating evolutionary rate, the highest value is obtained in the optimum alignment, if deletions are not taken into account.

For choosing an optimum path, the most simple method is to survey every possible paths in V matrix. However it is impossible to carry out the procedure, because the number of possible paths are numerous for long sequences. It is therefore necessary to use a more efficient procedure to choose an optimum path.

2.2 The Procedure for Obtaining an Optimum Alignment

At first, in order to obtain an optimum path(alignment), let us consider to survey every possible paths in the V matrix. In order to attain this, a path is constructed as follows: If there are not one but many candidates for (α_k,β_k), the (α_k,β_k) such that

$$\alpha_k = \min\{\alpha_k^1, \alpha_k^2,\ldots,\alpha_k^n\}, \tag{8}$$

$$\beta_k = \max\{\beta_k^1, \beta_k^2,\ldots,\beta_k^n\}, \tag{9}$$

is chosen, where $(\alpha_k^1,\beta_k^1),\ldots,(\alpha_k^n,\beta_k^n)$ are candidates for (α_k,β_k). The first path constructed in this way,

$$(\alpha_1,\beta_1),\ldots,(\alpha_L,\beta_L)$$

is the most upper right path in the V matrix. An example is shown by the solid line (a) in Fig.1. Next

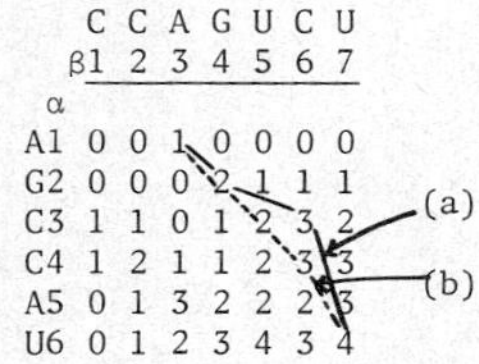

Fig. 1. An example of V matrix.

the second path is reconstructed from the first as follows: For i=L,L-1,...., it is examined if the (α_i,β_i) can be replaced by the (α_i',β_i') such that

$$V(\alpha_i',\beta_i') = V(\alpha_i,\beta_i), \tag{10}$$

$$\alpha_i' \geq \alpha_i,\ \beta_i' \leq \beta_i, \tag{11}$$

and $$\alpha_i' = \alpha_{i-1}-1 \text{ or } \beta_i' = \beta_{i-1}-1. \tag{12}$$

When the (α_i',β_i') is not determined uniquely, the (α_i',β_i') is to be choosen by (8) and (9). If (α_i,β_i) can be replaced by (α_i',β_i'), which is equal to (α_k,β_k), for the first time at i=K, the path is reconstructed from (α_K,β_K) such as the new partial path from (α_K,β_K) is the upper right partial path. In other word, in constructing a partial path, again, if (α_j,β_j) is not determined uniquely, it is to be chosen by (8)and (9). In Fig.1, the broken line (b) shows the second path in the V matrix. In the same way, the k-th path is reconstructed from the (k-1)-th path. After this procedure is continued until any other path can not be reconstructed, we can have every possible paths in the V matrix. It should be noted that the number of all the possible paths, which is constituted only by the cells above or on the k-th path,is k, if the paths are constructed one after another in the above procedure. In other word, there is no k'th path (k'>k) which is constituted only by these cells.

We can obtain an optimum path by comparing the lengths of every possible paths. However, if we employ the following procedure, we can get an optimum path without constructing every possible paths. If the length of the k-th path is not shorter than any other s-th path (s<k), then we do not have to take the cells above the k-th path in the V matrix into consideration, during the further paths construction. This means that if the s'-th(s'>k) which is constituted by the cells both below and above the k-th path is longer than the k-th path, we can obtain a much longer path which is constructed without the cells above the k-th path. In the following paragraphs, this assertion is proved.

For saving the number of constructions of paths, if the k-th path is the longest among the 1-st, 2-nd,..., and k-th paths, the cells above the k-th path in the V matrix are not taken into account in the further survey of paths. Such a procedure is permitted only when the longest path in V matrix does never disappear in doing so. There is no problem about the paths which are restricted only in the region above the k-th path or only in the region below the k-th path. Let us consider the path which begins in the region below the k-th path and ends in the region above the k-th path. Hereafter, in brief, a path is represented such as $P_1,,,P_i,,,P_p$ or $Q_1,,,Q_j,,,Q_q$, where p and q represent the lengths of the paths. In Fig.2 (case 1 or case 2), the path $P_1,,,P_i,,,P_p$ represents the k-th path and the path $Q_1,,,Q_j,,,Q_q$ represents the k'-th path (k'> k) which starts in the region below the k-th path and ends in the region above the k-th path. If we employ above procedure, the path, $Q_1,,,Q_j,,,Q_q$ is left out of consideration. In order to justify the above procedure, we need to show that the path $Q_1,,,Q_j,,,Q_q$ is not longer than a path which is constructed only by the cells below or on the k-th path.

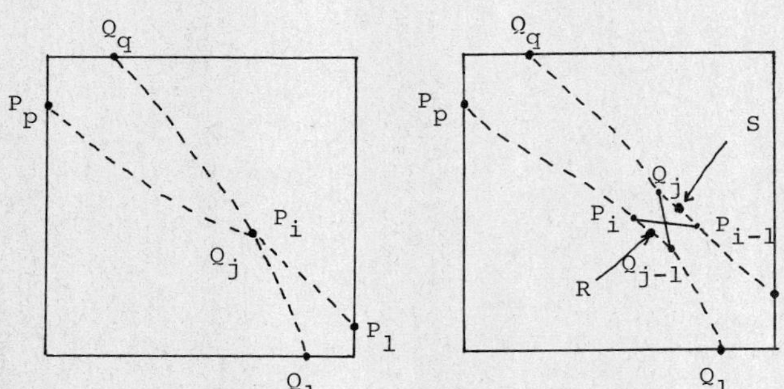

Case 1 Case 2

Fig. 2.

As shown in Fig.2, there are two cases when two paths cross at one point. In Case 1, two paths cross on a cell in the V matrix, whereas two paths do not cross on a cell but between cells in the V matrix in Case 2.

At first, let us consider Case 1, where it is easily verified that the path $Q_1,,,Q_j,,,Q_q$ is not longer than a path in the region below or on the k-th path. If the two paths $P_1,,,P_i,,,P_p$ and $Q_1,,,Q_j,,,Q_q$ cross at P_i(= Q_j), both $P_1,,,P_i(=Q_j),,,Q_q$ and $Q_1,,,Q_j(=P_i),,,P_p$ are obviously also paths in this case. The length of the path $P_1,,,P_i(=Q_j),,,Q_q$ is q-j+1, and that of $Q_1,,,Q_j$(= $P_i),,,P_p$ is p-i+j. Note that the path $P_1,,,P_i(=Q_j),,,Q_q$ is restricted only in the region above the k-th path $(P_1,,,P_i,,,P_p)$. Therefore the path $P_1,,,P_i,,,P_p$ is not shorter than the path $P_1,,,P_i(=Q_j),,,Q_q$;

$$p \geqq q-j+i \tag{13}$$

(13) is rewritten as

$$p-i+j \geqq q \tag{14}$$

The inequality (14) means that the path $Q_1,,,Q_j,,,Q_q$ is not longer than $Q_1,,,Q_j(=P_i),,,P_p$ which is constructed only by the cells below or on the k-th path.

Now, we consider Case 2.We will prove the followings,

1] When the two paths $P_1,,,P_{i-1}P_i,,,P_p$ and $Q_1,,,Q_{j-1}$ $Q_j,,,Q_q$ cross both between P_{i-1} and P_i and between Q_{j-1} and Q_j, the $Q_1,,,Q_{j-1}(R)P_i,,,P_p$ and $P_1,,,P_{i-1}$ $(S)Q_j,,,Q_q$ are also paths,

2] The path $Q_1,,,Q_{j-1}Q_j,,,Q_q$ is not longer than the path $Q_1,,,Q_{j-1}(R)P_i,,,P_p$.

The proof of 1]: Let the P_i be the cell of the α_{P_i}-th row and the β_{P_i}-th column in the V matrix, that is $(\alpha_{P_i},\beta_{P_i})$. In the same way, the Q_j means $(\alpha_{Q_j},\beta_{Q_j})$. It is shown that when $V(\alpha_{Q_j},\beta_{Q_j})$ is equal to n, $V(\alpha_{P_i},\beta_{P_i})$ is also n. If $V(\alpha_{P_i},\beta_{P_i})<n$, ",,,$P_{i-1}P_i$,,," is not a path, or if $V(\alpha_{P_i},\beta_{P_i})>n$, ",,,$Q_{j-1}Q_j$,,," is not a path. When $V(\alpha_{Q_j},\beta_{Q_j})=n$ and $V(\alpha_{P_i},\beta_{P_i})=n$, either $Q_1,,,Q_{j-1}P_i,,,P_p$ or $Q_1,,,Q_{j-1}RP_i,,,P_p$ is a path. Note that if $\alpha_{P_i}\neq\beta_{Q_{j-1}}-1$ and $\beta_{P_i}\neq\beta_{Q_{j-1}}-1$, ",,,$Q_{j-1}P_i$,,," is not a path by the definition of a path in the V matrix. In this case, if we consider the cell, R, which is $(\alpha_{P_i}+1,\beta_{Q_{j-1}}-1)$, then ",,,$Q_{j-1}RP_i$,,," is a path. This is easily understood. If $V(\alpha_{P_i}+1,\beta_{Q_{j-1}}-1)=n$, then ",,,$Q_{j-1}RP_i$,,," is a path. In fact, if

$V(\alpha_{P_i}+1,\beta_{Q_{j-1}}-1)>n$, it contradicts by (7) that ",,,$Q_{j-1}Q_j$,,," is a path and if $V(\alpha_{P_i}+1,\beta_{Q_{j-1}}-1)$ <n, it contradicts by (3) that $V(\alpha_{P_i},\beta_{P_i})=n$. It is also shown that $P_1,,,P_{i-1}(S)Q_j,,,Q_q$ is a path in the same way.

The proof of 2]: For simplicity let L_{QP} and L_{PQ} be the lengths of the paths $Q_1,,,Q_{j-1}(R)P_i,,,P_p$ and $P_1,,,P_{i-1}(S)Q_j,,,Q_q$, respectively. For the proof of 2], it is sufficient to show $L_{QP}\geqq q$. Clearly the following inequalities are hold.

$$L_{QP} \geqq p-i+j, \tag{15}$$

$$L_{PQ} \geqq q-j+i. \tag{16}$$

As the path $P_1,,,P_{i-1}P_i,,,P_p$ is not smaller than the path $P_1,,,P_{i-1}(S)Q_j,,,Q_q$, we have

$$p \geqq L_{PQ}. \tag{17}$$

From (16) and (17), we have

$$p \geqq q-j+i, \tag{18}$$

or

$$p-i+j \geqq q. \tag{19}$$

From (15) and (19), we have

$$L_{QP} \geqq q. \tag{20}$$

This means that the path $Q_1,,,Q_{j-1}Q_j,,,Q_q$ is not longer than the path $Q_1,,,Q_{j-1}(R)P_i,,,P_p$. In the above argument, we have dealt only with the situation that two paths cross at one point. There may be paths which cross at several points.In this case, it is also verified by the same argument that we do not miss the longest path by using our procedure.

In the actual program in use, we also use the similar procedure except that surveying the paths is started from the most lower left path. At first, the most upper right path is constructed and the cells above the path is substituted by a certain value (for example, -1) so that no path can be constructed in these regions. Next, the most lower left path is constructed and the cells below the path is substituted by the value. These are one cycle in the procedure. In the following cycle, only if the new reconstructed path is not shorter than any path already constructed, the cells above the path is substituted by the value. If we continue the cycles, we can find an optimum path without trying every possible paths.

As an illustrating example, let us consider two sequences "AGCCAU" and "CCAGUCU". In this case,V matrix is

```
     C C A G U C U
   β 1 2 3 4 5 6 7
 α
 A1  0 0 1 0 0 0 0
 G2  0 0 0 2 1 1 1
 C3  1 1 0 1 2 3 2
 C4  1 2 1 1 2 3 3
 A5  0 1 3 2 2 2 3
 U6  0 1 2 3 4 3 4 .
```

At first, the most upper right path is constructed and the cells above the path are substituted by blanks. Then, the V matrix becomes

```
   C C A G U C U
 A 0 0 1
 G 0 0 0 2
 C 1 1 0 1 2 3
 C 1 2 1 1 2 3
 A 0 1 3 2 2 2
 U 0 1 2 3 4 3 4 ,
```

where the solid line shows the most upper right path, and L_R, the length of the path, is 4. Next, the most lower left path is constructed and the cells below it are substituted by blanks. Then V matrix becomes

```
   C C A G U C U
 A 0 0 1
 G 0 0 0 2
 C 1 1 0 1 2 3
 C   2 1 1 2 3
 A     3 2 2 2
 U             4 ,
```

where the solid line shows the most lower left path, and L_L, the path length, is 4. This is the end of one cycle. As the next cycle, the second path is constructed from the first one (the most upper right path). As the length of this path(=5) is greater than L_R(=4),the cells above the path are substituted by blanks. Then,

```
   C C A G U C U
 A 0 0 1
 G 0 0 0 2
 C 1 1 0 1 2
 C   2 1 1 2 3
 A     3 2 2 2
 U             4 ,
```

where the solid line shows the second path constructed from the most upper right path and L_R is substituted by the length of this path, that is 5. Next, the second path is constructed from the most lower left path. As the length of the path(=5) is greater than L_L, the cells below the path are substituted by blanks. Then,

```
   C C A G U C U
 A     1
 G       2
 C         2
 C           3
 A
 U             4 ,
```

where the solid line shows the second path constructed from the most lower left path and corresponds to the optimum path. Since no further path can be constructed, it is the end of the procedure. In this example, the optimum alignment chosen by the procedure is

```
  * * A G C C A U
      :   :   :
  C C A G U C * U .
```

3. PHYLOGENETIC RECONSTRUCTION PROCEDURE FROM SEQUENCE DATA (M. Hasegawa)

A combination of computer programs (Fig. 3) is used to reconstruct phylogenetic trees from aligned sequence data of nucleic acids and proteins. Because of the complexity of protein phylogeny due to the degeneracy of the amino acid code, only nucleic acids, which are composed of four species of nucleotides, are treated in this section. Our program system is similar to that of Moore et al.[7] for the construction of protein phylogeny, and is named PARSIMON (the maximum parsimony method). It has been described briefly in ref.[8].

The first program DISMTX computes distances over all pairs of the aligned sequences being compared. Per cent difference between sequences is used as distance. The second program UWPGMC, executing the unweighted pair group method of clustering [9], produces a tree from the distance matrix. Since we do not consider ancestors or the evolutionary process in constructing the UWPGMC tree, it is not to be taken as the correct phylogenetic tree, but is used only as a first approximation in the search for it. Using the aligned nucleotide sequences and the UWPGMC tree, the minimum number of nucleotide replacements over the entire tree (MINCSP) is calculated by the third program MINCSP, executing the method provided first by Fitch [10] and proved later by Hartigan [11].

The search for trees with fewer total replacements is directed by those changes in topology which reduce the MINCSP value. At first, we restrict our search to the nearest neighbour 1-step changes in topology (NN1SC) [7]. In a tree of N exterior points or contemporary sequences, 2(N-3) topologies are examined. The change which results in the greatest lowering of the MINCSP is saved, and this new topology is taken as the start for the next cycle of NN1SC's. When the tree has been modified to a topology where the MINCSP value can no longer be lowered by the NN1SC procedure, then the iterative search by this procedure stops.

In order to represent the tree in graphical form, the length of each branch must be calculated. The maximum parsimony ancestral sequences for each interior point or node of the tree can be reconstructed by using the program MINCSP, and the solution is generally degenerate. The nucleotide replacement lengths of all the branches in the tree are calculated by averaging the numbers of nucleotide replacements among the set of reconstructed degenerate ancestors (Program BRNLGT). The program GRAPH display the tree graphically according to the topology and the branch lengths obtained by the above procedures.

The next step of our search for the most parsimonious tree is carried out by subjecting the tree to more extensive changes in topology and persist in searching until we are satisfied that a sufficiently wide range of phylogenetic possibilities have been examined. If there is a subtree with questionable or irrational location from a biological or biochemical viepoint, the subtree is cut off from the tree and is transplanted onto every possible branch of the remaining tree. The MINCSP value is calculated for every topology of the transplanted tree, and we continue to search for the most parsimonious tree until every questionable sub-

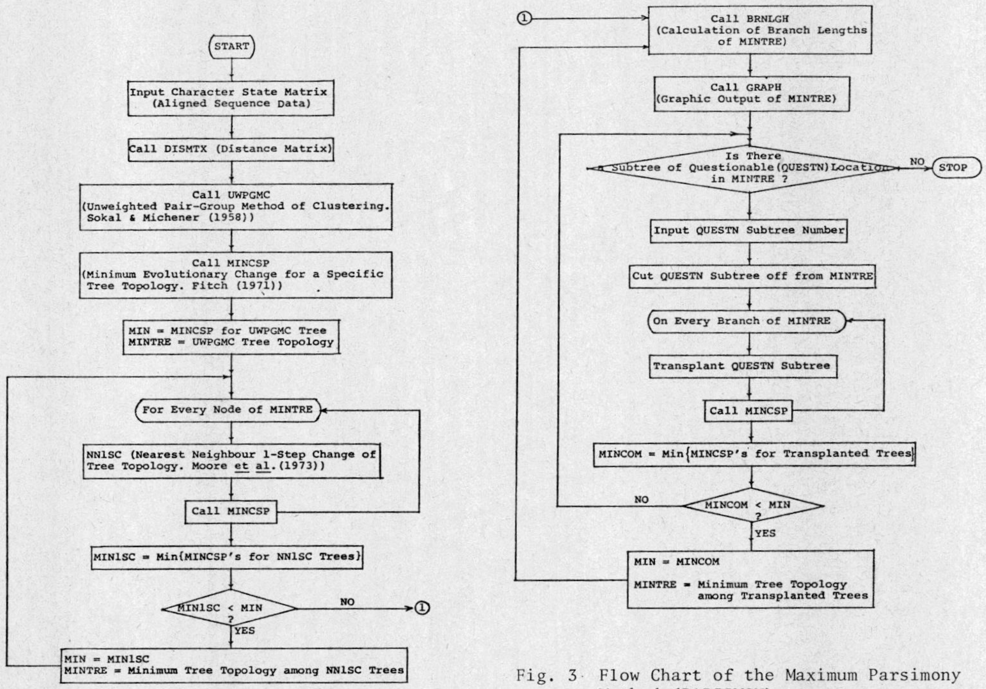

Fig. 3 Flow Chart of the Maximum Parsimony Method (PARSIMON)

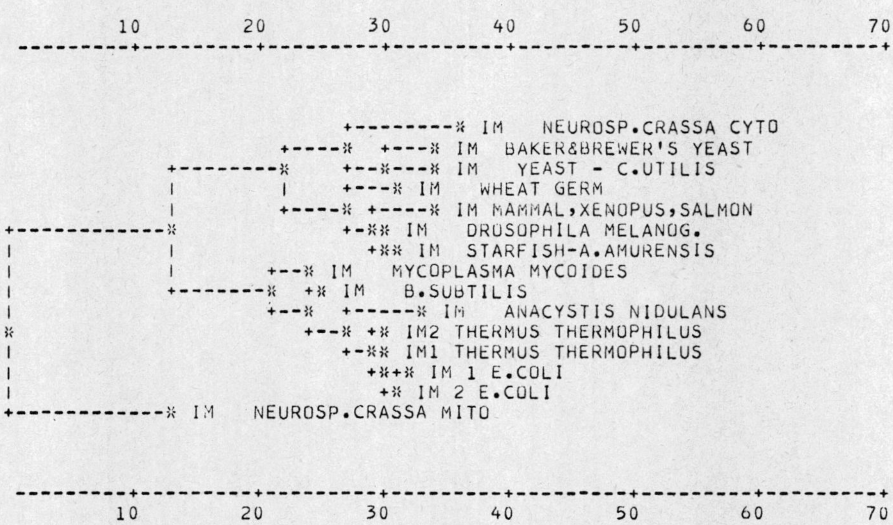

Fig. 4. The most parsimonious tree for initiator tRNAs. The abscissa shows the minimum number of nucleotide replacements along a branch. Data are from ref.[23].

tree has been examined.

Generally speaking, there are more than two topologies with the same MINCSP value. When we find more than two parsimonious trees with the same MINCSP value in a cycle of procedure, we choose the tree with the fewest numbers of nucleotide replacements per site (NRP) at the conservative sites and with the most NRP's at the variable sites among the alternative trees. Such weighting should be related to the coding property of the nucleotide sites and/or the spatial position of each nucleotide site in the secondary or tertiary structure of the nucleic acid molecule.

As an example, the most parsimonious tree for initiator tRNAs is shown in Fig. 4.

4. CONSTRUCTION OF MINIMUM PATHWAY EVOLUTIONARY TREES OF C3-RELATED COMPLEMENT COMPONENTS AND IMMUNOGLOBULINS BY EMPLOYING PRINCIPLE COORDINATE ANALYSIS

(K. Ohnishi)

The minimum pathway method(MPM) has recently been deviced for constructing evolutionary trees of homologous proteins in a multidimensional real number-valued Euclidean space. In the previous paper[12], the 'Euclidean space' was spanned by 18 orthogonal axes, each of which represents the molar ratio of the respective kind of amino acids, and thereby each protein was mapped, by its amino acid composition, as a sample point in this space. In this section, a method is proposed for constructing a multidimensional evolutionary tree based upon a difference matrix (for example, minimum mutation distances (MMD's)[13] among proteins) obtained by aligning homologous protein sequences. Firstly, homologies among three structurally similar complement components (C3, C5 and C4), β_2-microglobulins(β_2-μ's), immunoglobulin(Ig) constant(C) and variable(V) domains are presented. Secondly, these related proteins are mapped in a multidimensional pseudo-Euclidean space by employing principal coordinate analysis(PCOA)[14], and finally, MPM is employed for constructing evolutionary trees of these proteins in the pseudo-Euclidean space.

4.1 Alignment of C3a-related Protein Sequences

N-terminal sequences of C3a[15,16] and C5a[15] anaphylatoxins(N-terminal fragments of α chains of C3 and C5, respectively) and C4 α chain[17] were compared with one another and also with other sequences[18] shown in Fig. 5. Sequences were aligned so that smallest MMD's[13] could be obtained among these sequences. These alignments can be considered to represent true homology when tested by Moore and Goodman's method [19] of alignment statistics. Detailed discussions on the result of Moore-Goodman's test applied to these alignments will be published elsewhere(Ohnishi, in preparation). Consequently, these proteins can be considered to have evolved by gene duplication, and we here propose a term 'C3 family'[20] for homologous complement components, C3, C5 and C4.

4.2 PCOA: Theory

PCOA is a method for mapping n samples as n corresponding sample points in an (n-1)-dimensional character space so that given differences among samples may equal corresponding geometrical distances among sample points in this space[14,21]. As Gower[14] has described, PCOA has essential similarities to princip component analysis[12] when original samples can be mapped in an (n-1)-dimensional real number-valued

		1 5 10 15 20 25 30
C3a	human(1 -27)*	SVQLTEKRMNKVGKYPK-E---LRKCCEDGM
"	pig (1 - 27)	SVQLMEKRMQKLGQYSK-E---LRRCCEHGM
C5a	human(1 - 25)	--TLZKKIZZIAAKYKHSV---VKKCCYBG/
C4 α	human(1 - 21)	NVNFEKAINEKLGEYAS-P---TAK/
β_2-μ	human(1 - 25)	-IQRTPKI-QVYSRHPA-E---NGKSNFLNC
"	mouse(1 - 25)	-IQKTPZI-QVYSRHPP-E---NGKPNILNC
"	guinea pig(1-21)	-VLHAPRV-QVY--HPA-E---NG--NFLNC
"	rabbit(1 - 25)	-VQRAPNV-QVYSRHPA-E---NGKDNFLNC
"	dog (1 - 25)	-VQHPPKI-QVYSRHPA-Z---BGKPBFLBC
Cμ4	Ou (446-474)	GVALHRP--BVYLLPPARZZLNLRESATITC
Cα3	Bur (341-369)	SGNTFRP--QVHLLPPPSZZLALBZLVTLTC
Cγ3	Eu (341-367)	GQPR-EP--QVYTLPPSREEMTKNQ-VSLTC
Cε4	ND (427-454)	SGPRAAP--EVYAFATP-EWPGSRDKRTLAC
Cκ	Eu (107-134)	KGTVAAP--SVFIFPPSDEQLKSGT-ASVVC
"	MOPC 21(107-134)	KGADAAP--TVSIFPPSSEQLKSGG-ASVVC
"	rabbit4135(106-133)	KGDPVAP--TVLIFPPAADQVATGT-VTIVC
Cλ	Sh (108-135)	GQPKAAP--SVTLFPPSSEELQANK-ATLVC
"	MOPC 315(110-137)	GQPKSTP--TLTVFPPSSEELKENK-ATLVC
V_H	shark 225(1-22)	EVTLTZPZ-AZDSZP--------GGALTLTC
V_{HIII}	Nie (1-22)	ZVQLVESG-GGVVQP--------GRSLRLSC
V_L	shark 225(1-23)	AITMTESP-PVLSVG-------LGQTATITC
V_κ	Ti (1-23)	EIVLTQSP-GTLSLS-------PGERVTLSC

Fig.5. Alignment of C3a-related protein sequences. Abbreviations of amino acids are from ref.[1]. Sequence data are from refs.[15]~[18]. *Site of residues.

Euclidean space, $\underline{\underline{R}}^{n-1}$. But, when samples cannot be mapped in $\underline{\underline{R}}^{n-1}$, we can map them only in an (n-1)-dimensional space spanned by several real number orthogonal axe and several purely immaginary number ones, where the distance between two points, $P(x_1,x_2,..x_{n-1})$ and $P'(x'_1,x'_2,..,x'_{n-1})$, is given by a pseudo-Euclidean distance;

$$d_{pp'} = \left[\sum_{k=1}^{n-1} (x_k - x'_k)^2 \right]^{1/2} \tag{21}$$

This space is the so-called (n-1)-dimensional pseudo-Euclidean space[22], and here we denote it by

$$\underline{\underline{\Psi}}^{n-1}(p,\ n-p-1)$$

where p is the number of real number axes. It should be reminded that, in $\underline{\underline{\Psi}}^{n-1}$, the distance given by (21) is left invariant only by an (n-1)-dimensional orthogonal transformation, ${}^t(x_1,\ldots,x_{n-1}) \to \underline{\underline{L}}\,{}^t(x_1,\ldots,x_n)$, whose transformation matrix is given by $\underline{\underline{L}}=(l_{ij})_{n-1,n-1}$ where $\sum_{j=1}^{n-1} l_{ij}^2 = 1$.

We start from a difference matrix $\underline{\underline{\Delta}} = (\Delta_{ij})_{n,n}$ where Δ_{ij} is a difference between samples i and j. Consider an appropriate monotonously increasing function of Δ_{ij}, defined by

$$d_{ij} = f(\Delta_{ij}), \qquad (d_{ii} = 0), \tag{22}$$

then we have real symmetric matrices, $\underline{\underline{D}} = (d_{ij})_{n,n}$ and $\underline{\underline{B}} = (b_{ij})_{n,n}$, where

$$b_{ij} = \Big(\sum_{k=1}^{n} d_{kj}^2/n + \sum_{k'=1}^{n} d_{ik'}^2/n - \sum_{k=1}^{n}\sum_{k'=1}^{n} d_{kk'}^2/n^2 - d_{ij}^2 \Big)/2 \tag{23}$$

As we find that rank($\underline{\underline{B}}$) $\leq$ n-1, n eigenvalues of $\underline{\underline{B}}$ can be numbered such that

$$|\lambda_1| \geq |\lambda_2| \geq \ \ldots \geq |\lambda_{n-1}| \geq \lambda_n = 0 , \tag{24}$$

where $|\ \ |$ denotes absolute value.

Mapping of sample points $P_i(X_{i1},\ldots,X_{i,n-1})$ $(i=1,\ldots,n)$ is given by

$$X_{ik} = m_{ik}\sqrt{\lambda_k} \qquad (i=1,\ldots,n;\ k=1,\ldots,n-1), \tag{25}$$

where ${}^t(m_{1k},\ldots,m_{nk})$ is an eigenvector of $\underline{B}$, corresponding to λ_k. As given in Eq.(25), X_k axis is a real axis for $\lambda_k \geq 0$, and a purely immaginary one for $\lambda_k < 0$. If we denote, by p+1, the number of non-negative eigenvalues of $\underline{B}$, Eq.(25) gives a mapping in $\underline{\underline{\Psi}}^{n-1}(p,n-p-1)$. From (25) variance of $X_k(k=1,..,n)$ is

$$\mathrm{Var}(X_k) = \left(\sum_{i=1}^{n} X_{ik}^2\right)/(n-1) = \lambda_k/(n-1). \tag{26}$$

Thus, from the relation (24),

$$|\mathrm{Var}(X_1)| \geq |\mathrm{Var}(X_2)| \geq \ldots \geq |\mathrm{Var}(X_{n-1})| \geq \mathrm{Var}(X_n) = 0. \tag{27}$$

Accordingly, X_i might be called 'k-th principal coordinate', and then percent variance of the k-th principal coordinate axis is given by

$$V_k^{\%} = 100\cdot|\mathrm{Var}(X_k)|/\sum_{k=1}^{n}|\mathrm{Var}(X_k)| = 100\cdot|\lambda_k|/\sum_{k=1}^{n}|\lambda_k|, \tag{28}$$

and the accumulative percent variance of the 1-st to k-th axes is given by

$$V_k^{\%} = \sum_{k=1}^{n} V_j^{\%} \tag{29}$$

Further, we here define 'real part ratio' by

$$\mathrm{R.P.R.} = \sum_{\alpha=1}^{m}\lambda_{k_\alpha} / \sum_{\alpha=1}^{n}|\lambda_{k_\alpha}|. \tag{30}$$

where $\lambda_{j_1} \geq \lambda_{j_2} \geq \ldots\ldots \geq \lambda_{j_p} \geq \lambda_n = 0$.

4.3 PCOA of C3a-related Protein Sequences

In order to obtain a difference matrix $\underline{\Delta} = (\Delta_{ij})$ above mentioned, we here define MMD_{21} and mean MMD_{21} (m-MMD_{21}). In the alignments shown in Fig. 5, we consider 'gaps' as to be 21-st amino acids hypothetically coded by nonsense (termination) codon. MMD_{21} between two amino acids X and Y (X and Y are any two of the 21 kinds of amino acids) is defined as similarly as the original definition of MMD[13]. MMD_{21} between aligned sequence pair is counted by summing MMD_{21} values over aligned positions of amino acids. For those sequences including B(D or N) and/or Z(E or Q), m-MMD_{21} was defined as follows;

$$\text{m-}MMD_{21Z}^{\ \ B} = (MMD_{21E}^{\ \ D} + MMD_{21Q}^{\ \ D} + MMD_{21E}^{\ \ N} + MMD_{21Q}^{\ \ N})/4 = 1.75\ , \tag{31}$$

$$\text{m-}MMD_{21X}^{\ \ U} = (MMD_{21X}^{\ \ U'} + MMD_{21X}^{\ \ U''})/2, \tag{32}$$

where (U,U',U'') is (B,D,N) or (Z,E,Q), and X is any one amino acid other than B and Z.

Difference matrix, $\underline{\Delta} = (\Delta_{ij})$ where $\Delta_{ij} = (\text{m-}MMD_{21})_{ij}$ was counted for alignments over positions both 1 to 25 and 1 to 30 in Fig. 5. For Eq.(22), we select

$$d_{ij} = \Delta_{ij}^2 = (\text{m-}MMD_{21})_{ij}^2. \tag{33}$$

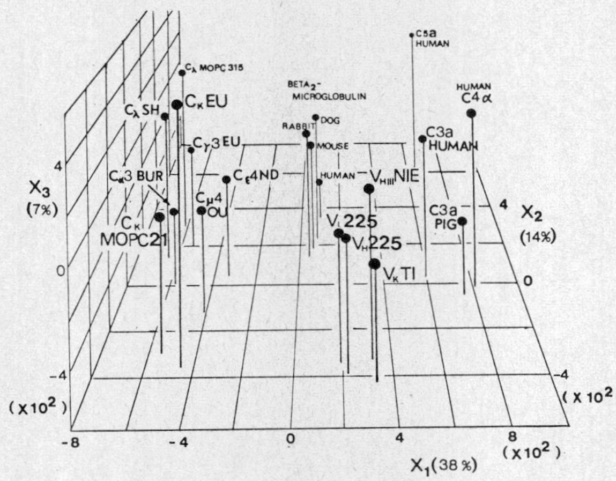

Fig. 6. Twenty C3a-related protein sequences projected by the 1st(X_1), 2nd (X_2), and 3rd(X_3) principal coordinate axes. $d = (\text{m-}MMD_{21})_{ij}^2$. See text for details.

Mapping of 20 C3a-related sequences is shown in Fig. 6. Here $V_1^{\%}$, $V_2^{\%}$, $V_3^{\%}$, $V_4^{\%}$ and $V_5^{\%}$ are 38, 14, 7, 6.2 and 5.7, respectively, and further, $V_{1-3}^{\%} = 59$, $V_{1-4}^{\%} = 65$, and R.P.R. = 0.79. As shown in Fig. 6, X_1 axis discriminates C3 family proteins from Ig C and V domains, with leaving β_2-μ's arround the mean value (zero). Thus, by employing PCOA, different protein classes are well-separately distributed in $\underline{\underline{\Psi}}^{19}$, with forming their respective clusters such as Ig C domains, V domains, β_2-μ's and C3 family complement components.

If we employ $d_{ij} = \Delta_{ij}$ and $d_{ij} = \sqrt{\Delta_{ij}}$ for Eq.(22), ($V_1^{\%}$, $V_2^{\%}$, $V_3^{\%}$) are (33, 16, 9.5) and (22, 15, 9.7), respectively, and we have 0.94 and 0.997 for the respective R.P.R. values.

4.4 Minimum Pathway Evolutionary Tree

The MPM[12] is now applied to the mapping in $\underline{\underline{\Psi}}^{n-1}$ obtained here by employing PCOA. Some comments and details of the method for applying the MPM in $\underline{\underline{\Psi}}^{n-1}$ will be published elsewhere (Ohnishi, in preparation). Fig. 7 shows three dimensional mapping of 8 representatives selected from Fig. 5, in 3-dimensional subspaces of $\underline{\underline{\Psi}}^7$. Here Eq.(33) was employed for PCOA by using alignments over positions 1 to 30 in Fig.5. ($V_1^{\%}$, $V_2^{\%}$, $V_3^{\%}$, $V_4^{\%}$) = (60, 15, 9.9, 9.8) and X_4 axis is purely immaginary one. Thus $V_{1-3}^{\%} = 85$, $V_{1-4}^{\%} = 95$, and R.P.R. = 0.88. The MPM using $d_{ij} = \Delta_{ij}$ and $d_{ij} = \sqrt{\Delta_{ij}}$ for Eq.(22), did not give good evolutionary tree. Therefore, of these three cases, $d_{ij} = \Delta_{ij}^2$ is the most suitable for constructing minimum pathway trees in $\underline{\underline{\Psi}}^{n-1}$. In Fig. 7, solid triangles (▲) indicate a common ancestral protein

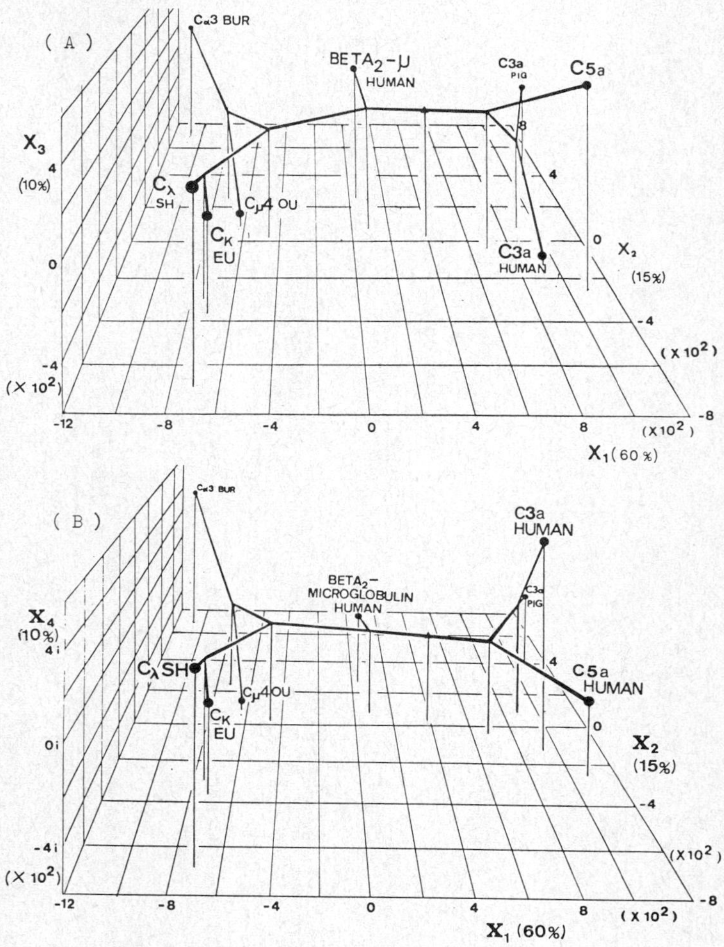

Fig. 7. A minimum pathway tree projected by the 1st, 2nd and 3rd (A), and the 1st, 2nd and 4th (B) principal axes. See text.

or the root of the minimum pathway tree. The resulting tree is considered to well indicate the evolutionary process of these proteins.

4.5 Summary

Alignments of C3a anaphylatoxin and other proteins in Fig. 5 are found to show true homology. PCOA and MPM have been shown to be useful to analyse their evolutionary relationships and to construct evolutionary trees of them.

REFERENCES

[1] M. O. Dayhoff (ed.), "Atlas of Protein Sequence and Structure", Vol. 5, Suppl. 3, National Biomedical Research Foundation, Washington, D.C. (1978).

[2] S. B. Needleman and C. D. Wunsch, J. Mol. Biol., 48, 443-453 (1970).

[3] D. Sankoff, Proc. Natl. Acad. Sci. USA 69, 4-6 (1972).

[4] D. Sankoff and R. J. Cedergren, J.Mol. Biol., 77, 159-164 (1973).

[5] T. C. Elleman, J. Mol. Evol., 11, 143-161 (1978).

[6] W. C. Barker and M. O. Dayhoff, in "Atlas of Protein Sequence and Structure" (M. O. Dayhoff, ed.), Vol. 5, pp. 101-110, National Biomedical Research Foundation, Washington, D.C. (1972).

[7] G. W. Moore, M. Goodman and J.Barnabas , J. theoret. Biol., 38, 423 (1973).

[8] M. Hasegawa, in "Origin of Life, Proceedings of 2nd ISSOL Meeting" (H. Noda, ed.), pp. 495-500, Center for Academic Publications Japan, Tokyo (1978).

[9] R. R. Sokal and C. D. Michener, Kansas Univ. Sci. Bull., 38, 1409 (1958).

[10] W. M. Fitch, Syst. Zool., 20, 406 (1971).

[11] J. A. Hartigan, Biometrics, 29, 53 (1973).

[12] K. Ohnishi, in "Evolution of Protein Molecules" (H. Matsubara and T. Yamanaka, eds.), p. 75, Japan Scientific Societies Press, Tokyo (1978).

[13] W. H. Fitch and E. Margoliash, Science 155, 279 (1967).

[14] J. C. Gower, Biometrika, 53, 225 (1966).

[15] H. N. Fernandez and T. E. Hugli, J. Biol. Chem., 252, 1826 (1977).

[16] N. C. Corbin and T. E. Hugli, J. Immunol., 117, 990 (1976).

[17] C. Bolotin, S. Morris, B. Tack and J. Prahl, Biochemistry, 16, 2008 (1977).

[18] B. A. Cunningham et al., Cold Spr. Harb. Symp. Quant. Biol., 41, 351 (1977); F. W. Putnam, in "The Plasma Proteins. Structure, Function and Genetic Control" 2nd ed., Vol. III (F. W. Putnam ed.), p. 1, Academic Press, New York (1977); E. S. Dugan, R. A. Bradshow, E. S. Simms and H. N. Eisen, Biochemistry, 12, 5400 (1973); C. Sledge, L. M. Clem and L. Hood, J. Immunol., 112, 941 (1974); M. O. Dayhoff (ed.) "Atlas of Protein Sequence and Structure" Vol. 5, Suppl.2, National Biomedical Research Foundation, Washington, D.C. (1976).

[19] G. W. Moore and M. Goodman, J. Mol. Evol., 9, 121 (1977).

[20] K. Ohnishi, Proc. of the Complement Symp. (Tokyo) (in Japanese) 14, 63 (1977).

[21] W. S. Togerson, "Theory and Method of Scaling" John Wiley & Sons, New York (1958).

[22] F. Klein, "Vorlesungen über Nicht-euklidische Geometrie" Springer, Berlin (1928); H. Weyl, "Raum-Zeit-Materie, 5. Aufl.", Springer , Berlin (1923).

[23] D.H.Gauss, F.Grüter and M.Sprinzl,Nucl.Acids Res 6,r1(1979); K.Watanabe et al.,J.Biochem.,86,893 (1979); S.Yamashiro-Matsumura and S.Takemura, J.Biochem.,86,335(1979); S.Silverman et al., Nucl.Acids Res.,6,421(1979); Y.Kuchino et al., Nucl.Acids Res.,6,3459(1979).

Scientific Information Systems in Japan
H. Inose (ed.)

INTERACTIVE C-13 NMR DATA RETRIEVAL SYSTEM; SPIRES

Shin-ichi SASAKI, Hidetsugu ABE*, Shizuko SASAKI*, Kazuo TANAKA*, Tohru YAMASAKI* and Masato SAKATA***

**School of Materials Science, Toyohashi University of Technology, Hibarigaoka, Tempaku, Toyohashi*
***Computer Center, Tohoku University, Katahiracho, Sendai*

A file search system of C-13 NMR spectra has been prepared based on 1435 spectra provided by NIH-EPA[1]. The queries used in the present system are chemical shifts, structure and molecular formula. In case of searching by the structure, not only full structure but also partial structure are used to search for the corresponding spectra. Similarly, it is possible to look for the structures consistent with a partial spectrum as well as a full spectrum. Moreover, the results given by such primary searches are treated with logic operation (AND, IOR, EOR and ANDNOT) for providing finer and fewer responses. A new notation method for structure representation prepared by the authors themselves, CANOST, are described.

1. INTRODUCTION

Among many spectroscopies utilized in organic chemistry, C-13 NMR (CNMR) spectroscopy is one of the most useful for identification of organic compounds[2], since the proton-noise decoupled spectra display single, well-separated lines, one for each carbon nucleus in the molecule. This inherent advantage of the spectra should be utilized in a CNMR search system. That is the system with which one can retrieve any compound containing the carbon atom associated with a distinctive chemical shift and also can retrieve the chemical shift data associated with a carbon atom in a specified partial structure. To realize this, we have developed an interactive system named SPIRES (SPectral Information REtrieval System) with which one can retrieve compounds by three features, i.e., partial structure, partial spectral data and molecular formula.

2. CANOST CODING SYSTEM

Structure representation in the system is performed by using a new notation system named CANOST (CANOnical STructure). The major objects of developing this new notation system are for simplification of coding work and for differentiation of distinct carbon atoms in a molecule. A CANOST notation for salicylic acid is shown in Fig. 1. As shown in the figure, the notation consists of two lines, the first line shows a string of codes and the second shows their connectivities. For the simplification of coding work, canonicalization task which is inevitable for any notation systems is performed by means of computer programs[3]. Thus a person at coding work is not required to learn troublesome canonicalization rules, but to learn only simple rule of translating organic structure into CANOST codes. The coding procedure for salicylic acid is also illustrated in Fig. 1.

At first, the structure is converted into graphical representation of CANOST codes according to Table I. Then, the codes are numbered arbitrarily, and arranged in that sequence, and their connectivities are represented as shown in the figure. The numerals arranged continuously mean the connection of adjacencies and zero means discontinuation of the connection. This non-canonical representation is then canonical-

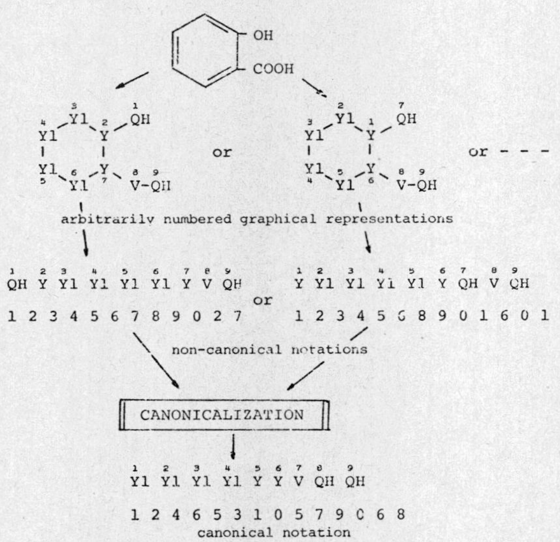

Fig. 1 Example of CANOST notation and its coding procedure

ized by means of the algorithm based on the idea of connectivity stack[4]. Using this representation, any carbon atoms in a partial structure can be specified simply by designation of their node numbers. Therefore, Chemical shift data associated with a specific carbon atom can be retrieved easily.

In SPIRES, structure of a compound is stored by the form of CANOST representation and chemical shift values of peaks of its spectrum are associated with corresponding nodes of the structure.

Table I CANOST CODE

No.	CODE		No.	CODE	
1	T	-C≡	18	H	-H
2	T1	≡CH	19	LF	-F
3	DD	=C=	20	LC	-Cl
4	DS	>C=	21	LB	-Br
5	D1	-CH=	22	LJ	-I
6	D2	CH_2=	23	XR	x⋮
7	C	>C<	24	XW	O=x=O
8	C1	>CH-	25	XD	x=
9	C2	-CH_2-	26	XX	=x=
10	C3	CH_3-	27	XT	x≡
11	Y	-C⋮	28	XP	xHp
12	Y1	CH⋮	29	SG	SINGLE BOND
13	V	-CO-	30	+	CATION
14	VD	=CO	31	-	ANION
15	Q	-O-	32	·	RADICAL
16	QH	-OH	33	/	CHELATION
17	QD	=O	34	X	OTHERS

Symbol x stands for elements excepting C, O and halogens.

Table II COMMANDS for SPIRES

command	appendant parameters		purpose
PEAK	a	b	by searching chemical shift
STRUCTURE	c		by searching partial structure
FORMULA	d		by searching molecular formula
LOGIC	e	f	for making Boolian operation
PSTRUCTURE	g	h	for displaying partial structure
SSPECTRA	i	j	for displaying partial spectrum
LIST	k		for displaying items specified by subcommands*
HELP	m		for requesting directions
RESET			for system initialization
END			for closing system

a and b are the chemical shift value and its allowance
c is the partial structure coded by CANOST notation
d is the molecular formula
e, f, g, i, and k are the set numbers
h is the size of partial structure
j is the vertex number of partial structure
m is command name or CANOST code
* subcommands for "LIST" are shown in Table III

3. COMMANDS FOR SPIRES

As SPIRES is operated interactively, user's instructions for the system are transmitted through several commands shown in Table II. Brief explanations are given below.

PEAK: As shown in the table, this command is used for chemical shift search. The searching result is a set of ID numbers corresponding to those compounds whose spectra contain signals between a - b and a + b ppm.

STRUCTURE: This command is for searching by partial structure. The input partial structure should be represented by means of CANOST notation. The searching result is also a set of ID numbers.

FORMULA: This is the command for searching by molecular formula. A complete molecular formula such as C3 H8 O is required for the appendant. The searching result is again a set of ID numbers

The sets of ID numbers which contain searching results of execution of above three commands are numbered serially and are stored temporarily for the following execution of the other commands.

LOGIC: This command is used with four subcommands, AND, IOR, EOR, and ANDNOT. They are used for Boolian logic operations, AND, inclusive OR, exclusive OR, and ANDNOT, among two ID number sets previously obtained. The result of execution of this command is also a set of ID numbers, and the set can be used for further logic operations and for execution of following commands.

PSTRUCTURE: This command is used for displaying a partial structure of a compound in the set obtained by PEAK search. Therefore, the partial structure contains a carbon atom which shows particular chemical shift value designated at PEAK search. The first appendant parameter of this command designates the set number and the second limits the range of output partial structure, for example, when two numerals 3 and 2 are inputted as appendant parameters, partial structures containing atoms up to β-positions for target carbons of the compounds in the set 3 are displayed

SSPECTRA: This is the command for displaying chemical shift data associated with the particular carbon atom of a compound in the set obtained by STRUCTURE search. The first appendant parameter designates set number and the second the carbon atom by vertex number.

LIST: This is the command for displaying all records associated to the compounds in a designated set. Displaying items are name, molecular formula, Wiswesser line notation, spectral data, solvent used, referred literature, and structure formula.

Table III SUBCOMMANDS for "LIST"

subcommand	function
NAME	compound name listing
SSHOW	full structure displaying
SPECTRA	full spectra listing
MOLECULAR	molecular formula listing
WLN	Wiswesser line notation listing
REFERENCE	referred literature listing
SOLVENT	solvent name listing

The items to be displayed can be independently designated with subcommands listed in Table III. Structural formulas displaing by the use of this command are not CANOST notations but two-dimensional figures using common symbols, such as C, N, O, and etc.[5]

RESET: This command is used for initializing the status of the system. Once this command is executed, all the temporary files containing sats of ID numbers of compounds are cleared off.

HELP: This is the command for requesting directions how to use commands or to decode CANOST symbols.

END: The command for closing SPIRES execution.

4. EXAMPLE OF EXECUTION

Examples of multiple search, search of partial structure by partial spectrum and search of partial spectrum by partial structure are illustrated in Figs. 2, 3, and 4. Question mark is the prompting symbol to open the gate of SPIRES. As the next on the mark, user can input his own data. A serial number is given to each set of the compounds resulted by the primary search and logic operation. To express chemical structure two-dimensionally, symbols, +, =, and # are used to indicate single bond, double bond and triple bond, respectively. Aromatic carbon is specified by adding (*) at the carbon atom.

```
 $$$CON,TSS
ACOS-6 TSS(R3.3) ON Ø1/19/79 AT 15.375 CHANNEL4431
USER ID -K2ØØ3NØ719
PASSWORD--
22L7 LLINKS FILE SPACE USED
SYSTEM ?RDB SPIRES

COMMAND ? PEAK 125.3Ø 1.ØØ
        ENTRY 161
        SET #    1
COMMAND ? PEAK 24.3Ø 1.ØØ
        ENTRY 174
        SET #    2
COMMAND ? LOGIC 1 2
:: SUBCOMMAND? AND
        ENTRY  13
        SET #    3
COMMAND ? STRUCTURE C3 DS D1 1 2 3
        ENTRY  14
        SET #    4
COMMAND ? LOGIC 3 4
:: SUBCOMMAND? AND
        ENTRY   1
        SET #    5
COMMAND ? LIST 5
:: SUBCOMMAND? NAME MOLECULAR SOLVENT REFERENCE SPECTRA;
        MORE? SSHOW

    *** REG.#  361 ***

    1  ISOPHORONE
    2  C9 H14 O
    3  DEUTEROCHLOROFORM
    4  J&J 357

    5  SHIFT      ASSIGNMENT
       199.ØØ      1
       125.3Ø      3
       159.7Ø      2
        45.1Ø      8
        33.3Ø      9
        5Ø.7Ø      7
        24.3Ø      4
        28.2Ø      5  6

    6  MOLECULAR STRUCTURE

                          C6
                         +
                   C7+++C9
               O    +   ++
                = +  C5 +
                C1      +
                  +     C8
                  C3   +
                    = +
                    C2
                     +
                     +
                    C4
```

Fig. 2 Example of Multiple search

```
COMMAND ? PEAK 5Ø.3Ø Ø.2Ø
        ENTRY  22
        SET #   6
COMMAND ? PSTRUCTURE
  SET # ? 6
   SIZE ? 1 2
        :: SUBSTRUCTURE (  1 EDGES AWAY FROM NODE 1)

        1 C1= 5Ø.1Ø PPM(REG.#   3 )

               +
              +
           CL+C+C3
             1  +
                 +

        2 C1= 5Ø.5Ø PPM(REG.#  59 )

           O+C+C##
             1 2

        3 C1= 5Ø.5Ø PPM(REG.#  98 )

           C4
            +
             C1+N+:
            +
           C3

        4 C1= 5Ø.4Ø PPM(REG.# 2Ø3 )

           +
            +
            C4      +
              +   +
              C1+C3
               +
               N

        5 C1= 5Ø.4Ø PPM(REG.# 246 )

               +
               +
              C4
               +
           ==C+C1
             2  +
                 C3
                  =
                   =
```

Fig. 3 Example of 'PSTRUCTURE' search

4-1 Search of the compounds with spectral signals at 125.30 ± 1.00 ppm and 24.30 ± 1.00 ppm and partial structure, CH_3-C=CH- (Fig. 2).

The first step of searching is carried out using the command "PEAK". That the compounds with a signal at 125.30 ± 1.00 ppm amount to 161 is responded, and the set of the 161 compounds is numbered 1.
Similarly, the set of 174 compounds having a signal at 24.30 ± 1.00 ppm is numbered 2. Consequently, by the command "LOGIC", Boolian logic operation AND of sets 1 and 2 is carried out to search for the compounds having both signals. The result is 13 compounds and the set is numbered 3.

On the other hand, the compounds with CH_3-C=CH- is searched using the command "STRUCTURE", and 13 compounds are picked up from the file (set number 4). As shown in the figure, the partial structure is expressed in a CANOST notation.

By Boolian logic operation AND of the sets 3 and 4, candidate compounds are squeezed into one compound, isophorone.
The command "LIST" functions to list up all the items designated with subcommands; name, molecular formula, solvent used, referred literature, full spectrum, and full structure for the finally responded compound. Numerals given for the carbon atoms in full structure correspond to numbers of individual signals of the spectrum. Full structure is displayed in the form of two-dimensional chemical structure without hydrogen.

4-2 Search of partial structure by partial spectrum

Structures up to α- and β-positions for the carbon with a signal at 50.30 ± 0.20 ppm are searched as shown in Fig. 3. Twenty-two compounds are extracted from the file by inputting 50.30 ± 0.20 ppm through the command "PEAK" (set number 6). The user can requset for the set what are the structures at α- and β-positions linked to the vertex with 50.30 ± 0.20 ppm by inputting 1 and 2, respectively, through the command "PSTRUCTURE". The output is the structures up to α- and β-positions for the vertex together with its chemical shift. Only five examples for α-positions are shown in the figure.

4-3 Search of partial spectrum by partial structure

Search of the chemical shifts of methyl group(1), quarternary carbon(2) and aromatic carbon(3) in partial structure a is shown in Fig. 4.

The partial structure a is inputted in the form of non-canonical CANOST notation. That the compounds with partial structure a amount to 3 is informed by the command "STRUCTURE" (set number 7). The chemical shifs of C3(methyl), C(quaternary carbon) and y(aromatic carbon) are informed by simply typing in 1, 2 and 3, respectively, from the set 7 through the command "SSPECTRA" as shown in the figure.

```
COMMAND ? STRUCTURE C3 C Y Y1 Y1 C3 C3 ;

   MORE ? 1 2 3 4 5 Ø 2 6 Ø 2 7

        ENTRY   3
        SET #   7

COMMAND ? SSPECTRA

  SET # ? 7
VERTEX #? 1 2 3

        ⁑ REG #   SHIFT (PPM)
            389     29.5Ø
            426     31.4Ø
            429     29.7Ø

        ⁑ REG #   SHIFT (PPM)
            389     34.4Ø
            426     34.2Ø
            429     34.4Ø

        ⁑ REG #   SHIFT (PPM)
            389    136.1Ø
            426    148.ØØ
            429    135.6Ø

COMMAND ? SSPECTRA 7 1 2 3

        ⁑ REG #   SHIFT (PPM)
            389     29.5Ø
            426     31.4Ø
            429     29.7Ø

        ⁑ REG #   SHIFT (PPM)
            389     34.4Ø
            426     34.2Ø
            429     34.4Ø

        ⁑ REG #   SHIFT (PPM)
            389    136.1Ø
            426    148.ØØ
            429    135.6Ø

COMMAND ? END

⁑ JOB END ⁑
```

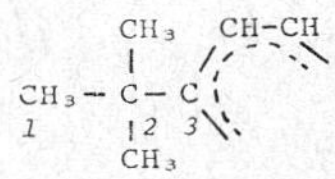

(a)

Fig. 4 Example of 'SSPECTRA' search

REFERENCES

[1] S.R.Heller, G.W.A.Milne, and R.J.Feldman, Science, 195, 253(1977).
[2] W.Bremser, M.lier, E.Meyer, Org. Magn. Reson., 7, 97(1975).
R.Scwalzenbach, J.Meili, H.Koenitzer and J.T. Clerc, ibid., 8, 11(1976).
B.A.Jezel, D.L.Dalrymple, Anal. Chem., 47, 203(1975).
W.Bremser, Z. Anal. Chem., 286, 1(1977).
[3] Details of CANOST system will be reported near future.
[4] Y.Kudo and S.Sasaki, J. Chem. Doc., 14, 200 (1974).
[5] R.E.Carhart, D.H.Smith, H.Brown and C.Djerassi, J. Amer. Chem. Soc., 97, 5755(1975).

Scientific Information Systems in Japan
H. Inose (ed.)

AN APPROACH TO INTEGRATING INFORMATION IN MEDICAL SCIENCE

Shigekoto KAIHARA

Hospital Computer Center, University of Tokyo Hospital, University of Tokyo, Bunkyo-ku, Tokyo 113

The efforts and problems towards integration of the medical science data were described. The medical information was classified into four types; namely, medical literature, drugs, clinical data and health statistics. Studies were performed on the drug information and clinical information. The development of database of test drugs as a pilot of general drug database was described. This database is now successfully used in multiple hospitals, but this study also clarified the problems to be solved when the database is used in larger scale. The problems of developing database of clinical data were identified, the methods to overcome these difficulties were discussed. The database system of clinical cases at University of Tokyo Hospital was described as one of the examples of such systems.

1. INTRODUCTION

In the field of medical science, there are four types of information to be taken into consideration for the study of "Integration of scientific information". They are 1) information on medical literature, 2) information on drugs, 3) information on clinical cases which are of value in medical science and 4) information of health statistics. These four informations are not yet integrated at present time except for limited part of medical literatures. Towards the integration of the informations in medical science, it is necessary to study and clarify in all these four informations, what type of data they are, how they are structured or what requirements are necessary for their information processing.

In this report, the efforts of research groups were described which dealt with medical data. The medical data studied by these research groups were mainly the second and the third type of informations among the above four, because of the framework where the research groups were placed. However, we also discuss briefly the other two types of medical informations here, whose integration seems to require bigger organizations than research groups of universities.

2. GENERAL CHARACTERISTICS AND REQUIREMENT FOR PROCESSING OF MEDICAL INFORMATION

The four types of medical information described above are different in their general characteristics and requirements for information processing. Table 1 summarizes the difference. In the information of drugs and literature, the structure of data is usually simple, although the volume of data very large. The requirement for information processing is data retrieval only, and the maintenance of data is not frequent and it can be performed by a small group of professional people. On the contrary the data of clinical cases or patients have very complicated data structure, which will be discussed in section 5. The number or volume of data is usually small, and rarely we encounter a large volume of database. Information processing required is not only data retrieval, but statistical analyses are required usually after retrieving data. Multiple people participate in the maintenance of the data and changes of data are frequent.

Table 1

	Clinical Data	Medical Literature & Drugs
DATA	many data items usually small number	few data items large volume of data
Processing required	complicated processing retrieval and analysis	retrieval only
Type of processing	real time & batch processing	real time
Cycle	inconstant once a month	any time

Considering these differences of characteristics the four data will be discussed separately in the following section.

3. MEDICAL LITERATURE

The database of medical literature available at present in Japan are those developed by foreign countries. MEDLARS, EXCERPTA MEDICA or BIOSIS are some representative examples of such database. There is no database of medical literature developed in Japan so far.

In this field, we need two endeavour towards the goal of integration of medical informations. Firstly, the establishment of more systematic plan and strategy to reach the goal by which medical researcher can access these database already in existence. Secondly, the development of new database of Japanese medical literature. At present, MEDLARS is available through the Japan Information Center for Science and Technology and EXCERPTA MEDICA and BIOSIS through some university computer centers or commercial networks of world wide. However, the access to these databases is geographically limited and also there are many more useful medical databases waiting for being served. As far as we understand, no systematic planning has been made for the access of these databases from all over the country, and the need for such plan seems to be obvious.

The second endeavour to be made is the development of Japanese database of Japanese medical literature. In other field of science and technology some databases have been developed in Japan, but no such databases are available for medical literatures. One technological difficulty will be the processing of Japanese kanji characters. However, since the technological progress in this field is very rapid, it will not long before kanji can be processed as easily as other alphabetical letters. Considering the fact that it takes long time to develop databases, we must start the development at any time now. To achieve this goal, bigger organizations than university research groups will be required. But more pioneering studies by research groups will also be necessary to realize the final goal.

4. INFORMATION ON DRUGS

4.1 WHAT IS DRUG DATABASE?

Information on drugs are unique in that the information is important from scientific standpoint as well as from practical standpoint in clinical medicine. In USA and Europe, databases of drugs have been developed and in practical service. RINGDOC, DRUGDOC TOXILINE are some of the examples of these systems. Although these databases are of some use even in Japan, the drugs used in Japan are not the same as those in other countries, which means that the database of drugs marketed in Japan must be developed separately.

Generally speaking, the database of drugs must include all the informations on the drug use such as name, commercial name, generic name, chemical structure, dosage, adverse effect, interaction references etc.. The number of drugs commercially available in Japan is more than several thousands and if the drugs under clinical tests are included the number will be easily doubled. This means that the database of general purpose cannot be developed on research base. Since efforts have been started to develop such database recently by non-profit organizations, we may be able to have such data available through terminals placed in hospitals in the near future, but we still do not know when.

Can the researchers in university hospitals or research institutions only wait for the development of such database by other organizations? Answer seems to be "no". General purpose database will be first aimed at practical use, and medical research institutions will require more detailed data on drugs, for their clinical studies. Pioneering studies to prove the usefulness of such systems must be performed also by researchers at universities. On the base of these small scale systems, the large system of practical use will eventually be developed.

Considering this, the group of University of Tokyo Hospital launched into this field and decided to develop a database of test drugs. "Test drugs" means the drugs which are not approved by the government committee yet, and are used on trial bases by clinical researchers. The trial use of drugs is required for the approval by the government committee and it is mostly performed by university hospitals or big teaching hospitals. The reasons for taking up "test drugs" for database are three folds; namely, the information of testdrugs is not available in other forms such as books, in spite of its importance especially at university hospitals. Secondly, the number of drugs to be included in this database is about 3000 to 4000 and this was considered to be the suitable scale for pioneering study. This study was also aimed to show the usefulness of this type of system. Accordingly, not only the development of database but also the practical service on limited scale was included in the study to find out the practical problems as well as its usefulness. This project was started in 1977, and it is now operational in a few hospitals. Since this is the only database of this type in real operation, I would like to describe more details of this database.

4.2 DATABASE OF TESTDRUGS

The TESTDRUG database includes the drugs, which are not yet commercially available, but are used on trial base. These data were searched from medical and pharmaceutical literatures, and if the drugs were identified the following items were recorded and stored in a computer, namely, name of the drug, reference number used for the identification, company which developed the drug, therapeutic effects an references. The drugs identified in the past 5 years were approximately 3000, and this data are maintained by the Drug Information Group of the Pharmacy of University of Tokyo Hospital every other week. Ten percent of the data requires the maintenance per month.

The data are now stored in a computer of the Hospital Computer Center of the University of Tokyo Hospital. The system used for the retrieval of the data is IDEAS 77 developed by Tsukuba University. It is now possible to access to this database by dialing telephone from any part of Japan. Figure 1 shows the examples of the retrieval of data using this system. The most frequent search is made using NUMBER to identify the drug as is shown in the first part of the figure. Here the drug of the number which starts with 37661 was searched. As is seen, even when the number is not complete it is possible to search the drug. The second part of the example shows the search using the name of the drug, which is not complete. Here the drugs were searched which have the data-string PENICILLIN in some part of its name. Seven drugs were identified and their names are displayed.

The limitation of the "TESTDRUG" database is that most of the data are written in English. It is possible to use katakana instead of English, but since the user of the systems are mainly in university hospitals or big teaching hospitals, English was preferred.

However, if more data are included such as adverse effects, the shortcoming of English may become more manifest even among clinical researchers. Telephone cost is also a problem when this system is accessed from the hospitals in far distance. Several centers which provide these services will be necessary all over Japan eventually.

```
?0001 DATABASE TESTDRUG
        SUB-DB (TESTDRUG)  :    2560
      USABLE RECORD COUNTS :    2560

      ** REQUEST COMPLETED **

?0002 SEARCH NUMBER 37661$
        SET NUMBER 001  CREATED
        *NUMBER OF RECORDS SEARCHED             :         1

      ** REQUEST COMPLETED **

?0003 DISPLAY

    ***** A LIST OF RETRIEVED RECORDS   (STANDARD FORMAT)   *****

RECORD =        1
  IDNO      : 79/07/13/053
  NAME      : BACAMPICILLIN HYDROCHLORIDE
  NUMBER    : 37661-08-8
  COMPANY   : PFIZER&YOSHITOMI
            : TAITO PFIZER
  THERAPY   : ANTIBIOTICS
  REF       : JAMA,237(17)1870(1977),CLIN.PHARMACOL.THER.,19(5)592(1976)

      ** REQUEST COMPLETED **


?0015 SEARCH THERAPY ANTIBIOTICS
        SET NUMBER 005  CREATED
        *NUMBER OF RECORDS SEARCHED             :       303

      ** REQUEST COMPLETED **

?0016 SELECT NAME $PENICILLIN$
        SET NUMBER 006  CREATED
        *NUMBER OF RECORDS SELECTED             :         7

      ** REQUEST COMPLETED **

?0017 DISPLAY NAME

    ***** A LIST OF RETRIEVED RECORDS   (STANDARD FORMAT)   *****

RECORD =        1
  NAME      : BISNOR PENICILLIN V

RECORD =        2
  NAME      : AZIDOCILLIN
            : AZIDOBENZYL PENICILLIN

RECORD =        3
  NAME      : SULFOBENZYLPENICILLIN EYE DROP

RECORD =        4
  NAME      : ISOPENICILLIN N

RECORD =        5
  NAME      : PENICILLIN N
            : ADICILLIN

RECORD =        6
  NAME      : PENICILLIN V BENZATHINE

RECORD =        7
  NAME      : PENICILLIN V HYDRABAMINE

      ** REQUEST COMPLETED **
```

Fig. 1 Data Retrieval of "TESTDRUG"

5. INFORMATION ON CLINICAL CASES OF SCIENTIFIC IMPORTANCE

5.1 PROBLEMS ASSOCIATED WITH CLINICAL DATA

It is needless to say that medical records are important for physicians to take care of their patients. But medical records are at the same time very important scientific materials for medical science. Most of the progress of clinical medicine has been based on the observations of patients by cautious physicians. Recently there is a trend that more data are required to draw any conclusions in the field of clinical medicine. Accordingly, the importance of accumulating medical records data and making it available to researchers systematically has increased tremendously.

However, this task is not so easy, because of the following reasons. The first difficulty is encountered in that there is no standard method of recording clinical cases. The medical record at present time is a kind of memorandum made by physicians, and terms used in medical records sometimes have different meanings from one physician to another. The structure of the data is not at all standardized either, and there is no common agreement how to categorize the contents of medical record data. The second difficulty is that the definition of scientifically important cases is varied and sometimes it is impossible to know beforehands the scientific importance of the clinical cases. If we try to accumulate all the data of all the patients treated in hospitals, the volume of the data will be astronomically large and most of the data are useless from scientific points of view. So when we start accumulating these data, we have to define the purpose of the data collection clearly and select the data to be accumulated from the routine medical records. This can only be done by experienced clinical researchers. Even when the above two difficulties were mostly overcome, there is the third difficulty. The method of management of clinical data as database is not established. The structure of medical data has some characteristics which are different from other data. The requirements for processing are also somewhat different, as will be discussed in the later section of this report. Accordingly, when we want to use the general purpose database management system, some modifications become necessary and in some cases it is desirable to develop a new database management system only for clinical purpose.

The above three represent the difficulties associated with clinical data. Many studies have been performed to overcome these difficulties, and there have been a number of successful cases even though in limited area. In this project, several groups dealt with these subjects. Prof. Abe and others set an objective to develop the database of patients with cardiovascular diseases at Osaka University Hospital. They used commercially available database management system, IDS, and found it applicable if the data are limited to the cases with specific diseases. Prof. Kudo and others of Kyushu University dealt with the hereditary diseases and developed a sophisticated method of inputting the data. This means a big progress in standardizing the data of this type of diseases. Dr. Nose and others of Kyushu University collected the data of electrocardiogram and clinical tests in the form of database using DL/I. Prof. Okajima and others of Nagasaki University developed a database of people exposed to atomic bomb using also DL/I. Since the data contained "privacy", the protection of the data was one of the main concerns of this group. Since this kind of database will never been available again, this is one of the most important database of clinical data in Japan. In developing these databases, standardization of the terminology and the structure of the data were mede in each group. Since their interests were concentrated in a single disease, this standardization seemed to work satisfactorily. However, at the moment, there has been no effort, to standardize them inter-institutionally.

The database management systems which each group used are varied. Some of them used network type database and other used hierarchical database. However, regardless of the database structure, these general purpose database management systems seem to have worked well, if the following conditions are fulfilled; namely, the purpose of the database is mainly retrieval of data and the computer is large enough to install these general purpose database management systems. There is strong need, however, in medical research for the analysis of the data in addition to data retrieval. Data retrieval is sometimes only the preprocessing for the analysis of the data. The analyses are performed often by "try and error", accordingly it is difficult to know beforehands what kind of analytical methods are required for which part of the data. To satisfy this need, the system must enable the researcher to retrieve the data and analyze them in sequence. The software which meets these requirements is now under development at University of Tokyo Hospital.

5.2 DATABASE SYSTEM FOR CLINICAL DATA AT TOKYO UNIVERSITY HOSPITAL

Fig. 2 shows the structure of clinical data in general. One of the characteristics is that it has three dimensional structure. Items in time sequence are the data for one patient and this unit is accumulated in multiple patients. The number of time sequence is not constant in each patient. Another characteristic of the clinical data is that these data are not filled completely; on the contrary more than half of the data are vacant. For the retrieval of data any item must become a key and in addition to this retrieval, statistical analyses must be performed from these retrieved data as was discussed in the previous section. The methods of analyses are varied from simple basic statistics as means and standard deviation to multiple variate analyses such as discriminant analyses.

To meet these requirements, at University of Tokyo Hospital, two batch systems were developed in the past named as "MERS" and "Patient oriented Database System". These systems have been used for the past 5 years, but considering the recent needs for on-line data analyses, a new system is now under development. The basic concept of the system is shown in Fig. 3. It utilizes two general purpose systems; one database management system and one statistical package. Because of the compatibility to the computer, the database management system used was IDS II and the statistical package SPSS. Since the data retrieved by the database system must be analyzed immediately, the interface program was developed which connects these two general purpose systems. The main function of the interface program is to change the structure of time-oriented three dimensional structure of clinical data into two dimensional tables for statistical analyses. There are several ways for this change, and some examples are shown in Table 2 a-c.

At University of Tokyo Hospital, more than ten databases exist which are accumulating data routinely. Database of obstetrical data, patients who had neurosurgical operations, in-patients at geriatric department, patients with idiopathic cytopenic purpura etc..

6. INFORMATION ON HEALTH STATISTICS

At present health statistics data are mainly collected by the central or local government and researchers who required these data use the data published by these organizations. Population statistics, birth rate, incidence rate of some diseases, morbidity rates and death rates are some of the examples of this type of data. These data are important not only for administrators but also for medical researchers. Published data are not sometimes sufficient for researches and more detailed data are required. Some method may have to be worked out for researchers to be able to access these data more

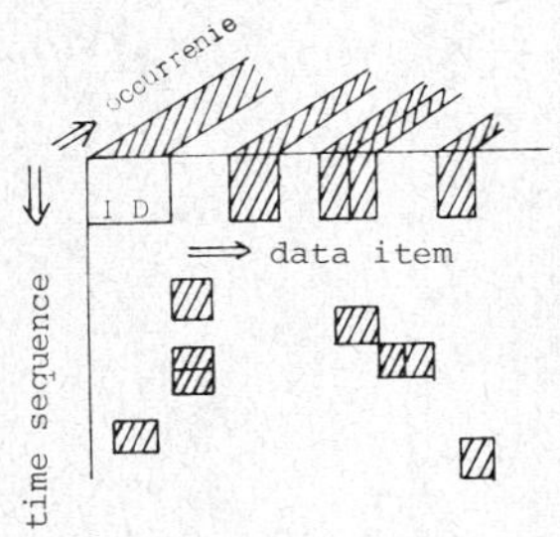

Fig. 2

Structure of Clinical Data

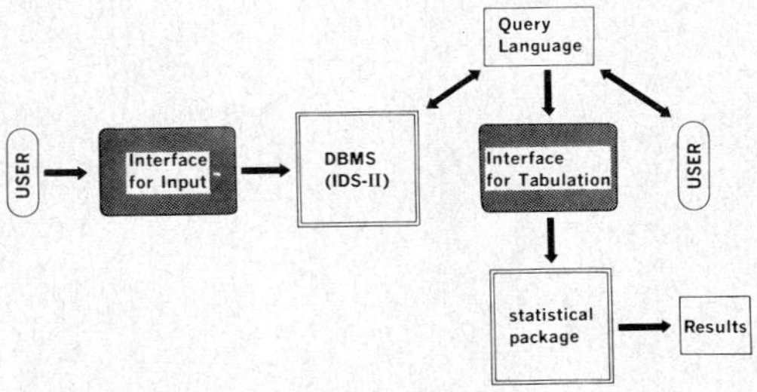

Fig. 3 Concept of Clinical Database System

Table 2-a original date

case number	name	age	sex	date of visit	blood pressure	urine protein
1	Taro TIDAI	50	M	1977. 1.15	160	+
				1977. 2.15	180	−
				1977. 3. 4	190	+
2	Jiro OSAKA	46	M	1978. 2.15	130	−
				1978. 4.20	140	−
3	Hanako HIROSHIMA	20	F	1978. 4.21	120	−

Table 2-b Data tabulated for analysis -- 1st type --

case number	age	sex	blood pressure	urine protein
1	50	M	177	+
2	46	M	135	−
3	20	F	120	−

Table 2-c Data tabulated for analysis -- 2nd type --

case number	age	sex	blood pressure	urine protein
1	50	M	160	+
1	50	M	180	−
1	50	M	190	+
2	46	M	130	−
2	46	M	140	−
3	20	F	120	−

easily. However, this problem is beyond the scope of this study.

7. PROBLEMS REMAINING TO BE INVESTIGATED

After several years of research by this study group "Integration of Scientific Information", it may be said that the integration of data took off in the field of medical science, too. A couple of databases which are used by multiple institutions now exist, although its scale is small and somewhat experimental. But we have to admit that there still remain many problems to be solved before larger databases are developed which lead to the integration of larger scale. Some of the technological problems are the management of the data which contain kanji characters and image data. These two types of data are unavoidable in medical science.

Maintenance of databases is also a big problem as is in other fields. In medical field, once database is developed, it will soon be used widely among the people in practice of medicine. If it is used outside the research environment, it will present some different problems, which are not fully known yet. For the solution of these remaining problems, the cooperation of the researchers in different field are mandatory and we hope strongly that some standing organizations which deal with these problems will be established in the near future.

REFERENCES

[1] S. Kaihara, H. Ueno and I. Wakai, "Database in Health Care," Information Processing, 17(10), 986-990 (1976).

[2] H. Ueno, M. Saito, M. Fukui and S. Kaihara, "POD: Patient Oriented Data Base for Medical Research Based on End-User Facility," Jap. J. Med Elect & Biol Eng., 17(1), 16-22, (1979).

[3] S. Kaihara and T. Koyama, "An Approach to General Purpose Medical Database System," Jap. J. Med Electr & Biol Eng 17(suppl), 306-307, (1979).

[4] S. Kaihara, "An Application of Computers for Selection of Drugs," Sogo Rinsho, 28(8), 1485-1490, (1979).

Scientific Information Systems in Japan
H. Inose (ed.)

ECONOMIC DATA SYSTEM AND MICRO DATA MATCHING FOR CREATING SOCIO-ECONOMIC DATA BASE

Yoshimasa KURABAYASHI and Yoshiro MATSUDA***

** Institute of Economic Research, Hitotsubashi University, 2-1 Naka, Kunitachi, Tokyo*

*** Documentation Centre for Japanese Economic Statistics, Hitotsubashi University, 2-1 Naka, Kunitachi, Tokyo*

Firstly the present situations of economic statistics data base compilation in Japan are reviewed. Secondly various experiments of micro data matching carried out by this project are summarized. The experiments cover the matching of the households, business establishment and others. Finally a short note on our programming system for the econometric analysis is followed.

1. INTRODUCTION

1.1 Objectives of the research projects

The recent development of economics in relation to information science is found in i) the bibliographic data base compilation and ii) the statistical data base compilation. The former aspect has been intensively examined by the another branch project organized by Professor Tairoku Kose. They compiled and analyzed the retrospective bibliographic data base for the publications from sixteenth century to twentieth century (1581-1977) of about 800 thousands records and also the current awareness search file based on the abstracts in the *Journal of Economic Literature* in 1975. The detailed report of their project in English will be published in another occasion. Here we will concentrate in the latter topics.

This research project has been organized by Professor Yoshimasa Kurabayashi. The emphasis was laid on i) the survey of the present situations of the statistical data base compilation in Japan, ii) the experiments of compiling micro data sets of current and historical economic systems and iii) the developing the application systems suitable to utilze such data base systems.

1.2 Macro econometric models and their data

The statistical data itself has occupied the prime importance in the analysis of macro econometric model building but this kind of analysis started a little bit earlier than the development of electronic computer, resulting in the neglect of the file construction for the common use or the data base compilation. The expansion of the size of the simultaneous macro econometric models, however, has required the systematic data file construction. Thus the new frontier of the information science has formed in the fields economics as an indispensable part of econometric model building.

On the other hand from the side of the data suppliers the schematization of statistical systems becomes necessary in order to get the well consistent data through statistical surveys. Thus to compile a well integrated data base becomes an urgent business for them.

The new SNA or Standard National Accounts system may be quite attractive for the model builders of macro econometric models but not so much attractive for the persons who are responsible to carry out statistical surveys. The actual statistical data system requires well articulated list of sampling and the well designed sampling. It must be admitted that the new SNA system contributes much to the unification and consistency of the concepts used but it is still necessary to compile a well consitent sampling frame for various sampling surveys. This kind of data base may be constructed on the philosophy of micro data sets compilation.

Before going further to the problem of micro data sets it is a good place to review the data base system in Japan.

1.3 The statistical data bank and the data base

Since nineteen-sixties the computerization of the econometric model building and sociological or political sceintific analysis using statistical data has attracted researchers' attention to the data bank. But soon they found it difficult to organize such data bank without having standardized data file system and so their interests have been shifted to the compilation techniques of the data base commonly shared.

The Bureau of Statistics of Prime Minister (hereafter abbreviated as BS) tried to organize a big statistical data bank system. The project, however, eventually retarded. They changed their policy to compile a data base system which will act as a frame of sampling such as households in the Population Census and the establishments in the Establishment Census and to create a document retrieval system for the statistical information.

The Economic Planning Agencies (abbreviated as EPA) has compiled a data base system for their own uses. Because they are to compile a SNA statistics and to estimate their econometric models for forcasting and planning.

Between these two extremes, one for conducting the statistical survey and the other for data reconstruction and application for the model building, the Ministry of International Trade and Industry (abb. as MITI) also has his own data base system for policy making and carrying out statistical surveys.

Many other ministries has their own data base system.

The recent point of issue is how to integrate these data base systems. But except for the data base by BS other data bases may be difficult to be integrated in one unified system because the decentralized system of collecting statistical information has stemmed from the begining of their statistical system formation in Meiji period or about one hundred years ago.

Other systems are organized by private institutions or semi-public organizations such as Japanese Economic Research Center supported by Nippon Keizai Shinbun-sha, economic newspaper publisher, which circulate NEEDS data base both on-line and batch systems and the Institute of Developing Economies which publishes various data bases on international foreign trade and inter-industry analysis of various developing countries for they have compiled them in collaboration with such countries.
But most of these data bank systems in Japan are policy orientated and thus their attention has been focussed in the current awareness services using mainly time series data for macro economic models. Thus the constructing the micro data sets and the retrospective data sets especially in the historical perspective has been left to the academic institutions. But on average they cannot afford to carry out such a project. Our project was started to fill this gap. But it is difficult to span every aspects of socio-economic data. We have been concentrated in the developing the micro data matching techniques. We expect that further extension of the data base may be realized by adopting the policy used by Inter-University Consortium for Political and Social Research (ICPSR) which consists of the data bases donated by the collaborating institutions.

2. MICRO-DATA MATCHING AND CREATION OF MICRO DATA SETS

2.1 Necessity and characteristics of micro data sets

Most of the data bases of economic time series data and also the census data are compiled by the vertical way. By the vertical compilation of data it is meant that statistical data are compiled and summarized into a number of classes according to a hierarchy of classification categories or attributes. It is quite natural by the formulation of the hierarchy of classification categories that it is typically represented by a tree structure composed by the classes of categories. Statistical data produced by the vertical compilation are displayed in a summary form of tables according to either one-way, two-ways or multi-ways classification. Although the consistencies between concepts used may be kept by introduction of new SNA standardization, the structure of data is rather collapsed into simplified tree structure in most of the Japanese data base systems. This stimulates the curiosity of researchers to compile a micro data sets.

The construction of the micro data bases differs from the traditional approach in several aspects.

First, the micro data bases refer to a set of statistical information for separate and distinguished agents, e.g. business establishments, government agencies, households and non-profit private institutions. This set of statistical information is amalgamated of different sources of statistical surveys,censuses and sample surveys. Second, this amalgamation and integration of different statistical sources is important to recognize the characteristics of the micro data sets. Statistical data for separate agents generated from different statistical sources are merged into the micro data bases by the help of the technique of data matching whose development is essentially owned to enlarging potentials of high-powered computers. Third, by the construction of the micro data bases the collection, processing, storage and retrieval of statistical information incorporated in them become much easier to be handled in the consistent and systematic computational processes. A series of operation on the micro data sets is particularly pertinent to obtain the statistical information which is cross-classified by multi-dimensional attributes. Fourth, it is worth noting that the multi-dimensional character of statistical data in the micro data sets has a great importance in the compilation of social data. The construction of the micro data sets is perhaps the most powerful apparatus for incorporating the social data into an integrated set of statistical information.

2.2 Micro data matching as compilation technique of integrated data base

There are two kinds of matching: exact matching (one to one correspondence of identical agency) and the statistical matching (one may be linked with another if they have shared the common attributes.)The former type of matching requires to show the identification code or name proper to that agencies. In most cases, the intelligence information utilizes this kind of information. But the statistical information need not be of this character. Thus the protection of personal data contained in each statistical survey straightly come to the point in this connection of the use of the technique of data matching. A great concern of our research in this category of topics is how and by what means the personal data compiled in a full-fledged data base be protected from the abuse and invasion of privacy.

The statistical matching needs no identification code or name proper to the enumeration unit. In stead of that the refinement of the categories of attriutes is indispensable. Among various separate and distinguishable agents which act as an enumeration unit of the statistical survey the household and the establishment are the most basic operationally defined concepts.

The statistical matching of various surveys should solve two basic problems. One is the linkage of the different surveys which may be regarded as surveyed at the same time and the other is the linkage over the surveyes carried out in different time. The attributes used for the linkage should satisfy these two requirements.

2.3 Business establishments data matching

Most of the statistical surveys on the business world depend on either establishment base or corporate base as an enumeration unit. Production statistics are in most cases surveyed by the establishment base but financial aspects are by corporate base. Thus the linkage of the statistics classified and surveyed by the establishment base with those of corporate base needs to compile a conversion table between establishment and corporate. But such compilation is practically quite difficult except the exact matching.

This kind of exact matching or uniting the different establishments into one owner or corporate has been tried to the establishment of all kind of industries by BS based on the Census of Establishments carried out every three years. Another matching of this kind is the yearly Manufacturing Census by MITI.

The matching of BS covers only the data of the categories of industries and the size of establishment by employee. While the MITI has used much wider informations such as size of employee and the capital which are linked with production data like amounts of products and value added.

1975 supplies us both BS matching results and MITI matching results. If we compare these results in the manufacturing industries, unmatched establishments in BS matching amount to 17.21 % but the figures of unmatched are not reported in the MITI matching. These unmatched establishments are concentrated in the smaller size or below 100 employee. The discrepancy ratio between the BS and the MITI is 10.18 % for below 20 employee. This may be attributed to the cutting off the tail of the distribution below the 20 employee in the case of MITI. In the case of over 5,000 employee the discrepancy becomes -25.34 %. This may be attributed on the classification principle of the establishments in the broad industrial categories. The MITI covers all the manufacturing establishments but the BS classified many to other industrial categories like commerce, transportations etc.

This fact finding implies the difficulties of linking other surveys which covers all industries with the Manufacturing Census by MITI. For the former may try to classify many establishments which are classified in manufacturing in MITI in other industries.

But the MITI matching is quite important because it will provide the conversion table between capital size and employee size because recent development of huge plant with smaller employee like the chemical plant has brought out the inadequacy of classification by the size of employee. Thus the classification by the capital size may gain the practical importance in compiling statitical data base.

The intertemporal comparison, however,is quite difficult in the case of capital size due to the inflation processes. Capital is defined by money term and so may not be comparable to the real term by employee.

2.4 Households data matching

Experiments are carried out in two aspects. One is to create a new data base by matching two different statistical surveys and the other is to make an intertemporal comparison.

Merging two surveys is tried through the matching of BS's Family Income and Expenditure Survey with BS's Basic Survey of Social and Community Life. They are abbreviated as FIES and SSCL.

SSCL contains abundant information for individual households and their members concerning their social and community life, encompassing in its scope such activities as domestic life, living for well-being, learning, earning and leisure, on the one hand. The detailed information on receipts and expenditure for urban households including transfer income and payments and some financial transactions, on the other hand, is available from FIES sources on monthly basis. It is quite interesting to construct a data base on the basis of separate households unit so that social data obtainable from SSCL may be merged into economic data derivable from FIES. Aforementioned experimental work refers to the statistical matching of SSCL with FIES data. The experiment is only preliminary to the extent that samples collected for the experiment are limited to those for Tokyo Metropolitan area, drawing approximately 1,000 samples from SSCL and 5000 samples from FIES.

This kind of matching is based on the various attributes of the households. But in most of the previous published statistics are mainly classified by the income class. Another experiment is for the intertemporal comparison of the distribution of the consumer durables ownership by income class. In this case the comparison between each survey of FIES conducted for every five year is tried through the computation of the Gini coefficients based on the decile income class data.

Even if the relative position of income class of each households may not be changed, the absolute income class varies due to the inflation. The decile data are used to avoid such influence of inflation.

The linkage of the yearly results through defining the quasi-stock which is derived from the remaining amounts of the consumer durables purchased in the previous years and still in use is quite satisfactory except for the discrepancies caused by the change of definition of income from monthly income to yearly income.

Another experiments are done for matching the data of Census of Agriculture and Forestly with those of Farm Household Economy Survey. This is also experimental; the samples drawn for the experiment are restricted in a particular county district near Asahikawa, Hokkaido. The final aim of this creating a data base is to construct a consolidated file which will contain the information concerning the management of production activities of farm households and their family composition based on the methodology of exact matching. The identification of a household is made according to i) the composition of family members , ii) the size and variety of management and iii) finaly the names of family members in Chinese characters.

3. RECONSTRUCTED SURVEY USING MICRO DATA MATCHING

3.1 Historical data base construction

Some words should be added to the project of creating the historical data base. As we have mentioned in the previous section to construct a historical data base by a single research project team is quite difficult task. In stead of that we prepared a core data base based on the *Estimates of Long-Term Economic Statistics of Japan since 1868* (LTES estimates) by Kazushi Ohkawa, Miyohei Shinohara and Mataji Umemura in 9 volumes. (The whole series will amounts to 13 volumes.)

This LTES data base will be regarded as macro economic data base. The basic philosophy of the estimation techniques is to inflate or deflate the basic time series data with appropriate ratio derived from the cross-section data of certain bench mark year.

The statistical surveys which will supply cross sectional informations are quite scarce before the world war second. Thus to improve the degree of accuracies should be realized through finding a new statistical source books hetherto unknown. But such a case must be a rather exceptional one. In stead of that we have tried to reconstruct a survey whose tabulation schemes were not yet well developed like those of later periods but contained the information of the codes or names proper for identification.

3.2 Reconstructed survey

It is possible to reconstruct a survey using the data published in one table formats which contains the proper name and their attributes like a list of factories or companies. In Japan we do not have a statistics of corporates and establishments in 1880s.

We have tried to construct a data base of the whole corporates combined with their factories using the list data found in the Who's Who directories.

The techniques for compiling such reconstructed survey data is the exact matching using the input of the Chinese characters.

4. PROGRAMMING SYSTEM FOR THE ECONOMETRIC STUDIES

The research activities under this heading are primarily taken up by Professors Ippei Sugiura and Kei Mori and their research groups. The interests of their research are directed to the development of those packaged programms which are particularly useful for econometric studies, forecasting and simulation.

Mori has developed a programming system called EMS (Econometric Modeling System) placing special emphasis on heuristic aspects of finding suitable econometric models from the point of estimation and simulation based on the data base included in their system. Sugiura's system is called as STEPS FOIL, an abbreviation of Simplified Techniques for Econometric Plannings and Simulations Function Orientated Interpretative Languages.

REFERENCES

[1] KURABAYASHI, Yoshimasa. *Studies in National Economic Accounting*. Tokyo, Kinokuniya Book Store, 1977.

[2] KOSE, Tairoku & KURABAYASHI, Yoshimasa (eds.) *Shakaikagaku ni okeru Johokensaku System no Genjo. [The Current of Information Systems for Social Sciences in Japan.]* Tokyo, Nippon-hyoronsha, 1978. (in Japanese)

[3] KURABAYASHI, Yoshimasa & MATSUDA, Yoshiro. "Ensogeijutsu no Juyo to Kyoku." [Demand for and Supply of Performing Arts.] *Hitotsubashi Ronso[the Hitotsubashi Review]* 79(1), 1978. (in Japanese)

[4] KURABAYASHI, Yoshimasa & MATSUDA, Yoshiro. "Service toshite no Ensogeijutsu to Orchestra no Choshu." [An Experiment of Compiling Synthetic Audience Data File for Orchestra Audience in Japan.] *Hitotsubashi Ronso[the Hitotsubashi Review]* 82(3), 1979. (in Japanese)

[5] MATSUDA, Yoshiro. *Data no Riron -- Tokeichosa no Data Kozo no Rekishiteki Tenkai --. [Treatise on Data -- Development of Data Structure in the Statistical Survey.]* Tokyo, Iwanami-shoten, 1978. (in Japanese)

[6] MATSUDA, Yoshiro & TERASAKI, Yasuhiro. "Economic Development and the Distribution of the Consumer Durables Ownership; Japan and Taiwan, 1954-1975." *Phillipine Economic Journal*. (1980, accepted to be published.)

[7] MIZOGUCHI, Toshiyuki. *Data base for the Studies on Income and Expenditure Distribution, an Example for the FIES Data*. Tokyo, Hitotsubashi University, 1978-1979. 2 vols.

Scientific Information Systems in Japan
H. Inose (ed.)

INFOLOGICAL APPROACH TO RESEARCH TREND ANALYSIS†

Hajime TANAKA and Michiyo NIKKUNI***

** Faculty of Science, Hokkaido University, Nishi 8-Chome, Kita 10-Jo, Kita-Ku, Sapporo*
*** Center for Information Processing Education, Hokkaido University, Nishi 5-Chome, Kita 10-Jo, Kita-Ku, Sapporo*

Categories exist by which sentences in documents are objectively classified. SCAT-IR system is produced where each sentence is retrieved with the scientific terms and the classification categories. The procedure of the clusterization for a set of documents is proposed. The useful information for grasping the research trend is found out with the mechanical procedure for it. The algorithms for advanced scientific information are presented.

1. INTRODUCTION

Infological research regarding scientific information hitherto has been mainly carried out on collections, store and utilization of numeric data, symbolic data and documents. In the case of scientific documents, it is seldom that work was extended to include the conceptional content of the document, and in the majority of cases, bibliography and key words in the title and the abstract alone were dealt with infologically.

However, in the course of scientific studies the information produced is highly diversified in addition to numerical data and symbolic data appearing in the documents. Semantic information expressed by a single sentence and research trends in various scientific fields, in other words advanced information are numerously produced as well as usual one. The research trend is produced after a long time through the course of research where the fundamental problem in each branch of science has been solved.

In this paper new attempts are developed for retrieving the content itself of documents. Particularly retrieval method of sentences included in each document is our most important proposal. The sentence retrieval is realized only when every sentence in each document may be objectively classified by any way. It is shown that the objective classification of sentences in documents is possible by adding information on the relationship of the sentences and the function thereof against the entire context. The twelve classification categories are introduced for that and the classification information of each sentences is chosen from twelve categories. Really each sentence is retrieved by two kinds of information, that is, above classification category and several key words which can be pointed out arbitrarily from all scientific terms in each branch of science. The retrieval system is produced and is called "SCAT-IR".

†) This is the co-work of the sentence retrieval group of which the members are Prof. Yoshinori Akaishi (Hokkaido Univ.), Prof. Jun Hiura (Iwate Univ.), Prof. Hiroharu Bandō (Fukui Univ.), Prof. Ryozo Tamagaki (Kyoto Univ.), Prof. Shinobu Nagata (Miyazaki Univ.) and the authors. The classification experiment of the sentence is also performed by Mrs. Yōko Kawai (Kyoto Univ.) and her group in the branch of solid physics.

Several algorithms can be shown for advanced scientific information, for example, the grasping of research trends and institution of research problems by using the above retrieval system and the information of documents citation. These algorithms are not extravagant but rather usual ones because they are brought just by replacement of several manual processes in the research course with mechanical ones where thinking steps in the research course are kept unchangeable. In other words several labored works finding out the necessary opinion from many documents are replaced with mechanical information processing and are connected with the usual thinking steps into whole courses of grasping research trends and instituting research problems. It may be said that the information processing can be extended to new regions, that is, to advanced scientific information.

2. CLASSIFICATION OF SENTENCES AND RETRIEVAL SYSTEM

2.1 Classification Category

Every sentence starts next with a symbol 'blank' and ends at a 'period'. They play a role in each document in their own way like as a 'title' expresses content of a document with a short phrase.

The following 12 categories will be introduced to classify individual sentences which make up the document. Since the classification categories are to be used for classification of information in the document, it is also necessary to make classification in accordance with the writer's way of handling of individual sentences. Each category will be described in the following with its number expression called category number.

General views (01) are defined as, views held by the author which are generally accepted in that given field, views that encompass other fields beyond the confines of the author's field or wide concepts which are dealt with the author.

Basic problems (02) are defined as problems which are to be solved over a considerable number of years in that field. Among such problems some require many years while some may be solved in a comparatively short period. Both types are written above will be dealt with as Basic problems.

Awareness of problems (03). Problems are taken up by

the author from a singular angle belonging to the author alone which the author attempts to solve. The angle is called Awareness of problems. Starting from the Basic problem, the sentences related to the course of formation of the Awareness of problems of the author may be divided into Research trend and Awareness of problems. When the Basic problem, Research trend and Awareness of problems can not be clearly differentiated, the classification should be made as Awareness of problems.

Analysis of the status quo (04). This deals with the course which leads to setting up of individual problems after starting from the Awareness of problems. The investigation in this course is the author's Analysis of the status quo of the problem. In this analysis, it is clear that recognition of the status quo of research related to the Awareness of problems is necessary. Thus recognition of the status quo will be included in the Analysis of the status quo.

Moment of development (Idea) (05). With Awareness of problems alone development of research cannot be expected, but a key or keys to open up a new means may be found, or in some cases certain points discovered previously may be recognized to be applicable to the problem. Thus the development of the problem may be made and led forth using such incidences as the turning point. We call such turning points Moments of development or Ideas. When a previously discovered fact is recognized to make the development of the problem possible, we may also call this a Moment of development or an Idea.

Individual problems (06). This is a research problem performed actually in the paper. To discover a turning point which may lead to the development of the problem is a non consecutive course. This is not always necessary. There are times when it is possible to set up problems with the continuity of numerous research. In such cases turning points which make the development of research problems possible, may have already been found as a result of other work. All of the above cases may be classified as individual problems. When it is difficult to classify the sentences as to whether they are Analysis of the status quo or Individual problems, they should be classified as Analysis of the status quo.

Development (07). After producing individual problems, the course of development to the results obtained by the above are referred to as Development. In actual documents, this portion becomes massive. Numerous facts and views are taken in this course and all of these may come under Development.

Results (08). Facts and graphs etc. obtained as a result of development may be classified as Results. Explanations of results not containing evaluation, may be classified as Results. Discussions of results involving evaluation should be classified under the following Evaluation. When differentiation is not clear such items should be included in Evaluation.

Evaluation (09) and Conclusions (10). When the results are viewed in the light of Awareness of problems or Individual problems, this course in its entirety is called Evaluation, and the results of such evaluations are called Conclusions. Since Conclusions are led forth in reply to Awareness of problems, in most cases it takes the form of conceptional statements. A repetition of Results is not Conclusion. When it is not clear whether a sentence should be included in Conclusions, this should be classified as Evaluation.

Awareness of problems following next (11). That which is led forth from Results and Conclusions is the Awareness of problems that follow.

2.2 Classification Experiment

Next the classification categories were used for classification experiments, and from the results, attempts were made to find whether or not sentence classification has sufficient objectivity for the above sentence classification categories.

The documents selected as the subject matter, were mainly short documents (letters or short notes) on theoretical research of nuclear physics and solid state physics.

The experiment was conducted using 3 groups of classifiers. Group I is the group which set up the sentence classification categories. Group II composed of graduate students majoring in nuclear physics. Group III is made up of a group of theoretical and experimental scholors in the field of solid state physics.

Classification experiments in each group were made roughly in the following manner. First documents were selected at random. The sentences of the documents were consecutively numbered. After each sentence was numbered, classification of the sentences was conducted independently, using the classification categories, then discussions were held in each group to determine a classification standard solution. When the classification results coincided in each group their classification was taken as the standard solution. When the results differed discussions were held. The discussions here were centered around understanding of the sentence classification category and the usage of the categories. Discussions were continued until an agreement was reached. The resulting time required for each document was 1-2 hours or thereabout.

Tables 2.1 and 2.3 show the results of Group I and tables 2.2 and 2.4 show the results of Group II. In tables 2.1 and 2.2 the classification results of each member and the number coinciding with the standard solution are listed, here A∿F are classified in each group. In tables 2.3 and 2.4 the rate of coincidences of the classification results and standard solution are shown in %. The order of the experiment is expressed by documents 1-6.

Table 2.5 shows a comparison of the results of Group I and Group III. This shows the experimental results after the understanding of the categories become common among the members. In spite of the difference in fields, namely experimental and theoretical workers in the field of solid state physics, against theoretical workers in the field of nuclear physics, the results of the average coincidence rates were astounding. In addition to that expressed in table 2.1 through table 2.5, sentence classification has been done on approximately 1116 documents. The results shown in tables 2.3, 2.4 and 2.5, indicate that the mean of the rate of coincidence with the standard solution shows a rapid rise as the experiment members become well aquainted with the method.

classifier / category NO.	A	B	C	D	E	F	standard solution
01			1	1	1		
02							
03							
04	1-4	1-4,9	2-4	2-4	2-4	1-4	1-4
05	16			5,16	5	5	
06	5	5	5	5	5	5	5
07	6-15 16-24	6-8,10-19 20-24	6-24	6-15 17-24	6-8,10-19 20-24	6-15 16-24,30	6-24
08	25,28	25,28	25	5,25,28	25,28	25,28	25,28
09	26,27,29 30	26,27 29,30	26-30	26-30	9,26,27 29,30	26,27,29	26,27 29,30
10	31	31	5,31	31	31	31	31
11							
rate of agreement	31/31	30/31	28/31	29/31	29/31	30/31	

Table 2.1 Frosch, R.: Shell Model Analysis of Elastic $\bar{e}$-^{4}He Scattering, Phys. Letters. 37B, 140 (1971).

classifier / category NO.	A	B	C	D	E	F	standard solution
01	1						
02							
03		5					
04	2-4	1-4,6-9	1-5	1-4 6-9'	1-4	1-4	1-4
05							
06	5		5	5	5	5	5
07	6-24	7,9,10-24	6-24	10-24	6-24	6-24	6-24
08	5,25,28 30	25,27,28 30	25,27 28	25,27 28	25,28 30	25,27 27	5 25,27,28,30
09	26,27 29	26,29	26,29 30	26,29 30	26,27 29	26,27 29,30	26,29
10	31	31	31	31	31	31	31
11							
rate of agreement	30/31	28/31	30/31	26/31	30/31	30/31	

Table 2.2 Frosch, R.: Shell Model Analysis of Elastic $\bar{e}$-^{4}He Scattering, Phys. Letters, 37B, 140 (1971).

As may be seen in tables 2.3 and 2.4, at the beginning of the experiments both the definitions of categories and their applications were a little hazy and the mean rate of coincidence was not necessarily high; later when the definitions of the categories became clear and the mode of application became established, a rapid rise in the rate of coincidence became apparent. Especially the coincidences with the standard solutions in tables 2.1 and 2.2 are surprisingly very close. Although Group I, II and III were composed respectively of members of different positions in different fields of research, the mean rate of coincidence showed an astonishing resemblance.

Now it can be concluded that our sentence classification categories give an objective sentence classification. The surprising result that the above

classifier / document consecutive NO.	A	B	C	D	E	F	average
1	93	74	81	78	78	52	76
2	100	81	70	78	100	67	83
3	96	86	80	84	91	84	87
4	57	60	67	67	67	80	66
5	78	84	-	88	81	88	84
6	100	97	90	94	94	97	95

Table 2.3 Rate of agreement by Group I. (%)

classifier / document consecutive NO.	A	B	C	D	E	F	average
4	72	66	83	83	72	72	75
5	73	90	73	80	80	83	80
6	94	88	94	81	94	94	91

Table 2.4 Rate of agreement by Group II. (%) (Graduate students of Hokkaido University)

Group I (%)	Group III (%)
66	73
84	83
95	94

Table 2.5 Comparative table between Group I and Group III.

category system has sufficient objectivity for the sentence classification may be originated to the fact that the categories are set up on the bases of stepwise characteristic of a research course.

Regarding the point of the present study on the course of scientific research, the course of research may be expressed as shown in Fig. 2.1. This course of study will be dealt with briefly but detailed description will not be made here. As to the individual categories, explanations will be made in §2.1. As shown in Fig. 2.1 the circle course depicted by a solid line is considered as a unit, and is on a scale which can be reported as one document. In Fig. 2.1, two research courses of such units are depicted. As a result of repetition of these individual courses, the basic problems are gradually solved.

Each function of the sentence in a document seems to correspond to each characteristic step of a research course. If the view is true, it can be said that the objectivity of sentence classification has its base on the structure of a research course.

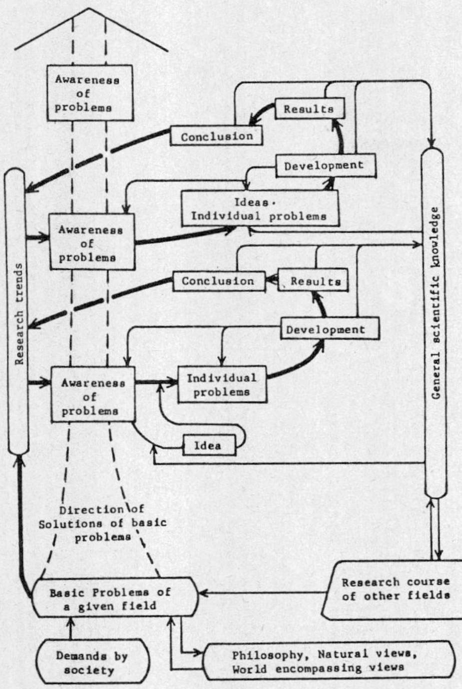

Fig. 2.1 An outline of research courses.

2.3 Sentence retrieval

Retrieval of the sentence is performable only by adding relevant information to all sentences before contents of the documents are stored. For that the classification categories and the scientific terms are used in our retrieval system. Of course the bibliographic information is available if necessary. Every sentence in the documents can be retrieved by two kinds of retrieval information, that is, the classification category, which shows the role of each sentence in the content in the document and the scientific terms which express the individual scientific content included in each sentence. The input-retrieval system called SCAT-IR is produced and all contents of 1116 documents are stored at present. The structure of SCAT-IR system is described in the 5th section as well as the application programs added to SCAT-IR.

Fig. 2.2 shows an example of retrieval by the usage of SCAT-IR. The set of documents is named as PROB34 which is defined so as to satisfy the following condition. The condition is given that the documents should include at least one sentence stating the statement, which is expressed by equating the set-definition statement DEF to the logical statement given by a logical product of scientific information following STW and a classification following to CAT. In this example the scientific information says that the document set should be made by the documents about the nuclei of three or four nucleons. A set of documents can be defined by bibliographic information, for example publishment year like as DEF YEAR70 =(YER=70∿); in Fig. 2.2. After the definition of the set of documents, the necessary output informations are given, that is, name of journal (JNL), volume (VOL), publishment year (YER), page (PAG), author (ATH) and title (TIL) in addition to the category numbers of the output sentences (SNT (02 06)). As already written, 02 and 06 mean the categories of Fundamental problem and Individual problem, respectively. The sentences satisfying these retrieval condition are outputed with each bibliographic information in Fig. 2.2. The set of documents can be defined at disposal as many statements of the set calculation are readily available in SCAT-IR.

```
DEF PROB34=(STW=(THREE+FOUR+ALPHA)*(BODY+PARTIC+NUCLEON))
        *(CAT=01∿03);
PRT (PROB34) CNT;
** PROB34 ** HITCOUNT=175
DEF YEAR70=(YER=70∿);
DEF RECENT=PROB34*YEAR70;
PRT (RECENT) CNT CSN SNT(02 06) JNL VOL YER PAG ATH TIL;
** RECENT ** HITCOUNT=79
 CSN=00111
 SENTENCE OF CAT NO=02
 SENTENCE NO=0001
   IN A RECENT LETTER PASK 1> HAS COMMENTED ON THE ROLE OF
   NON-CENTRAL INTERACTIONS IN LOW-ENERGY NUCLEAR PHYSICS,
   ESPECIALLY IN THE THREE-NUCLEON SYSTEMS.
 SENTENCE OF CAT NO=06
 SENTENCE NO=0013
   HOWEVER, THE PURPOSE OF THIS NOTE IS TO FIND A REASON
   FOR THE UNEXPECTED RESULTS FOR THE SQUARE WELL.
 JOURNAL=PHYS.LETTERS
 VOLUME=B 34
 YEAR=1971
 PAGE=184
 AUTHOR=R. VAN WAGENINGEN AND G. ERENS* (NATUURKUNDIG
        LABORATORIUM DER VRIJE UNIVERSITEIT, AMSTERDAM,
        THE NETHERLANDS)
 TITLE=SQUARE-WELL POTENTIAL AND THREE-NUCLEON BOUND-
       STATE PROPERTIES
```

Fig. 2.2 An example of retrieval by SCAT-IR.

3. CITED NUMBER AND CITED RATE

The next problem is how researchers with the intent of grasping research trend systematize effective information. A new method is given in this section by introducing cited number and cited rate. It is attempted for studies of three-body force in nuclear physics.

A total of 42 documents on nuclear three-body force published in main journals of nuclear physics from 1956 to 1977 were taken as examples, and document number n is given to each document in the order of publication date. The document n can be cited by documents which will be published later than it. It may be considered that set of documents L(n) is a collection of documents in principle possible for citing the document n. L(n) expresses a number of documents in the document set L(n) also. C(n) is a number of documents in L(n) actually citing the document n. C(n) is called 'cited number' and the ratio

R(n)=C(n)/L(n) 'cited rate'.

In Fig. 3.1 cited rate R(n) is taken as the abscissa and the year when each document is published is taken as the ordinate. In Fig. 3.1, 3 peaks appear in the graph. These peaks indicate that documents which are cited frequently, or documents which have a higher rate of citation, were published on 3 occasions. Such documents appear to give a new trend in the development of that research field. Therefore, Fig. 3.1 shows some research trends of three-body force problem and if the abstract of the document which is indicated by the position of the peak is outputed, such abstract may provide valuable information in finding the research trend of three-body force problem.

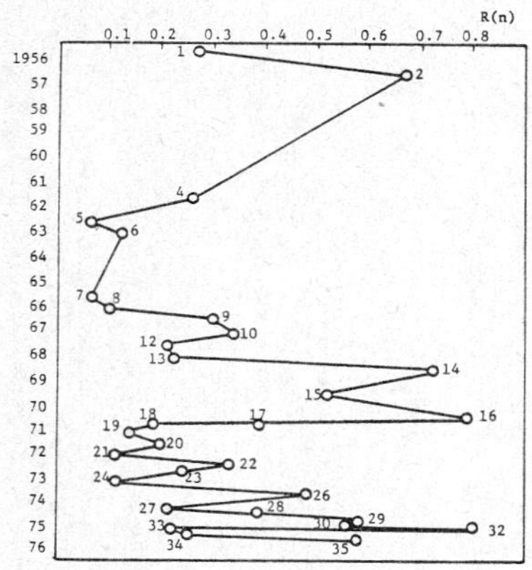

Fig. 3.1 Cited rate in the three body force problem

It should be remembered here that the process is mechanical in which the cited number C(n) and cited rate R(n) is found, the peak is determined and the abstract or necessary sentences of the document constituting the peak is outputed. This mechanical process is readily possible for SCAT-IR.

4. CLUSTERIZATION FOR SET OF DOCUMENT

The research theme of three-body force which was taken in the previous section is a comparatively independent field in nuclear physics and how to grasp the research trend of a basic problem may be typically shown in this example. Generally many research themes exist which are concerned with the different objects or different research methods. Therefore the method of the previous section may be applicable just when the mixture of many themes are clusterized into several independent or relatively independent problem. The set of documents may be roughly divided into several smaller sets when there are relatively independent problems. In this case each document cites many documents which belong to same subset as the document but total cited number between different subset of documents are not many. Inversely the clusterization procedure can be performed in various ways by the usage of citation relation and it should be noted that the resultant subset of documents corresponds to the research theme characterized by its object or its method.

Therefore it may be natural that a subset of documents or a cluster of documents is called a 'problem block' which means a problematic character appeared in a cluster of documents as well as an usual subset of documents.

Research activities may be represented by modeling them on the motion of many problem blocks, where many problem blocks interact with each other, a new problem block is produced and an old one disappears sometimes in a long research course. Shortly speaking the research trend may be grasped through the motion of a problem block.

Generally a clusterization procedure of documents can be performed on the bases of the strength of the relation between two individual documents. The citation relation is an easily manageable one for that but is not a unique relation. For example, we can use a number of common scientific words appearing in two sentences, which belong to different documents, with each other though two sentences are classified with the same classification categories. Here it is limited to the usage of the citation relation.

Kitagawa[1], Asano[2], Okino and Saito[3] proposed their methods to clusterize a set of documents by the use of the citation relation and obtained many interesting results. Also Yazima et al. introduced the manner to show the citation relation of many documents graphically. Here a simple procedure for the clusterization of documents will be shown. First a cited number c is counted for each document which belongs to the set L of documents. The documents which have a cited number over a critical value C_0 is named 'boss'. The value C_0 may be chosen suitably. The document which cites the boss is called 'follower'. Generally every document cites plural bosses and so arbitrary two bosses own one or several followers jointly in many cases. When a number of jointly owned followers is over a critical value C_1, two bosses and their followers are grouped into one cluster. If all bosses are indepedent of each other, that is, they do not own jointly any follower, each boss and its followers constitute each cluster.

For example the above method of clusterization is applied to 178 documents which are concerned with a few body problem in nuclear physics and make a subset produced by the adaptable retrieval condition from whole set of documents. The value C_0 is taken 15 times and it produces 15 bosses. C_1 is taken 10 times. The bosses of which C_1 is less than 10, are grouped in the same cluster when those Awareness of problems are similar. This procedure is not mechanically performed for 4 bosses by looking over the sentences of Awareness of problems. Fig. 4.1 shows the result where three clusters appear. There exist several documents which cite 2 or 3 bosses belonging to a different cluster of each other and so each pair or three clusters overlap with not so many documents. Some modification is applied to get the final result. A follower which cites 2 or 3 bosses in different clusters simultaneously, comes to be classified into the cluster in which documents cite the follower more than those in the other cluster. The final result is shown in Fig. 4.2, where the documents which belong to 2 or 3 clusters decrease rather than in Fig. 4.1. The resultant three clusters come to have a character on the physical problem and are named VAR, DIS and REAL, respectively. VAR means that each document of the cluster VAR is concerning with the variational method in quantum mechanics. Similarly DIS means a direct solution or some method to solve a physical system directly. REAL is a subset of documents where the realistic nuclear potential is used. Three problem blocks are derived by our method through the clusterization of 178 documents. Our method may be called 'territory procedure'. What results will be obtained when each cited rate of the documents in 3 clusters, VAR, DIS and REAL is plotted to the publishment year, as like section 3.2. Fig. 4.3 shows the cited number-publishment year curves for three clusters, where the publishment years of the cited documents are limited to 1958∿1964 but the citing documents were published from 1958 to 1976. The real line expresses the development of the whole research activity in a few body problem. However each peak is not always brought by the same cluster or the same problem block. The peak of 1958 indicates the activity of the problem block VAR and the one of 1962 indicates the development of the problem block DIS.

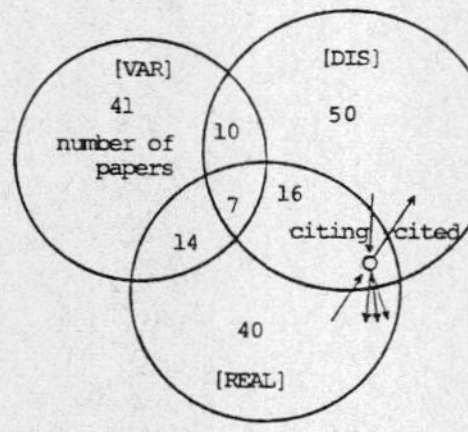

Fig. 4.1 The procedure of clustrization by territory method

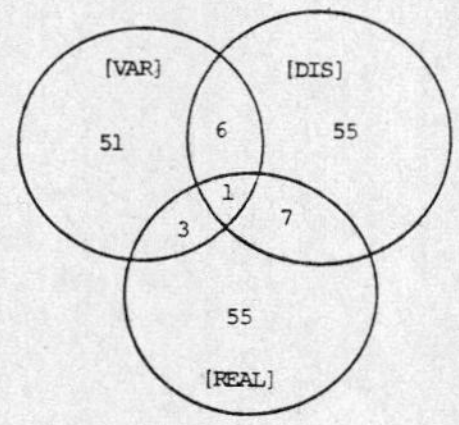

Fig. 4.2 The result of clusterization by territory method

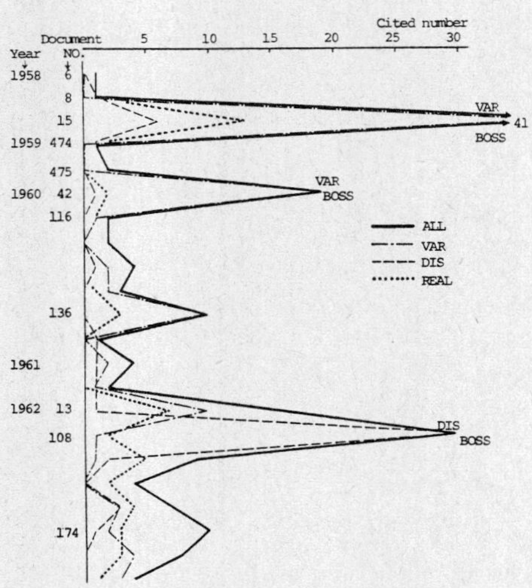

Fig. 4.3 Spectra-analysis

The method by which the development on the fundamental problem is analyzed through the clusterization of documents set and the cited rate-publishment year relation, may be called 'spectra analysis'.

SCAT-IR has application programs for the above procedures as well as the function of the sentence retrieval. In the next section the details on the architecture of SCAT-IR are described.

5. THE ARCHITECTURE OF SCAT-IR SYSTEM

5.1 Input Data

A document consists of "Bibliography", "Title", "Author", "Abstract", "Text" that consists of 'Sentences' given '12 Category Numbers' as shown in §2, "Reference" and "Comment". The ones shown by " " are called items of a document and the ones by ' ' are called sub-items. Individual information belonging to them is called an item occurrence. In a document every item occurrence varies greatly in length. In addition, the number of the occurrences belonging to each item is variable. For example, the length of a sentence is variable and the number of sentences belonging to a certain category is variable.

It is possible to distinguish such occurrences in a document by inserting some sort of delimiters as a sentence is done by a period. For example, the end of the occurrence of variable length can be given by a delimiter. In addition, suppose '/' is an item delimiter and '*' is an occurrence delimiter, a multi-occurrence can be shown by /abc*ahge*k*/ and a null-occurrence can be shown by //.

Here, an input data stream of a document to store is given as follows.

> Journal△volume△year△page/title/author$_1$..author$_2$../abstract/02△sentence$_1$..05△sentence$_2$..·····10△sentence$_n$../refno$_1$△reference$_1$..refno$_2$△reference$_2$..·····/comment$_1$..·····comment$_m$..&&

'/' is an item delimiter. '△' and '..' are sub-item delimiters. '&&' means the end of a data stream of a document as well as an item delimiter.

5.2 "Normalized Table" and "Mapping Language"

The data model of SCAT-IR is called "normalized table", which can represent multi-occurrences and null-occurrences. As the consideration of updating is unnecessary in a document, it can be generated by very simple normalization rules as follows.

1) To make a table where each item of a document is put in a row as Fig. 5.1. Fig. 5.1 shows a part of a paper.
2) To classify the items into A group having only one occurrence and B group having multi- or null-occurrences. In Fig. 5.1, A={JNL, VOL, YEAR, PAGE}, B={CAT$_1$, ···, CAT$_{12}$, REF}.
3) To classify the items of B by the type of the item occurrences. B is divided into B'={CAT$_1$, ···, CAT$_{12}$} and B"={REF}. The type of item occurrences of B' is a sentence and that of B" is a reference.
4) To give a name to the set of the items of each group classified in 3) and to the set of the item occurrences. The names given become new item names and the elements of both the sets become new

item occurrences. In B', {CAT_1, CAT_2, ·····, CAT_{12}}=CAT; {S_{ij}}=SENT. In B", {r_{ij}}=REF. CAT, SENT and REF are new item names and S_{ij} and r_{ij} are the item occurrences.

5) To make a table every group; A, B' and B" respectively. And to add an item for a document identifier. If necessary, to add an item for the identifier to each item occurrence such as a sentence number. Consequently three normalized tables are generated as shown in Fig. 5.2.

ID	JNL	VOL	YEAR	PAGE	CAT 01	CAT 02	CAT 03	---	CAT 12	REF
1	a	54	1976	125	s_{11}		s_{12}		s_{1n}	r_{11}
							s_{13}			r_{12}
										r_{13}
2	b	3	1975	64		s_{21}	s_{23}		s_{2m}	r_{21}
						s_{22}				r_{22}

Fig. 5.1 Table representation of a paper

ID	CAT	SENT
1	01	s_{11}
1	03	s_{12}
1	03	s_{13}
⋮	⋮	⋮
1	12	s_{1n}
2	02	s_{21}
2	02	s_{22}
2	03	s_{23}
⋮	⋮	⋮
2	12	s_{2m}

ID	JNL	VOL	YEAR	PAGE
1	a	54	1976	125
2	b	3	1975	64

ID	REF
1	r_{11}
1	r_{12}
1	r_{13}
2	r_{21}
2	r_{22}

Fig. 5.2 Normalized tables

A mapping language is devised to map input data streams into normalized tables. It interprets the delimiters inserted in input data streams. The syntax of the mapping language is shown in Fig. 5.3.

```
<delimiter>::= a character | some characters
<table delimiter>::='<delimiter>'|"<delimiter>"*1)
<item name>,<table name>::= a character string including English
                            letters, figures or symbols
<table element>::=<item name>|<item name><<delimiter>>|
                  <item name><<delimiter>>(length of fixed length data)
                  |<item name>(#(length) by <<delimiter>>)*2)|
                  <item name>(*)*3)
<repeating group>::=[<table element>,<table element>,······]
Mapping Language => <table name>=<table element>,······,'<delimiter>',
                    <table name>=<repeating group>,'<delimiter>',·····,
                     "<delimiter>"
*1) " " means the end of a data stream.
*2) (#(length) by <<delimiter>>) means that a consecutive number is
    generated every the appearence of the appointed delimiter.
*3) (*) means the item occurrence appeared or generated before.
```

Fig. 5.3 The syntax of a mapping language

The characteristics are as follows.

1) A repeating group, where the occurrences of some items are repeated as one group, can be dealt with.
2) Consecutive numbers can be generated to individual item occurrences.
3) It is possible to modify a delimiter when it appears as a character in a document.
4) A mapping language plays the role of data model definition, which can be derived by removing delimiters from it.

The mapping language of the input data stream shown in §5.1 is given as follows. P-ID means a paper identifier.

```
BIB=P-ID(#(4) by <&&>), JNL(△), VOL(△), YEAR(△),
PAGE'/', TIL=P-ID(*), AUTHOR=[P-ID(*),ATH<··>] '/',
ABSTRACT=P-ID(*), ABST '/', TEXT=[P-ID(*),CATNO<△>
(2), SENTNO(#(4) by <··>), SENTENCE<··>] '/',
REFERENCE=[P-ID(*), REFNO(*), REF<··>] '/', COM-
MENT=[P-ID(*), COM<··>] ''&&''
```

According to this, 7 normalized tables (BIB, TIL, AUTHOR, ABSTRACT, TEXT, REFERENCE, COMMENT) are generated.

5.3 Dynamic Generation

"Individual Problems" and "Conclusion" of cited papers can not be retrieved directly from normalized tables generated in §5.2, for such information is not contained in the references of a paper. However, it is possible to retrieve P-ID by matching the bibliography of the BIB table and that of the REFERENCE table. Consequently, a table, which consists of P-IDs of citing papers and P-IDs of cited papers, can be generated. As it is possible to retrieve cited P-IDs from this table, their "Individual Problems" and "Conclusion" can be retrieved easily from the TEXT table. In addition, it can support various retrievals in connection with the citation relation of papers. Thus when such a requirement can not be retrieved directly, to generate a table as mentioned above, it is called "Dynamic Generation". This table is deleted whenever it becomes unnecessary. By the device of "Dynamic Generation", it became unnecessary to consider the future requirements when a data base is designed. Normalization rules described in §5.2 are established without such considerations.

5.4 Retrieval

Every item occurrence can be regarded as a character string. For example, a sentence and a term given for retrieval can be regarded as a string having some words and some spaces. The basic retrieval method of SCAT-IR is a pattern matching as follows.

```
an item occurrence→·····abc△def△three△body△klb ···
a given term        →           three△body
   ( △ means a space.)
```

By the pattern matching, it is possible to retrieve sentences by not only words but also parts of a word, phrase, a compound word, or an arbitrary character string.

In document information, it must be considered to support mechanically a misspelling, an inflection and the different representation such as 'three body' and 'three-body'. For it, it is effective to give the number n of unmatched characters permitted on the pattern matching. For example, when n=1, 'three body' and 'three-body' are regarded as the same string. From the above a basic element of queries is denoted as ('a character string', n). Retrieval is executed intra or inter tables. The query expression is denoted as follows.

<element>::=('a character string', n)
<basic condition>::=<element>|<basic condition> OR <element>|<basic condition> AND <element>
<expression>::=Table name.item name=<basic condition>|<expression> OR <expression>| <expression> AND <expression>

A user language as shown in Fig. 2.2 is developed to support the researches described in §3 and §4. It has the following functions,

i) to name a set of resultant P-IDs of retrieval.
ii) to execute the boolean operation among the sets.
iii) to retrieve the relation of citation of papers and to output cited number and cited rate with a graph.

Usually it takes from 10 seconds to 3 minutes to scan sentences when a boolean combination of several words and a few category numbers are given. In the case of the first query in Fig. 2.2, it takes 25 seconds.

SCAT-IR system was developed on the HITAC 8250 computer of Hokkaido University with a main memoty of 96 Kbytes, one card reader, one line printer, two discs (100MB×2) and two MT drivers.

6. ADVANCED SCIENTIFIC INFORMATION

The procedure through which the research trend is grasped and the research theme is instituted may not be uniquely fixed and the algorithm is also diverse. However it does not mean that the constituents of the algorithm are also very diverse. Figs. 6.1 and 6.2 show the algorithm for grasping the research trend and instituting the research theme respectively where each algorithm is constituted by combining the process of thinking and the proposed procedure, that is, clusterization for a set of documents, territory procedure and spectra-analysis. These algorithms also can be said to be examples for obtaining advanced scientific information.

In the realistic activity of the research, the advanced scientific information can be obtained with both the process of thinking and the usage of necessary information. In our cases of Figs. 6.1 and 6.2, many kinds of information come to be cited much more heretofore. Really the retrieval procedure of various sentences with many classification categories, the method of clusterization for a set of documents alike the territory procedure and the spectra-analysis by the use of the citation relation, make it possible to increase the kind of advanced information which is mechanically manageable.

As mentioned above, the natural activity in the scientific research, for example grasping the research trend or instituting the research theme, includes two important parts, the process of thinking and the process of retrieval of many kinds of scientific information. However the latter part is not always adequate in both the kind and the quantity of information.

These manual retrieval procedures confined in the usual research activity are replaced with the above many mechanical procedures in Figs. 6.1 and 6.2. Then the processes of thinking included in the courses of the research may go more deeply after they are combined in the adaptable manner with the mechanical procedure of retrieval for the advanced scientific information.

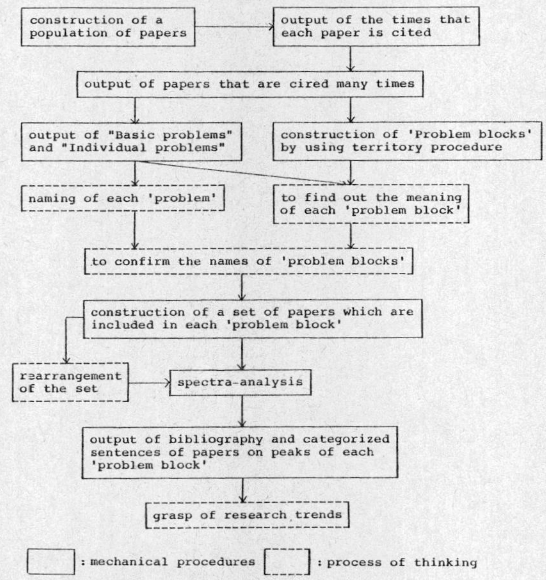

Fig. 6.1 Flow diagram of grasping of research trends.

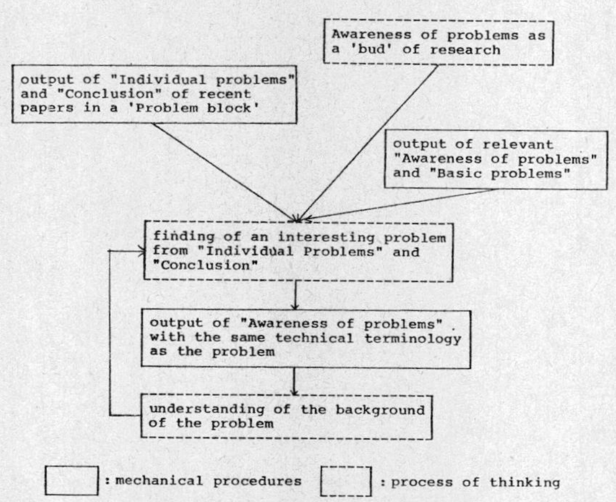

Fig. 6.2 Flow diagram of instituting of research themes.

7. RESULTS AND CONCLUSION

Our results are summarized as follows.

1. Categories exist with which each sentence in documents can be objectively classified.
2. The retrieval system for the sentence is produced, where the scientific term and the classification category are used as the retrieval information.
3. The sentence retrieval system SCAT-IR is produced and the whole content of documents are already accumulated.
4. A new clusterization procedure 'territory one' for a set of documents is proposed, where the citation relation is used. Each cluster is obtained from the method correspondence to some research problem.
5. The relation between the publishment year of a document and its cited number or cited rate brings the useful information for grasping the research trend.
6. The algorithm for the advanced scientific information can be constituted by combining the process of thinking and the mechanical procedure such as sentences retrieval, territory procedure and spectra-analysis.

Advanced scientific information, for example the research trend or the research theme, has not been investigated infologically until now. However, our results seemingly suggest that the new branch of information science may be opened to the infological research for advanced scientific information, that is to say, to the infological research for the research course itself.

ACKNOWLEDGEMENT

Prof. T. Kitagawa and Prof. M. Kotani have had their strong interest in our problem and encouraged us since the research started. The authors heartily appreciate them for their encouragement. Further the authors are grateful to Prof. T. Kitagawa, Prof. S. Asano, Prof. N. Okino and Mr. T. Saito for their stimulating discussions. The authors also thank Mrs. T. Matsumoto for her help in producing the SCAT-IR system and Miss U. Kohata and Mrs. H. Yamamoto for their kind help in punching and typing.

REFERENCES

[1] T. Kitagawa, private communication.

[2] C. Asano, "Cluster Analysis for Many Variables", Research Report, Research Institute of Fundamental Information Science, Kyushu University, Japan. (in preparation)

[3] T. Saito and N. Okino, "A Mathematical Model and a Cluster Analysis for Seeing Scientific Research Trends and Retrieving Scientific Articles", Research Report of 'Working Group for the Grasping of Research Trends' in Japan (1979) (in Japanese).

Scientific Information Systems in Japan
H. Inose (ed.)

KNOWLEDGE REPRESENTATION SCHEME FOR QUESTION-ANSWERING SYSTEMS & IMAGE INTERPRETATION SYSTEMS

Kokichi TANAKA and Hitoshi OGAWA
Faculty of Engineering Science, Osaka University, Toyonaka, Osaka 560

Knowledge representation has recently become one of the crucial issues in artificial intelligence. In fact, whichever system we might consider, how to represent knowledge along with how to utilize it on a computer is a key problem to realize advanced information systems such as natural language processing, image or speech understanding, machine vision, intelligent information retrieval, and intelligent man-machine communication. As for knowledge representation scheme, there have been made several proposals such as conceptual dependency, production system, semantic network, KRL and FRL based on frame theory, and so on. This paper presents a knowledge representation using micro-actors, a version of actors originated by Hewitt, C., which are beneficial to interaction as well as modularity and distinguished from actors in such points that LISP is used for the script instead of PLASMA and frame concept is introduced into the acquaintance. Through the demonstration of a question-answering system and an image interpretation system, the present authors have manifested tractability as well as feasibility of the knowledge representation scheme employing micro-acotors proposed here.

1. INTRODUCTION

In conjunction with organizing scientific information in order to establish the Scientific Information System, there are keenly requested researches and/or developments on more intelligent and convenient input/output systems, structural analysis and cognition systems relevant for a diverse kinds of information, more efficient and tractable information storage and retrieval systems, and their related methodologies and techniques. All of these research subjects may seem to be involved into the so-called artificial intelligence research in a wide sense. In fact, taking up document retrieval, there lie such crucial problems to be developed as automatic indexing, automatic abstracting, natural language inquiring and so on. As to fact retrieval, the more sophisticated techniques will be needed especially in question-answering systems.

In case of human search of a literature ever seen before, one will often employ such an ambiguous memory as; that literature was probably published in such-and-such a year, or such-and-such a picture was possibly on such-and-such a page. To simulate the same as the above human behavior by a machine, it should be implemented with a sort of associative memory to facilitate inference or association based on, for example, frame theory or n-ary fuzzy relation*, or other relevant techniques. Thus we face to the study on what is called key problems in artificial intelligence.

Actually, as one of the crucial issues in artificial intelligence, we have to study three themes as follows: (a) how to represent expertise and experience in a specific world as well as general knowledge and common sense, (b) how to implement the above knowledge on a computer, and (c) how to utilize acquired knowledge. All of these techniques are eventually to be integrated to realize the more intelligent information processing than usual. This integrated discipline is named as Knowledge Engineering after Feigenbaum, E.A.

This paper describes a work on knowledge representation scheme and its application to such systems with inference and understanding as question-answering systems or image interpretation systems.

In this discussion, "Knowledge does not mean only the facts about a world state but also the techniques dealing with the facts". The artificial intelligence system obtains information from the scene and/or the statements in natural language. The system, then, acquires new knowledge by comparing the obtained information with the knowledge the system has. Therefore, the knowledge representation is requisite, in which it is easy to specify knowledge, modify it and refer to it. There have been two types for the representation of knowledge: the declarative one and the procedural one. The modularity is emphasized in the former; e.g., the logical formulas. On the other hand, the interaction is stressed in the latter; e.g., PLANNER, QA4, CONNIVER and so on. The declarative representation has several benefits: The separation of knowledge into data and procedures brings the flexibility and economy of the declarative representation. The independence of the axioms or facts leads to the understandability and learnability of them. On the contrary, in the procedural representation, a frame axiom need not be specified. Furthermore, the heuristic knowledge can be treated in it. To utilize both benefits of the declarative representation and the procedural one, it is necessary to develop the method in which the modularity is established without sacrificing the possibilities for interaction.

KRL(Bobrow, D.G. & Winograd, T.) and FRL(Roberts, R.B. & Goldstein, I.P.) are developed as the approaches to the problem mentioned above. Both KRL and FRL are based on frames, and attempt to integrate procedural

* The n-ary fuzzy relation has been defined and formulated by the present author in 1971 [IEEE Trans. SMC-1, No.1, pp.61-66]. Fuzzy reasoning using fuzzy logic including fuzzy relation has also been studied by the present author [Proc. IJCAI-79, pp.589-591 & pp.917-919].

Knowledge with a broad base of declarative forms. On the other hand, there is some knowledge about the action; for example, the control of the reasoning mechanism, the data base processing and so on. Such knowledge is preferable to be expressed by the procedures rather than the declarative forms. If declarative knowledge is integrated with the procedures which deal with it, and the procedures have the modularity, then we can freely use both declarative knowledge and procedural one.

In this paper a micro-actor* is employed as a method for knowledge representation. The script of a micro-actor contains procedural knowledge to deal with only the acquaintance of the micro-actor. The acquaintance has declarative knowledge in the form of frame.

2. REPRESENTATION OF WORLD-STATE[1]

2.1 Frame Representation In Acquaintance

A micro-actor is composed of a script and an acquaintance. The script describes the behavior which the micro-actor should take when it receives a message. The acquaintance is the memory space for the micro-actor. The acquaintance stores the data that the micro-actor knows; for example, its native values, its attributes, the names of the other micro-actors it directly knows about, and the relation with them.
The following are the reason why the concept of Frame Theory[2] is used for construction of an acquaintance:
(1) A frame is a data-structure representing a stereotyped situation.
(2) Several kinds of information can be attached to each frame, and some of this information is about how to use the frame.
(3) Frames enable us to represent knowledge in a hierarchical structure.

Fig.1 shows a slot array representation to indicate a frame intuitively. A frame is, however, represented by the nested association lists as shown in Fig.2 in its implementation. The respective substructures of a frame are named; Slot, Facet, Datum and Comment.
A slot shows which property the data in it are connected with. A facet shows kinds of the data; e.g., values, constraints, procedures and so on. A datum is elementary information associated with the facet, and a comment is additional information for the datum.

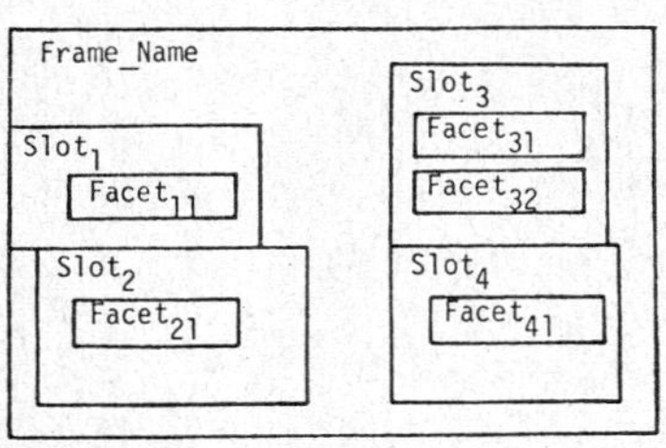

Fig.1 Slot Array Representation of a Frame.

* The micro-actor was proposed as the implementation of some of abilities of actor(Hewitt, C.)[3, 4] in 1977[5]. A new version of a micro-actor described in this paper has similar abilities of actor. Parallel process and coroutine control are possible in a micro-actor system.

(Frame_Name

$(S_1\ (F_{11}\ (D_{111}\ C_{111})\ \cdots\ (D_{11h}\ C_{11h})))$

$(S_2\ (F_{21}\ (D_{211}\ C_{211})\ \cdots\ (D_{21i}\ C_{21i})))$

$(S_3\ (F_{31}\ (D_{311}\ C_{311})\ \cdots\ (D_{31j}\ C_{31j}))$

$\quad (F_{32}\ (D_{321}\ C_{321})\ \cdots\ (D_{32k}\ C_{32k})))$

......

......

$(S_s\ (F_{s1}\ (D_{s11}\ C_{s11})\ \cdots\ (D_{s1m}\ C_{s1m})))$

Fig.2 Nested List Representation of a Frame.
S: slot, F: facet,
D: datum, C: comment.

2.2 Effect On Other Frames

An addition of procedures to a slot of a frame enables the frame to acquire information from other frames and to change the contents of the other frames. This may make it possible to combine the declarative knowledge and the procedural knowledge; that is, the modularity and the interaction. However, a question still remains to be solved.

A frame is freely dealt with using three basic functions; obtaining something from the frame, putting something to it, and removing from it. The use of only the three functions is, however, inconvenient and uneconomical when a frame is dealt with in several ways. New functions should be made either in advance or afterward. We will try to put box A on box B in a block world.

Assume that there are frames which correspond to box A and box B. To put box A on box B is executed by the following process; If box A and box B are a kind of Block and Cleartop**, and box B is Supportable***, then let box A be supported on box B and let box B support box A. The knowledge concerning the process mentioned above is represented in three methods: (1) Function ON is made in order to put an object on another, and is added to the basic functions. (2) The process is distributed to the frames corresponding to box A and box B. (3) A new frame is created corresponding to the function ON.

In (1), there exists the chunk of procedures dealing with frames. A slight change of a procedure often affects the other procedures. As a result, (1) is not good to represent the knowledge about actions in A.I. system. General functions can be used in both (1) and (3). On the contrary, a procedure depends on a particular frame in (2). (2) is convenient to describe the procedure closely depending on the frame, and to establish the modularity among procedures.
It is, however, wasteful when several frames have the similar function. (3) is equal to (1) in essence.
So there remains the possibility of the unlimited and unrestricted interactions. However, the application of the idea of actor to(3) yields modularity for representing knowledge with proper interaction method.

The actor model is suitable to control and describe

** Cleartop means the state in which the box supports nothing.
*** The box which is supportable can support another box.

the interaction between the modules. Namely, the action of an actor is defined in terms of only one kind of action; sending messages to the other actors, and any other side effect is not permitted. When the actor receives a message, it becomes active and tries to play its role. The authors think that the actor model is useful for parallel processing.

3. REPRESENTATION OF ACTION[1]

The chapter 2 stated that a frame is good for representing the knowledge about a world state, but not so good for representing the knowledge about an action. On the other hand, a micro-actor is very suitable for representing the action. The micro-actor, which is a procedure, is a basic module in order to construct the knowledge base. This chapter describes the form of script which represents the behavior, and illustrates how to deal with the knowledge about the world state using message passing.

3.1 Micro-Actor

The script of the micro-actor precisely represents the message passing, the parallel computation and the coroutine control. The form of the script is similar to Yonezawa's notation[6]. But the micro-actor is based on LISP whereas his notation is based upon PLASMA.

(i) Script

The script is specified with a set of pairs of message patterns and rules as follows:

((message_pattern$_1$ rule$_{11}$ ... rule$_{1m}$)
(message_pattern$_2$ rule$_{21}$ ... rule$_{2n}$)
⋮
(message_pattern$_p$ rule$_{p1}$ ... rule$_{pq}$)).

(ii) Message Pattern

There are two types for pattern:
(1) (MES: ?%N *content* TO: ?CONT).
(2) (RE: ?%N *content*).

?%N is only matched with the message number %n, where n is a positive integer. A message number is assigned to an input sentence or a message for distinguishing it from the others as a tag. The message number is also used to distinguish a datum from the others stored in the same micro-actor. The atoms prefixed by "?" (e.g., ?CONT) are the variables for pattern matching. The prefix "!" is used instead of the prefix "?" to refer to the pattern matching variables for their values; for example, !CONT. "MES:" means a message and "RE:" means a reply. "TO:" means that the continuation is !CONT. The continuation is the micro-actor which is sent a reply. Assign "NO-ONE" to !CONT in the case that the answer is not required. Assign "ME" to !CONT when the reply is demanded. This is equal to the subroutine control.

(iii) Rule

A rule is a set of programs, which carries out a job. The rule consists of three parts as follows:

((P-C: program$_{p1}$ ... program$_{pq}$)
(N-C: program$_{n1}$ ... program$_{nr}$)
(C-E: program$_{c1}$ ... program$_{cs}$)).

"P-C:" is an abbreviation for pre-condition. "N-C:" means next-condition which is an option. "C-E:" is an abbreviation for caused-events in which the messages are sent to the other micro-actors when both the pre-condition and the next-condition are satisfied. The form of the rule is given by Dr. Yonezawa, and it is suggested that the form of the rule is useful to deal with the parallel computation. The programs should be carried out in order of the appearance of them in the pre-condition and the next-condition. On the other hand, the programs in the caused-events can be executed concurrently.

(iv) Program

A program is written in the restricted LISP notation. The user can freely use the LISP functions which don't have influence upon the pointers to lists and atoms; e.g., MEMBER, ASSOC and so on. The variables are restricted to the form !m or !!n (Both m and n are positive integers) for the LISP functions which have influence upon the pointers; e.g., SETQ and so on. This restriction prevents the behavior of a micro-actor from giving unitentional effect to other micro-actors. The variable !m is used as well as PROG variable, which !!n can be used in a special way as mentioned in the next section.

A specification for message passing is as follows:
(=> *the_name_of_a_micro-actor a_message*).
The user uses the basic function (FGET, FPUT and FREMOVE) to deal with the acquaintance. These functions differ from those of FRL in point of the necessity of the frame name. The functions in this paper do not need the frame name because the micro-actor deals with only its acquaintance. Therefore, the execution of the basic function does not affect the other micro-actors.

(v) Behavior of Micro-actor

A micro-actor tries to find a message pattern which is matched with a message M_1 when it receives M_1. If the message pattern P_1 is found, a rule R_1 is searched for, which is one of rules making a pair with the pattern P_1, and whose pre-condition should be satisfied. When there is the next-condition in R_1, all programs in the next-condition should be evaluated. If the next-condition is satisfied, the casued-events are executed. If there is no rule whose pre-condition is satisfied concerning the pattern P_1, the micro-actor tries to find another message pattern which is matched with M_1.

Since the system holds the COMMON-SCRIPT which is necessary for all micro-actor, the users may specify the micro-actor by the peculiar script. The COMMON-SCRIPT has the message pattern (MES: ?%N ?X TO: ?CONT) and acts as follows:
(a) If the variable !X (corresponding to ?X) indicates one of the basic functions (FGET, FPUT and FREMOVE), then it is put into practice.
(b) If !X shows one of the extension functions (i.e., there exists a micro-actor corresponding to it), the rules are obtained from the micro-actor corresponding to the extension function, and they are performed. The variables !!n (n = 1, ..., 5) are corresponded to the arguments; for example, !!1 means the first argument.
(c) If !X indicates a set of rules, then !X is accomplished.

3.2 Example

A block world gives us good examples to illustrate the representation of the script and the acquaintance of a micro-actor, and message passing. The following discussion concerns with the micro-actor PROPERTY. It is used in the implementation of putting an object on another. The script of the micro-actor Boxl is NIL, because it needs just only the COMMON-SCRIPT. The acquaintance is expressed by one slot as follows: ((AKO (:VALUE (BOX)). This means that boxl is a kind of box. The AKO slot has a higher level concept of boxl as the value marked by the :VALUE facet. Concerning the micro-actor BOX, the script is NIL as well as boxl, and the acquaintance is as follows:

```
((AKO       (:VALUE (BLOCK)))
 (INSTANCE (:VALUE (BOX1 BOX2 BOX3)))
 (PROPERTY (:VALUE (SUPPORTABLE)))  ).
```

This shows the following facts: (1) Box is a kind of block. (2) The instances of box are boxl, box2 and box3. (3) The native property is supportable which means that the box can support another box.

Fig.3 gives the script and the acquaintance of the micro-actor PROPERTY which deals with the rules about the properties. In the rest of this paper, a variable prefixed by "!" indicates its value until it is quoted. The script has a rule to ask the micro-actor !B if it has the property !A. According to the rule, PROPERTY sends !B the rules in which !A is substituted for "!!1". The AKO slot means that PROPERTY is a kind of E-FUNCTION (which is an abbreviation of extension function). The CONTENT slot shows the rules which contain the following algorithm. If the micro-actor which executes the rules has the value SUPPORTABLE in its PROPERTY slot, then it sends the reply OK to !CONT. Otherwise, if it has the value A in the AKO slot, then A is sent the message to ask if A has the property SUPPORTABLE. If A has it, A sends the reply OK to !CONT; otherwise, A sends NIL to !CONT. Note that A need not answer the micro-actor which sent the message to A. This means the possibility of a coroutine control.

```
((MES: ?%N ?B PROPERTY ?A TO: ?CONT)
 ((P-C: (SETQ !1 (FGET CONTENT :VALUE)))
  (N-C: (SETQ !1 (*SUBS !1 (LIST (CONS
                  (QUOTE !!1) !A)) )))
  (C-E: (=> !B (MES: !1 TO: !CONT)))
))
```

(a)

```
(AKO      (:VALUE (E-FUNCTION)))
(CONTENT (:VALUE
  ((P-C: (SETQ !5 (FGET PROPERTY :VALUE))
         (ASSOC !!1 !5))
   (C-E: (=> !CONT (RE: !%N OK)))
  )
  ((P-C: (NOT (ASSOC !!1(FGET PROPERTY :VALUE)))
         (SETQ !4 (FGET AKO :VALUE))
         (SETQ !4 (CAAR !4)))
   (C-E: (=> !4 (MES: !%N PROPERTY !!1
                          TO: !CONT)))
  )
  ((P-C: (NULL (FGET AKO :VALUE)) )
   (C-E: (=> !CONT (RE: !%N NIL)))
)))
```

(b)

Fig.3 (a) The Script of the Micro-actor PROPERTY.
(b) The Acquaintance of PROPERTY.

Fig.4 shows the diagram of the message passing for the process to see if boxl is supportable. The user can communicate with micro-actor system using the special micro-actor * as shown in Fig.4. For example, when the user wants to know whether boxl is supportable or not, he may write as follows:

(* PROPERTY BOX1 PROPERTY SUPPORTABLE).

The first argument of * is the target of the message whose content follows the first argument. The micro-actor * composes the message and sends it to the micro-actor PROPERTY (See linel and line2 in Fig.5).

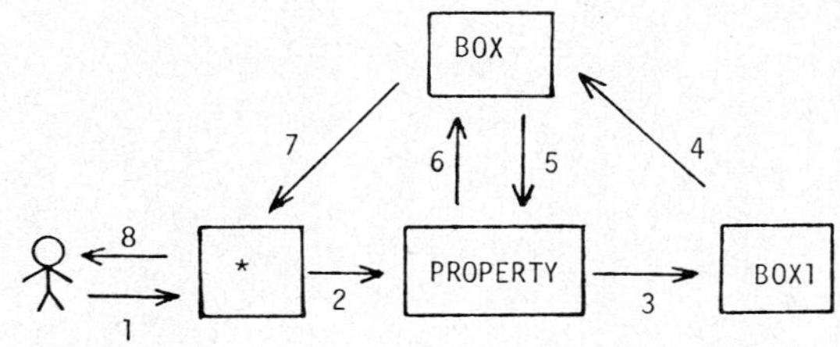

Fig.4 The Diagram of the Message Passing.

Fig.5 gives us the messages and the targets illustrated in Fig.4. Each line in Fig.5 shows the message passing which is represented by the arrow attached the same number that the line has. The front atom in each list (i.e., *, *+, etc.) illustrates the level of message passing. The symbol "*" means the subroutine control, and "+" means the coroutine control. Each line will be explained.

Linel: The user communicates the message (* PROPERTY BOX1 PROPERTY SUPPORTABLE) to the system.

Line2: The micro-acotr * composes and sends it to the micro-actor PROPERTY.

Line3: The micro-actor PROPERTY sends the BOX1 the message including the rules to see if the target has a datum SUPPORTABLE in its property slot. The continuation of the message on the line3 is the same one on the line2. In other words, "ME" means the micro-actor *.

Line4: BOX1 puts the rules into practice according to COMMON-SCRIPT. As Boxl does not have a property slot, BOX1 sends the message to the BOX which is the higher level concept of boxl. The continuation "ME" is BOX in this case, and this means the subroutine control.

Line6: PROPERTY returns the rules to BOX.

Line7: BOX carries out the rules. As BOX has the value SUPPORTABLE in the property slot, BOX sends OK to the micro-actor *.

Line8: * replies OK to the user.

This example shows the method of the coroutine control and the subroutine control, and explains how the extension function defined as the micro-actor is used and is executed. Furthermore, it gives the preferred inference about the knowledge represented with frames.

```
1 (* PROPERTY BOX1 PROPERTY SUPPORTABLE)
2 (* PROPERTY <= (MES: %2 BOX1 PROPERTY SUPPORTABLE TO: ME))
3 (*+ BOX1 <= (MES: %2 (((P-C: (SETQ !5 (FGET PROPERTY :VALUE))
                               (ASSOC SUPPORTABLE !5))
                          (C-E: (=> !CONT (RE: !%N OK))) )
                         ((P-C: (NOT (ASSOC SUPPORTABLE (FGET PROPERTY :VALUE)))
                                (SETQ !4 (FGET AKO :VALUE))
                                (SETQ !4 (CAAR !4)) )
                          (C-E: (=> !4 (MES: !%N PROPERTY SUPPORTABLE TO: !CONT))) )
                         ((P-C: (NULL (FGET AKO :VALUE)))
                          (C-E: (=> !CONT (RE: !%N NIL))) ) )
                 TO: ME))
4 (*++ BOX <= (MES: %2 PROPERTY SUPPORTABLE TO: ME))
5 (*++* PROPERTY <= (MES: %2 FGET CONTENT :VALUE TO: ME))
6 (*++* REPLY: (RE: %2 (((P-C: (SETQ !5 (FGET PROPERTY :VALUE))
                               (ASSOC !!1 !5))
                          (C-E: (=> !CONT (RE: !%N OK)) ))
                         ((P-C: (NOT (ASSOC !!1 (FGET PROPERTY :VALUE)))
                                (SETQ !4 (FGET AKO :VALUE))
                                (SETQ !4 (CAAR !4)) )
                          (C-E: (=> !4 (MES: !%N PROPERTY !!1 TO: !CONT)) ) )
                         ((P-C: (NULL (FGET AKO :VALUE)) )
                          (C-E: (=> !CONT (RE: !%N NIL))) )
                  )))
7 (* REPLY: (RE: %2 OK))
8 (* RE: %2 OK)
```

Fig.5 Message and Targets shown in Fig.4.

4. A DIALOGUE IN QUESTION-ANSWERING SYSTEM REGARDING GEOMETRICAL FIGURES[1]

In the question answering between a machine and a human, one of most important things is a topic of conversation and a context to understand what the partner said. We assume that the speaker and the listener have the common knowledge about the discourse. Consequently, the inference plays a great role to understand the partner's intention. The micro-actor can get a good policy from the acquaintance in which the facts of the world state are expressed by a frame.

4.1 Mapping From World

The authors consider that the knowledge about a world is a "snapshot" of world state. The followings show the difficulty of mapping data from a world state to a knowledge state. What part of a house one directs his attention to when he sees it? Someone might direct his attention to the door. A gardener might be interested in the trees and flowers in the garden. An architect might look the style of the house. Consider another example. When we hear "John is like a flog", what individuality do we assume about John? The answer depends on the context and the topic of the discourse. If one heard that Mike was a handsome boy just before, then he considers that John is ugly. If one heard that Mary was gentle then he thinks that John is noisy.

The frame representation is suitable for extracting information from the world. The mapping method using the frame is to move a part of the world state to a slot of the frame. The data acquired by the mapping must satisfy the condition of the slots which depend on the topics and the context. So, we will use the frame called P-Frame (Prototype Frame), which has the slots to fill out. The P-Frame is used to compose an I-Frame (Instance Frame) which represents the knowledge state. The P-Frame has the default values which are communicated to the I-Frame if necessary. In the same way, using a Q-Frame (Query Frame) which corresponds to "question", the I-Frame for the answer obtains the data from the I-Frame for the knowledge. A Q-Frame means such an action as human observes a world and gets the data (See Fig.6). In Fig.6 the knowlege is indicated by a frame for brevity. The knowledge is, however, represented by several frames in practice.

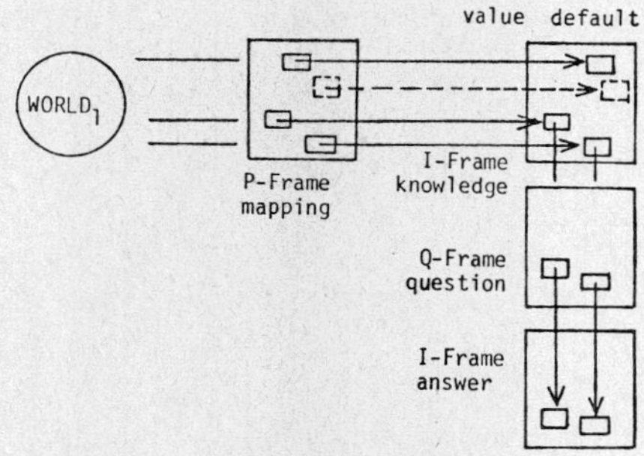

Fig.6 The Observation of a World.

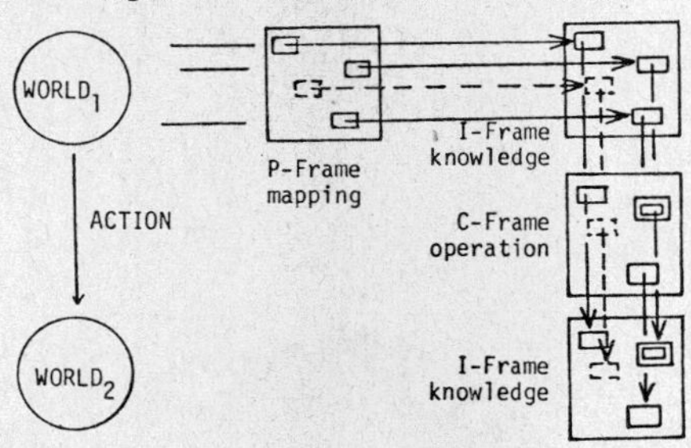

Fig.7 The Change of a World.

Fig.7 gives two I-Frames which correspond to the world1 and the world2 changed from the world1. A C-Frame(Change Frame) represents the action which affects the world, or which changes the observer's situation (In the latter case, the world is not changed.). A C-Frame transfers the data from an I-Frame to another frame, and changes the data if necessary.

4.2 Conversation Examples

The question answering system about the geometrical figures is constructed with the micro-actors according to the consideration in 4.1. The user communicates with the system in the limited Japanese for the present. The authors are developing a Japanese analyzer. After it is accomplished, the user will be able to converse with the system in Japanese. Fig.8 presents the examples in the conversation. The sentences in the figure are written in English for the readers to understand. The user's sentences are marked by "*", and the replies are marked by "!".

The system has the geometrical world as the context in advance. The context, of course, is changed during the conversation. The system also has the knowledge about several kinds of triangles, rectangles and squares. The system tries to associate the input sentence with the knowledge. If it fails, the system asks the user about the sentence as shown in line2.

When the micro-actor build receives the sentence (line1), it tries to make the micro-actor house. As "house" is still undefined and the other parts of line1 are not associated with the knowledge, the system asks the user about a house (line2). The user defines the house as shown in line3, then build can create the micro-actor house (line4). The context plays a great role to decide kinds of slot in the acquaintance of house and to fill the slot with the data. In this case, build is regarded as P-Frame. As the world "roof" on line5 is not defined yet, the micro-actor prefer will associate an isosceles triangle with the knowledge as follows. Since the topic is about the house, prefer investigates the relation between the house and the isosceles triangle. The triangle is in the AKO slot of the micro-actor isosceles triangle. The house is in the PART-OF slot of triangle. Then prefer infers that the roof is a part of the house, and asks the user if the result of the inference is correct (line6). If the user answers yes (line7), prefer creates roof and adds the PREFER facet into the PART slot of house (line8). In this case, prefer acts as a C-Frame. The above process is smoothly performed using message passing between the micro-actors as mentioned in the chapter 3.

There have been several studies of metaphors using transfer frames. Our system makes use of these studies (line9, 10, 11 and 12). The like behaves like a P-Frame. That is, like receives the slot which has an extremal value from the tower, and communicates it to house.

5. IMAGE INTERPRETATION USING KNOWLEDGE BASE[7]

In chapter 4 dealt is the fact in a specific world, that is, the story concerning what is called a house as an example. As is stated above, micro-actor is also capable of dealing with knowledge pertaining to action, movement, change, and behavior. As an illustrative instance of image understanding using knowledge base, the present chapter treats a tracing technique of blood vessels in stereoradio-graphic images of a cerebrum-vascular system. In a usual tracing technique, deadlock or mistracing will often happen when a tracer is going to approach to an apparently crossing point of two or more blood vessels.

This difficulties are coped with by refering to the knowledge base implemented beforehand in the form of micro-actors system in which stored are graphic data concerning a cross point and its preassigned neighbouring region (cross-region for short) and furthermore processing procedures to interpret which is the right way to trace when a tracer reaches a cross-region. The image interpretation system proposed here is composed of two subsystems: blood vessel tracing system (BVTS) and knowledge base system (KBS). There exist four tracing modes in the BVTS: normal, left-blind, right-blind, and both-blind mode. They are corresponding in this order to the next four states, respectively, where both of the tracer on the left- and right-hand image of a pair of stereoradiographic images are not within the respective cross-regions; the tracer on the left-hand image is within a cross-region and it on the right-hand image is not within another cross-region; the inverse case of the above; and both of the tracer are within the respective cross-region.

The KBS assignes which mode is the present tracer in whenever the BVTS asks. For example, in case of left-hand mode, regarding the present tracing on the left-hand image as unreliable, the KBS instructs the BVTS to repeat tracing once more only on the right-hand image with extending the search domain a little more steps ahead. Refering to the above tracing result on the right-hand image, the KBS assignes the tracing direction of the left-hand image. In case of both-blind, the KBS deduces the tracing direction on both of the right- and left-hand images from the tracing pattern obtained hitherto. In this conneciton, the BVTS assumes the following three conditions: (i) blood vessels curve smoothly and do not have acute curvatures, (ii) consequently tracing directions do not change abruptly, (iii) and the tracing results do not contradict with the binocular principle that the parallax between a point on the left-hand and its corresponding point on the right-hand image is inverse proportional to the depth between them.

Fig.9 shows an example of the correct results of blood vessel tracing interpreted by making use of the KBS in the case of stereoradiographic images of a cerebrum-vascular system. From this figure we can observe that the apparently crossing point does not intersect, but is apart each other in reality.

6. CONCLUSION

As a scheme for knowledge representation establishing modularity without sacrifice of interaction proposed are the micro-actors which are suitable to represent knowledge regarding to not only facts but also actions.

The micro-actors mainly consists of the message patterns and their corresponding actions in the script. The micro-actor conducts according to the script, refers to and modifies the acquaintance, and sends the message to the other micro-actors if necessary. Since only message passing is the interaction method between the micro-actors, the modularity is established. The message can communicate enough information to the micro-actor. Furthermore, the continuation in a message makes both the subroutine control and the coroutine control possible. This is very useful for obtaining the data from another micro-actor, in particular, acquiring information from the higher level concept.

The acquaintance is a memory space which stores the data and the names of the other micro-actors. Since the frame is employed to represent the acquaintance, the data in the acquaintance is used smoothly. The acquaintance is treated by only its related micro-actor. This prevents the acquaintance from being affected by another micro-actor.

The chapters 4 and 5 presented examples of the question-answering system and the image interpretation system, respectively. The dialogues in the question-answering system manifest the following abilities: creation of new micro-actors, association using higher level concept, comprehension of metaphors, and recognition of the constructed figures. These abilities greatly depend upon the frame representation.

The interpretation system of blind phenomena in the blood vessel tracing of stereoradiographic images shows the abilities of obtaining the correct results with avoiding the deadlock or mistracing in the apparently cross-region.

The recent advance in LSI technology makes micro processors and micro programming technique available for implement the knowledge representation scheme using micro-actors and consequently leads to speed-up of the operating time of the system to the extent of reasonable value in practice. The present authors are now in the course of constructing what is called the actor-machine.

REFERENCES

[1] Ogawa, H., Nanba, H. and Tanaka, K., "An Active Frame for The Knowledge Representation," Proc. IJCAI-79, 668-675 (1979).

[2] Minsky, M., "A Framework for Representing Knowledge," The Psychology of Computer Vision, McGraw-Hill (1975).

[3] Hewitt, C., Bishop, P. and Steiger, T., "A Universal Modular Actor Formalism for Artificial Intelligence," Proc. IJCAI-73, 248-249 (1973).

[4] Hewitt, C. and Smith, B., "Towards a Programming Apprentice," IEEE Trans. on Sof. Eng., 26-45 (1975).

[5] Tanaka, K., "A Knowledge Representation Structure for Natural Language Processing," Preprints IFAC Workshop on Control of Management Systems (October 10-13, 1977, Bulgaria, Varna).
Ogawa, H. and Tanaka, K., "A Structure for the Representation of Knowledge — A Proposal for A Micro-Actor —," Proc. IJCAI-77, 248-249 (1977).

[6] Yonezawa, A., "A Specification Technique for Abstract Data Types with Parallelism," Res. Rep. C-17, Dep't of Information Science, Tokyo Institute of Technology (1978).

[7] Nishi, J., Kitahashi, T. Tanaka, K., "An Application of Knowledge Base to Interpretation of stereoradiographic Images," The 20th Annual Convention Record of Information Processing Society of Japan, 1E-3, 551-552 (1979).

1. *Build a house.
2. ! What is a house?
3. *It is the composition in which a triangle is on a rectangle.
4. ! OK.
5. *I prefer an isosceles triangle for the roof.
6. ! Is the roof the triangle of the house?
7. *Yes, it is.
8. ! OK.
9. *The house is like a tower.
10. ! Do you mean the house which is now on display?
11. *Yes, that's right.
12. ! OK.
13. *What kind of house is it?
14. ! The house is composed of the isosceles triangle-4 and the rectangle-3, and is probably a tall house.
15. *Explain in general what is called a house.
16. ! A house is composed of a triangle and a rectangle, and its instance is the house-1.

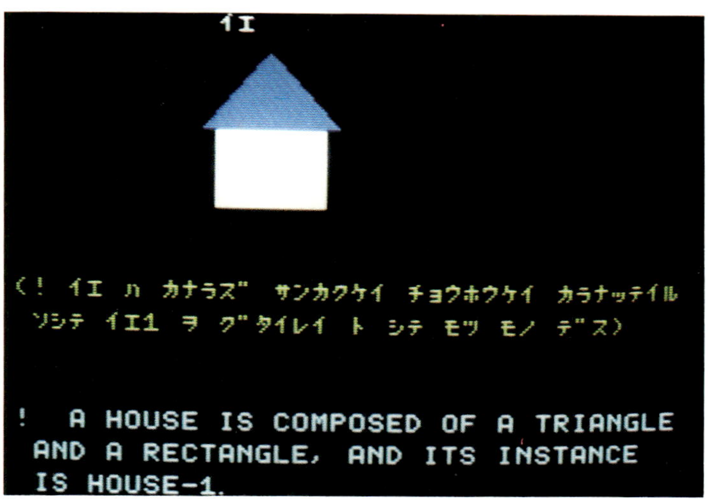

Fig. 8 Example of the Conversation (*: user, !: computer).

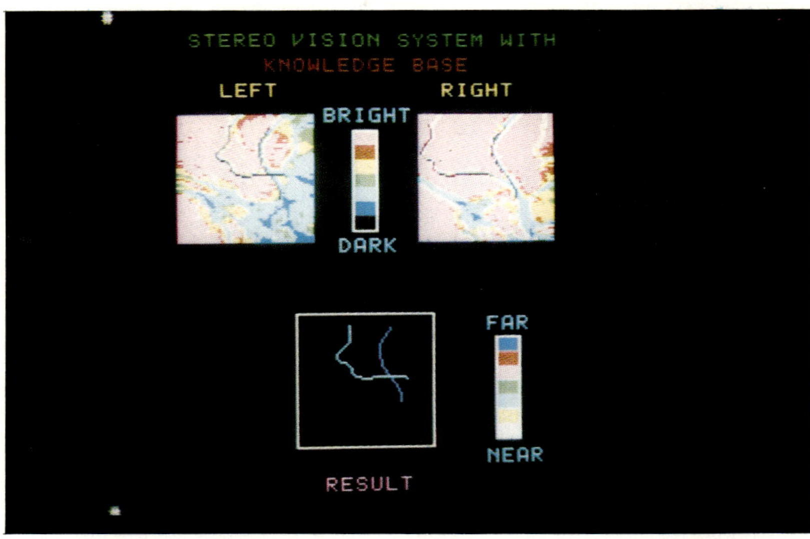

Fig. 9 Blood Vessel Tracing using Knowledge Base.

Scientific Information Systems in Japan
H. Inose (ed.)

COORDINATE RESEARCH ON SYSTEMS SCIENCE APPROACH TO UTILIZATION OF SCIENTIFIC INFORMATION

Y. SAWARAGI, H. KUMAMOTO** and K. INOUE***

** Department of Applied Mathematics & Physics, Faculty of Engineering, Kyoto University*
*** Department of Precision Mechanics, Faculty of Engineering, Kyoto University*

This project was started in 1977 by a group of systems scientists in Japan, intending broader applications of systems approach to today's problems which were typified by a large amount of scientific information or data. The present article summarizes a project report which was published in Japanese in 1979. After a brief description of what necessitated the project, basic approaches such as DEMATEL, ISM, and CIM are described briefly, using materials in the project report. Then, papers in the report are tabulated from a viewpoint of data items, item relations, methods, results, etc. Finally, special topics are described in more details.

1. INTRODUCTION

Our group of 37 systems scientists headed by the first author of this article and sponsored by Japan Ministry of Education have been working for two years since 1977 on the project, "Coordinate Research on Systems Science Approach to Utilization of Scientific Information". A project report containing 26 papers written in Japanese was released early 1979 [1-27]. The present article overviews the activities of our group, hoping to inform people outside Japan of present state of the art of systems approach to problems typified by a large amount of scientific information or data.

With the possible exception of information science and computer technology, no other branch of applied science has developed and broadened so dramatically in a few decades as systems science. There are a seemingly endless number of sophisticated theories and algorithms in this field, and one may wonder whether this degree of sophistication and complexity is justified, or whether this is a purely academic indulgence. Theories and algorithms must be developed not as an academic exercise, but because we need them to solve problems. Our activities were motivated by a recognition that truely fruitful development of systems science became possible only through tight linkages with scientific information or actual data.

Most of problems facing mankind today are consequences of the success of modern science and technology, and thus are flooded with a large amount of scientific information. For example, present-day energy saving problems must be analyzed through data on energy consumptions in all sectors of human activities; we would also have to organize good curricula in universities to stimulate new developments of interdiciplinary science and technologies to solve problems of concern to mankind. The systems approach which contributed to sending men to the moon must now be linked with these information or data.

Systems approach is a systematic collection of procedures or modules for solving problems according to a purpose. Solutions are more important for some people than problem solving procedures. For systems scientists, however, the procedures are as important as, or sometimes are more important than the solutions. They have investigated the procedures, and formulated them in terms of general languages not specific to particular problems. The resulting set of modules, i.e., the systems approach, can apply to various, new problems through formulations by the general languages.

The today's problems are difficult to solve because these are new and there is a great paucity of experts in a conventional sense. One of the greatest features of the systems approach is that people, if they work with the approach, can solve problems in a way as good as the conventional experts could do. Even foreigner could drive successfully in Tokyo to his destinations if the roads had sufficient guideposts written in English; the systems approach is a kind of a good map.

A large number of modules are now available in systems approach. These are, for example, ISM (Interpretative Structural Modeling), DEMATEL (Decision Making Trial and Evaluation Laboratory), CIM (Cross Impact Method), Multivariate Analysis, Delphi Method, IR (Information Retrieval), LP (Linear Programming), DP (Dynamic Programing), FA (Fuzzy Algebra), GMDH (Group Method of Data Handling). The reader will find in the project report many applications of these modules to today's problems. We now start the overview of the project with a brief description of DEMATEL, ISM, and CIM, using materials from our report[1].

2. BRIEF DESCRIPTION OF DEMATEL, ISM, AND CIM.

2.1 DEMATEL

DEMATEL was developed by Battelle Institute in Geneva [28]. Given a set of objectives, this method quantifies influencing-influenced relations among the objectives, using matrix manipulations. The objectives are, for example, curricula in universities, journals of science and engineering, and alternatives in a city planning problem. An interesting application of DEMATEL is in Ref. [6] in our project report.

Consider a set of 7 journals in the field of control engineering: (1) IEEE Trans. on Automatic Control, (2) International Journal of Control, (3) SIAM Journal of Control, (4) Automatica, (5) Information and Control, (6) Automation and Remote Control (USSR), (7) Automatisme. DEMATEL starts with a matrix like Table 1. Here, the matrix entry, $a_{12} = 303$, indicates that papers in Journal 2 cited those in Journal 1 303 times, while the entry, $a_{21} = 58$, shows that Journal 1 cited Journal 2 58 times. Self-citations are neglected.

JOURNAL	1	2	3	4	5	6	7
1	0	303	19	153	13	0	0
2	58	0	8	27	0	0	0
3	77	63	0	13	10	0	0
4	65	55	0	0	0	0	0
5	21	11	0	0	0	0	0
6	8	0	0	33	0	0	0
7	0	0	0	0	0	0	0

Table 1. Frequency Distribution Of Citations Among Seven Journals In Control Engineering.

The normalization of the matrix by the maximum value of the row sums yields matrix D:

$$D = \begin{bmatrix} .000, & .620, & .039, & .313, & .027, & .000, & .000 \\ .119, & .000, & .016, & .055, & .000, & .000, & .000 \\ .158, & .129, & .000, & .027, & .020, & .000, & .000 \\ .133, & .113, & .000, & .000, & .000, & .000, & .000 \\ .043, & .025, & .000, & .000, & .000, & .000, & .000 \\ .016, & .000, & .000, & .068, & .000, & .000, & .000 \\ .000, & .000, & .000, & .000, & .000, & .000, & .000 \end{bmatrix}$$

Entry d_{ij} of this matrix is a measure which quantifies a degree of Journal i influencing Journal j. Thus, Journal 1 influences Journal 2 by the intensity, 0.62, while Journal 2 influences Journal 1 by 0.119.

Matrix D is called a direct influence matrix since it gives only the degrees of the direct influence. We should note that Journal i might influence Journal j through Journal k. Such combinations of direct and indirect influences can be quantified by a formula in DEMATEL:

$$F = D + D^2 + D^3 + \dots = D(I - D)^{-1}.$$

Matrix F is called a direct-and-indirect influence matrix. Matrix F after the normalization by the sum of all entries is:

$$F = \begin{bmatrix} .049, & .241, & .018, & .127, & .010, & .000, & .000 \\ .047, & .034, & .008, & .034, & .001, & .000, & .000 \\ .065, & .085, & .004, & .033, & .008, & .000, & .000 \\ .053, & .071, & .003, & .021, & .001, & .000, & .000 \\ .016, & .018, & .001, & .006, & .000, & .000, & .000 \\ .010, & .009, & .001, & .025, & .000, & .000, & .000 \\ .000, & .000, & .000, & .000, & .000, & .000, & .000 \end{bmatrix}$$

The influencing-influenced relation represented by matrix F can be visualized by the diagram in Fig. 1, where ordinate and abscissa are a row sum and a column sum of matrix F, respectively. We observe:
(1) IEEE Trans. on Automatic Control has the greatest influence upon the other journals. This is a leading journal in control engineering, and control engineers can find innovative articles in this journal.
(2) International Journal of Control is most influenced by the other journals. Engineers who want to find developed articles may be referred to this journal.
(3) SIAM Journal of Control has a unique status; it influences the other journals, but is not influenced. This result is reasonable because SIAM Journal is directed to mathematicians.
(4) Automatica locates in between IEEE Trans. and Int J Control. This journal influences the other journals as much as it is influenced by the others. This shows a healthy development of Automatica.
(5) Information and Control is isolated from the other journals. Actually, the journal is biased to information science.
(6) Automation and Remote Control is also isolated. This journal is written in Russian.

The present approach can be extended to cases containing more journals from different fields. Results may be useful for us to retrieve scientific documents from a set of unfamiliar journals. DEMATEL is akin to sensitivity analysis, and thus useful for decision makings through importance rankings.

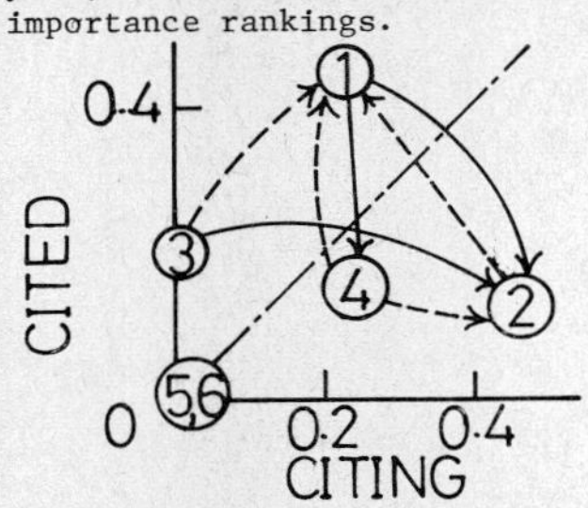

Fig. 1 Utilization structure of the journals.

2.2 ISM

ISM was developed by Warfield in Ref. [29], which also contains interesting and imaginative applications of systems structure concepts to decision makings. Given a set of objectives, the method clarifies structural relationships among them, using digraphs and matrix manipulations. Various things such as modules in curricula and alternatives can be regarded as the objectives in ISM, similarly to DEMATEL.

To illustrate the method, consider an example in Ref. [27] in the project report. This paper analyzes why there exist in control engineering discrepancies between theory and practice. Twenty one problems causing the discrepancies were picked up.
(1) There exist discrepancies between theory and practice; the original problem.
(2) Paucity of case studies.
(3) Weak cooperations between universities and industries.
(4) Inappropriate goals of theory.
(5) Industries hesitate to pay money for software.
(6) Insufficient budget for research.
(7) Few control engineers are engaged in national projects.
(8) Modern control theories are too academic to apply to practical problems.
(9) Practical problems are too complicated to be solved by existing theories.
(10) Poor demand for theories.
(11) Societies of control engineers are stereotyped.
(12) Inappropriate management of research activities.
(13) Scarcity of simple methods or hardware.
(14) Industries want to solve problems by themselves.
(15) A lot of copies and little originarity.
(16) Few chances for continueing educations.
(17) Curricula in universities are biased to theory.
(18) Research activities are evaluated by volumes of academic papers.
(19) Few people act as go between for theory and practice.
(20) Plenty of academic information and little industrial informaticn.
(21) Practical experiences are not well organized.

Consider a binary relation:

R: Solving Problem i contributes to settlements of Problem j.

The following zero-one matrix is obtained by a questionnaire.

	1	2	3	4	5	6	7	8	9	10	11	12	13	14	15	16	17	18	19	20	21
1																					
2	1																		1		1
3	1	1				1														1	
4																					
5													1								
6																					
7		1	1																		
8	1	1								1			1				1				
9	1																				
10	1	1	1		1		1														
11																					
12																					
13	1	1								1											
14		1	1							1										1	
15																					
16		1								1									1		
17																					
18																			1	1	
19	1	1	1							1			1							1	
20		1																			
21	1																				

ISM yields the hierarchical digraph in Fig. 2 which shows how the problems should be solved to decrease discrepancies between theory and practice.

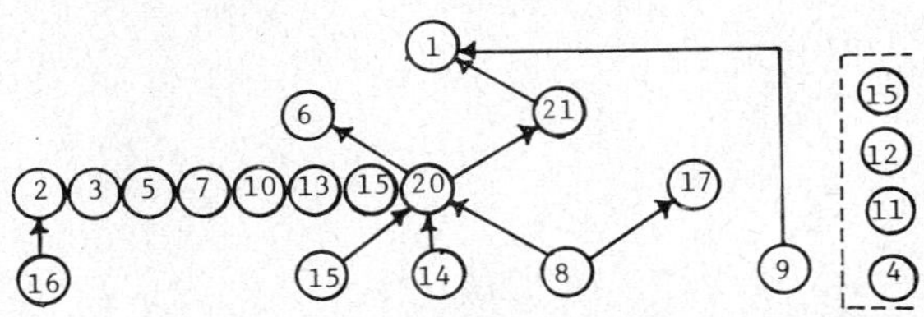

Fig. 2 Hierarchical structure of 21 problems.

2.3 CIM

CIM was first developed by Gordon & Hayward in 1968 when they intended to write scenarios for predictions [30]. The prediction is based on probabilities which are estimated by experts for occurrences of events in future. Dupperin & Godet advanced the method by introducing a second order non-linear programming [31]. Another improvement using a linear programming is proposed by Kaya & Ishikawa [32].

Consider a set of n events in future. Let

$$x_i = \begin{cases} 1, & \text{if event } i \text{ occurs,} \\ 0, & \text{otherwise.} \end{cases}$$

CIM calculates simultaneous probabilities $P(x_1,\ldots,x_n)$, based on marginal probabilities $P(x_i)$ and conditional probabilities $P(x_i|x_j)$ that are estimated by a group of specialists. The conditional probabilities represent cross impacts between event i and event j.

Let us illustrate CIM by a paper in the project report. The article predicts evolution of Japan up to 2000 A.D. [16]. It is pointed out that CIM can be used successfully with ISM; the prediction would become easier if the evolution is divided into several sectors:
(1) Change of ways people work.
(2) Change of values in life and labour.
(3) Change of status of women and families.
(4) Change of social welfare systems.
(5) Change of life style of older people.
(6) Change of relationship between academic careers and jobs.
ISM yields causal relations among these sectors, as is shown in Fig. 3.

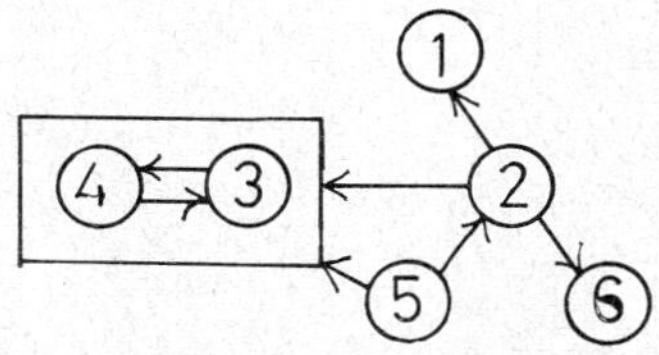

Fig. 3 Relation among sectors.

Several events are introduced into each sector. Sectors 3 and 4, for example, include the events:
1) Less discriminations of women against men in occupation.
2) Increase of employment in social welfare business.
3) More tight cooperation between professionals and volunteers of social welfare business.
4) More jobs open to handicapped persons.
5) Recurrence to old Japan. ...
Similar events are introduced into the other sectors. Marginal probabilities and conditional probabilities are evaluated by a group of specialists. CIM yields three different scenarios which forecast the evolution of Japan. The scenarios are summarized as:
(1) Movement to an individualistic society like those in Europe and USA.
(2) Recurrence to the old Japanese society.
(3) Continuation of present trends.

3. BRIEF SUMMARY OF PAPERS.

Most of the articles in our project report are tabulated in Table 2 which uncovers a close linkage between the systems approach and scientific information. Readers may overview in top-down the four columns in this table; data items, item relations (or viewpoints), methods, and results.

We give here several comments on papers which are not listed in the table. Ref.[1] is the project report containing 26 papers [2-27]. Papers [10-14] are comments on the articles for a thesaurus construction [9]. Paper [21] discusses present status of international relations, intending to apply military technologies to organization of an urban accident management system. Paper [23] developes a new method for finding subsystems in a directed graph. A data processing system is discussed in paper [25]. Papers [9] and [23] are described in Sections 4.3 and 4.6, respectively, of the present article.

4. SPECIAL TOPICS.

4.1 ISM And Curriculum Analysis [2].

Department of Applied Mathematics and Physics of Kyoto University listed 88 modules in curricula of engineering mathematics. Part of the list is shown in the first column in Table 3. The second column lists modules which are prerequisite for the modules in the first culumn. The prerequisite relations were obtained by a questionnaire which had been sent to professors of the engineering mathematics.

ISM yields a hierarchical structure of the 88 modules. Part of the structure is shown in Fig. 4. We see, for example, that "Integral And Differential Calculus" and

Table 2 Tabulation of Papers in the Project Report.

Paper	Data Items	Item Relations (Viewpoints)	Methods	Results
[2]	Modules in curriculum for engineering math. Engineering departments.	Prerequisite relations. Utilization relations.	ISM, Cluster Analysis, Questionnaire	Hierarchical structure of modules. Clusters of departments from a viewpoint of utilizing modules.
[3]	Modules in curriculum for electrical dept. Grades in entrance examinations.	Prerequisite relations among modules. Correlations of modules and grades.	ISM, DEMATEL, Principal component analysis.	Hierarchical structure of modules. Influencing-Influenced relations among modules. Basic factors affecting the students' grades.
[4]	Keywords in scientific documents.	Mapping between keywords and documents.	Fuzzy algebra.	Improvement of the mapping via a posteriori information.
[5]	Grades in examinations and modules in curriculum in department of mechanical engr.	Correlations of students' grades.	Multivariate analysis.	Several basic factors affecting the grades. Influencing-influenced relations between modules and grades.
[6]	Journals listed in JCR.	Citing-cited relations.	DEMATEL	Utilization structure of journals of control engineering. Influence relations among fields relevant to control engineering.
[7]	Modules in curriculum. Students' grades in examinations. Jobs after graduation.	Prerequisite relations among modules. Correlations between students' grades and jobs.	DEMATEL, Tabulation method, Questionnaire.	Influencing-influenced relations among modules. Relationships between students' grades and jobs after graduation.
[8]	Problems in a city planning.	Influence relations.	DEMATEL ISM	Hierarchical structure and importance ranking of problems.
[9]	Vocabulary of systems science.	Part-whole relation. Class inclusion relation....	Brainstorming	Thesaurus of systems science.
[15]	Address, name, title, field, career, etc., of systems scientists.	Correlations of data items.	Tabulation method.	Clarification of present status of systems scientists.
[16]	Events in future in Japan.	Probabilities of occurrences of the events in future.	CIM	Three probable scenarios of Japan up to 2000 A.D.
[17]	Land-mesh-data on population and social capital investments.	Average access time to public facilities.	Parameter identification method. NLP.	Extended version of Newling model. Optimal solution for social capital investments.
[18]	Land-mesh-data on population, industries, fuel consumptions, etc.	Time series trends.	Information retrieval, Data bank.	Method of predicting numbers of employees in industry sectors, etc.
[19]	Social indicators	Status of utilizations of social indicators	Graphs, Tables, Questionnaire.	Several problems specific to the social indicators.
[20]	Economic statistics.	Discrepancy among statistics.	Inspection	Causes of discrepancy.
[22]	Treaties and laws for utilization of space	Historical developments.	Survey.	Retrospective and perspective issues.
[24]	Six methods for displaying multivariable data	Relative merits of methods.	Multivariate analysis.	Suggestions for designing display pannels.
[26]	Control engineers. Scientific information.	Utilization of information by control engineers.	Questionnaire, Multivariate analysis.	Present status of utilization of scientific information. Four groups of control engineers.
[27]	21 problems to be resolved to decrease discrepancy of control theory from practice.	Prerequisite relations	ISM, DEMATEL	Hierarchical structure and importance ranking of the problems.

"Matrix And Determinant" form basic branches of the curriculum. Following the diagram, a student can select a sequence of modules by himself, depending on his current knowledge and goals.

College of Engineering of Kyoto University consists of about 14 (groups of) departments. A total of 13 departments answered what modules of the engineering mathematics were prerequisite for classes in the departments. The results are shown in Table 4, where the first column gives 17 groups of the modules. The entry at the second column and the second row is the number of answers saying that a module in Group 1 is prerequisite for a class in department B-1.

A cluster analysis was applied to the data in Table 4, yielding Fig. 5. There are 14 clusters of the depart-

Table 3 Prerequisite Relations of Modules.

	MODULES	PREREQUISITES
A.	Analysis	
A.1	Calculus	
1	Elementary Calculus	
2	Calculus (Fundamentals)	1
3	Differential Calculus (One Variable)	2
4	Integral Calculus (One Variable)	3
5	Sequence of Functions	4
6	Differential Calculus (Multivariable)	3,4,51
7	Integral Calculus (Multivariable)	4,6
A.2	Differential Equations	
.....	....	....

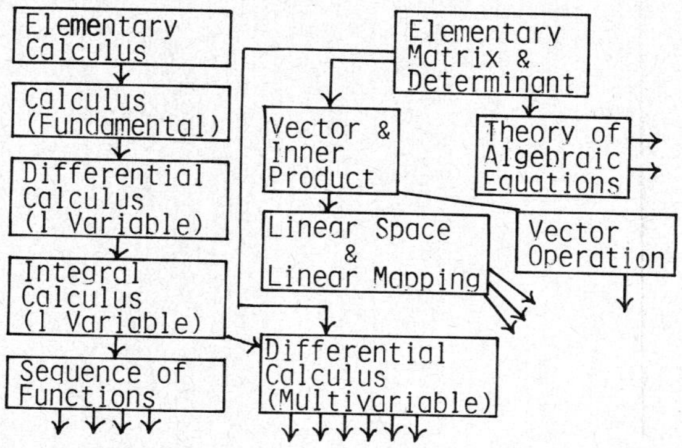

Fig. 4 Hierarchical structure of modules.

ments. Cluster 1 consists of departments D-1 and D-2, while Cluster 9 is composed of C-1, D-4, and A-6.

Some modules of engineering mathematics are taught at classes for students from several different departments. These classes are called "Common Class". Figure 5 helps us orgainze such classes, uncovering common features of physically different departments. Curriculum analyses similar to this one are also found in Papers [3] and [5] in the project report.

Table 4 Utilization Distribution of Modules by Departments in College of Engineering.

	B-1	A-1	C-1	C-2	D-1	D-2	C-3	D-3	----
1	78	109	34	146	67	89	77	72	
2	36	49	12	51	27	23	28		
3	44	75	19	46	33	25			
4	21	26	12	27	17				
5	9	13	0	10	9				
6	26	36	16	29	25				
7	9	9	7	13					
8	12	13	6	13					
9	2	1	0	15					
:::	:::	:::	:::	:::					
17	30	19	9	31					

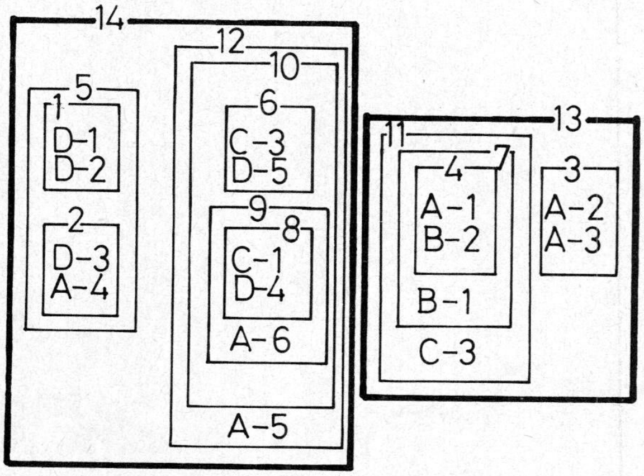

Fig. 5 Clusters of departments.

Table 5 MODULES IN DEPT. OF INFORMATION SCI.

	MODULES
A	ELEMENTARY CALCULUS
B	DIFFERENTIAL EQUATIONS
C	LINEAR ALGEBRA
D	PROBABILITY & STATISTICS
E	COMPLEX-VALUED FUNCTIONS
F	LAPLACE TRANSFORM
G	VECTOR & TENSOR
H	NUMERICAL ANALYSIS
I	SYSTEMS ENGINEERING
J	MULTIVARIATE ANALYSIS
K	MATHEMATICAL PROGRAMMING
L	INFORMATION THEORY
M	PROGRAMMING LANGUAGE
N	ELECTRICAL CIRCUIT
O	DIGITAL CIRCUIT

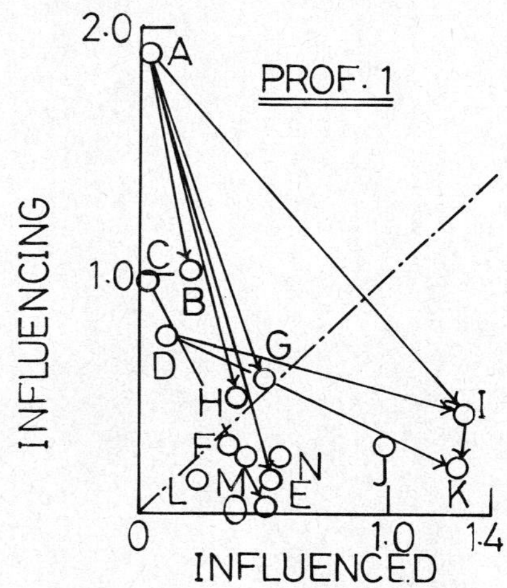

Fig. 6 Diagram by Prof. 1.

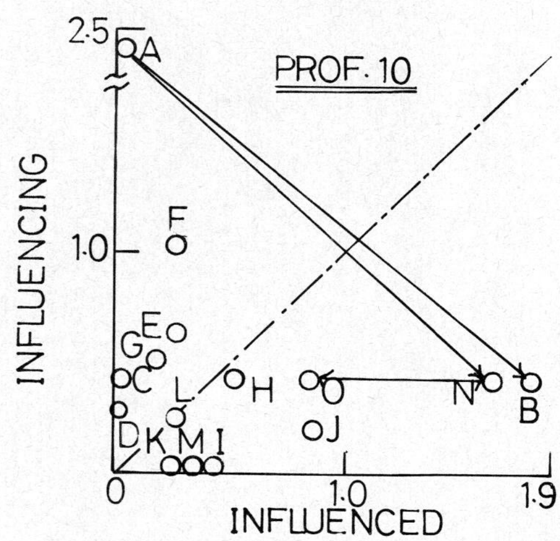

Fig. 7 Diagram by Prof. 10.

4.2 DEMATEL And Curriculum Analysis [7].

Table 5 shows part of modules in curricula of Department of Information Science of Tokushima University. Ten direct influence matrices similar to the one in Section 2.1 were obtained for the modules, through a questionnaire to 10 professors. DEMATEL was applied to the 10 matrices, and it yields 10 diagrams which quantified influencing-influenced relations among the modules. Figures 6 and 7 show 2 sample diagrams. We observe that the influencing-influenced relations considerably vary, depending on the professors who gave the direct influence matrices.

Consider now obtaining a consensus of the professors in the influence relations. The simplest method is to use the average direct-influence-matrix:

$$\bar{X}_{ij} = \sum_{r=1}^{10} X_{ij}^{(r)}/10,$$

where

$X_{ij}^{(r)}$: matrix entry given by professor r.

A better consensus can be obtained by iterating revisions of the matrices. Each matrix r is associated with it a standard deviation from the average matrix:

$$d_{r\bar{X}} = \sqrt{\sum_i \sum_j (X_{ij}^{(r)} - \bar{X}_{ij})^2/N^2}, \quad r = 1,\ldots, 10.$$

N: total number of modules = 15.

The deviations prior to the first revision is given in Table 6. Professors with large deviations are asked to revise their direct influence matrices, considering the average influence relation, and the relations by other professors. The process of consensus during three iterations is shown in Fig. 8. This diagram uncoveres basic, advanced, and more advanced curricula, for the new interdiciplinary department.

Table 6 Deviations Of Answerer From Average Matrix.

ANSWERER r	1	2	3	4	5	6	7	8	9	10
$d_{r\bar{X}} \times 1000$	29	35	22	21	25	24	21	32	21	39

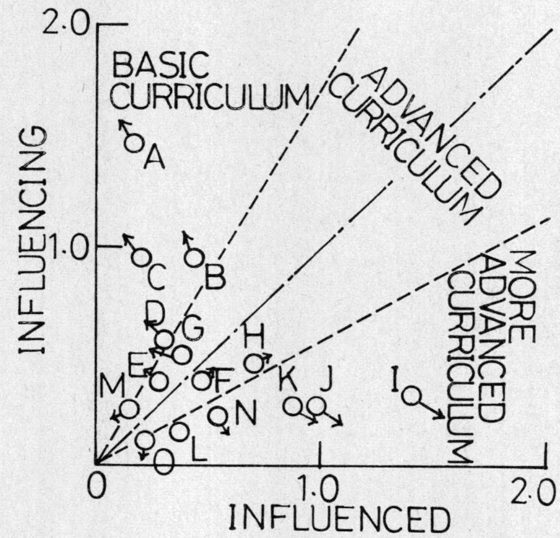

Fig. 8 Movement from the first to the third consensus.

Convergence of the consensus process can be quantified by measures such as mean deviation $\bar{D}$ of the 10 matrices from average matrix $\bar{X}$.

$$\bar{D} = \sum_{r=1}^{10} d_{r\bar{X}} /10.$$

A total of 4 measures similar to $\bar{D}$ were introduced. The convergence of the consensus process is shown in Fig. 9.

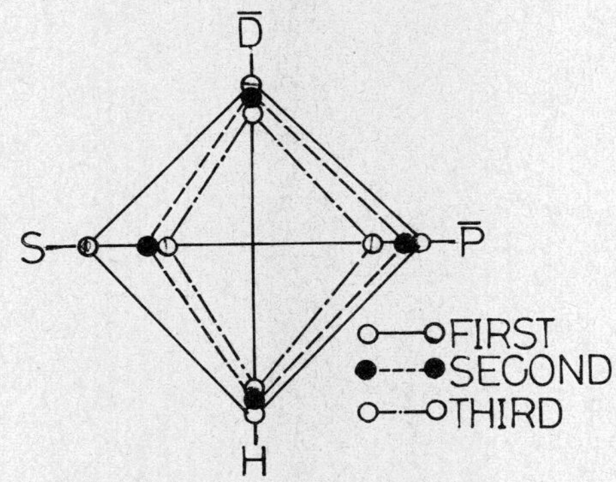

Fig. 9 Convergence of four measures.

4.3 Thesaurus Construction for Systems Science [9].

A present status of the systems science is clarified by constructing a thesaurus of relevant vocabulary. Part of the thesaurus is listed below:

- SYSTEMS SCIENCE
  - SYSTEM METHODOLOGY
    - AXIOMATIC APPROACH
      - FORMAL APPROACH
      - DESCRIPTIVE APPROACH
    - PRAGMATIC APPROACH
      - EXPERIMENTAL APPROACH
      - EMPIRICAL APPROACH
  - SYSTEM THEORY
    - GENERAL SYSTEM THEORY
      - META MODEL
        - WHITE BOX
        - BLACK BOX
          - ENVIRONMENT
          - INPUT.....OUTPUT
          - STATE
          - BEHAVIOR
      - MODELLING
        - STRUCTURE
          - ELEMENT
          - INTERACTION
        - DUALITY
        - ANALOGY
    - ....
  - SYSTEM ENGINEERING
    - PROBLEM DEFINITION
      - BRAINSTORMING
      - :::::::: :::::::: ::::::

It is suggestive to look back upon procedures of the thesaurus construction. Similar procedures could apply to thesauruses for other fields.

(1) 12/1/1976. A purpose of the thesaurus construction, "Clarification of Present Status of Systems Science", was determined.
(2) 2/22/1977. "Classification of Knowledge through Trial and Error" was adopted as an approach to the construction.
(3) 7/8/1977. Survey of papers dealing with thesaurus constructions.
(4) 10/13/1977. A total of 283 terminologies were picked up from "Unified Systems Engineering (Battell Memorial Institute)" and "An Approach to General System Theory (G. Klir)". These terminologies were classified into four categories: systems concept; general system theory; individual system theory; procedures & algorithms.
(5) 11/28/1977. Revision of the categories. Addition of new terminologies, resulting in 324 words and phrases. An automated method using a connectivity matrix was tried, and it failed because the relations among the terminologies were too complicated to be represented by the matrix.
(6) 1/14/1978 - 2/18/1978. Several tree heads of the thesaurus were proposed and revised. One of the tree heads is shown below:

```
|SYSTEMS SCIENCE|

        SYSTEM THEORY
             |GENERAL SYSTEM THEORY
             |INDIVIDUAL SYSTEM THEORY
             |SYSTEMS APPROACH
        SYSTEM METHODOLOGY
             |AXIOMATIC APPROACH
             |PRAGMATIC APPROACH
```

(7) 4/7/1978. Several difficulties came out of the thesaurus construction. These are:
1) A terminology is recognized differently by persons.
2) Many contradictions and troubles appear when the terminologies are forced to arrange in a hierarchical order.
3) New terminologies which have never appeared in any documents must be invented to maintain completeness of the hierarchical structure.
(8) 5/11/1978 - 8/26/1978. The construction continued, alleviating the difficulties by intensive discussions. Revisions of the tree heads were repeated, developing the heads into lower levels step by step.
(9) 10/6/1978. The tree head was completed.
(10) 12/16/1978. The thesaurus was completed by developing the tree head into the lowest level.

The thesaurus would be a good communication tool for people who try to solve problems by the systems approach.

4.4 Land-Mesh-Data And City Planning [17].

Part of Japan is subdivided into squares with the area of 1 km^2. The land-mesh-data includes information on population, industries, etc, for each square. An extended version of Newling population model was obtained by identifying constant a, b, and c by the data:

$$P(S) = a + b\cdot\exp(cS),$$

S: cummulative area measured from city center,
P: population density.

The mesh data can be used efficiently via this analytican model; for example, interpolation and extrapolation of the data become possible.

Let us consider allocating schools, hospitals, etc., in such a way as to minimizing the total access time to these public facilities. Let

$f(x)$: population density at location x,
$g^2(x)$: density of social capital investment.

Assume that the access time at x is inversely proportional to the capital density. Then, the allocation problem is formulated as a non-linear optimization:

$$\text{Minimize total access time } J = \int f(x)g^{-1}(x)dx,$$
$$\text{subject to } \int g^2(x)dx = \text{total capital.}$$

The solution is

$$g(x) = k\cdot f(x)^{2/3} = k\cdot[a+b\cdot\exp(cS)]^{2/3}.$$

This could apply to:
(1) optimal allocation of social capital,
(2) evaluation of present status of the investments.

4.5 Land-Mesh-Data And Data Base [18].

A data base was constructed for utilizing the land-mesh-data. Various types of data can be retrieved via simple commands to a computer. The data types include:
(1) Number of Employees (Industry, Region, Year),
(2) Population (Region, Year),
(3) Total Population (Year),
(4) Fuel Consumption (Industry, Fuel, Region, Year).
Graphs, charts, histrograms, and tables are available through utility programs. Routines such as correlation analysis and principal component analysis are also available.

The data base can apply to predictions of a city. As an example, consider predicting the numbers of employees in industrial sectors of Osaka prefecture. The past data, i.e., "Number of Employees (Industry, Year), $1966 \leq \text{Year} \leq 1979$", are retrieved from the data base. Average growth rates $\gamma(i)$ of the employees during several subintervals in future are estimated for each industry i. The number of employees of industry i is predicted similarly to computing compound interests. If there are some inconsistensies in this prediction, then the growth rates are modified. The prediction is revised in a conversational mode with a computer. It was turned out that this approach using the data base yielded a good prediction.

4.6 Directed Graph And Subsystems [23].

Structural model using directed graphs like Figs. 2 and 4 are useful tool for analyzing complicated systems. Interpretations of the model, however, are not easy. This difficulty can be alleviated if subsystems are identified in the model. A new method of finding the subsystems is developed by using pricipal component analysis. PCM requires measures quantifying closeness of nodes in the directed graph. Three measures are introduced:
(1) Connectivity;

$$a_{ij} = \frac{1}{2}(\text{number of branches from node i to node j} + \text{number of branches from node j to node i}).$$

(2) Likelihood of existing at the same level.

$$hij = \frac{\text{number of branches from group } \{i,j\}}{\text{number of branches to } \{i,j\} + 1}$$

(3) Human judgement factor c_{ij}.
These measures are combined, yielding

$$d_{ij} = \alpha\cdot a_{ij} + \beta\cdot b_{ij} + \gamma\cdot c_{ij},$$

where
α, β, γ: weighting factors.

Matrix $\{d_{ij}\}$ can be regarded as a correlation matrix. PCM then applies to finding clusters of the nodes in the directed graph; the second and the third eigen vectors are used to find the subsystems.

Consider as an example the directed graph in Fig. 10. This is a structural model for a city planning problem. The blocks in this figure were obtained via analysis of practical meanings of the nodes. Assume $\alpha = \beta = 1$ and $\gamma = 0$. The result of PCA is given in Fig. 11. Five subsystems similar to those in Fig. 10 are identified. The automated method using PCM yields as good result as the non-automated syntactical approach.

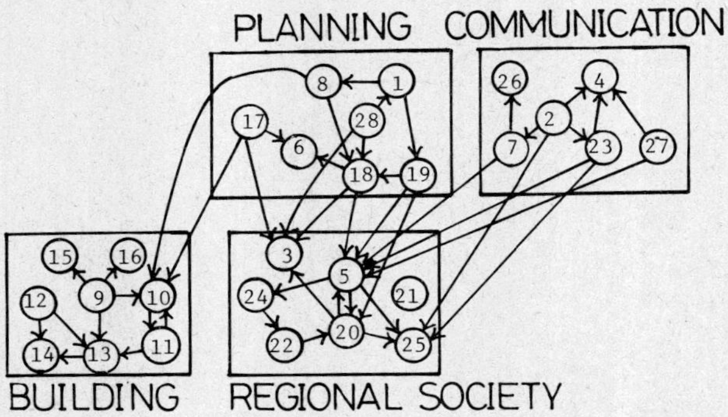

Fig. 10 Subsystems of a model.

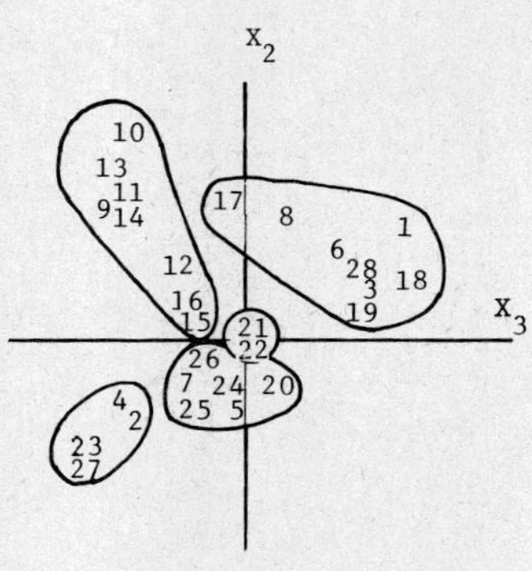

Fig. 11 Clusters by PCA.

REFERENCES

[1] Group D, "Coordinate Research On Systems Science Approach To Utilization Of Scientific Information", Report of Japan Ministry of Education Special Research, February, 1979, pp. 1-252, (in Japanese).

[2] Y. Sawaragi, T. Hasegawa, H. Tokumaru, "Systematization Of Educational Information", in Report Of Japan Ministry of Education Special Research, February, 1979, pp. 10-27.

[3] Y. Nishikawa, T. Udo, "Curricula For Under Graduate School And Entrance Examinations To Post Graduate School In Department Of Electrical Engineering Of Kyoto University", ibid., pp. 28-37.

[4] S. Iwai, K. Nakamura, "Adaptive Literature Search By Fuzzy Algebra", ibid., pp. 38-45.

[5] S. Iwai, K. Nambu, S. Tumtavitikul, "Multivariate Analysis Of Curricula In Department Of Mechanical Engineering", ibid., pp. 46-53.

[6] K. Inoue, T. Kimura, "Macro Analysis Of Utilization Of Scientific Documents By Journal Citation Index", ibid., pp. 54-61.

[7] Y. Sawaragi, T. Soeda, T. Shoman, H. Takeda, "Application Of DEMATEL To Structural Analysis Of Educational Information Systems", ibid., pp. 62-75.

[8] H. Takeda, "A City Planning Using ISM And DEMATEL", ibid., pp. 76-81.

[9] S. Kobayashi, A. Ichikawa, "How A Thesaurus Of Vocabulary Of Systems Science Was Constructed", ibid., pp. 85-99.

[10] H. Eto, "Thesaurus Of Systems Science From A Viewpoint Of Historical Development Of Science", ibid., pp. 100-105.

[11] T. Kitamori, "Comment On Structuralization Of Systems Science", ibid., pp. 106-110.

[12] M. Kimura, "Structure Of Systems Science", ibid., pp. 111-118.

[13] E. Shimemura, "Structure Of Terminologies Of Systems Science", ibid., pp. 111-118.

[14] Y. Takahara, "Comment On Structuralization Of Systems Science From A Viewpoint Of General System Theory", ibid., pp. 121-133.

[15] S. Kobayashi, A. Ichikawa, "Census Of Systems Scientists", ibid., pp. 134-144.

[16] Y. Kaya, J. Kawakami, S. Mori, M. Toda, M. Ishikawa, "Modeling Of Social Systems By Cross Impact Method", ibid., 149-154.

[17] H. Ishitani, H. Ina, N. Hyodo, "Land-Mesh-Data, Population Distribution Model, And Social Capital Investments", ibid., 155-159.

[18] Y. Suzuki, H. Pak, H. Kin, H. Kamitani, "On-Line Processing System For Regional Information Data", ibid., pp. 160-166.

[19] Y. Harashina, K. Shinozaki, "Social Indicators And Political Decision Making", ibid., pp. 167-172.

[20] H. Koshihara, "Particularities Of Economic Statistics And Non-statistical Discrepancy", ibid., pp. 173-177.

[21] S. Aida, "International Relations And Systems Science", ibid., pp. 182-189.

[22] T. Kosuge, "International Laws Of Activities In Space", ibid., pp. 190-196.

[23] T. Terano, "Aggregation Of Nodes In Directed Graph For Interpretation Of Structural Model", ibid., pp. 197-201.

[24] N. Honda, "Comparison Of Several Methods Of Displaying Multivariable Data", ibid., pp. 202-207.

[25] H. Kodama, "Improvement Of Data Processing System For Decision Making", ibid., pp. 208-213.

[26] T. Terano, "Scientific Information For Measurement & Control Engineering", ibid., pp. 214-235.

[27] T. Terano, et al., "ISM - Theory And Practice In Control Engineering", ibid., pp. 236-252.

[28] E. Fontela, "Perceptions Of The World Problematique, Report No. 2", DEMATEL 1973 Reports, Battelle Geneva, October, 1973.

[29] J.N. Warfield, "Societal Systems - Planning, Policy And Complexity", John Wiley & Sons, 1976.

[30] T.J. Gordon, H. Hayward, "Initial Experiments With The Cross Impact Matrix Method Of Forecasting", Futures, Vol. 1, No. 2, 1968, pp. 100-116.

[31] J.C. Dupperin, M. Godet, "SMIC74", Futures, Vol. 7, No. 4, 1975, pp. 302-312.

[32] Y. Kaya, M. Ishikawa, S. Mori, "A Revised Cross Impact Method", Tech. Forecasting and Social Change, Vol., 11, 1979, (to appear).

Scientific Information Systems in Japan
H. Inose (ed.)

SOFTWARE ENGINEERING APPROACH TO SCIENTIFIC INFORMATION SYSTEMS

Yutaka OHNO

Department of Information Science, Kyoto University, Yoshida Hon-machi, Sakyo-ku, Kyoto

A computer aided design system for scientific information systems, named IDEA (Integrated Design and Evaluation Aid) system is presented. The IDEA system makes it possible to verify and evaluate the design result before its implementation. It is based on two concepts; a flexible and extensible description language, and the verification by observing the behaviours of the system described.
A software engineering approach for the DB design is also presented. It includes three steps; to define the requirements of the target system, to derive the requirements conceptual schema and the requirements external schemata and to calculate an actual weight of an access path to optimize the DB system. The requirements descriptions about an application program of the DB system are treated as a requirements external schema. The system designer will read these descriptions and merge the several terms which mean the same concept. It is regarded as a requirements conceptual schema. With the procedure description, he can get the frequency of the data access through a certain path to design the optimal DB system.

1. INTRODUCTION

It has been widely recognized that a systematic methodology supporting the whole life cycle of a software system is needed to make it high reliable. The life cycle consists of a requirements (or needs) analysis, a specification, an implementation and a maintenance. So much software development has been done without the efforts of the consistent specification or the valid design, then some people are interested only in programming technique and confuse it with a designing one.

We have investigated several computer-aided systems to support the development activities in the ealier stages of a software life cycle. Because a programming will be an easy job, we think, if a valid and precise description of the design result is afforded to the programmers. There are two important concepts to check the validity of the description, the observation of the system behavior and extensible description language.

The system developed is named IDEA system which is the abbreviated form of Integrated Design and Evaluation Aid system. The activity's results at each stage of the software development should be validated and evaluated. Up to now, the requirements specification and the design descriptions are checked only after the implementation of the system. It may be one of the reason for making the cost and period of the system development more than scheduled. These problems can be solved with the integration of the description and its validation / evaluation.

The fundamental ideas on which IDEA system is based and the brief explanation about IDEA subsystems are presented in this paper. The IDEA system can be utilized during the formation process of information systems. Moreover, the database system will play an important role in the organization of scientific information. Then, the example of the software engineering approach to the design of the data base system is also mentioned.

2. DESCRIPTION AND VERIFICATION

To explain the designer's intention clearly, the suitable language to its application field is required. If the description language has a flexible and extensible language structure, we can use it in any stage of the system development. Then the close communication between the requirement definer and the system designer can be expected.

The functions required in the target system are realized with a sequence of the actions of the computer. The gap between the static description and the dynamic system causes some trouble. The validation of the description will be done with the observation of the system behavior.

2.1 Necessity of a Flexible Description Language

The different system may include the different concepts. We need the common concepts to communicate to each other. In the software life cycle, the communication is required at each transition between two stages. There is a design result from the preceeding stage view and a design specification from the succeeding one. If a description language which can be used at any stage is available, it is very easy for a designer to look over all descriptions and to get more precise information about the specifications.

The natural language daily used has such features as a flexibility, extensibility and understandability. Daily conversations often include ambiguous meanings which we can suppose with their context. We intend to utilize the computer to aid the designers' activities. Since the ambiguity of the language may cause some trouble,we need the artificial language whose semantics are well defined and in which the designer can write his description easily.

A concept is often represented with one term or one symbol among the members of the restricted group. In this case, the communication depends mainly on the term used in the description, not on the syntax of the language. Then we can use a same language at various stages of the software development with a corresponding level of abstraction, when a proper language is provided. In a very early stage, the communicability is seemed most important. In a latter stage, the rigorus validation or formality may be most significant. We have proposed the language definition system which allows a designers' group or a project to define their own language favourable to their purposes. The description evaluator of the language derived from the definition is automatically constructed and used for the verification of the design descriptions.

2.2 Verification Based on Operational Semantics

The information system in some development phase is faced to the corresponding level of the nested abstract machine. When we suppose that the ultimate center of this onion is a conventional computer hardware, it is an adequate way to explain or to define the required function as a sequence of the operations of the abstract machine. To observe the behavior of the abstract machine is the best way to check whether the designers' intension is completely described or not. Consequently, it is a recommended way to avoid the designing error that the gap between the static description and its dynamic behavior of the system implemented is reduced (narrowed)

The check of the dynamic behavior is done by observing how the inputs or events of the system is processed and through which path of the system description the output is acquired. However, it is difficult to make the description be executed without the resource allocation. In the IDEA system, such problem is resolved with the man-machine interaction. That is, when the evaluation system meets the description which it cannot perform without the activities of the detailed design, the designer should tell what to do.

The descriptions of the earlier stage of the system development, e.g. the requirement description or the system specification, are too much abstract to be executed. Therefore, we provide the facility to describe the dynamic properties of the system. With the comparison of the static descriptions and the dynamic ones, the definer can verify the consistency, the completeness, the traceability and so on of the description more precisely.

In the latter stage, the designer can describe two aspects of the system, i.e., the control structure and the data flow. The evaluation system will check the contradiction between two aspects to find the designer's error.

3. COMPUTER AIDED SYSTEM FOR SYSTEM DEVELOPMENT

The IDEA system has been developed to support and to maintain the design activities of whole life cycle of the information systems. As the subsystems, it inculdes the verification system of the requirements descriptions, the automatic module construction and evaluation system, the universal interpretation system based on operational semantics, and the software description and evaluation system.

3.1 Verification System for the Requirements Descriptions

The requirements description is the result of the analysis of the function and performance required in the target system. This description is quite high level in the abstraction. Then, if the system cannot recognize the meaning or concepts of the terms used, the precise verification by the computer may not be

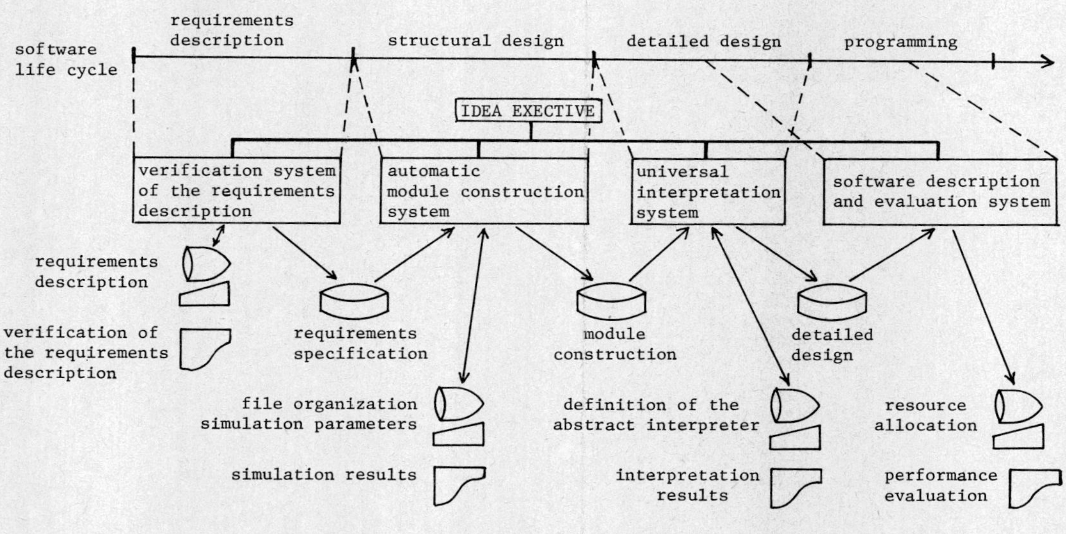

Fig.1 The organization of the IDEA system.

expected. In the state of art of the computer understanding of the natural language, it is very difficult for the computer to get full meaning of the sentence which has high level abstraction.

In many support system of this stage, only the syntactical errors and the undefined input-output relations are tested. To check the semantics of the description, ASL and its analyzer (Assertion Statement Lanruage/Analyzer) have been provided in our case. They allow the requirements definer to describe the dynamic aspects of the system and to analysis his description from the dynamic view point. It is not the description for defining the target sytem. Therefore, it is treated as an assertion and does not affect to the requirements descriptions directly. If needed, the definer should modify his descriptions.

3.2 Module Design Aid System

The Module Design Program(MDP) is developed to construct a design proposal from a requirements specification. MDS produces module structure concurrent-executable maximally and the execution paths of the transactions (or inputs). It allows for the definer to simulate his descriptions to evaluate its performance. The results of the computer simulation can be relied on because it is obtained using the module design in some sence.

In many cases, the requirements description does not include the definition of the dynamic flow structure but only the definition of the functions and the performance requirements. It may be recommended since the requirements description should not impose a particular design selection and leave free as much as possible If possible, the flow structure of the target system is remained undefined. However, the undefined flow structure makes the simulation difficult.

MDP can help it. That is, it derives the set of partial orders between two processes based on input-output relations appeared in the requirements description. The second, it calculates the the maximally parallel process structure. It determines a total ordering of processes according to the similarity of the preceding structure of a process and to the localization of the access of a particular data. The simulator is activated to evaluated the performance of the target system whose structure is derived as mentioned above.

3.3 Universal Interpretation System based on Operational Semantics

It is desirable that a system designer can use any design language suitable for his target system on his abstraction level. It is also needed for him to know the dynamic feature of his system on his abstraction level before its implementation.

We propose the universal interpretation system which consists of a definition language for a design language (ODL: Operational Definition Language) and its interpreter. The design language, which contains abstract concepts to be considered primitive for the target system on his abstraction level, is defined in terms of ODL primitives and user-defined abstract primitives. The user- defined abstract primitives represent "well-known and common concepts among designers".

The language definition written in ODL is the definition of an abstract interpreter, which includes abstract syntax of the language, state syntax of the abstract interpreter, initial and final states of the interpreter, and instruction definitions of the interpreter. The ODL interpreter accepts definition texts, interprets a system written in the defined language, and reports the dynamic feature of the system.

3.4 Software Description and Evaluation System

The time and conceptual gaps between the design activity and the evaluation activity cause many troubles. For example, whether the model used for the evaluation corresponds correctly to the design description or not is very difficult to check. The language which makes for designers to integrate two

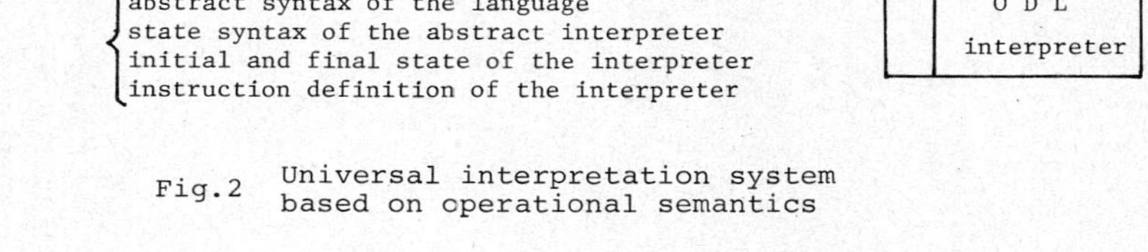

Fig.2 Universal interpretation system based on operational semantics

activities has been proposed and implemented.

The two aspects of the system are described in one description, i.e., active entity which processes the data, and passive entity which is processed by the software system. The designer first describes the structure of the target sytem with active entity , i.e., a process view. Then he overlaps the behavior of the passive entities on the structure. The unit which represents a passive entity and traces its behaviour is named "traverser". The description of a passive entity means the assignment of the paths and attributes of a traverser. The evaluation using a traverser is called a traversing method.

The designer can utilize this integrated description and a traversing method to define the software system and to validate the functions and the performance.

4. SOFTWARE ENGINEERING APPROACH FOR DATABASE DESIGN

Today, it is recognized that all stages of the life cycle of the information system including the requirements describing stage should be properly maintained to get the reliable system. If the system now under designing is used by only designers and implementers, the developed system will work as intended. In usual cases, however, its user is neither the designer nor a specialist about the information system. A close communication between a system designer and users is required to realize such a system. Almost all DB systems belong to the class of such systems.

It is easier for non-specialists to express their intension about a target system with their mother languages, i.e., natural languages than with a special language such as a programming language. Daily conversations often include ambiguous meanings which we can suppose with its context. Since this ambiguousness should be deleted for the system specifications, we use an artificial language whose meanings are well defined.

Several languages for requirements descriptions are proposed. For instance, PSL(Problem Statement Language) of Michigan University [ISDOS(1975), Teichroew(1977)], RSL(Requirements Statement Language) of TRW [Bell(1976)], BDL (Business Definition Language) of IBM [Goldberg(1975)] etc. are known well. Moreover, there are several papers about the formal specification for Database systems [Smith(1977), Lockemann(1979)]. However, a few papers are presented about the Database System Description using a Requirements Engineering approach[Kahn(1976)]. In this paper, we propose how to describe the requirements of the DB systems and how to utilize them to design the optimal systems. We adopted PSL as its language. However, we assign some special semantic to get the design parameters with a computer.

To define DB systems, their two aspects should be described, that is, data structures and data utilizations. We call the former aspect static requirements and the latter dynamic requirements.

The requirements descriptions which represent a direct interface between the system and users are treated as an external schema in the ANSI/SPARC DBMS model [Jardine(1977)]. There are many external schemata corresponding with DB applications. Each application has its own view of the DB correspond- ing with static requirements and procedures defined as dynamic requirements. The requirements description of DBMS should clarify what kinds of data are stored, retrieved and updated, how often such processes are activated, and how the data in the DB relate to each other. Some relations among the data are so inherent that they may be regarded as a fundamental common sense in the DB system. Then such relations derived from the requirements descriptions will be regarded as a conceptual schema, more precisely said, a requirements conceptual schema of data structures.

For designing the DB systems, the system parameters such that process operations, their frequencies and the cardinalities of relations should be also defined to determine how often a certain access path is used. The requirements about typical processes can indicate a dynamic mapping ratio between two relations.

Since such a ratio varies from process to process, we provide several data structures which are suitable for each process. These data structure can be corresponds to an external schema. Then we call it a requirement external schema.

4.1. CORRESPONDENCE OF REQUIREMENTS TO DBMS LAYER

4.1.1 DBMS layer

In this section, we present the DBMS layer which is studied by ANSI/SPARC and approved widely as a guideline for the future investigation. The DBMS architecture is partly based on the concept of "nested machine". Users of DBMS access the data through several layers which called an external schema, a conceptual schema and an internal schema. Users can define an external schema as own model which is relevant to their particular needs. Some users use only a subset of the whole provided model. This subset of models is described as an external model.

The conceptual schema represents the current "best model" or "way of doing business" for DBMS establishment. It is treated as a common denominator between the data structure installed and several views of particular users. The internal schema is an intermost schema and related closely to machines. It defines how the data are stored. However it doesn't define its actual storage strategy. One of the benefits of this three layers architecture is that we can modify an internal schema without affections to the relation between a conceptual schema and many external schemata. As against this, the mapping between a schema and subschema should be redesigned in DBTG mapping mechanism when the schema is changed.

4.1.2 DB requirements layer

Now, we follow the approach of the ANSI/SPARC framework. We consider three aspects of requirements descriptions which correspond to each layer of a DB system. Requirements descriptions represent the interface between the DB system and the real world where it operates. Then, the most important aspect represented in the requirements description is an external schema since it faces directly to users. However, if there is no common concept among the application programs using the DB, the designer or implementer will lose sight of the target system. It means that the requirements descriptions about a conceptual schema also should be represented in some form. With regard to the principle that we should leave as much room as possible for designers' choice, the user need not to describe about an internal schema. Since an internal schema have close relation not only to the conceptual schema but also to the system configuration, it seems reasonable that the internal schema is left undefined until the latter phase of the system development.

4.1.3 Requirements description for external schema

The requirements description can be regarded as an external schema since it defines the interface between a real world and the DB system. It expresses a static data structure and a dynamic one.

The static data structure are defined with the descriptions of the relation between two entities and the inclusive relation between the entity and the elements. We call it a static requirements. In the requirements description, we also define the processes which shows how the data in DB are used through the user's view, how frequently it will occur and the the effective cardinality of the entity determined dynamically and so on. We call it a dynamic requirements.

It is impossible that the requirements definer describes all procedures accessing to the DB which will appear in the real operations. Then, he will write only the typical or most frequently used procedures. It is true that there may be sometimes the procedure which is frequently used and he passes over. But it has no problem if it accesses to the DB legally. Generally speaking, there is no existency without the perception in the requirements description. The designer will make use of this static and dynamic requirements to get the suitable conceptual schema and its internal schema. He can add some function and schema which, he thinks, will be needed in future or in the extended use. It can help to recover some portion of the requirements definers' overlooks.

In this paper, these static data structure are defined mainly in the ENTITY section with attribute statements or the RELATION section. Since some part of the requirements external schema is defined from a dynamic view point, it is depicted in a procedure definition. In our case, it is described in a PROCESS section. With their inputs and outputs, we can know which data is retrieved or updated, which access path is selected to get the data etc. If there is more than one access path defined by a pair, that is an input of a particular process to an output, the requirements definer can specify one of them using a MAINTAIN statement.

4.1.4 Requirements description for conceptual schema

The conceptual schema is a fundamental relation of entities and doesn't vary with each application view. The example of these requirements conceptual schema is the relation between an entity and its inherent attributes. A man has his name, address, years, height,...etc. If we deal with individual people, we can not say about a man without his name and address. It doesn't depend on its application.

There may be several types of a data structure. It seems natural that we provide three types; relational structure, hierarchy structure and network structure. These types are well known as typical data structures in DB system. However, it does not mean that the requirements definer should declare what type of them is used in his description. He can define his data structure model as a relational one partially and as a network one in another part. The system designer or implementer is held responsible for what type of a data structure is adopted in the target system.

The requirements definer does not need to state the conceptual schema. It may be specified by the designer or implementer. However, it is better to provide the function to derive the candidate of the conceptual schema with a computer, if possible. We show the correspondence of the ANSI/SPARC DB layer with the requirements layer in Fig. 3.

It is a very difficult job for a computer to decide what is a fundamental relations in the requirements descriptions. The trivial method to derive a conceptual schema is to make a union of all external schemata defined as the static requirements. We should utilize the information defined in the requirements de- scription as much as possible to get a suitable system. Then we calculate the usage frequencies of the relations and attach them to the union as their weights.

4.2. REQUIREMENTS DESCRIPTION LANGUAGE FOR DB SYSTEM

We adopt PSL to define DB systems, since it has an enough capability from the syntactic view. We add some semantics to it to get the external schema and the conceptual schema of the target DB system. The addition of semantics means the restrictive use of some reserved terms such as HAPPENS, CARDINALITY, RELATION etc.

PSL is one of the languages which support to maintain the documentation describing the target system, the environments where it operates, and the project organization. One of the most important features of PSL is a computer-aided management of the documentations. PSL will give full play when combined with its analyzer PSA (Problem Statement Analyzer).

4.2.1 PSL AS DB requirements description language

PSL is invented as a definition language of the requirements of arbitrary kinds of information systems. The concept appeared in the target system may vary from each other. To accept all varieties of the system concepts, PSL has a simple and extensible language structure. Sometimes it is too simple to analyze its semantics in detail. The meaning of the descriptions is left undefined until the common concepts or mutual agreements are founded between the definer and the reader. For example, the inputs of a process are defined with USES statement in PSL. We can define the data and/or the control as its inputs. It is very difficult to distinguish them without other information, that is the meaning of the system and the terms, the environments, the definer's intensions, etc. However, with a careful investigation of PSL, it comes to light that almost all concepts appearing in DB system requirements can be defined using PSL syntactically.

4.2.2 Correspondence of PSL statements to DB design concepts

We add some functions which derive the useful parameters for the DB system design to PSA. To utilize these additional functions, the requirements definer should be careful to use some special reserved words of PSL. We show the correspondence of PSL sections and statements to the terms in DB fields.
(a) SET: A SET means one DB. That is a collection of ENTITYs. Where ENTITYs may be thought as logical records, a SET may be thought as a logical file.
(b) ENTITY: An ENTITY is a record type of CODASYL or a table of relational DB.
(c) ELEMENT: An attribute of a record are defined as an ELEMENT of PSL. Notice that an ATTRIBUTE statement of an ELEMENT section shows a domain of the attribute of DB. That is, there is no description of each occurence in DB.
(d) RELATION: An RELATION defines a relationship between two ENTITYs. This correpondence a set, i.e., the owner-member relationships of CODASYL. This section defines the type of DB. That is, if there is no RELATION description, it means a relational DB. If there is, a hierarchical DB or a network type depending on the existence bi-directional relation.
(e) USES/DERIVES: In PSL, a PROCESS USES some data and DERIVES another data. If such data is an ELEMENT or an ENTITY defined as an attribute or a table of a DB, we assume that this PROCESS will access this DB. Moreover the access is assumed to be executed through the path defined in RELATION sections. If there is more than one path, we can define explicitly one of them using MAINTAIN statements.
(f) others: The frequency of the activation of a process is defined with HAPPENS statement. The cardinality of ENTITYs or ELEMENTs is defined CARDINALITY statements.

4.3. REQUIREMENTS SCHEMA CONSTRUCTION

The requirements descriptions written in PSL may be used to support the whole development of the target system. However, we will treat only the part of it to get the DB schema of the requirements phase.

The requirements descriptions of DB systems include the data descriptions and the process descriptions. If the designer or implementer of the DB systems defines the requirements. They shall be regarded as an conceptual schema of DB. On the other hand, if the application system designer or the user, they will be external schemata. Here, we consider the latter case, that is, in the case that the requirements is defined by users.

Each description defines the data structure derived from the static requirements. The weights of its structure are attained by the calculation of the utilizing frequency of the data.

4.3.1 Amalgamation of data

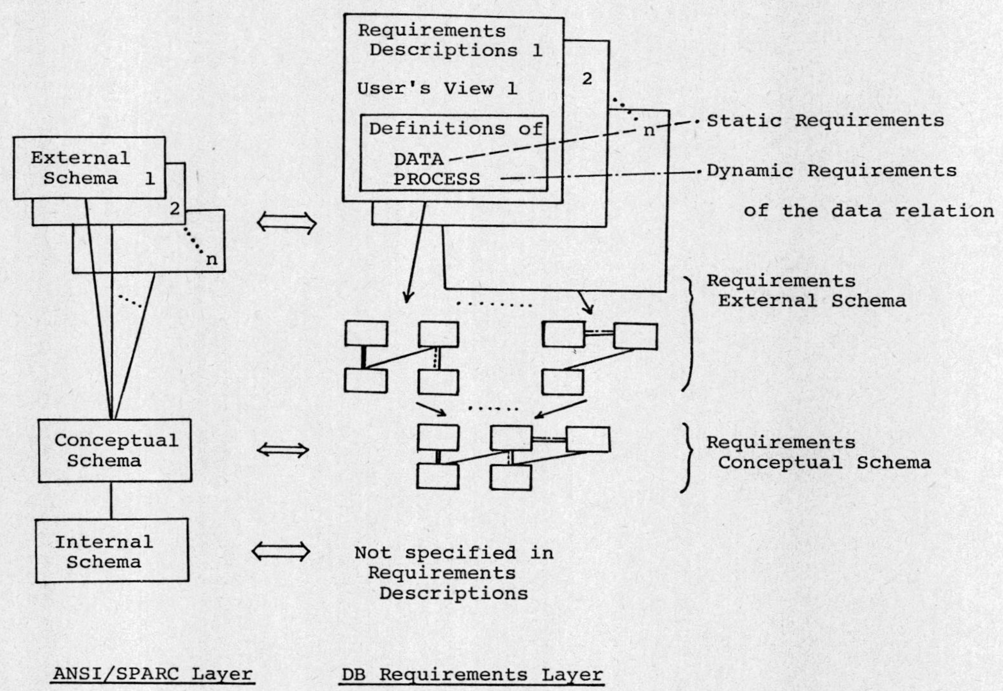

Fig.3 The Correspondce of the ANSI/SPARC Layer with the Requirements DB Layer.

To design a DB system, we need a certain conceptual schema which represents a common concept in user's view. The requirements in PSL is easy to read and understand because of its formatted outputs. However, some terms in one description may be used to express different meanings in other one. In other case, the different terms are used for the same meaning. It is very difficult to solve these problem with a computer. Some conversation between the DB design support system and the designer is required to combine the descriptions written by more than one requirements designer.

There is no section to combine the descriptions of different DBs of requirements description into one in PSL. However, the COMBINE command is provided in PSA to correct errors in the descriptions of a DB. For example, a requirements definer remembers a spell of some word incorrectly and have written his descriptions with the erroneous spell. To correct all errors is a tedious job. If he can find the use of DATABASE and DATA-BASE as a same meaning and wants to merge them, then he will type in a following command.

```
COMBINE      CNAME=DATABASE    DNAME=DATA-BASE
```

As a result of the execution of this command, DATA-BASE is deleted and both informations are amalgamated in DATABASE. It is true that it is not a recommanded way to use a COMBINE command, but a DB designer can use it to derive a conceptual schema. If some data or its structure is seemed to be natural, common or inherent, he can describe his own ENTITY or RELATIONs between them and combine such definers' data into his data. If needed, he should combine also the ELEMENTs which appear in CONSISTS statement in the ENTITY section.

The requirements conceptual schema is defined as a union of the external schemata combined by the DB designer.

4.3.2 Weight of relations

Here, we will consider the dynamic behaviour of the DB system written as the dynamic requirements. When a process P_i recieves an input I_i whose cardinality is C_i, uses it as an access key of a DB and derive an output O_i, what is the cardinality of O_i? The cardinality of O_i may be defined in the description of O_i. In this case, however, the cardinality of O_i depends on the number of the occurrences which are retrieved from the DB with the corresponding key. Of course, it depends on also the task of the process. It is very difficult to calculate its dynamic cardinality using the task of the process without understanding the meaning of the description. In the RELATION section, the definer can define its connectivity N in the CONNECTIVITY statements. We use the ratio of two cardinalities of the ENTITYs for the CONNECTIVITY if there is no description of the RELATION like in the case of a relational DB. If there is more than one RELATION, we can chose one of them using MAINTAIN statements. This facility can help to get more precise value when the definer assigns the individual value to the different process.

We use this value to calculate the dynamic cardinality, e.g., C_i N in the above case. This is illustlated in Fig.4. If there is a feedback loop, we can assume that the input cardinality is $C/(1-\rho)$ where C means the input cardinality defined as the static requirements, and ρ means the probability of the feedback from the output to the input. If there is more than one DB access in a PROCESS description, the dynamic cardinality of the output can not be estimated so easily. For the simplification, however, we define it with the maximum number among them. These values are regarded as the weights of the relations between the data. It is easy to get these values in a

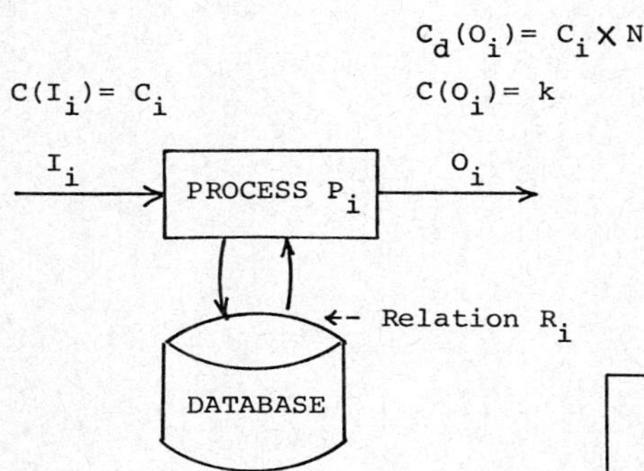

C(a): the cardinality of a.

C_d(a): the dynamic cardinality of a.

```
PROC Pi;
  USES  Ii;
  DRVS  Oi;
REL  Ri;
  BTWN Ii AND Oi;
  CONN IS 1 TO N;
ENT  Ii;
  CARD Ci;
```

Requirements Description

Fig.4 The Calculation of the Dynamic Cardinality.

computer-aided way. The designer can decide which relation should be maintained by the inverted files with them. Moreover, he can add the shortcut access path from a ENTITY to another if all calculated weights between them are significantly large.

The dynamic aspect of the conceptual schema is obtained in the same way. That is, the designer combines several PROCESSes into one, if needed. Next, the activation frequencies of the application programs are multiplied to the weights. Then we sum up all of them. It is regarded as the weight of the relation in the requirements conceptual schema.

Consequently, we can show the equation of the weight W of the relation as follows;

$$W = \sum_i P_{ix} \max(C_{ij} \prod_k n_{ik})$$

$i \in$ all requirements external schemata

P_{ix}: the frequency of the activation of the process i corresponding one of the requirements external schema.

C_{ij}: the cardinality of j which is an input of the process i.

n_{kj}: the connectivity of the relation. k is defined with the path along which the global inputs reach to its subprocess.

5. CONCLUSION

The IDEA system which will help the designer to develope the scientific information systems. It provides the powerful description languages and the verification tool based on operational semantics. However, the current state of the IDEA system is not completed one. We are now researching what is most important in the man-machine system for the development of information systems. Moreover, the fluent transition between two subsystems of the IDEA system is also under the investigation.

The development of structured or automated DB design methodology is one of the main subjects in the DB research fields. The formal description of the requirements of the target DB system is a first step of it. It is said that the requirements description should have the property called design free, that the requirements description does not impose the particular design selection. However, the approach described in this paper is introduced to propose the particular design selection for an automated DB design. Of course, some parts of design selection are left to be decided by the designer, e.g., what should be seemed as a conceptual schema, what is most important in the DB application, security, privacy or others, etc.

The DB designer should check all external schemata described in the requirements with PSA commands. Since the command to get the information along the particular pathes not provided, he must calculated the access frequency with PSA reports. We have proposed ASA (Assertion Statement Analyzer) to verify the requirements descriptions described in PSL [AGUSA(1979)]. It has the base on the path analysis of the system structure or the data derivation of the system aspects. We will extend our ASA to enable to output some reports about the dynamic data utilization. Moreover, a facility to help deriving a conceptual schema from requirements descriptions is now under the implementation.

REFERENCES

ISDOS Project. (1975). "Problem Statement Language Version 3.0 Language Reference Manual". ISDOS Working Paper, No.68. May 1975.

Teichreow D. and Hershey E.A. (1977). PSL/PSA:A Computer-aided Technique for Structured Documentation and Analysis of Information Processing System. IEEE Trans. on Software Engineering, Vol.SE-3, No.1. January 1977, pp 41-48.

Bell T.E. and Bixler D.C. (1976). A Flow-oriented Requirements Language. Proc. of MRI Symposium on Computer Software Engineering, April 1976, pp 109-121.

Goldberg P.C. (1975). Structured Programming for Non-Programmer. IBM Report RC-5318. March 1975.

Smith J.M. and Smith D.C. (1977). Database Abstraction: Aggregation and Generalization. ACM Trans. on Database Systems, Vol.2, No.2, June 1977, pp 105-133.

Lockemann P.C., et al. (1979). Data Abstraction for Database Systems. ACM Trans. on Database Systems, Vol.4, No.1, March 1979, pp 60-75.

Kahn B.K. (1976). A Method for Describing Information Required by the Database Design Process. ACM SIGMOD 1976. pp 53-64.

Jardine D.A. (1977). "The ANSI/SPARC DBMS Model". North-Holland Publishing Company. 1977.

Agusa K., et al. (1979). Verification of Requirements Descriptions. Proc. of 12th HICSS, vol.1, January 1979, pp 131-139.

Itoh k. et al. (1978). An Evaluation System for Concurrent Processes by the Traversing Method. Proc. of 3rd UJCC, October 1978, pp 41-45.

Hirota T. et al. (1978). Computer Aided Design of Software Modeules. Proc. of 3rd UJCC, Ocober 1978, pp 220-224.

Tabata k. et al. (1978). A DBMS for a Composite Type of Distributed Data Base. Proc. of 3rd UJCC, October 1978, pp 187-191.

Scientific Information Systems in Japan
H. Inose (ed.)

AN UNDERSTANDING SYSTEM OF NATURAL LANGUAGE AND PICTORIAL PATTERN IN THE WORLD OF WEATHER REPORT

Tuneo TAMATI, Eiji KAWAGUCHI*, Masao YOKOTA** and Tsutomu ENDO**

** Graduate School of Engineering Sciences, Kyushu University, 6-10-1 Hakozaki, Fukuoka 812*

*** Information Science Laboratory for Bio-Medicine, Kyushu University Hospital, Fukuoka 812*

An information understanding System Of BAsic weather Report(ISOBAR) is introduced first. It can adapt to both linguistic and pictorial input, and produce each output. The central part of this system is semantic processing, but ISOBAR's world is limited to weather report of Japan and Far East Asian area in terms of Japanese and corresponding weather chart. Acording to experimental results, the performance of the system proved to be satisfactory, except the computing time of picture processing procedures. Finally the problems for the future research are pointed out.

1. INTRODUCTION

The present paper is based on the experimental research for understanding system of natural language and pictorial pattern. The prototype system is called ISOBAR(Information understanding System Of BAsic weather Report). The model was designed and developed after the following phylosophy:

1. In order for human/machine to understand natural language, each word must be associated in some way with the human/machine knowledge about a external world, or their experience in it. The external world in our case is mainly the visual world.
2. Understanding of pictorial pattern can not be achieved without extracting and recognizing such specific evidences or features in the given pictorial information that show clear correspondence to each concept in human mind.
3. Through these interactive mechanism, man or machine can explain the pictorial pattern by using natural language and vice versa.

The ISOBAR's world is limited to only Japanese sentences of weather report and weather chart of Japan and Far East Asian areas. But from the view point of semantics, the fundamentals of processing algorithm or mechanism of this limited world is applicable to more general world.

ISOBAR has two operating mode in principle. One is accumulation of meteorological database both from linguistic and pictorial inputs. The other is its retrieval. In the accumulation mode, man presents weather report sentences and charts to the machine. The machine interprets and understands the input information semantically by the aid of general knowledge about meteorology. It transforms the input into suitable data format and adds to the database if and only if the input is meaningful and conveying new information. If not, the machine either inquires about ambiguous points, or rejects the total input.
The semantics in the final version is designed to have such hierarchy as in Fig.1. GKS is the symbol for General Knowledge about Semantics, and GKM for General Kowledge about Meteorology. SWM and KMP are Semantics inthe World of Meteorology and Knowledge about Meteorological Phenomena respectively. Generally speaking, it is difficult to distinguish GKM from GKS strictly.

The present ISOBAR is not completed as a total system with full semantic components as in Fig.1. SWM is already implemented in the present system, because it is the core of the semantics in the system, and fortunately it has rather simple structure. In the meantime, GKS must cover all field of our linguistic activities in Japanese. So in our case, GKS is limited only to general linguistic knowledge which is essential to the field of weather report. Present ISOBAR is partly containing GKS in the form of "Event of Constituent" which will be introduced in the next chapter. KMP is regarde as the total contents of an "Encyclopedia of Meteorology", but this part is fairly limited and is not explicitly separated from SWM in the present system.

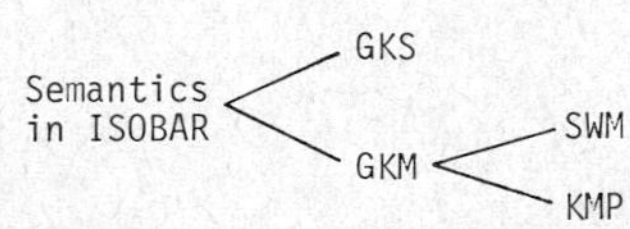

Fig.1 Hierarchy of semantics in ISOBAR

2. OUTLINE OF ISOBAR

First, the weather report sentences and weather charts for ISOBAR are outlined. Then some important semantic notions are presented here.

2.1 Sentences of weather report

We limited the words and sentence patterns for ISOBAR only to those of weather reports in Japan which are broadcasted every day on NHK NETWORK 2. We termed them basic weather reports. NHK weather reports consists of four parts:

- Part 1 General weather conditions
- Part 2 Local weather condirions
- Part 3 Reports from vessels
- Part 4 Information for fishermen

Among these, we excluded the sentences from Part 1, and all other parts were taken into our consideration. All the materials were collected from real NHK program between June 1976 and May 1977, and these were stored in the database of ISOBAR. Examples of sentences from Part 2 to Part 4 are shown in Fig.2.

2.2 Weather charts

Weather charts for ISOBAR are illustrated in Fig.3. The meaning of each pictorial component will be found in Table 1.

Part 2)
O-SAKA DEWA NNE NO KAZE , FU-RYOKU 3 TEN-KI WA HARE DE , KIATSU 998MB , KION- 23°C . (OOSAKA area, NNE wind, wind force 3, fair, pressure 998MB and temperature 23°C.)

Part 3)
KAN-TO-TO-HO- NO N35°E146° DEWA W NO KAZE , FU-RYOKU 4 , TEN-KI WA HARE DE , KIATSU 05MB . (At N35°E146° east of the KANTO district, west wind, wind force 4, fair and pressure 1005MB.)

Part 4)
(a) KAN-TO-HIGASHIKAIJO- NO N36°E149° NIWA , 1004MB NO TEIKIATSU GA ATTE , E NI 75KM DE SUSUN-DEIMASU . (A low pressure of 1004MB at N36°E149° east of the KANTO district, proceeds east at 75KM/hr.)
(b) NIHON-HUKIN- O TO-RU 1020MB NO TO-ATSUSEN- WA , N60°E140° , N56°E136° , ··· , N19°E113° NO KAKUTEN-O TO-TTEIMASU . (An isobar of 1020MB in and around Japan passes through N60°E140°, N56°E136°, ··· and N19°E113°.)

Fig.2 Examples of weather report sentences

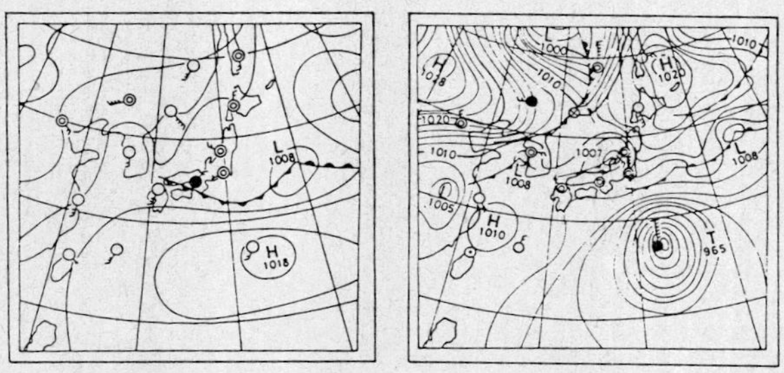

Fig.3 Examples of weather charts

2.3 Notions in the semantic structure of weather report[1],[2],[3]

We introduce three basic semantic notions to describe the semantic contents of the world of weather report. The first is "Constituent" of meteorological world in ISOBAR. The second, "Attribute" of Constituent, and the last, "Event" in Attribute-space. "Attribute-space" in this context means an abstruct space whose bases are all Attributes essential to that Event.

After careful analysis of weather report sentences and weather charts, we postulated 17 Constituents (denoted by C_1, C_2, ···, C_{17}) in all. These are tabulated in Table 2. Attributes of constituents are extracted to be 14 different kinds (A_1, A_2, ···, A_{14}) as in APPENDIX I. Each Attribute is, in other words, a semantic phase or point of view of Constituents. Relation between Constituents and Attributes makes an "Cross-relation Table" of Constituents and Attributes. "Semantic Table" is defined to be such Cross-relation Table that is filled with each "Attribute Value". A set of weather reports at a special date and time, or a sheet of weather chart, makes complete Semantic Tables, if all input sentences or pictorial components are meaningful, and all are recognized correctly. APPENDIX II shows an example of Semantic Tables.

"Event" in an Attribute-space is introduced as a semantic contents of "verb", "adjective", "adverb", etc.; those are all representing the conditions of Constituents or relation among Constituents. We visualize each event as a locus of some Constituent in an Attribute-space. There are many patterns of Events in the meteorological world in ISOBAR. Some are temporal patterns, others are spatial patterns. In the linguistic analysis of weather report, we need an explicit expression of each event pattern. ISOBAR has all list of variations of event patterns, which appear in the world of weather report. These event patterns work as both GKS and SWM in the present system. Semantic Tables, in the mean time, exclusively work as SWM.

Table 1 Picture components in weather chart

Component	Pictorial pattern	Informantion from the component
Weather symbol		Location, variety of weather
Wind symbol		Location, direction, force
High pressure	H	Location, central pressure
Low pressure	L	
Tropical depression	TD	
Typhoon	T	Number, location, central pressure
Front line		Location, variety of front
Isobar		Locus, pressure value
Coast line		No information (constant components)
Latitude, longitude		
Numeral	01 9	Typhoon number, pressure value

Table 2 List of constituent

Symbol	Constituents	Linguistic Expression	Pictorial expression
C_1	Low pressure	TEIKIATSU	L
C_2	High pressure	KO-KIATSU	H
C_3	Typhoon	TAIFU-	T
C_4	Tropical depression	NETSUTAISEITEIKIATSU	TD
C_5	Warm front	ON-DAN-ZEN-SEN-	
C_6	Cold front	KAN-REIZEN-SEN-	
C_7	Stationary front	TEITAIZEN-SEN-	
C_8	Occluded front	HEISOKUZEN-SEN-	
C_9	Isobar	TO-ATSUSEN-	
C_{10}	Ridge	KIATSUNOMINE	No direct correspondence
C_{11}	Wind	KAZE	
C_{12}	Fog	KIRI	
C_{13}	Surge	NAMI	
C_{14}	Storm	BO-FU-U	
C_{15}	Local weather	KAKUCHINOTEN-KI	
C_{16}	Weather at sea	KAIJONOTEN-KI	No direct correspondence
C_{17}	Time of observation	KAN-SOKUJIKAN-	

2.4 General performance of the system

The total configuration of ISOBAR is illustrated in Fig.4. It consists of seven processing units and three databases. Linguistic input, i.e., weather report sentences in the accumulation mode, and questions in the retrieval mode, are sent to STA (SynTactic Analyzer) unit first. If the input is syntactically correct, it is stored in the LDB(Linguistic DataBase) only in case of weather report sentences, and at the same time, it is transfered to SSG(Semantic Structure Generator) unit. SSG generates the semantic structure of the input sentence. Then SAS(Semantic Analyzer and Synthesizer) analyzes the structure. The output from SAS makes a Semantic Table which is to be stored in SDB(Semantic Database) in the accumulation mode, or to be hold in SAS itself in the retrieval mode.

On the other hand, pictorial input, i.e., the weather chart, both in the accumulation and retrieval mode, are given to PPP(Pictorial PreProcessor) first. Pictorial preprocessing consists of digitization of analog image, noise elimination, data compression, etc.. In the accumulation mode, all original data are stored in PDB(Pictorial DataBase) in a compressed form according to a special coding scheme. Output from PPP is sent to PR(Picture Recognizer) unit in every mode, where each component in the chart is recognized. Then its output is fed to SAS unit. SG(Sentence Generator) and CG(Chart Generator) work only in the retrieval mode. That is, linguistic answer to the question is synthesized by SAS using SDB and generated by SG, while pictorial answer is synthesized by SAS using SDB and PDB, then generated by CG. More details about these units and databases will be presented in Chapter 3 and 4.

Fig.5 shows the hardware system of ISOBAR, The main system has special I/O devices, all of which are peripherals of either T-40(TOSBAC-40C) mini-computer or O-4300(OKITAC-4300b) multi-CPU mini-computer system.

3. SENTENCE UNDERSTANDING

The outline of processing in ISOBAR for linguistic input is presented in this Chapter.

3.1 Restriction for input sentences

Original weather report sentences have mostly regular sentence patterns. In ISOBAR we admit any sentence that satisfies the following restrictions.

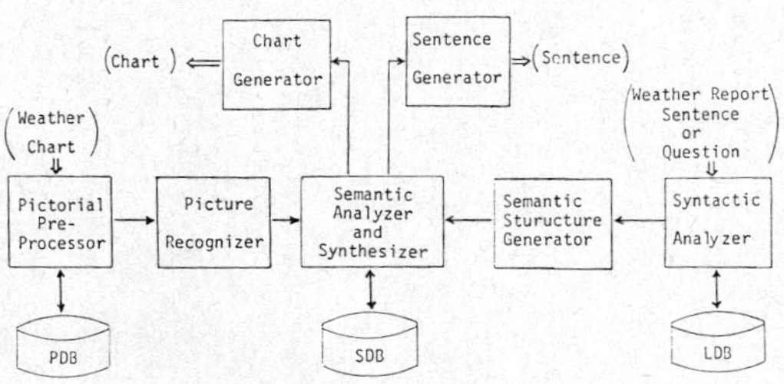

Fig.4 Configuration of ISOBAR

[Restriction 1] The vocabulary for input sentence is about 200 words, most of which is adopted from original weather report sentences, and some specific words are supplemented for interrogative and imperative sentences, e.g., "ITU(when)", "DOKO(where)", etc..

[Restriction 2] The forms of input sentences are either (1) imperative form like, "···GATA NO TEN-KIZU O SHUTSURYOKUSEYO(Output all weather charts of ··· type)", or (2) such forms that are arbitrarily transformed from original kernel sentences in weather report by the optional rules R1 - R5.

R1: Alternation of words in kernel sentences.
R2: Addition to such a word-group as "1976 NEN- 9 GATSU 11 NICHI 12 JI (WA)", which denotes a special date and time.
R3: Substitution of interrogatives to words or word groups of kernel or transformed sentences.
R4: Abbreviation of words.
R5: Addition of interrogative particle "KA".

[Restriction 3] Input Japanese sentences are segmented word by word with spaces.

3.2 Syntactic analysis, semantic structure generation and semantic analysis

The functions of STA and SSG in ISOBAR are as follows.

Input sentences are transformed in STA from surface structure into dependency structures by steps (1)-(4) below, and then into semantic structures in SSG by step (5).

Step (1) Parsing into indivisual words.
Step (2) Forming word groups.
Step (3) Construction of inter-word dependencies within each word-groups.
Step (4) Construction of dependencies among word-groups.

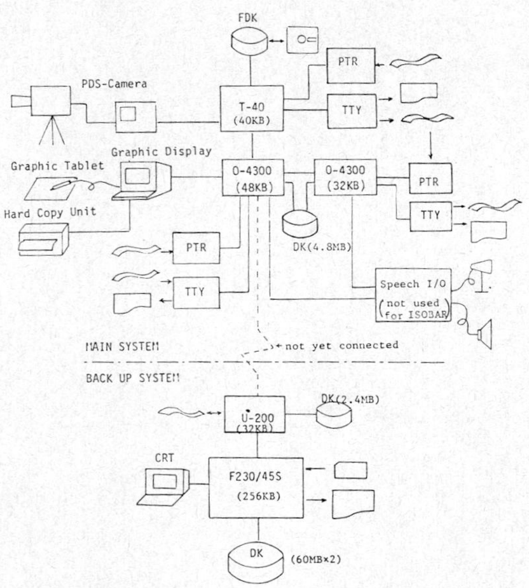

Fig.5 Illustration of the hardware system

Step (5) Transforming dependencies into semantic structure(i.e., Semantic Table) by referring the Dictionary of Conceptual Connection(DCC).

At the end of Step(4), each input sentence is identified with one of the template patterns, where verbs are essential keys. DCC as above is the most important dictionary that indicates which word-groups are representing "Constituents" or "Attribute Values", and in which slots in the Table they are to be stored.

In the course of above steps, syntactically anomalous sentences are rejected in STA. Several types of anomaly are detected in SSG, and in this case no semantic structures are generated. The semantic structures just generated are analyzed by SAS, and semantic contradictions are detected which can not be interpreted or understood in the world of weather report. This is realized as follows. SAS examines every slot of the Semantic Table whether stored Constituents and Attribute Values are correct. If not, the semantic structure is contradictory and rejected.

3.3 Access to SDB and output generation

Response of the system to a legal input sentence is detected according to kinds of input sentences. It is as follows. Case(1) is the only case that open the accumulation mode of ISOBAR, while Case(2)-(4) open the retrieval mode.

Case(1) Declarative sentence

Every declarative sentence is assumed to be asserting some facts about weather. SAS compares the generated semantic structure of the input with every structure that was already stored in SDB, where comparison is executed on every Attribute. If the input is conveying new information to SDB, it is stored there, otherwise SG comments that it is already stored or contradictory.

Case(2) YES-NO question

Case(3) WH question

Case(4) Imperative sentences

If an imperative input sentence demands all charts of a specific type of pressure configuration, the system interprets the input as "When did the pressure configuration of ···type occure?". Examples are shown in APPENDIX III.

4. PICTURE PROCESSING

In this chapter we describe the operation of PPP and PR in ISOBAR.

4.1 Pictorial preprocessor(PPP)

An analog image is converted into binary-valued picture with 1,024×1,024 pixels by PDS-camera. The noises caused by quantizing operation are eliminated by simple maskes. The preprocessed pictorial data are sent to PR. At the same time, they are stored in PDB. In this case, we took a strategy to reduce the amount of storage. In our working system, a picture is represented by a series of black or white square resions of verious sizes. The square size, its location, and the gray level(either black or white) are encoded in the structural information of "0", "1" and "(" series. We call this expression DF-expression[4],[5]. DF-expression of original chart, together with their second data, are stored in PDB. The second data include such items as complexity of picture, spectrum of primitives, headings of weather pattern, etc.. These are used for chart searching operation in the retrieval mode.

4.2 Picture recognition(PR)

The major function of this unit is to extract and recognize each picture component as shown in Table 1. We have some a priori knowledge about weather chart in PR. They are as follows.

(1) Every weather symbol is enclosed with a circle of a constant size. The variation of symbol is limited only to its inside pattern.

(2) Other components are not connected to the circle except wind synbol.

(3) There are 15 weather symbols in a chart. The location of respective symbol is fixed.

(4) The location of coast lines, latitudes and logitudes are fixed, but they are partly erased or interrupted by other components.

(5) Only front symbols and a few kinds of weather symbols have areas. The others are line drawings.

(6) Numerals are located near the characters or isobars. They specify the number of Typhoon or pressure values.

(7) The shape and size of numerals and characters are normalized.

The flow diagram of PR is shown in Fig.6, where each number within parentheses corresponds to above a priori knowledge which is used there. The final output from PR makes the Semantic Tables. Thus all information from a weather chart is recognized as the semantic content of the world of weather report at the given date and time. We illustrate the results of 1 - 5 processing steps in APPENDIX IV. Those experiments were carried out on the full picture, that is, the picture with 1,024×1,024 pixels. Though, the display terminal covers 780×1,024 pixels. So the upper and lower most part is not displaied in each picture.

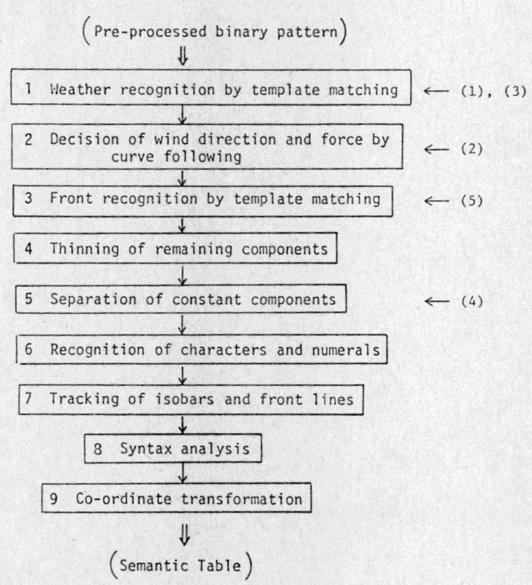

Fig.6 Flow diagram of PR

5. DISCUSSION

According to our experience, the performance of ISOBAR in the accumulation mode for linguistic input was very satisfactory. Also in the retrieval mode for linguistic input, the system worked successfully. In other words, we can say that the present restrictions for input sentences are rather severe. So do the vocaburary and sentence patterns. We should remove such restrictions as much as possible in the next version of ISOBAR.

On the other hand, its performance for pictorial input was not so much successful. There are no fundamental problems left for the performance of pictorial output, that is, the retrieval operation. But in the accumulation mode, or in the recognizing operation of the system, some problems are left unsolved. For instance, computation time is the most untolerable problem. This is partly because of our small (in terms of the size of the main memory) and slow (cycle time) computer system, and partly because of large size of input pictures. Another problem left are improvement of curve following algorithm. Present ISOBAR works in the man-machine interactive mannar, but we must find some good idear of automatic gap tracer. Any solution, if ever, of those and some other unsolved problems seemed to require a high performance machine.

6. CONCLUSION

The conclusions obtained from this research are as follows.
(1) The semantic model in ISOBAR was proved to be appropriate.
(2) The system performance for linguistic input was satisfactory.
(3) The present ability of ISOBAR for pictorial input is not necessarily sufficient, because of the machine size and some unsolved problems.
(4) In order to make ISOBAR more flexible and powerful, it is desired to remove restrictions for input sentences as much as possible.
(5) All picture processing procedures should be improved to be more time saving ones. Authors have already started such improvement.

ACKNOWLEDGEMENT

The authors are deeply indebted to Mr. Y.Ogawa, R.Taniguchi, K,Tsuzuki and K.Noda for their considerable assistance with the experimental work.

REFERENCES

[1] YOKOTA,M. and TAMATI,T.,"A Study on Formal Description of Word Meaning(I)",Technology Rep. of Kyushu Univ., Vol.48, No.6, pp.809-814, Dec. 1975.

[2] YOKOTA,M. and TAMATI,T.,"Description of Word Meaning and Its Application", Tech. Rep. of IECE Japan, AL77-41, 1977.

[3] YOKOTA,M. and TAMATI,T.,"On Syntactic and Semantic Analysis of Weather Report Sentences", Tech. Rep. of IECE Japan, AL77-78, 1978.

[4] ENDO,T. and KAWAGUCHI,E.,"Some Properties of DF-Picture Expression and Its Application to Data Compression", Trans. IECE Japan, Vol.J62-D, No.2, Feb. 1979.

[5] KAWAGUCHI,E. and ENDO,T.,"On a Method of Binary Picture Representation and Its Application to Data Compression", IEEE Trans. on PAMI, to be published.

APPENDIX I

LIST OF Attributes

Symbol	Attribute	Symbol	Attribute
A_1	Location	A_8	Temperature
A_2	Pressure	A_9	Wind Direction
A_3	Number of Typhoon	A_{10}	Time
A_4	Direction	A_{11}	Degree of Influence
A_5	Area	A_{12}	Velocity of Motion
A_6	Wind Force	A_{13}	Wind Speed
A_7	Weather	A_{14}	Visibility

APPENDIX II

Examples of Semantic Tables

Constituent	Time	A_1	A_2	A_3	A_4	A_5	A_6	A_7	A_8	A_9	A_{10}	A_{11}	A_{12}	A_{13}	A_{14}
$C_1(x)$	t_0	β	γ			α									
$C_1(y)$	t_0	p			α_1								β_1		
	t_1	q			α_1								β_1		
$C_6(z)$	t_0	α_1													
	t_0	α_2													
	t_0	α_3													

N.B. All above are the semantic structures of the following sentences (i.e. S1-S3) which appear in the weather report at t_0.
S1:KO-KAI(α) NO N34°E122°(β) NIWA 1000MB(γ) NO TEIKIATSU($C_1(x)$) GA ARIMASU.
S2:TEIKIATSU($C_1(y)$) GA MAIJI 20KM(α_1) DE HIGASHI(β_1) SUSUN-DEIMASU.
S3:KAN-REIZEN-SEN-($C_6(z)$) GA N34°E107°(α_1) , N35°E115° (α_2) , N37°E122°(α_3) O TO-TTEIMASU.
The tense of S2 is "Progressive present tense", which we denote by the pair of times, i.e. t_0(the present) and t_1(some immediate future).

APPENDIX III Examples of Q-A

1978/12/30/12/ WA NIHON-KAIGAWA DEWA TEN-KI WA DAITAI HARE DESU KA. (Was the weather on the side of the Japan Sea almost fair at noon on Dec. 30, 1978?)
SPATTERN=7 QUEST=1 PLACE=12 WEATHER=2 DEGREE=2
HAI , NIHON-KAIGAWA DEWA TEN-KI WA DAITAI HARE DESU. (Yes, the weather on the side of the Japan Sea was almost fair.)

1977/1/12/12/ WA NAHA DEWA TEN-KI WA S KA. (In NAHA area at noon on Dec. 12, 1977, was the weather south?)
ZOKUSEI(TEN-KI) TO ZOKUSEICHI(S) TOGA HUTEKIGO-. (The weather has not the attribute of direction.)

1976/9/12/12/ WA L NO CHU-SHIN- KARA ON-DAN-ZEN-SEN- GA DOKO NI NOBITEIMASU KA. (Where did a warm front extend from the center of the low pressure at noon on Sep. 12, 1976?)
SPATTERN=4 QUEST=1 TIME=1976/9/12/12/ ORG=L FNAME=5 LOC=#
L NO CHU-SHIN- KARA ON-DAN-ZEN-SEN- GA N26DE140D NI NOBITEIMASU. (A warm front extended to N26°E140° from the center of the low pressure.)

ITSU SEIKO-TO-TEI NO KIATSUHAICHI TO NATTEIMASU KA. (When did the pressure configuration of typical winter type occur?)
SPATTERN=6 QUEST=1 TIME=# TYPE=1
1976/11/29/18/ WA NIHON- HUKIN- WA SEIKO-TO-TEI NO KIATSUHAICHI TO NATTEIMASU. (At 6:00p.m. on Nov. 29, 1976, the pressure configuration of typical winter type occured around Japan.)

APPENDIX IV

Results of each processing in PR

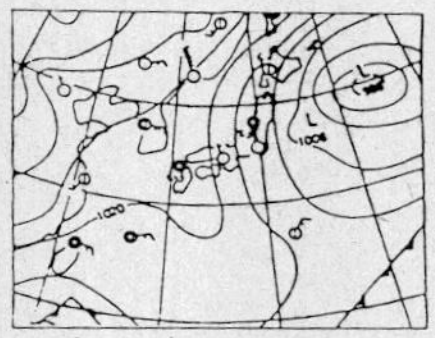

Preprocessed pattern

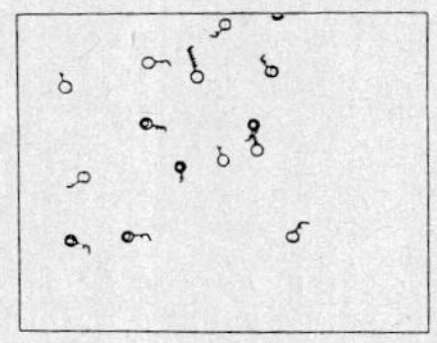

Extracted weather symbols

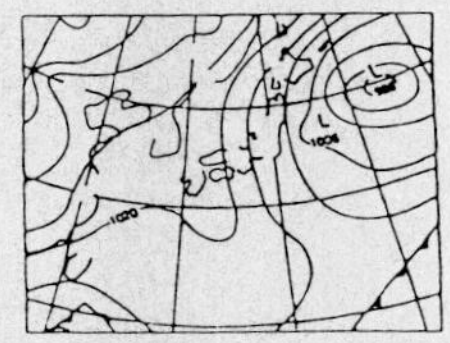

Residual pattern

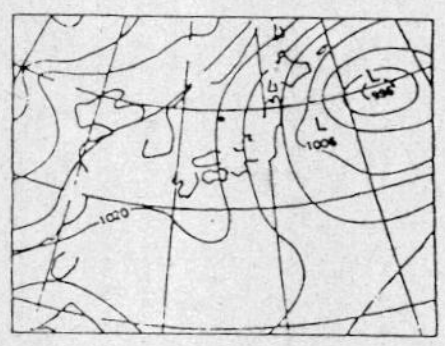

Detection of front symbols

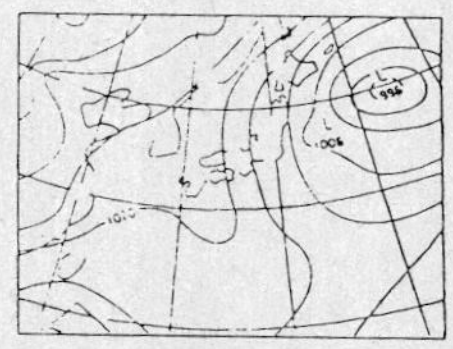

Result of thinning oreration

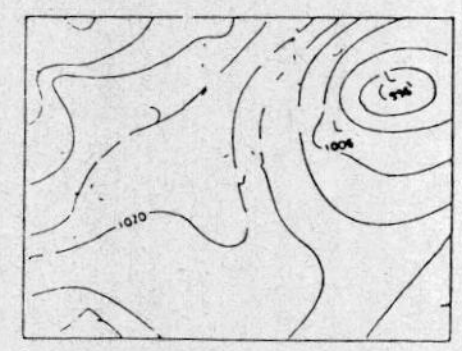

Elimination of constant component

Scientific Information Systems in Japan
H. Inose (ed.)

METHODOLOGIES OF JAPANESE LANGUAGE TREATMENT BY COMPUTER FOR INFORMATION AND DOCUMENTATION SCIENCES

Makoto NAGAO, Jun-ichi TSUJII and Takashi MATSUYAMA
Dept. of Electrical Engineering, Kyoto University, Sakyo, Kyoto 606

Abstract: Computer processing of natural language is studied for use in information and documentation sciences. Utilization of ordinary Japanese dictionary by computer, morphological analysis of Japanese sentences, machine translation of document titles from English into Japanese, text editor of Japanese, and a model of natural language question-answering system are the main topics which are explained in this paper.

1. ROLES OF NATURAL LANGUAGE IN INFORMATION AND DOCUMENTATION SCIENCES

There are varieties of information in science and engineering fields. Many are well organized and well formatted numerical and character data, which fit to the present day data base systems. However, there are vast amount of non-formatted, not well-organized natural language data, which are hard to be handled by the ordinary data base systems.

By the information retrieval system of science and engineering documents, the author's name, journal name, publication year, key words and so forth, can be handled in well formatted structures, but the document's titles and abstracts are not. It is very difficult to utilize the information contained in the titles and abstracts because some linguistic analyses are required.

In English there are a few studies which extract key words by utilizing a set of stop words, inflection information of word endings and so on. In Japanese there are very few studies of this kind. Computer processing of Japanese language is especially difficult because words are not segmented by spaces in Japanese sentences, and the word order is rather free except the main predicate comes to the last.

For the more advanced future information systems, the treatment of natural language is essential not only because the stored information is by natural language, but also the access to the system is to be by ordinary natural language. International exchange of documentation information requires the language translation, and it is to be by machine because the information amount to be translated is too huge to be done by human translators. Therefore the establishment of computer techniques of natural language processing has become an urgent need of the whole world.

Our main research objectives are, (1) computer analysis of Japanese sentences and automatic extraction of effective information from Japanese sentences for the information and documentation sciences, (2) machine translation of document titles between English and Japanese for the international exchange of documentation information, and (3)intelligent man-machine interface by using natural language.

This paper is the summary of what we have achieved by the three years' research project (1975-1978) under the title of "Analysis and Model Building of Linguistic Information Structure of Scientific Documents". We have studied the following topics.

(1) Flexible computer utilization of ordinary Japanese dictionary
(2) Collection of various linguistic data for use in the analysis of Japanese sentences
(3) Morphological analysis and segmentation of Japanese sentences into word sequences
(4) Machine translation of titles of scientific documents from English into Japanese
(5) Development of text editor for I/O device of Japanese characters
(6) Fundamental study of a model for intelligent question-answering system to complicated knowledge base

2. COMPUTERIZED JAPANESE DICTIONARY AND ITS ACCESS METHOD

Algorithms for natural language processing must be verified by a large corpus of text data from various different fields. The dictionary for the language processing programs, therefore, must be large enough to cover vocabularies in the varieties of fields. Texts in a certain field usually contain both the technical terms, which are specifically used in the field, and the common words, which are used in almost every field. A conventional dictionary which is currently published and daily used almost consists of the common words. We have completed to computerize such a dictionary. The dictionary we adopted is "Shinmeikai Kokugojiten" published by Sanseido Publishing Co., which contains about 70,000 lexical entries.

Because of the particular Japanese writing method, special techniques are required to utilize the dictionary that are not necessary for English and other European languages. We have computerized these techniques, which are also useful for the dictionaries of technical terms and terminology data banks in Japanese.

The main difficulty we encountered in developing the dictionary consultation system is from the fact that lexical entries in Japanese usually have more than one spelling. They have basically two different forms of spellings, a Kana-spelling and Kanji-spellings. Moreover, we often find mixed forms of these two spellings in ordinary texts (Fig. 1). We must be able to retrieve the lexical description of an entry by using any one of these spellings.

In our computerized dictionary system, the intermediate indexing structures are provided for Kana-spellings and Kanji-spellings as shown in Fig. 2. The other spellings, mixed spellings, can be normalized into one of these two basic spellings by utilizing a certain auxiliary information. We can transform a mixed spelling into Kana-spellings by systematically changing the Kanji-character in the spelling into the corresponding Kana-strings. However, each Kanji-character corresponds to three or four (or more) different Kana-spellings. If the resultant Kana-spelling is found in the dictionary, it will be the correct one. We can also obtain Kanji-spellings from a mixed spelling by changing a Kana-string in the mixed spelling into the corresponding Kanji-characters. Then we can utilize the index for Kanji-spellings to retrieve the lexical entry which has the given mixed spelling.

Another problem is the incorporation of the morphological analysis component into the dictionary consultation system. Because the inflection system for Japanese words is much richer than English, particularly because the predicative words have many inflectional variants, the morphological analysis component is indispensable for the dictionary consultation system of Japanese. By this study we completed the system of utilizing computerized Japanese dictionary flexibly. Another ongoing project of ours is to computerize English-Japanese dictionary, 'New Concise English-Japanese Dictionary', which contains about 90,000 lexical entries. This will be completed in 1980.

Kanji-Spelling	Kana-Spelling	Mixed Spelling	Meaning
繰返す	くりかえす	くり返す	to Repeat
繰り返す		繰かえす	
		繰りかえす	

Fig. 1. An Example of Kana-, Kanji-, and Mixed Spellings.

3. LINGUISTIC DATA FOR JAPANESE LANGUAGE PROCESSING

One of the important issues in natural language processing is the robustness of a system. We understand the concept, robustness, is how huge or large language texts can be processed by the system. In order to facilitate natural language processing in a large scale, we should provide comprehensive linguistic data. The Japanese dictionary and the English-Japanese dectionary are such linguistic data. Besides these, we should also collect and computerize other linguistic data appropriately from computational points of view. Such linguistic data are as follows:

(1) Terminology data banks that contain specific terms in specific fields
(2) List of inflectional suffixes of verbs, auxiliary verbs, and adjectives
(3) List of function words in Japanese
(4) List of suffixes and prefixes
(5) List of complex Kana-strings which often appear in Japanese texts
(6) List of sentential patterns

(2),(3) and (4)are important for morphological analysis programs. We have currently listed up 254, 150 and 70 items for these respectively. We also have gathered another linguistic data to treat some exceptional cases in morphological analysis. In Japanese, content words such as nouns, verbs, adjectives and so on, are usually written in Kanji-characters. On the other hand, function words such as postpositions, conjunctives, inflectional suffixes etc., are usually written in Kana-characters. These character types give us useful information to identify and segment words in written Japanese texts which are strings of continuously juxtaposed words without spaces. There are, however, many exceptions to this rule. Some function words are written in Kanji-characters and some content words are written in Kana-characters. To treat these exceptional cases, we should tabulate these exceptional words, and utilize them in morphological analysis program. We have about 700 such exceptional words.

The list of complex Kana-strings now contains 2500 different Kana-strings. The units of words in different languages often disagree with each other. That is, a concept which is expressed by a single word in one language may often be expressed by a sequence of words which constitute a complicated syntactic construction in another language. This tendency is

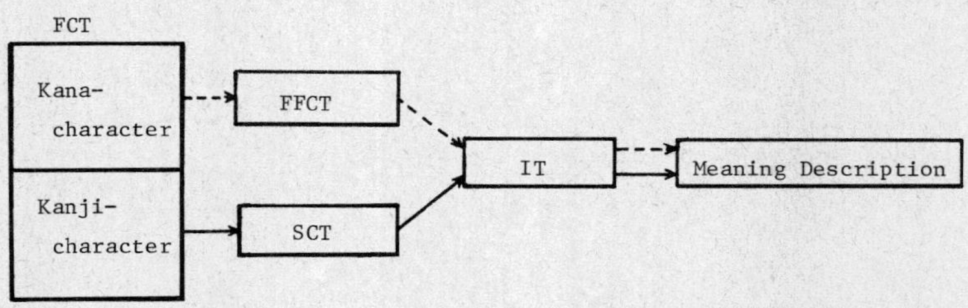

FCT : First Character Table
FFCT : First Five Characters Table
SCT : Second Kanji Character Table
IT : Item Table

Fig. 2. Overall Construction of the Dictionary System.

further emphasized if the two languages belong to quite different language families such as Japanese and English. There are many English single words which are to be expressed by phrases or other constructions in Japanese and vice versa.

Modals and aspects are usually expressed in English by auxiliary verbs such as 'can', 'may', 'must', 'have -ed', 'be -ing' and in some cases, by certain specific constructions such as 'it is impossible that S', 'it is necessary to VP'. Even some adverbs such as 'necessarily' etc. can express modalities.

These expressions for modals and aspects are highly dependent on languages. In Japanese, they are usually represented by complex Kana-strings which follow predicative words (verbs, adjectives, and so on) of sentences. We can separate these Kana-strings into several component words by applying morphological rules of Japanese. For example, the complex Kana-string 'なければならない' ('NAKEREBANARANAI') which marks the modality 'necessity' or 'obligation' can be analysed into a sequence of words as shown in Fig. 3.

The literally synthesized meaning of this complex Kana-string may be approximately "If not (s), (it) will not {work, succeed, go well}". But the real meaning of the whole expression is "should", or "must", and therefore, it is almost impossible to synthesize the whole meaning from the combination of the component meanings. There are lots of examples of this kind. This shows that it is much better to handle a longer Kana-string as a whole rather than dividing it into parts of speeches.

Another complex Kana-string in Japanese is a string that indicates 'case'. The case relationships between noun phrases and a verb are usually marked by a relatively small set of postpositions in Japanese. However, complicated constructions which contain certain specific verbs sometimes represent case marks. The expression によって (NIYOTTE), for example, contains the inflected form of the verb 'YORU'. The verb 'YORU' has lost its original meaning in this string, and the string as a whole plays the role of a case postposition which marks 'instrumental case'.

We can find out many such complex Kana-strings in ordinary Japanese texts. Currently, we have listed up about 2,500 of such expressions. By treating these complex expressions as the units we can simplify the subsequent syntactic and semantic processings.

4. MORPHOLOGICAL ANALYSIS OF JAPANESE

In English and some other European languages, words can be easily recognized, because they are usually separated each other by spaces or punctuation marks in texts. However, there are many languages in the world in which there are no such definite separators, and words in texts are simply juxtaposed. Japanese is one of such languages.

The procedure for morphological analysis of Japanese consists of the following components:

(1) Detection of boundaries between pause groups
(2) Analysis of inflections or consecutive kana-strings
(3) Testing the grammaticality of the connection of adjacent words in pause groups
(4) Dictionary look-up
(5) Segmentation of compound words composed of Kanji characters

Morphological analysis is usually performed as the first step of the whole language processing task. Errors in this step may have damaging effect on the subsequent language processing. Therefore, our morphological analysis procedure gives all possible interpretations, if they exist. This procedure utilizes various kinds of linguistic data listed in Section 3.

The Table 1 shows the results of the morphological analysis when it is applied to the sentences from an elementary chemistry text book.

A set of utility programs are also provided to augment the procedure. By using these utility programs, we can obtain various summary information of the analysis results. For example, we can obtain the following:

(1) list of the failures: The snap shots of the analysis can also be obtained.
(2) list of independent words (nouns, verbs, adjectives etc.) which are not registered in the lexicon of independent words: Because many independent words are domain dependent, it is impossible to have a comprehensive dictionary of independent words. Our morphological analysis procedure can work without such a comprehensive dictionary. In fact, it can even work without any dictionary of independent words. Our algorithm can estimate the part-of-speech of an unknown word from the kana-string which follows it (most of the independent words in

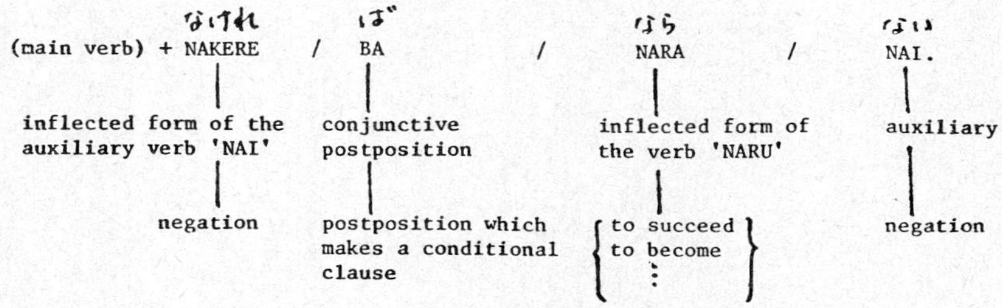

Fig. 3. An Example of Complex Word Sequences Expressing Modals in Japanese.

Japanese are written in Kanji-characters).
(3) Frequency of each words in the given text.

We can improve the morphological analysis procedure by adding new entries to the dictionary or by revising the morphological rules from the information of these summaries.

5. MACHINE TRANSLATION SYSTEM FOR TITLES OF SCIENTIFIC DOCUMENTS

By starting the research of machine translation systems, we first selected an easy domain, that is, the titles of scientific documents. They have no contextural information at all, and the words in this domain are usually less ambiguous than those in other domains. We restrict our domain in a more restricted area of electrical engineering documents. We collected one hundred thousand titles in this area. Our ultimate goal is to translate them correctly. We guess that the amount of one hundred thousand texts are large enough to check the robustness of the system. After one year's research effort, we have completed a prototype system. Now, we are going to develope the second, the augmented version of this prototype system.

One of the important research objectives of this project is how to develop a practical system which is feasible from the economical point of view. The system should be able to treat a large numbers of vocabularies. Therefore, we can not incorporate into our system a sophisticated semantic analysis component which requires detailed semantic descriptions for the vocabularies. The dictionary which we have at present contains 9,800 items in the field of electrical engineering. The outline of the system is as follows:

Step 1. Morphological Analysis of Words in English Titles.

Step 2. Dictionary Look Up: Because many technical terms or concepts are expressed by a set of words, not by a single word, we listed up those expressions in the dictionary. By looking up this dictionary, certain sequences of words are reduced into single words and treated in the next stage as if they were single lexical units. Some idiomatic expressions are also processed at this step.

ex. settling time ⟶ (n)
time varying ⟶ (adj)
perpendicular to ⟶ (prep)
based on ⟶ (prep)

Step 3. Processing of Local 'AND' Construction: In general, it is very difficult to determine the scope of the conjunctive phrases. However, a certain specific conjunctive constructions are easily analyzed by considering only the local syntactic environment. At present, only such conjunctive phrases are processed in order to make the succeeding processing easier.

Step 4. Processing of Consecutive Noun-Noun Constructions: In scientific documents, especially in their titles, many complex concepts are often expressed by long concatenated noun-noun sequences. In some cases, adjectives or adverbs are also imposed in this sequence. It may require a very sophisticated semantic analysis to determine the internal structures of such sequences. Therefore, in our present system, the internal structures of such noun-noun sequences are not analyzed. We only identify that such constructions exist. We provided a transition network only to accept such noun-noun constructions, and not to produce the corresponding internal structures. In the generation stage, such noun-noun constructions in English will be translated into corresponding noun-noun constructions in Japanese. Fortunately, the word orders in such constructions are almost the same in English and Japanese.

ex. extremely thin sio films ⟼ films (n)
high current arc discharges ⟼ discharges (n)

Correct Rate of Analysis (%)

	Single Correct Result		Plural Results are obtained. Correct one is contained.		Failure	
	Word	Pause Group	Word	Pause Group	Word	Pause Group
Before Consultation of Dictionary	73.0	61.4	20.9	33.2	6.1	5.4
After Consultation of Dictionary	87.0	74.8	9.5	20.0	3.5	5.2

(*) 500 PG's are processed by the procedure.

Average Processing Time (second)

	per a sentence	per a PG	per a word
Without Dictionary Consultation	4.6	0.43	0.17
With Dictionary Consultation	35.3	3.32	1.30

(*) The processing time includes the times for MT I/O.

Table 1. Scores of Morphological Analysis.

Step 5. Determination of Japanese Word Orders by Skelton Patterns: As mentioned before, English and Japanese have almost the same word orders in long noun-noun constructions. However, the global surface word orders are completely different. For example, prepositional phrases in English usually modify the preceding nouns or verbs. In Japanese, these modifying phrases should precede the words which are modified. Moreover, in some cases the same word orders in English should be translated into different word orders in Japanese, depending on the semantic contents of the words. An example is,

measuring temperature ⟼ 温度 (temperature) 測定 (measuring)

measuring device ⟼ 測定 (measuring) 装置 (device)

The determination of an appropriate word order is very difficult.

The input titles have been systematically reduced into simpler forms by Steps 1, 2, 3 and 4. This simplified forms, called 'Skeltons,' consist only of the words that are relevant to the determination of the global word order.

ex. Input: Industrial and scientific techniques for measuring field effect mobility
Skelton: Techniques for measuring mobility
Skelton pattern: (n) (prep) (ing) (n)

Input: Evaluation of high quality phosphor screens for image tubes
Skelton: Evaluation of screens for tubes
Skelton pattern: (n) (prep) (n) (prep) (n)

Input: An automated general purpose test system for solid state oscillators
Skelton: System for oscillators
Skelton pattern: (n) (prep) (n)

The skeltons thus extracted from the titles will be matched against the skelton patterns in the dictionary. Corresponding to each skelton pattern, several different word orders are prepared to generate different Japanese title structures. Which order is suitable for a title is determined by considering the following two factors:

1. flexible idiom, if any. Typical examples are:

   Application of N to N
   Technique for -ing N

2. Semantic contents of words: Very shallow semantic descriptions are given in the dictionary. Nouns are classified into 5 groups, Verbs are accompanied with their case patterns, and so on. These semantic descriptions are used to calculate semantic preference of each word order.

Step 6. Morphological Generation of Japanese Titles

At present, about 70% of English titles can be translated into Japanese, and 80% of translated Japanese titles are evaluated as good translation. Some of the results are shown in Fig. 4. Most of the failures are caused by lack of skelton patterns or lack of lexical entries. We are now accumulating skelton patterns which appear in the titles of scientific documents, and augmenting the system by adding the patterns and the lexical items.

6. INPUT/OUTPUT OF JAPANESE CHARACTERS

6.1 Input/Output Terminal for Kanji-Characters

Rapid progress of recent hardware technologies has

GENERATION OF STANDARD EM FIELDS USING TEM TRANSMISSION CELLS.
ＴＥＭ伝送セルを用いた標準ＥＭ場の生成

TRANSIENT ELECTROMAGNETIC PROPERTIES OF TWO, INFINITE, PARALLEL WIRES.
二無限平行ワイヤの過度電磁性質

A MULTIPROBE MINIATURE THERMINTOR SYSTEM FOR THE MEASUREMENT OF TEMPERATURE PROFILES.
温度プロフィールの？測定のための？多プローグ小型？系

A HIGHLY STABLE AND SENSITIVE TEMPERATURE REGULATOR FOR THE RANGE 1.5 350 K.
？範囲Ｋのための？高安定及び高感度温度調節器

STABILITY ANALYSIS OF INTERCONNECTED SYSTEMS.
相互連結系の安定性解析

CONTROLLABILITY OF NONLINEAR SYSTEMS WITH RESTRAINED CONTROLS.
拘束制御による非線形系の可制御性

METHOD FOR MEASURING MAGNETIC MOMENTS WITH PRECISION.
精密による磁気モーメント測定のための手法

THE SIGNAL NOISE RATIO OF ASTATIC MAGNETOMETERS WITH NEGATIVE FEEDBACK.
負フィードバックによる無定位磁力計の？信号雑音比率

INFRARED LASER INTERFEROMETER FOR MEASURING AIR BEARING SEPARATION.
空気軸受分離測定のための赤外レーザ干渉計

Fig. 4. Examples of Translation Results.

enabled us to input/output kanji-characters comparatively easily. Our research group has such a terminal connected to the central computer of Kyoto University.

Main Characteristics

(1) Display area $250^{mm} \times 125^{mm}$

(2) Number of characters on the screen
    40 characters × 20 lines (for kanji)
    80 characters × 20 lines (for alphabets)

(3) Size of display memory
    40 characters × 50 lines (kanji)
    80 characters × 50 lines (alphabet)

(4) Each kanji-character is displayed by 24 x 24 dot pattern.
    Each alphabetic character is displayed by 16 x 10 dot pattern.

(5) Number of character patterns in the ROM : 2560 characters
    Number of character patterns in the floppy disc: 6000 characters

(6) Transfer rate from/to the main computer: 9600 baud (max)

(7) Display mode: normal mode, reverse mode, flicker mode, blink mode,

(8) Laser beam printer: the content of display memory is printed out on ordinary paper sheet.

Because a microprocessor is built in this terminal, many flexible editing facilities are prepared in the local mode. A file system is also provided on the floppy disc. Therefore this terminal can also be used as a stand-alone system for producing private documents. We are writing a Japanese text editor for it. The terminal will become a good word processor for Japanese.

6.2 Software for Manipulating Kanji-characters

Until recently it was not easy for computer systems to input/output kanji-characters. Most computer systems have been designed to manipulate only alpha-numeric characters. Especially, high level languages such as PL/1, COBOL, LISP etc. have no provisions for manipulating kanji-characters. We had to revise such programming languages to accept kanji-characters. Now we have a kanji-LISP system, that was originally implemented at Utah University and revised by us to handle kanji-characters. Because we use PL/1 for morphological analysis and low level processings of Japanese, we have developed PL/1 software packages to manipulate kanji-data.

We have also developed a flexible editor to store, revise and manipulate kanji-data. Especially, this editor has a facility of defining the record format of stored data. Many useful commands are provided for this purpose. For example, SORT is the command to sort the records according to the values in the specified colums, and KEYFILE is for extracting records that have specified keys in the specified colums and for creating new file of them. These commands are very useful to manage various linguistic data described in Section 3.

7. SEMANTIC NETWORK AS A BASIS FOR NATURAL LANGUAGE CONVERSATION BETWEEN MAN AND MACHINE

Many researchers convince that various kinds of knowledge, which are not linguistic ones, play important roles in the process of natural language conversation between man and machine, and language understanding. We have chosen elementary chemistry as the micro-world of our conversation situation, and we provided our system with the real world knowledge of this field. Here we will describe the detailed constructions and the logical properties of the semantic network S.N. which we have developed for this purpose. Then we will show how to implement and use it in the problem solving processes. This is a fundamental study for natural language conversation between man and machine, which will become important in the future information retrieval and documentation systems.

Our contention is that, in order to represent knowledge in a complicated domain such as chemical field, the programming system must have not only certain innate inference abilities but also the ability to allow the programmer to embed various forms of control mechanisms (typically they are expressed by procedures) in appropriate locations.

Our semantic network S.N. can be seen as such an internal data structure. Various different kinds of knowledge, whichever it may be procedural or declarative, can be naturally embedded in this network. By traversing the network, the system not only retrieves relevant knowledge but also performs a certain kinds of inference at the same time. It has the following characteristics.

(1) The S.N. is much richer index structure for retrieving relevant knowledge than the uniform structures employed in the AI languages such as hash arrays.

(2) In the S.N., we have the following basic node types.
    (a) Predicate Node, (b) Function Node
    (c) Variable Node, (d) Constant Node
    (e) Logical Connective Node
    The predicate node is further subdivided into the following five node types according to the functions which the predicate performs in the problem solving processes:
    (a1) Concept Node
    (a2) Arithmetic (or Procedure) Node
    (a3) Relation Node (a4) Boolean Node
    (a5) Event Node

(3) An arithmetic node shows that there exists a certain computable relationship between the arguments. A relation node is used to show that the relationship between the arguments is stored in an external data base.

(4) In the problem solving process, the dependency relationships shown by Arithmetic nodes and Relation nodes are utilized to determine the unknown values of certain arguments from the known values of the other arguments. A boolean node is mainly evaluated when the values of the all arguments are determined.

(5) Generalization hierarchy of concept nodes is a useful technique for clustering relevant knowledge. In order to extend the expressive power of the hierarchy, we have introduced two new types of nodes, that is, SELF nodes and DISJOINT nodes.

(6) Because the representation in the S.N. is highly descriptive, the same knowledge can be used for different purposes.

(7) There is an important distinction of nodes in the S.N.. That is the distinction between the generic nodes and the occurrence nodes. The generic nodes for a predicate contains general information or definition of the predicate. An occurrence node is the node which indicates that the predicate is used in a certain statement.

We have developed a description language for the S.N.. The language can be seen as a programming language for problem solving and knowledge representation in general. The details of the system language are described in the literatures (22) $\sim$ (24) and (30). The internal data representation in the form of the S.N. provides a clear and systematic notion of how one description can describe another, and how its meaning can be used by the interpreter. It allows the language to be described in terms of a clear semantics, because the S.N. itself has a clear semantics. Note that each piece of knowledge in the S.N. is indexed by the content of the knowledge. Therefore, the programmer need not supply any external handles for each knowledge such as invocation patterns. This is one of the desirable characteristics of the S.N., because the programmer need not pay any irrelevant attention to the consistency between the external handles and the internal knowledge bodies.

The current version of the problem solving program is the primitive one which uses a simple problem reduction method as the basic strategy and depth-first method as the search strategy. However, it can be expected that we can develop much more powerful problem solvers by utilizing the characteristics of the S.N..

PUBLICATION LISTS

(1) M. Nagao and J. Tsujii: Analysis of Japanese Sentences by Using Semantic and Contextual Information, Proc. of 6th ICCL, Ottawa, Jul. 1976.

(2) J. Tsujii: Problem Solving Techniques in Robotics, Special Issue for Robot, Bit, Jul. 1976.

(3) M. Nagao and J. Tsujii: Survey of Current Computational Linguistics, Technical Report of the Professional Group on Computational Linguistics of IPS Japan, CL7-1, Sept. 1976.

(4) M. Nagao, J. Tsujii and A. Terada: Internal Data Representation of a Question-Answering System, Technical Report of the Professional Group on Automata and Language of IECE Japan, AL76-45, Oct. 1976.

(5) J. Tsujii: Computers and Language, Mathematical Sciences, No. 168, Jun. 1977.

(6) M. Nagao and J. Tsujii: Survey of Current Computational Linguistics and Natural Language Understanding Researches, Jour. of IPS Japan, Vol. 18, No. 1, 1977.

(7) M. Nagao, Y. Ikeda: Support System for Natural Language Processing I, Technical Report of the Professional Group on Automata and Language of IECE Japan, AL77-25, Jul. 1977.

(8) M. Nagao, J. Tsujii, A. Yamagami and S. Tatebe: Support System for Natural Language Processing II, Technical Report of the Professional Group on Automata and Language of IECE Japan, AL77-25, Jul. 1977.

(9) M. Nagao, J. Tsujii, A. Yamagami and S. Tatebe: Dictionary Organization for Morphological Analysis of Japanese Written Texts, National Convention Record of IECE Japan, Aug. 1977.

(10) M. Nagao, J. Tsujii, A. Yamagami, S. Tatebe and M. Hoda: Analysis of Japanese Long Compounds, National Convention Record of IECE Japan, Aug. 1977.

(11) M. Nagao, J. Tsujii, A. Yamagami, S. Tatebe and M. Hoda: Morphological Analysis of Japanese Written Texts, National Convention Record of IECE Japan, Aug. 1977.

(12) M. Nagao and J. Tsujii: Algorithms for Natural Language Understanding, Joint Convention Record of Four Institutes of Electrical Engineers of Japan, Oct. 1977.

(13) M. Nagao, J. Tsujii and A. Terada: Expressive Power of a Semantic Network and Problem Solving Techniques by Using it, Technical Report of the Professional Group on Automata and Language of IECE Japan, AL77-43, Oct. 1977.

(14) M. Nagao, J. Tsujii and A. Terada: Deduction Process Based on a Semantic Network, National Convention Record of IPS Japan, Nov. 1977.

(15) M. Nagao: Historical Survey and Future Trend in Computational Linguistics, Journal of IPS Japan, Vol. 19, N . 2, Feb. 1978.

(16) M. Nagao, J. Tsujii, A. Yamagami and S. Tatebe: Dictionary Organization and Morphological Analysis, Journal of IPS Japan, Vol. 19, No. 6, 1978.

(17) M. Nagao, Y. Ikeda: Input and Output System for Special Characters and Language Map, Journal of IPS Japan, Vol. 19, No. 6, 1978.

(18) M. Nagao, J. Tsujii, and H. Isahara: Natural Language Conversations in CAI Environment, 1978, National Convention Record of IPS Japan, 1978.

(19) M. Nagao and J. Tsujii: Natural Language Understanding Systems, Journal of IPS Japan, Vol. 19, No. 10, 1978.

(20) J. Tsujii: Dictionary Organization and Morphological Analysis for Japanese, 1978 Joint Convention Record of Four Institutes of Electrical Engineering in Japan, Oct. 1978.

(21) M. Nagao and J. Tsujii: Dictionary Organization and Morphological Analysis, in Proc. of COLING 78, 1978.

(22) M. Nagao and J. Tsujii: Semantic Network as an Internal Knowledge Representation, in Proc. of COLING 78, 1978.

(23) J. Tsujii: Internal Knowledge Representation S-Net and a Problem Solving System Using It, Technical Report of Professional Group on Artificial Intelligence of IPS Japan, 1978.

(24) M. Nagao, J. Tsujii, et.al.: Semantic Network S-Net and a Problem Solving System Using It, 1978 National Convention Record of IPS Japan, 1978.

(25) M. Nagao, J. Tsujii: Knowledge Representation and Data Base, Journal of the Society Instrument and Control Engineers, Japan, Vol. 18, No.1, 1979.

(26) M. Nagao: Information Science and Natural Language, Journal of IPS Japan, Vol. 20, No. 3, 1979.

(27) J. Tsujii: Production System and Its Application, Journal of IPS Japan, Vol. 20, No. 8, 1979.

(28) M. Nagao, J. Tsujii and S. Tatebe: An Approach to Machine Translation for Titles of Scientific Papers, Technical Report of Professional Group on Computational Linguistics of IPS Japan, 1979.

(29) J. Tsujii: Machine Translation for Titles of Scientific Papers, 1979 Joint Convention Record of Four Institutes of Electrical Engineers of Japan, 1979.

(30) M. Nagao and J. Tsujii: S-Net: A New Foundation for Knowledge Representation Languages, in Proc. of 6th IJCAI, 1979.

(31) J. Tsujii: Semantic Description — From the Computational Point of Views, Journal of IPS Japan, Vol. 20, No. 10, 1979.

(32) M. Nagao: Machine Translation, Journal of IPS Japan, Vol. 20, No. 10, 1979.

Scientific Information Systems in Japan
H. Inose (ed.)

STATISTICAL INFORMATION PROCESSING SYSTEM FOR PREDICTION AND CONTROL

Hirotugu AKAIKE
The Institute of Statistical Mathematics, Minami-Azabu, Minato-ku, Tokyo

A computer program package TIMSAC-78 is developed to provide an automatic procedure of modeling time series for the prediction and control of a stochastic system with gradually changing spectral characteristics. The procedure is realized with the aid of an information criterion (AIC) for the evaluation of possible models and the necessary least squares computations are efficiently performed by using the Householder transformation.

1. INTRODUCTION

Due to the recent development of efficient measurement techniques and equipments there is a tremendous growth of the amount of scientific information given in the form of the records of observational data. As the amount of records increases it becomes practically impossible to retrieve by human eyesight the part of data which is of particular importance for the study of the subject under consideration. One possible way out of this difficulty is to prepare a "condensed" record of the original data and realize the retrieval of the original data through this "condensed" record. This "condensation" of information is often realized by developing an appropriate statistical model for the data. In particular, we may consider the problem of predicting the future behavior of the object under consideration and develop a statistical model for this purpose. This usually leads to a characterization of the data sufficiently sensitive to the change of the characteristic of the system which generated the data. The model will be specified by a set of parameters and by recording these parameters for each local span of the original data we can get a condensed record of original observations.

The purpose of the present research was to develop a computer program package for the prediction and control of time series as a particular example of general statistical information processing system.

2. THEORETICAL DEVELOPMENTS

The basic element in realizing an efficient procedure for statistical model determination is the criterion for the evaluation of various possible models. In our approach to the modeling, each model is parameterized by a finite number of parameters and the parameters are determined by the method of maximum likelihood. The criterion, which is a measure of badness of such a model, is given by AIC which is by definition

$$\text{AIC} = (-2)\ \ln(\text{maximum likelihood}) + 2(\text{number of parameters}).$$

A simple minded, but often quite useful, procedure for the determination of a model is the minimum AIC procedure which chooses the model with the minimum AIC value. One of the significant results obtained during the process of the present research is the clarification by Shimizu [11] of the difficulty inherent in the minimum AIC procedure. This study shows that when AIC attains a spuriously low value for an excessively over parameterized model it does mean that the fitted model is actually a bad one.

The above observation suggests the necessity of paying attention to the risk, in choosing a model by AIC, due to its increased statistical variability when a model with a large number of parameters is fitted to a set of data. A reasonable procedure to avoid the difficulty is to use AIC_α which is defined by increasing the coefficient of the term (number of parameters) in the definition of AIC from 2 to α which is greater than 2, so that a model with a larger number of parameters will be penalized more heavily. Unfortunately this idea cannot be made practical unless we develop a procedure for the choice of α.

Since AIC was originally defined as an estimate of minus twice the expected log likelihood of a model specified by the parameters determined by the method of maximum likelihood, it seems natural to define the "likelihood" of the model by $\exp(-0.5\text{AIC})$. If we replace AIC by AIC_α we get $\exp(-0.5\text{AIC}_\alpha) = \exp(-0.5\text{AIC})\exp[-0.5(\alpha-2)k]$, where k denotes the number of parameters. This result shows that to consider AIC_α is equivalent to replace the "likelihood" $\exp(-0.5\text{AIC})$ by $\exp(-0.5\text{AIC})\rho^k$, where $\rho = \exp[-0.5(\alpha-2)]$. This is equivalent to introducing prior weight ρ^k to modify the original "likelihood" and this observation naturally leads to the introduction of a Bayesian modeling procedure developed by Akaike [1].

In this Bayesian procedure, rather than choosing a model with the minimum "posterior" probability, we choose a model which is determined by taking the average of the models with respect to the "posterior" probability proportional to $\exp(-0.5\text{AIC})\times$(prior probability of the model). The choice of the prior probability can be developed based on some theoretical analysis and empirical testing of the performance of the resulting procedure. A procedure thus obtained constitutes the basis of our program package TIMSAC-78, where TIMSAC stands for Time Series Analysis and Control, and allows an automatic modeling of locally stationary time series.

3. TIMSAC-78

The computer program package TIMSAC-78 (Akaike, Kitagwa, Arahata and Tada [4]) is designed mainly for the efficient modeling of locally stationary time series and its use for the realization of a control of a system operating under a stochastic environment with a gradually changing spectral characteristic.

The main features of TIMSAC-78 are as follows:

a. For the least squares type computation required for the fitting of autoregressive type models the Householder transformation is applied directly to the data matrix and the computation of sample covariances is avoided. This allows simple revision of models when a new additional set of observations is added to the original one (Kitagawa and Akaike [5]).

b. As is discussed in the preceding section, a new procedure of autoregressive model fitting by a Bayesian modeling is introduced.

c. Very practical procedures for estimating a gradually changing spectrum is realized.

d. Some predictors which are linear in parameters but not linear in terms of the past history are also considered.

e. The programs for the exact Gaussian likelihood computations of autoregressive and autoregressive-moving average models for scalar stationary time series are included.

The programs included in TIMSAC-78 are as follows:

CODE	NAME	DESCRIPTION
78.1.1.	UNIMAR	(UNIVARIATE CASE OF MINIMUM AIC METHOD OF AR MODEL FITTING)
78.1.2.	UNIBAR	(UNIVARIATE CASE OF BAYESIAN METHOD OF AR MODEL FITTING)
78.1.3.	BSUBST	(BAYESIAN TYPE ALL SUBSET ANALYSIS OF TIME SERIES BY A MODEL LINEAR IN PARAMETERS)
78.2.1.	MULMAR	(MULTIVARIATE CASE OF MINIMUM AIC METHOD OF AR MODEL FITTING)
78.2.2.	MULBAR	(MULTIVARIATE CASE OF BAYESIAN METHOD OF AR MODEL FITTING)
78.2.3.	PERARS	(PERIODIC AUTOREGRESSION FOR A SCALAR TIME SERIES)
78.3.1.	MLOCAR	(MINIMUM AIC METHOD OF LOCALLY STATIONARY AR MODEL FITTING; SCALAR CASE)
78.3.2.	BLOCAR	(BAYESIAN METHOD OF LOCALLY STATIONARY AR MODEL FITTING; SCALAR CASE)
78.3.3.	MLOMAR	(MINIMUM AIC METHOD OF LOCALLY STATIONARY MULTIVARIATE AR MODEL FITTING)
78.3.4.	BLOMAR	(BAYESIAN METHOD OF LOCALLY STATIONARY MULTIVARIATE AR MODEL FITTING)
78.4.1.	NADCON	(NOISE ADAPTIVE CONTROLLER)
78.5.1.	EXSAR	(EXACT MAXIMUM LIKELIHOOD METHOD OF SCALAR AR MODEL FITTING)
78.5.2.	XSARMA	(EXACT MAXIMUM LIKELIHOOD METHOD OF SCALAR AR-MA MODEL FITTING)
78.5.3.	TSSBST	(TEST OF SUBROUTINE SUBSET)
78.5.4.	TSWIND	(TEST OF SUBROUTINE WINDOW)
78.5.5.	TSCHCK	(TEST OF SUBROUTINE CHECK)
78.5.6.	TSROOT	(TEST OF SUBROUTINE CHROOT)
78.5.7.	TSTIMS	(TEST OF SUBROUTINE EDTIMS)
78.5.8.	TSCANC	(TEST OF SUBROUTINE CANCOR)

4. EXAMPLE OF APPLICATION OF TIMSAC-78

Fig.1 represents the results of applying the programs MLOCAR(MINIMUM AIC METHOD OF LOCALLY STATIONARY AR MODEL FITTING; SCALAR CASE) and BLOCAR(BAYESIAN METHOD OF LOCALLY STATIONARY AR MODEL FITTING; SCALAR CASE) to an artificially generated time series with gradually changing spectrum. The result was obtained by developing an AR model for each local span of data of length 200. The theoretical spectrum given in Fig.1-A is defined as the spectrum of the series when the system characteristics are fixed at the values given at the middle point of the local span. The result demonstrates the practical utility of the program for the purpose of estimation of a gradually changing spectrum.

Fig.2 represents the result of applying the program NADCON(NOISE ADAPTIVE CONTROLLER) to a set of simulated data of the yawing of a ship. The result shows that by this program the change of the environment is detected and the controller adapts to the change properly. This program has already been used to implement an auto-pilot of a ship (Ohtsu, Horigome and Kitagawa [7]) and produced good results in the testing on the real sea.

5. PLAN FOR FURTHER RESEARCH

For a stationarily fluctuating time series the power spectrum provides a fairly useful description. Since the programs within TIMSAC-78 provide various procedures of automatic determination of spectra it is quite possible to get a condensed representation of a time series in terms of the parameters of autoregressive models. This condensed information will be useful for the retrieval of the necessary part of the original data for some specific purpose of analysis. The task of realizing this idea in the form of a machine system for the information retrieval of a massive data is now being undertaken at the Institute of Statistical Mathematics.

When a time series contains trend and seasonal components, ordinary spectrum analysis is not directly applicable. Recently Akaike [2] developed a Bayesian model for the decomposition of a time series into trend, seasonal and irregular components. It is hoped that by improving this procedure we will be able to get a reasonably automatic procedure of trend and seasonality elimination. The improvement of the necessary computational procedures is now under way with a partial support by a grant from Asahi Shimbun Publishing Company.

One particularly interesting aspect of the above Bayesian modeling of trend and seasonality is that it can be applied easily to a time series with missing values (Akaike and Ishiguro [3]). This suggests that if we can develop a reasonable procedure for the detection of abnormal observations we can develop a trend and seasonality estimation procedure for data with abnormal observations simply by declaring them as missing values.

The areas of research of growing importance related with the stochastic process modeling are the point processes and non-linear, or non-Gaussian, processes. Some progress in these areas has been obtained in

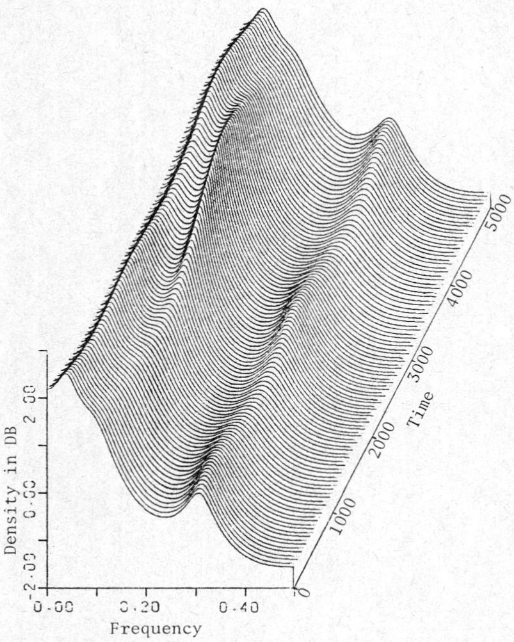

Fig. 1-A.

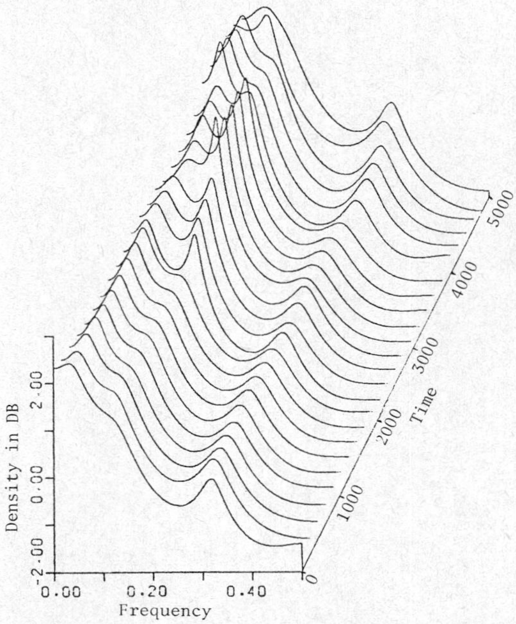

Fig. 1-B.

Fig.1-A Theoretical spectrum.

Fig.1-B Estimated spectrum obtained by MLOCAR.

Fig.1-C Estimated spectrum obtained by BLOCAR.

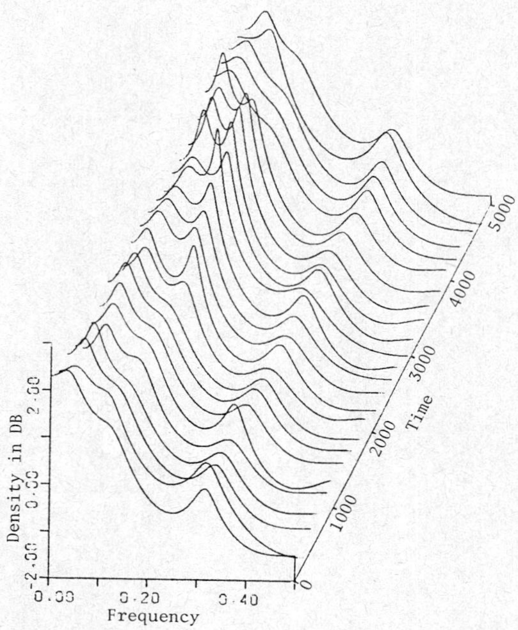

Fig. 1-C.

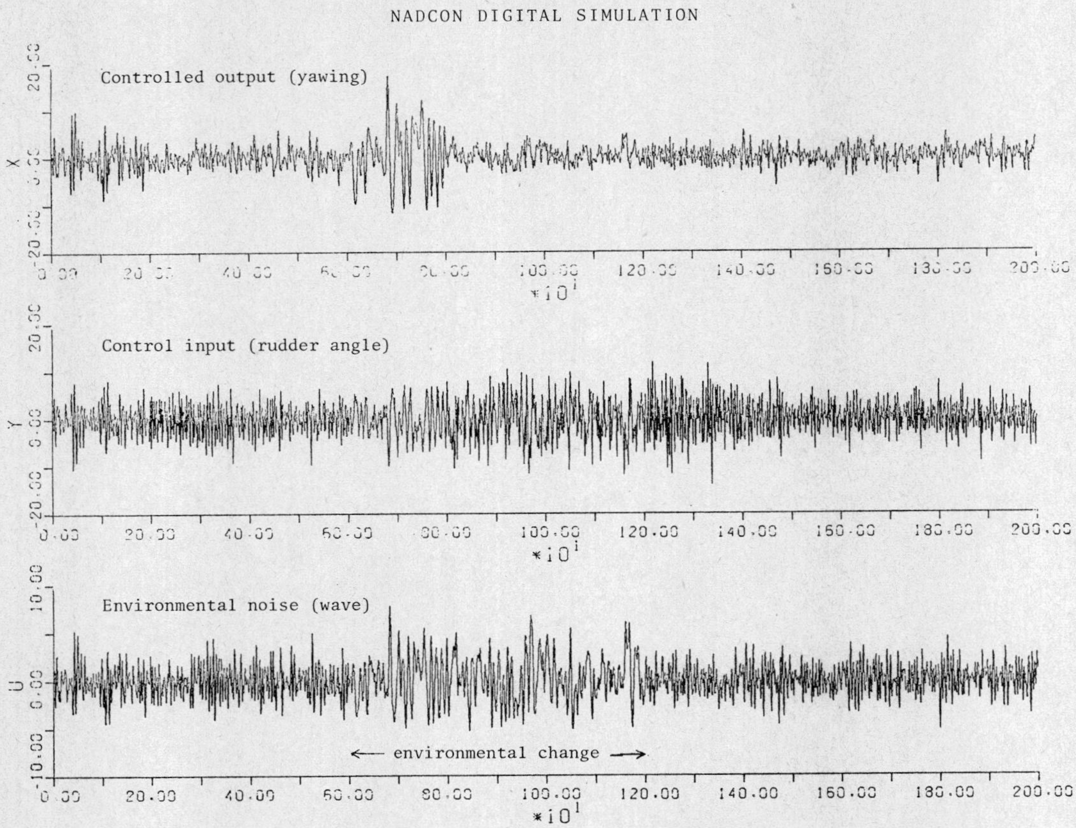

Fig.2 Simulated ship yawing under control by NADCON.

(Length of local span = 200)

conjunction with the present research; see, Ogata [6] and Ozaki [8], for point process modeling, and Ozaki [9] and Ozaki and Oda [10], for non-linear process modeling. It is hoped that the results of these researches originated within the present work on statistical information processing system for prediction and control will eventually lead to an extended version of TIMSAC-78.

ACKNOWLEDGEMENTS

It is a pleasure for the author to acknowledge the intellectual and psychological support by those who participated in the research project reported in this paper. He is indebted to G. Kitagawa for producing the numerical results illustrated in Fig's.1 and 2.

REFERENCES

[1] Akaike, H. A Bayesian extension of the minimum AIC procedure of autoregressive model fitting, Biometrika, 66, pp.237-242 (1979).

[2] Akaike, H. Likelihood and the Bayes procedure, A paper presented at the International Meeting on Bayesian Statistics, May 28 - June 2, 1979, Valencia, Spain.

[3] Akaike, H. and Ishiguro, M. Trend estimation with missing observations, A paper presented at the International Conference on Recent Developments in Statistical Inference and Data Analysis, November 28-30, 1979, Tokyo.

[4] Akaike, H., Kitagawa, G., Arahata, E. and Tada, F. TIMSAC-78, Computer Science Monographs, 11, The Institute of Statistical Mathematics, Tokyo (1979).

[5] Kitagawa, G. and Akaike, H. A procedure for the modeling of nonstationary time series, Ann. Inst. Statist. Math., 30, pp.351-363 (1978).

[6] Ogata, Y. The asymptotic behavior of maximum likelihood estimators for stationary point processes, Ann. Inst. Statist. Math., 30, pp.243-261 (1978).

[7] Ohtsu, K., Horigome, M. and Kitagawa, G. A robust autopilot system against the various sea conditions - Noise Adaptive Autopilot -, A paper presented at the third International Symposium on Ship Operation Automation, November 28 - December 1, 1979, Tokyo.

[8] Ozaki, T. Maximum likelihood estimation of Hawkes' self-exciting point processes, Ann. Inst. Statist. Math. (1979) To appear.

[9] Ozaki, T. Nonlinear time series models for non-linear random vibrations, J. Appl. Prob., (1979) To appear.

[10] Ozaki, T. and Oda, H. Nonlinear time series model identification by Akaike's information criterion, Information and Systems, B. Dubuisson (ed.), Pergamon Press, pp.83-91 (1978).

[11] Shimizu, R. Entropy maximization principle and the selection of the order of an autoregressive Gaussian process, Ann. Inst. Statist. Math., 30, pp.263-270 (1978).

Scientific Information Systems in Japan
H. Inose (ed.)

STRUCTURAL DESCRIPTION, ENCODING AND RESTORATION OF CONTOUR MAPS -TOWARDS THE CONSTRUCTION OF WEATHER MAP DATABASE-

Teruo FUKUMURA
Faculty of Engineering, Nagoya University, Furo-cho, Chikusa-ku, Nagoya

A class of contour map patterns extracted from weather maps is an important class of pictorial data because its description and analysis have serious implication in the organization of an efficient and flexible image data bank of weather maps. In this context, a structural tree description of contour map is proposed after first introducing a model for contour map. From this tree description, a string representation that contains the global features of the corresponding contour map is derived, and an attempt is made to utilize this representation for determining the storage structure of contour maps. Furthermore, a method to restore the line drawing of contour map in the retrieval phase is presented with some experimental results.

1. INTRODUCTION

A class of contour map patterns is an important class of pictorial data because its description and analysis have serious implication in the flexible and effective image data bank organization. In this context, this paper presents methods for the structural description of contour map, for the formation of the storage data structure, and for the restoration of the original map from the description, including applications of the methods to a set of weather maps.

The first four sections, from Section 2 through Section 5, are devoted to describe the structuring method of a contour map and its application. In Section 6, by exploiting a description scheme proposed in the previous sections, the binary encoding for a set of contour maps extracted from weather map data is attempted to obtain the data structure in a storage system. In Section 7, a formal description scheme for an individual contour is presented with the aim of associating with a retrieval system a function of restoration of the original contour, and some experimental results are also presented. The final section is a conclusion.

To facilitate the understanding of the following description scheme, an example of a weather map and its isobar map (contour map) is given in Fig. 1 (where the numeral attached to each contour is its contour number and has no particular significance other than being its label).

2. BASIC IDEAS AND DEFINITIONS

2.1 Contour Maps and Contours

A contour map is a collection of contour lines inscribed in a rectangular boundary called frame. The four sides of the boundary are labeled in clockwise order as S_1, S_2, S_3, and S_4. (See Fig. 2.)

〔Def. 1〕 Contour Types
The 4 types of contour are defined as follows (See Fig. 2.):

(1) Contour C is of type 1 iff C is cut by the same side and denoted by C_1;
(2) Contour C is of type 2 iff C is cut by adjacent sides and is denoted by C_2;
(3) Contour C is of type 3 iff C is cut by opposite sides and is denoted by C_3; and
(4) Contour C is of type 4 iff C is a circuit and is denoted by C_4.

〔Def. 2〕 Enclosing Region of Contours
The region enclosed by a contour is defined as follows (See Fig. 2.):

(1) If contour C is of type 1 or type 2, then its enclosing region is the area bounded by the contour and the cutting side(s);
(2) If contour C is of type 4, then its enclosing region is

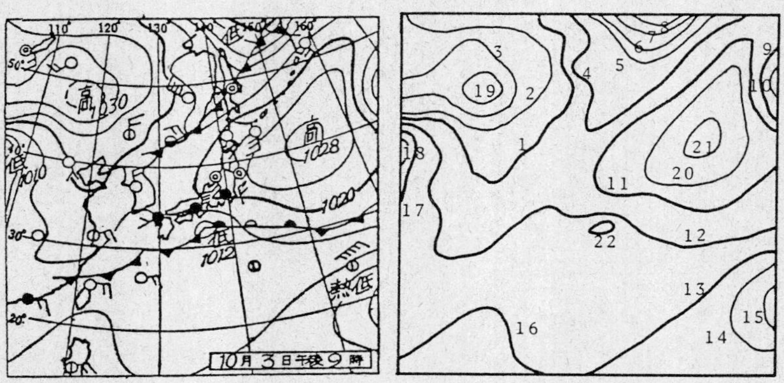

(a) A weather map (b) An isobar map

Fig. 1 A weather map and its isobar map.

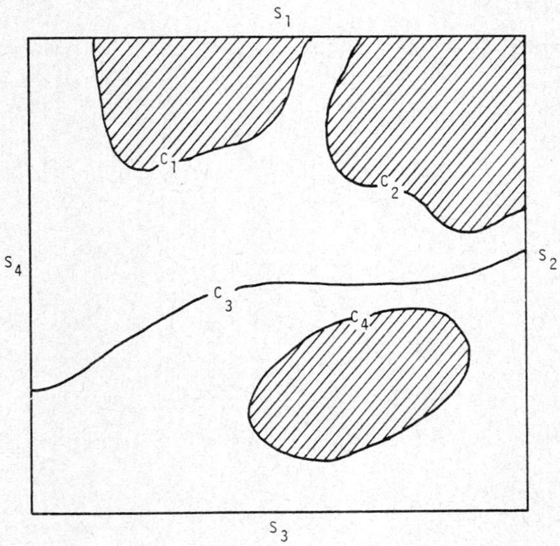

Fig. 2 Sides of frame, types of contour (C_i is of type i), and enclosed regions.

the area enclosed by itself; and

(3) If contour C is of type 3, then its enclosing region is not defined.

[Def. 3] Relations between Two Contours

(1) If contour C_i is in the region enclosed by contour C_j, then this fact is denoted by $C_i <\cdot C_j$ or $C_j \cdot> C_i$.

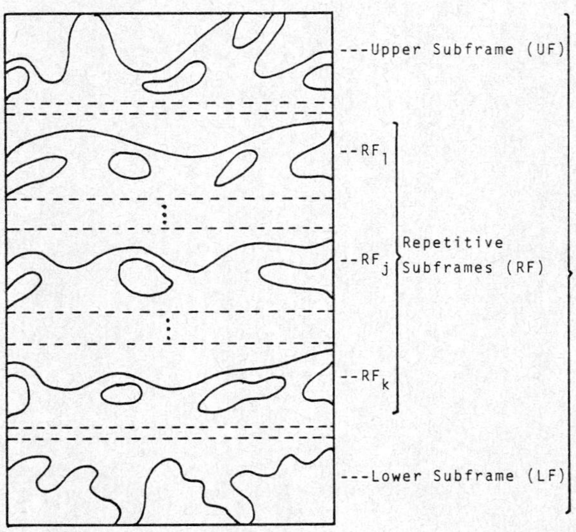

Fig. 4 Frame, subframes, and labels of subframes.

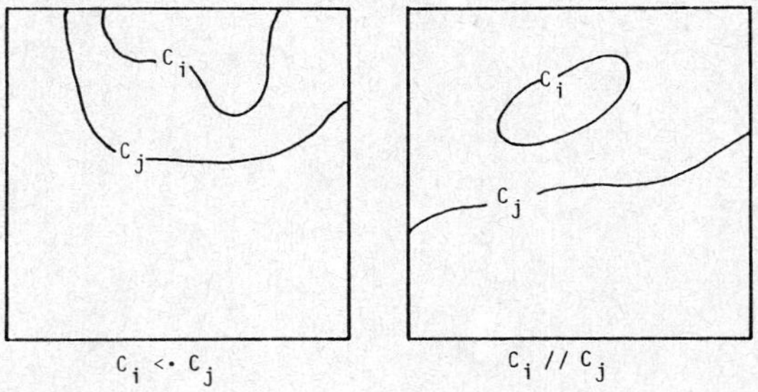

Fig. 3 Relations of two contours, C_i and C_j.

(2) If neither $C_i <\cdot C_j$ nor $C_j <\cdot C_i$ holds, then this fact is denoted by $C_i // C_j$.

Then for any two distinct contours C_i and C_j, only one of these relations is true. Some examples of these relations are shown in Fig. 3.

2.2 Subframe

It is known from the above definitions that type 3 contours do not enclose any area, thus, any contour; but they can be considered to partition a set of contours into several subsets of contours. An imaginary boundary can be considered to enclose each of these subsets and each boundary is called a subframe. Three kinds of subframe are defined as shown in Fig. 4. Of course, the number of repetitive subframes is determined by that of the contours of type 3. The lower subframe is defined conveniently so that all the repetitive subframes have the same components. (For the sake of simplicity, only contour maps with horizontal contours of type 3 are considered here.) Then for any given frame, there are at most three different kinds of subframes. Unique label is associated with each subframe as shown also in Fig. 4, where UF, LF, and RF denote the upper subframe, the lower subframe, and the repetitive subframe, respectively.

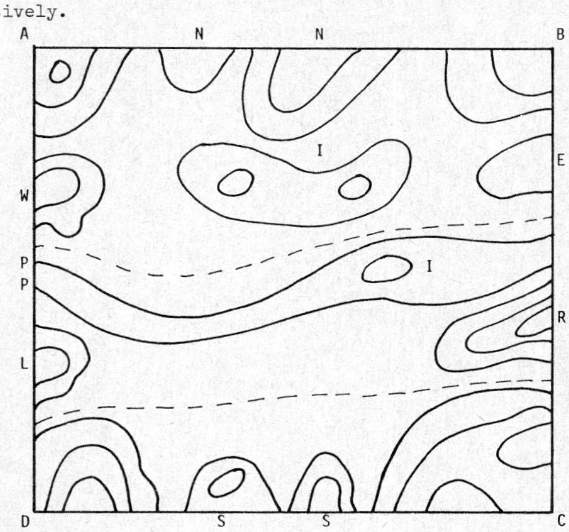

Fig. 5 Types of cluster.

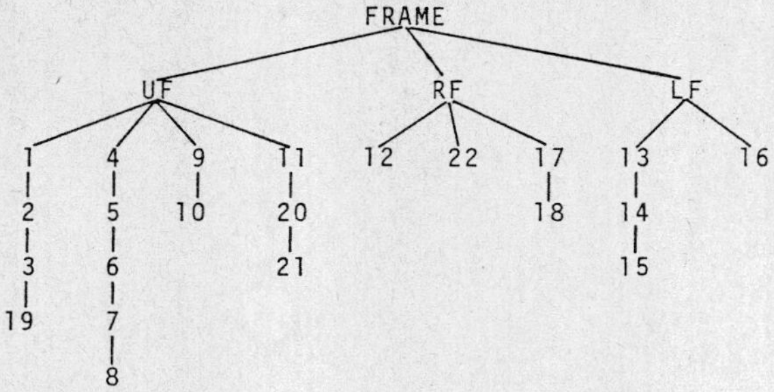

Fig. 6 The tree description of the isobar map in Fig. 1.

2.3 Maximal Covering Contour Set

Given a set of all contours **c** in a frame, its Maximal Covering Contour Set (MCCS), $\hat{\mathbf{c}}$, is defined as the largest subset of **c**, all elements of which are pairwisely in relation //.

The MCCS of the set of contours in a subframe is defined similarly to that of the frame. It is then obvious that the MCCS of the set of contours in a subframe is a subset of that in the frame. For example, the MCCS of the contour map shown in Fig. 1 (b) is {1, 4, 9, 11, 12, 13, 16, 17, 20}, where each of the elements is indicated by a thick line in the figure; the MCCSs of the upper subframe, the lower subframe, and the repetitive subframe are {1, 4, 9, 11}, {13, 16}, and {12, 17, 22}, respectively.

2.4 Contour Map Clusters

It is known from above that the cluster forms an important feature of a contour map. The various types of cluster in a subframe are defined below.

[Def. 4] Cluster Types

(1) A corner cluster is a set of a type 2 contour in the MCCS and all the contours enclosed by it. The upper left corner cluster is denoted by A; the upper right corner cluster is denoted by B; the lower right corner cluster is denoted by C; and the lower left corner cluster is denoted by D.
(2) A side cluster is a set of a type 1 contour in the MCCS and all the contours enclosed by it. A side cluster in the upper subframe is denoted by N when it is on side S_1, by E when it is on side S_2, and by W when it is on side S_4. The side cluster in the lower subframe is denoted by S. A side cluster in the repetitive subframe is denoted by R when it is on side S_2 and by L when it is on side S_4.
(3) A parallel cluster is a set of only type 3 contours which are not interleaved by contours of any other types. This cluster is denoted by P.
(4) An interior cluster is a set of a type 4 contour in the MCCS and all the contours enclosed by it. This cluster is denoted by I.

These cluster types and their respective symbols are illustrated in Fig. 5.

3. DESCRIPTION OF CONTOUR MAP

In this section, an overall description of a contour map called tree description is developed. From it a string representation which captures the global features of contour map is derived.

3.1 Tree Description

The tree description of a contour map is formed with the concepts of subframe and the relation $<\cdot$ in the following manner:

(1) The frame is the root of the tree;
(2) The subframes are the immediate sons of the root;
(3) The elements of the MCCS of a set of contours in a subframe are the immediate sons of that subframe; and
(4) In a subtree of a covering contour, a node corresponds to one of the contours enclosed by the covering contour. And given two nodes C_i and C_j, C_i is the immediate son of C_j iff $C_i <\cdot C_j$ and there exists no C_k such that $C_i <\cdot C_k <\cdot C_j$, where i, j, and k are all distinct.

An example of a tree description is shown in Fig. 6. In this figure, the numerals representing the nodes of the tree are the corresponding contour numbers used in Fig. 1 (b), and the positional relation among the nodes which are brothers to one another is arbitrarily determined.

This tree description has a property that the structure of its upper levels describes the global features of the contour map, while the structure of its lower levels describes the local features of contour clusters. In the following section, the string representation is derived from this structural property.

3.2 String Representation

The string representation of contour map's global features, called Global Description of Contour Map (It is abbreviated as GDCM and will be used to denote either the description itself or the description rule described below.), is formulated with the cluster symbols defined in Definition 4 as follows:

$$\text{GDCM} \triangleq \langle A|\lambda\rangle\langle B|\lambda\rangle\langle C|\lambda\rangle\langle D|\lambda\rangle\langle N|\lambda\rangle\langle E|\lambda\rangle\langle S|\lambda\rangle\langle W|\lambda\rangle\langle I|\lambda\rangle[\langle P|\lambda\rangle\langle I|\lambda\rangle\langle R|\lambda\rangle\langle L|\lambda\rangle]^* \quad (1)$$

where

(1) symbols, $\langle$ and $\rangle$, are the meta-symbols used for delimiting cluster symbols;
(2) $\langle\alpha\rangle\langle\beta\rangle$ means $\langle\alpha\rangle$ concatenates $\langle\beta\rangle$;
(3) $\langle\alpha|\lambda\rangle$ means α if at least one cluster with label α is present, and null otherwise; (Because of this rule, the number of clusters

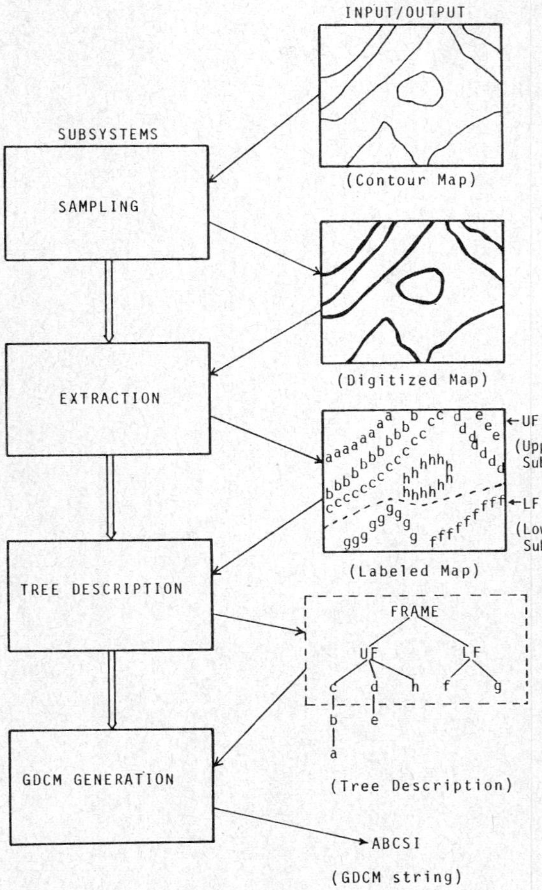

Fig. 7 Block diagram of the subsystems and their respective inputs and outputs.

Table 1 Typical processing time

EXTRACTION	31 min.
TREE DESCRIPTION	44 sec.
GDCM GENERATION	20 sec.

contained in a map is greater than or equal to the length of the GDCM.) and

(4) [ ]* means the occurrence of the contents within the brackets for a finite number of times, including no occurrence.

For example, the GDCM of the map shown in Fig. 1 (b) is ACDNEPIL.

4. SYSTEM IMPLEMENTATION

A system which accepts the image data of contour maps and output their corresponding GDCMs is implemented. This system consists of four major subsystems: SAMPLING of input image, EXTRACTION of contour lines, TREE DESCRIPTION, and GDCM GENERATION. Its block diagram is shown in Fig. 7. The first subsystem is implemented on a mini-comupter system, while the remaining subsystems are implemented on the computer system FACOM 230-38 (Memory Cycle Time: 960 ns, Gibson Mix: 3.8 μs, and Main Memory Capacity: 256 kB).

The SAMPLING subsystem converts a contour map into a computer-readable form. Then the EXTRACTION subsystem performs, on the output of the previous subsystem, the extraction of contour lines and the analysis of their properties. The TREE DESCRIPTION subsystem structures these informations into its corresponding tree description. The GDCM GENERATION subsystem then generates the GDCM corresponding to the input contour map by structuring the upper level informations of the tree description.

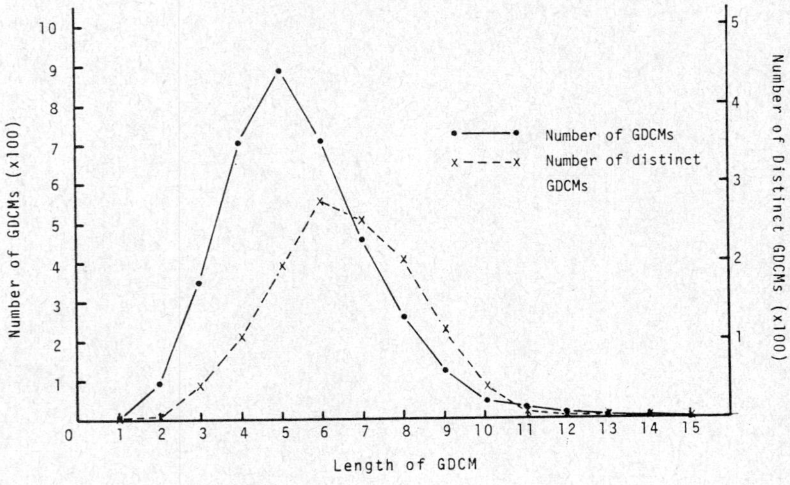

Fig. 8 Frequency distributions of the length of the GDCM and that of the distinct GDCM for the isobar map data.

Table 2 Marginal relative frequencies of the GDCM symbols

Symbol	A	B	I*	D	E	C	W	N	S
Relative Frequency	0.92	0.89	0.53	0.43	0.37	0.32	0.25	0.24	0.13

* Only interior clusters appearing in the upper subframe.

Table 3 Correlation coefficients of symbol pairs in GDCM strings

Symbol	I*	W	S	E	N	D	C	B
A	-0.0393	-0.2994	-0.0037	0.0270	-0.2053	-0.0476	0.0183	-0.0085
B	-0.0304	-0.0216	0.0065	-0.1656	-0.1801	0.0594	-0.0924	
C	0.0171	0.0061	-0.1322	-0.0968	0.0213	-0.0420		
D	-0.0020	-0.0642	-0.1550	0.0013	0.0505			
N	0.1098	0.1785	0.0082	0.0100				
E	0.0383	0.0164	0.0158					
S	0.0080	0.0181						
W	0.1587							

* Same as that of the Table 2.

In our experiment, the isobar maps, where isobars are manually traced from ordinary weather maps, are our contour map data. An isobar map is first digitized into 900 x 900 pixels by the FSS (Flying Spot Scanner). Before the EXTRACTION phase is started, the scanned picture, which consists of a density data of 8-bit gray level, is converted into a binary pattern picture and is then thinned. Each isobar line is identified by a labeling algorithm, and then its properties, such as endpoints and chain code sequence are extracted in the EXTRACTION phase. These and other derived properties, i.e. the subframe and the contour type, are used in the TREE DESCRIPTION phase first, to describe the inclusion relation of the isobars and then, to form the tree description. Furthermore, the informations about cluster type is extracted in the GDCM GENERATION phase to derive the GDCM of the input isobar map.

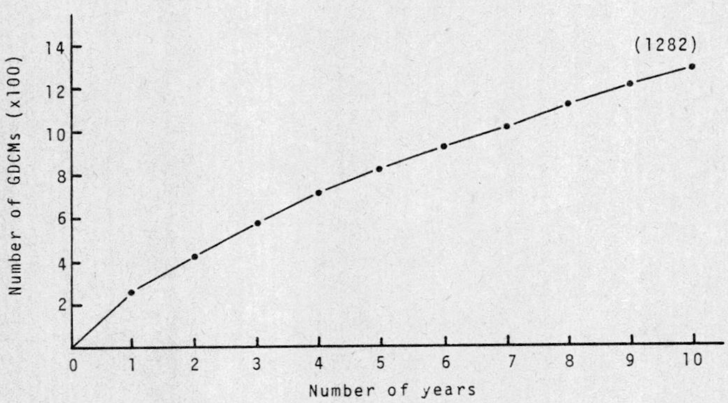

Fig. 9 The number of distinct GDCMs versus the number of years.

From the result of a preliminary test run of the system on some sample data of isobar maps, a rough estimate of the processing time (CPU) of each subsystem, except that of the SAMPLING, is obtained as shown in Table 1. It should be noted that converting a contour map into a binary array accounts for most of the total processing time. The main reason is that because of the limited memory capacity, it is necessary to divide the input image into some segments with overlapping boundary areas and then to process each of them while ensuring the continuity of the isobar lines along the dividing boundaries.

The program size of the system is about 3000 Fortran statements.

5. APPLICATION TO THE STRUCTURING OF WEATHER MAP DATA

The system developed in the previous section is applied to a set of isobar maps traced out of the weather map data (an example is given in Fig. 1) from 1969 through 1978 [1]. For each day, only one representative map indicating that day's weather condition at 9 PM (a fixed time for all the data) is chosen. The description of these maps by the GDCM are then used to investigate the properties of the GDCM itself and to analyze the underlying structure of the weather map. (Because of the substantial amount of processing time needed in the EXTRACTION phase,

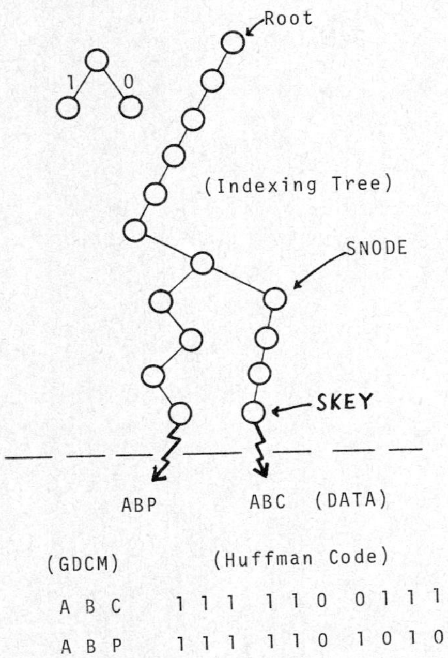

Fig. 10 Storage structure formed by encoding the GDCM.

most of the GDCM data used in this section are manually coded.)

Some statistics of isobar maps can be obtained in a straightforward manner through their description by the GDCM. The frequency distribution of the length of the GDCMs in 10 years of data is given in Fig. 8, where the distribution curve of the length of the distinct GDCMs in 10 years of data is also presented. Table 2 and 3 show two kinds of statistics obtained from the same isobar map data. The former one gives the marginal relative frequencies of some symbols, i.e. clusters, having relatively high frequency, and the latter one gives the correlation coefficients for some symbol pairs, i.e. cluster pairs, having relatively high frequency. The small values of the coefficients are noticeable here.

It is interesting to have a statistics indicating how many distinct GDCMs appear in these ten years of data. Figure 9 shows the cumulative increase of the number of distinct GDCMs as the sample data increases in number. In the figure the cumulative number grows almost linearly and reaches 1282 at the end of ten years. The ratio of this number to the total sample size (3652) is 0.35. That is, on the average, about three isobar maps are transformed to one and the same GDCM. This reduction ratio is very much smaller than that observed in, for example, the recognition of hand-written characters.

But, at present, the GDCM does not deserve any treatment of the category name of patterns. It can rather be considered as a simple description facility for the isobar map data. The detail of this section is contained in the reference [2].

6. ENCODING OF THE GDCM

As the GDCM has a property of being a simple code as described in the previous section, the statistical nature of the description by the GDCM will be an immediate property derived from the original isobar map data. Thus a set of the GDCMs obtained for the isobar maps should be treated as an information source in the information theory. If symbols and strings of the GDCM are considered as the messages and the outputs of the source respectively, a set of isobar maps described by the GDCM can be considered as a nearly memoryless information source as can be inferred from Table 3.

A frequency distribution of the symbols, a part of which is already shown in Table 2, is utilized in the binary Huffman coding of the GDCM. The storage structure using this code is illustrated in Fig. 10, where SNODE is an intermediate node of a binary tree and SKEY is a leaf node representing a distinct GDCM, which usually corresponds to a set of weather maps. A determining factor of the complexity of the storage-retrieval system is the number of SNODEs in the structuring tree. If the whole GDCM data are encoded directly, this number grows with the increase of the data size at nearly constant rate as shown in Fig. 11. Of course this fact results in a great difficulty in the estimation of the necessary memory capacity.

The difficulty can be overcome by introducing an idea to partition a GDCM into two parts and encode them separately. The first part is the non-repetitive part in the GDCM description (i.e., from

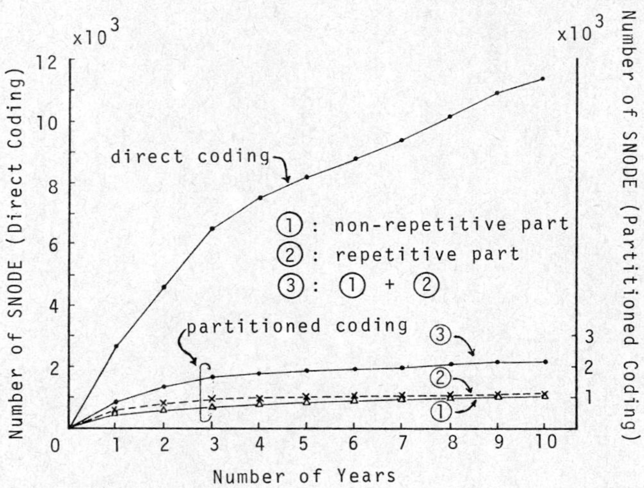

Fig. 11 Growth of the number of SNDDE with the increase of data size.

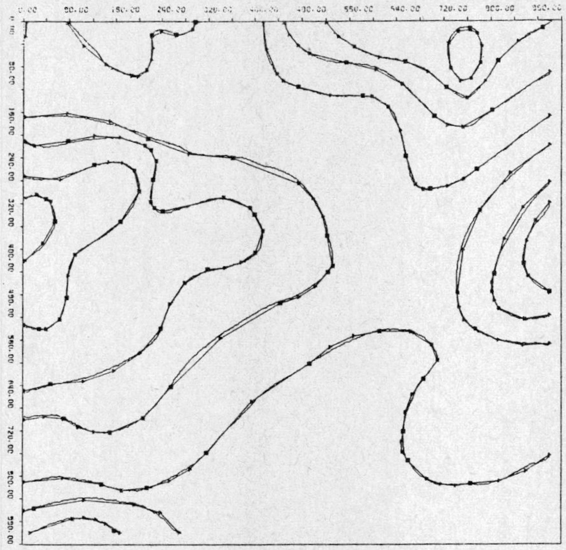

Fig. 12 An example of restoration of an original contour map. Zigzag lines show the original curves and piecewise-linear curves are restored curves. Dots on the curves are the feature points.

A to I in Eq. (1)) and the remaining part is the second one. The encoded resuts are shown also in Fig. 11, where saturation tendency of curves is clearly observed. Thus the feasibility of the file formation by using the GDCM as a code can be confirmed.

7. FORMAL DESCRIPTION OF CONTOUR MAPS AND THEIR RESTORATION

In this section, an outline of a method to formally describe a contour and to restore the original contour from the description is presented [3]. First, it is assumed that every contour is originally coded by the chain code, each symbol of which represents one of the eight directions.

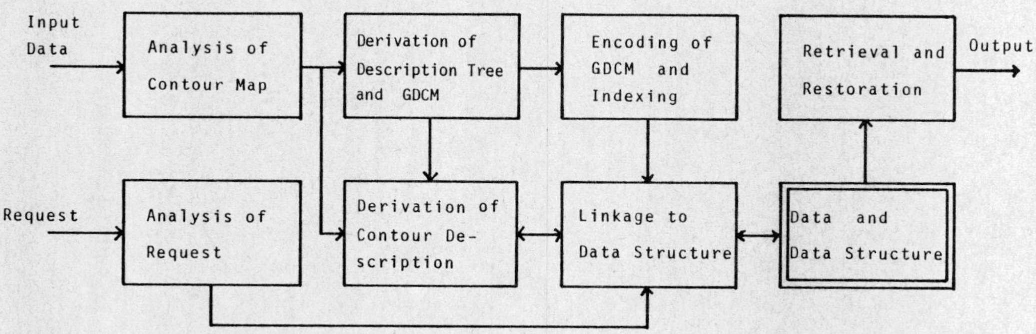

Fig. 13 Storage and retrieval system of weather map data, which has the restoration capability of contour maps.

(1) linear approximation
By using some criterions, every contour is transformed to a chain of vectors with arbitrary length.

(2) extraction of feature point
Every point of inflection is detected on a vector chain, and every vector sub-chain between two adjacent points of inflection is subjected to one of the following processings according to its curvature.

(i) To be reduced to a sub-chain consisting of one vector connecting the two points (smoothing).
(ii) To be divided to more than two sub-chains.
(iii) To be kept unchanged.

All points delimiting sub-chains are called the feature points, and a vector connecting two adjacent feature points is defined as the principal vector.

(3) definition of primitives
A set of 24 vectors, each having unit length and a proper direction and being called the unit vector, is defined. A primitive is defined as a sequence of unit vectors, which are distinct to each other and selected from the set so as to obtain a sequence with unticlockwise progressing. The selection is arbitrary except the starting vector, which is predetermined and fixed for all the primitives. Each vector sub-chain obtained in (2) is transformed to a primitive. This corresponds to the normalization of vector sub-chain with respect to the length and the angle, meaning an approximation by an arc. A set of primitives is obtained through the learning

process for the given contour data.

(4) description of contour

Formal description of a contour is defined as a quadruple

$$D_c = (\mathbf{P}, \mathbf{V}, \theta, s) \quad (2)$$

where $\mathbf{P}$ is a sequence of primitives, $\mathbf{V}$ is a sequence of principal vectors, θ is a normalized angle and s is the starting point of a contour. All these parameters are extracted from the given contour through the processing steps (1), (2) and (3).

(5) restoration

The restoration of an original contour from D_c given by Eq. (2) can be done by applying to $\mathbf{P}$ restoring operations with respect to size and angle. The necessary informations for these operations can be obtained from $\mathbf{V}$ and θ. The line drawing on the plotter starts at a point s. An example of the restoration is given in Fig. 12.

Finally a conceptual diagram of the storage and retrieval system of weather map data, which has the restoration capability of contour maps, is shown in Fig. 13.

8. CONCLUSION

In this paper, first a method of structuring a contour map, which is motivated by the final objective of constructing an image data bank of weather maps, is proposed. A fundamental description called the tree description of contour map is defined, and then the definition of the global description of contour map (GDCM) is derived. The algorithms for realizing these descriptions are implemented on computer systems and are applied to describe structurally isobar maps extracted from the weather maps. The results are successful except that substantial amount of time is needed in the EXTRACTION phase because of the limited memory capacity of the computer.

An exploitation of the description derived previously is an encoding of the GDCM to determine the data structure in a storage system. By introducing an idea to partition the GDCM string, in other words, to divide a structuring tree through selection of appropriate GDCM symbols as the roots of sub-trees, it is confirmed that the data structure can be formed within a reasonable amount of complexity.

One more idea is to restore a line drawing of contour map in the retrieval phase, instead of a real weather map. For this aim, a formal description of a contour and a restoring method are presented with successful experimental results.

The system organization of the partially developed subsystems described above and the performance evaluation of a realized system are left for future works.

ACKNOWLEDGEMENT

The author would like to thank Associate Professors Jun-ichiro Toriwaki and Yuji Yoshida of Nagoya University, members of his research group, for their intensive promotion of this research activity. He also thanks Messrs. Delfin Y. Montuno and Akio Okazaki, graduate students, for their much collaboration throughout the project. Further gratefulness must be expressed for the useful discussions of all the members of the A-group of this long term research project titled "Formation Process of Information Systems and Organization of Scientific Information". The author also would like to acknowledge the support of the Ministry of Education through the Grant-in-Aid for Scientific Research.

REFERENCES

[1] "Tenmon Tenkizu Nikki (Meteorological Weather Map Diary)," in Space Science, Tokyo: Chigin Shokan, 1968-78, vol. 34-44, (in Japanese).

[2] D. Y. Montuno, Y. Yoshida and T. Fukumura, "Structural description of contour maps and its application to weather maps", submitted to the Transaction of the Institute of Electronic and Communication Engineers of Japan (Section E).

[3] A. Okazaki, Y. Yoshida, and T. Fukumura, "Automatic description and restoration of curved line drawings and their applications to contour lines," Technical Report of the Pattern Recognition and Learning Group of the IECE of Japan, PRL79-41, Oct. 1979, (in Japanese).

Scientific Information Systems in Japan
H. Inose (ed.)

Appendix A

List of research groups and their leaders

A Group (Input Information Processing)

A-1 Coordinated research on input systems of scientific data and information
Toshiyuki Sakai, Kyoto University

A-2 Research on interactive graphic entry system
Tohru Moto-oka, University of Tokyo

A-3 Development of data base of standard images
Morio Onoe, University of Tokyo

A-4 Research of statistical program packages
Masashi Okamoto, Osaka University

A-5 Statistical information processing system for prediction and control
Hirostugu Akaike, The Institute of Statistical Mathematics

A-6 Dynamically structured image processing of input data
Teruo Fukumura, Nagoya University

A-7 Application of an on-line intelligent terminal to the organization of scientific information
Takeshi Kiyono, Kyoto University

A-8 A pre-processing system of patterns with the facility of structural description
Yasuhisa Niimi, Kyoto Technical University

A-9 Investigation on the input, storing and retrieval system for the chemical spectroscopic data
Shizuo Fujiwara, University of Tokyo

A-10 Research on automated analysis and processing of microscopic image information
Mikio Takagi, University of Tokyo

A-11 A study of computer input/output methods for medical image data base system
Chiyoshi Yoshimoto, Hokkaido University

A-12 Analysis of moving images for data base
Saburo Tsuji, Osaka University

A-13 Processing of X-ray image obtained by two-dimensional X-ray detectors
Yoichi Iitaka, University of Tokyo

A-14 Processing system of meteorological information
K. Gambo, University of Tokyo

A-15 Pattern presentation of solidification morphology of metals
Terutaka Ueda, University of Tokyo

A-16 Improvements of efficiency and compatibility of symbol manipulation
Eiichi Goto, Institute of Physical and Chemical Research

A-17 A system for on-line spike selection of brain waves and electrical recordings from neurons
Masao Nakamura, Fukui University

B Group (Information Structure Recognition)

B-1 Integrated researches of formation process of scientific information systems
Tosio Kitagawa, Kyushu University

B-2 Research on man-machine interactive problem solving system
Mamoru Hosaka, University of Tokyo

B-3 Processing and retrieval system of image structure
Hajime Enomoto, Tokyo Institute of Technology

B-4 Towards computer-based research in discrete systems
Shuzo Yajima, Kyoto University

B-5 Database architecture
Tosiyasu L.Kunii, University of Tokyo

B-6 Computer-aided design and evaluation of information systems
Yutaka Ohno, Kyoto University

B-7 Analysis and model building of information structure of scientific documents
Makoto Nagao, Kyoto University

B-8 Recognition of semantic structure of natural language and relational information structures between language, picture and speech in scientific field
Tsuneo Tamachi, Kyushu University

B-9 Studies on inference mechanism in pattern understanding system and its application
Kokichi Tanaka, Osaka University

B-10 Studies in symbolic and algebraic processing systems
Hisao M. Yamada, University of Tokyo

B-11 Development of retrieval system of literatures by their abstracts
Shuji Doshita, Kyoto University

B-12 Data base management system with performance evaluation software
Yoshikazu Tezuka, Osaka University

B-13 On firmware for information retrieval
Hiroshi Hagiwara, Kyoto University

B-14 Basic research on information-network systems
Shoichi Noguchi, Tohoku University

B-15 Reliability and security in formulation process of information systems
Tadao Saito, University of Tokyo

B-16 On the optimal network construction of input information processing system
Hisashi Mine, Kyoto University

B-17 Parallel accessing and processing of structured data by a multi-microprocessor system
Takao Tsuda, Kyoto University

B-18 Information storage and retrieval systems for scientific research
Sumiyasu Yamamoto, Hiroshima University

B-19 Research on scientific data base management system
Setsuo Ohsuga, University of Tokyo

B-20 Research on effective computer communication network via a satellite for wide-area information service system
Juro Oizumi, Chiba Institute of Technology

B-21 Researches on computer network architecture
Hiroyoshi Ohara, Waseda University

C Group (Storage and Retrieval)

C-1 Coordinate research on the organization of scientific information
Takehiko Shimanouchi, University of Tsukuba

C-2 Formation and administration of a large-scale scientific information retrieval system
Takeo Yamamoto, University of Tokyo

C-3 Organization of scientific information including patterns and figures
Hiroshi Inose, University of Tokyo

C-4 Studies on review literature and guantitative data in organizing scientific information
Masao Kotani, Science University of Tokyo

C-5 Accumulation and interpretation of chemical structure information
Shinichi Sasaki, Toyohashi University of Technology

C-6 A study of compatibility problems in scientific information systems
Yoichi Terashita, Kanazawa Institute of Technology

C-7 System of science information in nuclear
Yoshinori Akaisi, Hokkaido University

C-8 Development of on-line information retrieval system on protein structure data
Masao Kakudo, Osaka University

C-9 Information system for the study of molecular evolution
Haruhiko Noda, University of Tokyo

C-10 Development of database systems for disciplinary researches and their applications to geosciences and environmentology
Kiyoshi Wadatsumi, Osaka City University

C-11 Man-machine system for the alloy design
Yoshitsugu Mishima, University of Tokyo

C-12 Information retrieval system on finite element method, theory of plates and mechanical vibration
Masao Naruoka, Nagoya University

C-13 Storage and retrieval of medical information
Hiroshi Abe, Osaka University

C-14 Methodology of collection and management of human anomalies data with related pedigrees
Akio Kudo, Kyushu University

C-15 Practice and development of a clinical database
Yoshiaki Nose, Kyushu University

C-16 Database system for medical records of atomic bomb disaster
Shunzo Okajima, Nagasaki University

C-17 Studies for data accumulation and retrieval system of Japanese classical texts and documents
Kiyohiko Furukawa, National Institute of Japanese Literature

C-18 A research on information management system in ethnological studies
Komei Sasaki, National Museum of Ethnology

C-19 Basic study on the utilization of archaeological information
Migaku Tanaka, Nara National Institute of Cultural Property

C-20 Dynamical representation of molecules and its mechanical processing
Haruo Hosoya, Ochanomizu University

C-21 Study on the processing of medical data as database
Shigekoto Kaihara, University of Tokyo

C-22 Construction and evaluation of the IR system for social sciences
Tairoku Kose, Tohoku University

C-23 Designing and compilation of the economic data base through matching several micro data sets
Yoshimasa Kurabayashi, Hitotsubashi University

D Group (Systems Science Approach)

D-1 Coordinated research on systems science approach to utilization of scientific information
Yoshikazu Sawaragi, Kyoto University

D-2 Organization of scientific information in systems science
Atsunobu Ichikawa, Tokyo Institute of Technology

D-3 A study on systematization of information on land use
Yoichi Kaya, University of Tokyo

D-4 A study of structural analysis of policy science including international relations
Shuhei Aida, The University of Electro-Communications

E Group (Methodology of Finding Research Trend)

E-1 Comprehensive study of infological research of research trend
Hajime Tanaka, Hokkaido University

E-2 The establishment of techniques for the utilization of a database for the purpose of grasping the trends of research
Kazuhiko Nakayama, University of Tsukuba

E-3 Clustering of research trend using relational graph made by quoted references in scientific information
Norio Okino, Hokkaido University

E-4 Basic study of utility of scientific information system from a standpoint of researcher
Toshiro Terano, Tokyo Institute of Technology

Scientific Information Systems in Japan
H. Inose (ed.)

Appendix B

Development Committees and Their Members

Data Base Management System Development Committee (M-Committee)

Tosio Kitagawa	Kyushu University	Professor Emeritus
Setuo Osuga	University of Tokyo	Institute of Space and Aeronautical Science
Setsuo Arikawa	Kyushu University	Research Institute of Fundamental Information Science
Seiichi Utinami	Osaka University	Faculty of Engineering
Takuya Katayama	Tokyo Institute of Technology	Dept. of Information Science
Joji Kariya	Tokyo Metropolitan Institute of Medical Science	Clinical Epidemiology Information Processing
Yahiko Kambayashi	Kyoto University	Dept. of Information Science
Tosiyasu Kunii	University of Tokyo	Dept. of Information Science
Kaname Kobayashi	Fujitsu Limited	
Tadao Saito	University of Tokyo	Dept. of Electrical Engineering
Yuzuru Tanaka	Hokkaido University	Dept. of Electrical Engineering
Jun-ichi Tsujii	Kyoto University	Dept. of Electrical Engineering
Masamitsu Negishi	University of Tokyo	Research Center for Laboratory and Information Science
Kouichi Furukawa	Electro-Technical Laboratory	
Ryosuke Hodaka	Tsukuba University	Dept. of Social Engineering
Kiyoshi Wadatsumi	Osaka City University	Dept. of Geosciences
Takeo Yamamoto	University of Tokyo	Computer Centre

Network Development Committee (N-Committee)

Hiroshi Inose	University of Tokyo	Dept. of Electronic Engineering
Toshiyuki Sakai	Kyoto University	Dept. of Information Science
Syoichirou Asano	University of Tokyo	Institute of Space and Aeronautical Science
Noriko Iida	Kyoto University	Computer Centre
Jiro Iimura	Nippon Telegraph and Telephone P.C.	Musashino Electronic Communication Laboratory
Haruhisa Ishida	University of Tokyo	Computer Centre
Hirotaro Ohba	Nippon Telegraph and Telephone P.C.	Data Communications Bureau

Masanori Kanazawa	Kyoto University	Computer Centre
Hajime Kitagawa	Kyoto University	Computer Centre
Tadao Saito	University of Tokyo	Dept. of Electrical Engineering
Toshiharu Takatsuki	Nippon Telegraph and Telephone P.C.	Engineering Bureau
Hidehiko Tanaka	University of Tokyo	Dept. of Electrical Engineering
Koichi Tabata	Kyoto University	Dept. of Information Science
Akihiko Hashimoto	Nippon Telegraph and Telephone P.C.	Yokosuka Electronic Communication Laboratory
Kigen Hasebe	University of Tokyo	Computer Centre
Hisashi Yasunaga	University of Tokyo	Earthquake Research Institute

Scientific Information System Development Committee (O-Committee)

Hajime Tanaka	Hokkaido University	Dept. of Physics
Choichiro Asano	Kyushu University	Dept. of Fundamental Information Science
Atsunobu Ichikawa	Tokyo Institute of Technology	Dept. of Systems Science
Syu Ono	University of Tokyo	Faculty of Culture
Tairoku Kose	Tohoku University	Faculty of Economics
Shuji Takagi	Osaka University	Dept. of Fundamental Engineering
Yoshinosuke Terashima	Nagoya University	Institute of Plasma Physics
Masahiro Fujiwara	Ministry of Health and Welfare	Waste Management Division
Yoshiro Matsuda	Hitotsubashi University	Institute of Economic Research
Mitsuichi Yamanaka	National Institute of Japanese Literature	
Yasuhisa Yugawa	Science University of Tokyo	Dept. of Organic Chemistry
Haruhisa Ishida	University of Tokyo	Computer Centre

Input-Output System Development Committee (I-Committee)

Mamoru Hosaka	University of Tokyo	Institute of Space and Aeronautical Science
Toshiyuki Sakai	Kyoto University	Dept. of Information Science
Katsuo Ikeda	Tsukuba University	Dept. of Information Science
Morio Onoe	University of Tokyo	Institute of Industrial Science
Saburo Tsuji	Osaka University	Dept. of Control Engineering
Shizuo Fujiwara	University of Tokyo	Dept. of Chemistry
Tohru Moto-oka	University of Tokyo	Dept. of Electrical Engineering
Chiyoshi Yoshimoto	Hokkaido University	The Research Institute of Applied Electricity
Mikio Takagi	University of Tokyo	Institute of Industrial Science

Development Planning and Promotion Committee

Hiroshi Inose	University of Tokyo	Dept. of Electronic Engineering
Tosio Kitagawa	Kyushu University	Professor Emeritus
Takehiko Shimanouchi	Tsukuba University	Dept. of Information Science
Mamoru Hosaka	University of Tokyo	Institute of Space and Aeronautical Science
Toshiyuki Sakai	Kyoto University	Dept. of Information Science
Yoshikazu Sawaragi	Kyoto University	Dept. of Applied Mathematics and Physics
Hajime Tanaka	Hokkaido University	Dept. of Physics
Atsunobu Ichikawa	Tokyo Institute of Technology	Dept. of Systems Science
Haruhisa Ishida	University of Tokyo	Computer Centre
Takeo Yamamoto	University of Tokyo	Computer Centre
Hisashi Yasunaga	University of Tokyo	Earthquake Research Institute
Koji Maeda	Nippon Telegraph and Telephone P.C.	Engineering Bureau
Setsuo Osuga	University of Tokyo	Institute of Space and Aeronautical Science